PROBABILITY
AND STATISTICS
For Engineering and the Computing Sciences

J. Susan Milton

Radford University

Jesse C. Arnold

*Virginia Polytechnic Institute
and State University*

Liu Kwong Ip

Hong Kong Baptist University

McGraw Hill

Singapore • Boston • Burr Ridge, IL • Dubuque, IA • Madison, WI
New York • San Francisco • St. Louis • Bangkok • Kuala Lumpur
Lisbon • London • Madrid • Mexico City • Milan • Montreal
New Delhi • Seoul • Sydney • Taipei • Toronto

The McGraw-Hill Companies

Mc Graw Hill **Higher Education**

10 9 8 7 6 5 4 3 2 1
CTP SLP
20 15 14 13 12

When ordering this title, use **ISBN 978-007-108785-8** or **MHID 007-108785-0**

Printed in Singapore

About The Authors

J. Susan Milton is Professor Emeritus of Statistics at Radford University. Dr. Milton received the B.S. degree from Western Carolina University, the M.A. degree from the University of North Carolina at Chapel Hill, and the Ph.D. degree in statistics from Virginia Polytechnic Institute and State University. She is a Danforth Associate and is a recipient of the Radford University Foundation Award for Excellence in Teaching. Dr. Milton is the author of *Statistical Methods in the Biological and Health Sciences* as well as *Introduction to Statistics, Probability with the Essential Analysis,* and *A First Course in the Theory of Linear Statistical Models.*

Jesse C. Arnold is Professor of Statistics at Virginia Polytechnic Institute and State University. Dr. Arnold received the B.S. degree from Southeastern State University, and the M.A. and Ph.D. degrees in statistics from Florida State University. He served as head of the statistics department for ten years, is a fellow of the American Statistical Association, and is an elected member of the International Statistics Institute. He has served as President of the International Biometric Society (Eastern North American Region) and Chairman of the Statistical Educational Section of the American Statistical Association.

Liu, Kwong Ip is a Lecturer in the Department of Mathematics at Hong Kong Baptist University. Dr. Liu received the BSc. degree from the City Polytechnic of Hong Kong, the MSc. degree from the University of Hong Kong, and Ph.D. degree from Hong Kong Baptist University.

TO
PEGGY ARNOLD

AND IN LOVING MEMORY OF ENID K. AND GEORGE A. MILTON

Contents

Preface

Interpretation of much of the research in the engineering and computing sciences increasingly depends on statistical methods. Furthermore, the practicing engineer will be expected to understand and help implement statistical quality control techniques in the workplace. For these reasons, it is essential that students in these fields be exposed to statistical reasoning early in their careers. This text is intended as a first course in probability and applied statistics for students in the engineering and computing sciences. It is hoped that this first course will occur on the undergraduate level. However, the text can be used to advantage by graduate students who have little or no prior experience with statistical methods.

This text is not a statistical cookbook, nor is it a manual for researchers. We attempt to find a middle road—to provide a text that gives the student an understanding of the logic behind statistical techniques as well as practice in using them. A one-year course in elementary calculus should provide an adequate background for understanding everything presented here.

We chose the examples and exercises specifically for the student in the engineering and computing sciences. Most data sets are simulated. However, the simulation was done with care, so that the results of the analysis are consistent with recently reported research. References to reports upon which the data are based are given whenever possible. In this way, the student will gain some insight into the types of engineering problems that can be handled statistically. Many exercises are left open-ended in hopes of stimulating some classroom discussion.

It is assumed that the student has access to some type of electronic calculator. Many such calculators are on the market, and most have some built-in statistical capability. The use of these calculators is encouraged, for it allows the student to concentrate on the interpretation of the analysis rather than on the arithmetic computations.

We should point out that many of the data sets are rather small so that the student will not be overwhelmed by the computational aspects of statistics. We do not intend to imply that very small data sets are routinely used in the engineering fields. In fact, most major research projects involve a tremendous investment in time and money and result in a large body of data. New to the fourth edition, we have added some large data sets to better reflect the reality students will encounter after graduation.

Such data lend themselves to analysis by computer. For this reason, we include some instruction in the interpretation of statistical packages. The packages chosen for illustrative purposes are SAS and MINITAB. This was done because of their widespread availability and ease of use. We do not intend to imply that they are superior to other well-known packages such as SPSS (Statistical Package for the Social Sciences) or BMD (Biomedical Computer Programs, University of California Press).

Each chapter ends with a chapter summary that is intended to remind the student of the major topics presented in the chapter. This chapter summary also includes a list of important terms. A set of exercises is provided for each section of each chapter. In addition, each chapter has a set of review exercises in which the problems are presented in random order. It is hoped that this will help the student develop the ability to recognize the appropriate analysis. The appendices include statistical tables, selected derivations, and answers to all odd numbered and review exercises.

A number of different courses can be taught from this book. They can vary in length from one quarter to one year. It is difficult to determine exactly what material can be covered in a given time, since this is a function of class size, academic maturity of the students, and inclination of the instructor. However, we do offer some guidelines for the use of this text. In particular, the type of course presented can vary from one whose chief aim is to familiarize the student with the computational aspects of probability and the handling of data sets to one of a more theoretical nature. In many cases we include the proof or derivation of theorems in the text labeled as such. If an instructor wants to deemphasize theory, these proofs can be skipped easily with no loss of continuity.

CHANGES TO THE U.S. EDITION

At the suggestion of users of the first three editions of the text, some changes have been made to enhance the fourth edition. New exercises have been added throughout. A data disk containing all data sets that appear in the text as part of examples or exercises is provided with the Instructor's Solution Manual. At the suggestion of the reviewers, some of the data sets are rather large so the student can learn to manipulate such data via computer. The SAS computer supplements that appeared in earlier editions have been deleted. However, more discussion of the interpretation of computer output is now included in the text. Some of the more difficult derivations have been placed in an Appendix. This gives the text a more applied flavor while preserving the material for those who are particularly interested in the mathematical foundations of the statistical concepts presented. The discussion of the F distribution and comparison of two means has been simplified by making use of the folded F test for comparing variances. Other new material includes a discussion of Tukey's method of paired comparisons and a section on the use of tolerance limits in quality control.

Chapter 1 This chapter provides an introduction to probability and counting.

Chapter 2 The study of probability is continued. The laws governing probability are presented, and the notions of conditional probability and independence are introduced.

Chapter 3 The notion of random variables is introduced. General properties of discrete distributions are discussed. The notion of expected value is introduced, and the idea of the mean and variance of a distribution is developed. The moment generating function is presented as a means of finding the first two moments of a distribution. Important discrete distributions are studied in detail. The chapter closes with an optional section on simulating discrete distributions.

Chapter 4 parallels Chap. 3 with an emphasis on continuous distributions.

Chapter 5 discusses joint distributions of both the discrete and continuous types. The notions of covariance, correlation, and regression are introduced in the theoretical sense.

Chapter 6 is the link between the more theoretical concepts of statistics and the methods of data analysis. Here we present an introduction to classical data-handling techniques and descriptive statistics. We also introduce some of the newer techniques of exploratory data analysis.

Chapter 7 considers the notion of point and interval estimation of population parameters. Method of moments, maximum likelihood, and unbiased estimators are considered. Some distribution theory is also discussed. In particular, the distribution of $\overline{X}$ is investigated. The moment generating function is used as a fingerprint to help pinpoint the distribution of some important random variables that will underlie the statistical methods developed in later chapters.

Chapter 8 begins the study of the classical methods of data analysis. The topic of interest is inferences on the location and variability of a distribution based on a single sample. Both estimation and hypothesis testing are discussed and the T distribution is introduced. A full discussion of significance testing is included. The methods presented assume that sampling is from a normal distribution. The chapter closes with a section on nonparametric tests for location. These tests are especially useful when the normality assumption appears to be violated.

Chapter 9 In this chapter inferences on a single proportion are considered. The study of two sample problems is begun by showing how to compare two proportions based on independent random samples.

Chapter 10 is concerned with methods used to compare two variances and two means. The F distribution is introduced as a means of comparing variances. Means are compared first when variances are assumed to be equal. The Smith-Satterthwaite procedure is used to compare means when variances appear to be unequal. These procedures all assume independent sampling. A procedure for comparing means based on paired data is presented. The chapter ends with a section on nonparametric two-sample tests for location.

Chapter 11 studies simple linear regression and correlation. The least-squares method is given for estimating parameters in the regression model. Estimation and hypothesis testing is presented. Development for the simple linear regression model is quite thorough as preparation for the more general regression cases discussed in Chap. 12. The bivariate normal distribution is presented as needed for estimation and testing for product-moment correlation. A new section on the analysis of residuals is included.

Chapter 12 The simple linear regression model is extended to multiple and polynomial models. The methods of Chap. 11 are extended in matrix form. Variable selection procedures are discussed along with examples.

Chapter 13 The analysis of variance procedure is studied for various one-factor experimental designs. This chapter includes a discussion of randomized complete blocks, and some results on the effectiveness of blocking are given. A section on Latin squares is included as well as material on Bonferroni-type and Tukey-type multiple comparisons. Variance component estimation in random effects models is discussed.

Chapter 14 This chapter discusses factorial experiments and contains material on fractional factorials.

Chapter 15 is an introduction to the study of categorical data. Chi-squared goodness of fit tests are presented. Contingency table tests for independence and homogeneity are discussed in both the 2×2 and $r \times c$ cases.

Chapter 16 discusses the basic concepts of statistical quality control. Process control is discussed using control charts, and basic ideas of acceptance sampling are presented. The relationship of acceptance sampling with usual hypothesis testing is presented. Taguchi methods are discussed briefly. A new section on tolerance limits is included.

You should be aware that statistics is an art as well as a science. For this reason, there is always room for debate on how to properly analyze a given data set. We have presented in this text methods that have stood the test of time as well as some that are relatively new. In many cases we have intentionally left to you the decision of whether or not to reject a particular null hypothesis. The reason for this is simple: No one can really say how small a probability must be in order to claim that it is too small to have occurred by chance. You might disagree with our conclusions at times. Feel free to do so!

CHANGES TO THIS EDITION

The changes have been made so that the book would be suitable for one semester introductory Statistics courses. The contents of this edition focus on the applications of Statistics, rather than the mathematical details. Several sections and chapters were thus removed. For example, the topics about *moment generating functions* have been removed from the chapters of *Discrete Distributions* and *Continuous Distributions*. For the same reason, the chapters of *Joint Distributions*, *Multiple Linear Regression Models*, *Factorial Experiments*, and *Statistical Quality Control* have been removed. It does not mean that these topics are not important. In fact, they are excellent topics in a book for the second course of Statistics.

Furthermore, a section titled Real World Application has been added to three topics–*Comparing Two Means and Two Variances*, *Analysis of Variance*, and *Categorical Data*, for students to better relate the key concepts learned, to real cases.

Another significant change of this edition is the presentation of hypothesis testing. The conclusion of hypothesis testing is drawn by comparing the value of test statistic and the critical value(s), rather than the estimated P value. Moreover, hypothesis testing is presented in a "4-step format":

[Hypotheses]   . . .
[Test statistic]   . . .
[Critical value]   . . .
[Conclusion]   . . .

Hopefully, students will find that the presentation is easier to follow. Nevertheless, P value is still estimated. As stated in the preface of the earlier edition, statistics is an art as well as science. Students might disagree with the conclusion at a particular level of significance. Please feel free to do so!

Finally, for this new edition, for Instructors, they can access supplements to the text–which include Solutions Manuals, powerpoint lecture slides and additional reference materials–at the text Online Learning Centre website: www.mheducation.asia/olc/milton

Acknowledgements

We wish to thank the Chemical Rubber Company, Bell Laboratories, and the American Society for Testing and Materials for use of statistical tables. Special thanks go to SAS Institute for permission to use their package for illustrative purposes.

A particular thanks goes to Dr. Jill Stewart for her many hours spent checking answers and writing the solutions manuals.

We would like to thank David Dietz, Peter Galuardi, Joyce Watters, and the rest of the McGraw-Hill staff for their support and advice. Very special thanks are offered to the following reviewers for their many helpful suggestions during the preparation of this, and the previous three editions:

Lynne Billard, University of Georgia; Ahankar P. Bhattacharyya, Texas A & M University; Martha L. Bouknight, Meredith College; David C. Brooks, Seattle Pacific University; Saibal Chattopadhyay, University of Nebraska–Lincoln; Daren B. H. Cline, Texas A & M University; Michael W. Ecker, Pennsylvania State University; Peter G. Furth, Northeastern University; David Groggel, Miami University of Ohio; Robert Lacher, South Dakota State University; Chand K. Midha, The University of Akron; H. N. Nagaraja, The Ohio State University; Roxanne Peck, California Polytechnic and State University; Larry G. Richards, University of Virginia; Don Ridgeway, North Carolina State University; Thomas N. Roe, South Dakota State University; Paul Speckman, University of Missouri–Columbia; Larry Stephens, University of Nebraska; Harrison M. Wadsworth, Georgia Institute of Technology; and Vasar Waikar, Miami University of Ohio.

From J.C.A., special thanks for the unfailing inspiration and support of my wife Peggy and for the love and encouragement of our children Christa and Chuck. I would also like to thank my colleagues at Virginia Polytechnic Institute and State University for their helpful and enlightening discussions.

J. Susan Milton
Jesse C. Arnold

CHAPTER 1

Introduction to Probability and Counting

What is "statistics," and why is its study important to engineers and scientists? To answer this question, let us describe an aspect of the work of a scientist known as "model building."

Basically, the job of a scientist is to describe what he or she sees, to try to explain what is observed, and to use this knowledge to predict events in the world in which we live. The explanation often takes the form of a physical model. A *model* is a theoretical explanation of the phenomenon under study and, at the outset, is usually expressed verbally. To use the model for predictive purposes, this verbal description must be translated into one or more mathematical equations. These equations can be used to determine the value of a specific variable in the model based on the knowledge of the values assumed by other model variables. For example, the Perfect Gas Law states that the pressure and volume of a gas may both vary simultaneously when the temperature of the gas is changed. This verbal model can be translated into a mathematical equation by writing

$$\text{Perfect Gas Law: } PV = RT$$

where P is the pressure of the gas, V is its volume, T is its temperature, and R is a constant, called the *gas constant*. The numerical value of the gas constant depends on the physical units chosen for the other terms in the model. Once we know the values assumed by two of the three variables P, V, or T, we can calculate the value of the third via this mathematical model. For example, under a pressure of 760 mm mercury and a temperature of 273 kelvins, a mole of any gas is thought to have a volume of 22.4 liters. The gas constant in this case has a value of approximately 62.36. Based on the Perfect Gas Law, a gas with a volume of 5 liters at a temperature of 100 kelvins has pressure P given by

$$PV = RT = 62.36T$$

or
$$P(5) = 62.36(100)$$
$$P = 1247.2 \text{ mm mercury}$$

That is, our model leads us to expect the pressure to be 1247.2 mm mercury. A model such as the Perfect Gas Law is said to be "deterministic." It is deterministic in the sense that it allows us to determine an exact value for the variable of interest under specified experimental conditions. The Perfect Gas Law does describe *some* real gases at moderate temperatures and pressures. Unfortunately, many real gases cannot be described by this or any other deterministic model, especially at extreme temperatures and pressures! Under these circumstances we must find another way to predict the behavior of the gas with some degree of certainty. This can be done with the aid of statistical methods.

What do we mean by statistical methods? These are methods by which decisions are made based on the analysis of data gathered in carefully designed experiments. Since experiments cannot be designed to account for every conceivable contingency, there is always some uncertainty in experimental science. Statistical methods are designed to *allow us to assess the degree of uncertainty present in our results*. These methods can be classed roughly into three categories: descriptive statistics, inferential statistics, and model building. By descriptive statistics we mean those techniques, both analytic and graphical, that allow us to describe or picture a data set. Inferential statistics concerns methods by which conclusions can be drawn about a large group of objects, based on observing only a portion of the objects in the larger group.

This idea leads to the following definition:

> **Definition:** The overall group of objects about which conclusions are to be drawn is called the *population*. A subset or portion of the population that is actually obtained and that is used to draw conclusions about the population is called a *sample*.

Model building entails the development of prediction equations from experimental data. These equations are called statistical models; they are models that allow us to predict the behavior of a complex system and to assess our probability of error. These categories are not mutually exclusive. That is, methods developed to solve problems in one area often find application in another. We shall be concerned with all three areas in this text.

A statistician or user of statistics is always working in two worlds. The ideal world is at the population level and is theoretical in nature. It is the world that we would like to see. The world of reality is the sample world. This is the level at which we really operate. We hope that the characteristics of our sample reflect well the characteristics of the population. That is, we treat our sample as a microcosm that mirrors the population. This idea is illustrated in Fig. 1.1.

Figure 1.1
The sample is viewed as a miniature population. We hope that the behavior of the variable under study over the sample gives an accurate picture of its behavior in the population.

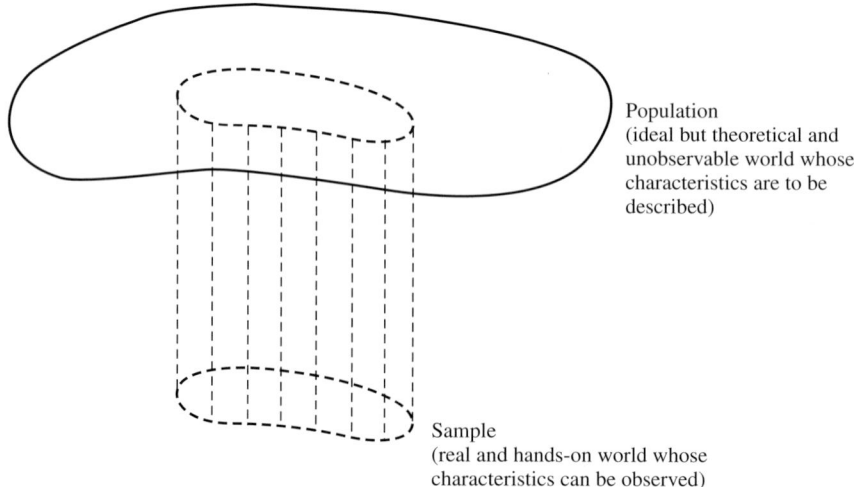

Population
(ideal but theoretical and unobservable world whose characteristics are to be described)

Sample
(real and hands-on world whose characteristics can be observed)

The mathematics on which statistical methods rest is called probability theory. For this reason, we begin the study of statistics by considering the basic concepts of probability.

1.1 INTERPRETING PROBABILITIES

When asked, "Do you know anything about probability?" most people are quick to answer, "no!" Usually that is not the case at all. The ability to interpret probabilities is assumed in our culture. One hears the phrases "the probability of rain today is 95%" or "there is a 0% chance of rain today." It is assumed that the general public can interpret these values correctly. The interpretation of probabilities is summarized as follows:

Interpretation of Probabilities

1. Probabilities are numbers between 0 and 1, inclusive, that reflect the chances of a physical event occurring.
2. Probabilities near 1 indicate that the event is extremely likely to occur. They mean not that the event will occur, only that the event is considered to be a common occurrence.
3. Probabilities near zero indicate that the event is not very likely to occur. They do not mean that the event will fail to occur, only that the event is considered to be rare.
4. Probabilities near 1/2 indicate that the event is just as likely to occur as not.
5. Since numbers between 0 and 1 can be expressed as percentages between 0 and 100, probabilities are often expressed as percentages. This is particularly common in writings of a nontechnical nature.

These properties are guidelines for interpreting probabilities once they are available, but they do not indicate how to assign probabilities to events. Three methods are widely used: the *personal* approach, the *relative frequency* approach, and the *classical* approach. These methods are illustrated in the following examples.

Example 1.1.1. An oil spill has occurred. An environmental scientist asks, "What is the probability that this spill can be contained before it causes widespread damage to nearby beaches?" Many factors come into play, among them the type of spill, the amount of oil spilled, the wind and water conditions during the clean-up operation, and the nearness of the beaches. These factors make this spill unique. The scientist is called upon to make a value judgment, that is, to assign a probability to the event based on informed *personal opinion*.

The main advantage of the personal approach is that it is always applicable. Anyone can have a personal opinion about anything. Its main disadvantage is, of course, that its accuracy depends on the accuracy of the information available and the ability of the scientist to assess that information correctly.

Example 1.1.2. An electrical engineer is studying the peak demand at a power plant. It is observed that on 80 of the 100 days randomly selected for study from past records, the peak demand occurred between 6 and 7 p.m. It is natural to assume that the probability of this occurring on another day is at least *approximately*

$$\frac{80}{100} = .80$$

This figure is not simply a personal opinion. It is a figure based on repeated experimentation and observation. It is a relative frequency.

The relative frequency approach can be used whenever the experiment can be repeated many times and the results observed. In such cases, the probability of the occurrence of event A, denoted by $P[A]$, is approximated as follows:

Relative Frequency Approximation

$$P[A] \doteq \frac{f}{n} = \frac{\text{number of times event } A \text{ occurred}}{\text{number of times experiment was run}}$$

The disadvantage in this approach is that the experiment cannot be a one-shot situation; it must be repeatable. Remember that any probability obtained this way is an approximation. It is a value based on n trials. Further testing might result in a different approximate value. However, as the number of trials increases, the changes in the approximate values obtained tend to become slight. Thus for a large number of trials, the approximate probability obtained by using the relative frequency approach is usually quite accurate.

Example 1.1.3. What is the probability that a child born to a couple heterozygous for eye color (each with genes for both brown and blue eyes) will be brown-eyed? To answer this question, we note that since the child receives one gene from each parent, the possibilities for the child are (brown, blue), (blue, brown), (blue, blue) and (brown, brown), where the first member of each pair represents the gene received from the father. Since each parent is just as likely to contribute a gene for brown eyes as for blue eyes, all four possibilities are equally likely. Since the gene for brown eyes is dominant, three of the four possibilities lead to a brown-eyed child. Hence the probability that the child will be brown-eyed is 3/4 = .75.

The above probability is not a personal opinion, nor is it based on repeated experimentation. In fact, we found this probability by the *classical* method. This method can be used *only* when it is reasonable to assume that the possible outcomes of the experiment are equally likely. In this case, the probability of the occurrence of event A is given by the following classical formula:

Classical Formula

$$P[A] = \frac{n(A)}{n(S)} = \frac{\text{number of ways } A \text{ can occur}}{\text{number of ways the experiment can proceed}}$$

One advantage to this method is that it does not require experimentation. Furthermore, if the outcomes are truly equally likely, then the probability assigned to event A is not an approximation. It is an accurate description of the frequency with which event A will occur.

1.2 SAMPLE SPACES AND EVENTS

To determine what is "probable" in an experiment, we first must determine what is "possible." That is, the first step in analyzing most experiments is to make a list of possibilities for the experiment. Such a list is called a *sample space*. We define this term as follows:

> **Definition 1.2.1 (Sample space and sample point).** A sample space for an experiment is a set S with the property that each physical outcome of the experiment corresponds to exactly one element of S. An element of S is called a sample point.

When the number of possibilities is small, an appropriate sample space usually can be found without difficulty. For instance, we have seen that when a couple heterozygous for eye color parents a child, the possible genotypes for the child are given by

$$S = \{(\text{brown, blue}), (\text{blue, brown}), (\text{blue, blue}), (\text{brown, brown})\}$$

As the number of possibilities becomes larger, it is helpful to have a system for developing a sample space. One such system is the *tree diagram*. The next example illustrates the idea.

Example 1.2.1. During a space shot the primary computer system is backed up by two secondary systems. They operate independently of one another in that the failure of one has no effect on any of the others. We are interested in the readiness of these three systems at launch time. What is an appropriate sample space for this experiment?

Since we are primarily concerned with whether each system is operable at launch, we need only find a sample space that gives that information. To generate the sample space, we use a *tree*. The primary system is either operable (yes) or not operable (no) at the time of launch. This is indicated in the tree diagram of Fig. 1.2(*a*), where yes $= y$ and no $= n$. Likewise the first backup system either is or is not operable. This is shown in Fig. 1.2(*b*). Finally, the second backup system either is or is not operable. The tree is completed as shown in Fig. 1.2(*c*). A sample space S for the experiment can be read from the tree by following each of the eight distinct paths through the tree. Thus

$$S = \{yyy, yyn, yny, ynn, nyy, nyn, nny, nnn\}$$

Once a suitable sample space has been found, elementary set theory can be used to describe physical occurrences associated with the experiment. This is done by considering what are called *events* in the mathematical sense.

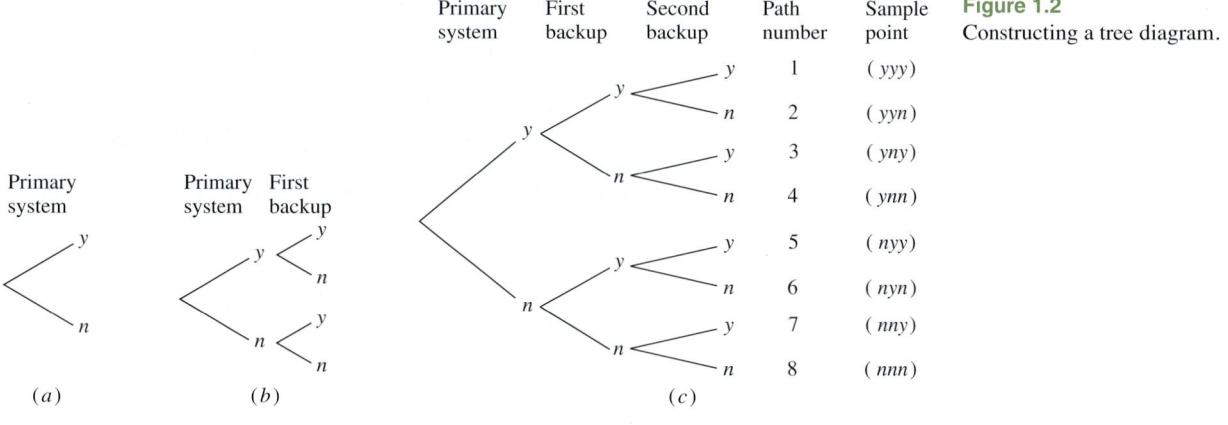

Figure 1.2
Constructing a tree diagram.

Definition 1.2.2 (Event). Any subset A of a sample space is called an event. The empty set $\varnothing$ is called the *impossible* event; the subset S is called the *certain* event.

Definition 1.2.3 (Union of events). The union of two events A and B, denoted by $A \cup B$, consists of all the sample points that belong to A or B or both. The union of events $A_1, A_2, A_3, \ldots, A_k$, denoted by $A_1 \cup A_2 \cup A_3 \cup \ldots \cup A_k$, consists of all the sample points that belong to any one or more A_i for $i = 1, \ldots, k$.

Definition 1.2.4 (Intersection of events). The intersection of two events A and B, denoted by $A \cap B$, consists of all the sample points that belong to both A and B. The intersection of events $A_1, A_2, A_3, \ldots, A_k$, denoted by $A_1 \cap A_2 \cap A_3 \cap \ldots A_k$, consists of all the sample points that belong to all A_i for $i = 1, \ldots, k$.

Example 1.2.2.   Consider a space shot in which a primary computer system is backed up by two secondary systems. The sample space for this experiment is

$$S = \{yyy, yyn, yny, ynn, nyy, nyn, nny, nnn\}$$

where, for example, *yny* denotes the fact that the primary system and second backup are operable at launch, whereas the first backup is inoperable (see Example 1.2.1). Let

A: primary system is operable

B: first backup is operable

C: second backup is operable

The mathematical event corresponding to each of these physical events is found by listing the sample points that represent the occurrence of the event. Thus we write

$$A = \{yyy, yyn, yny, ynn\}$$
$$B = \{yyy, yyn, nyy, nyn\}$$
$$C = \{yyy, yny, nyy, nny\}$$

Other events can be described using these events as building blocks. For example, the event that "the primary system *or* the first backup is operable" is given by the set $A \cup B$, the union of set A with set B. Recall from elementary mathematics that the *union of A with B consists of all sample points that are in set A or set B or are in both.* Thus

$$A \cup B = \begin{array}{c} \text{primary } or \text{ first} \\ \text{backup is operable} \end{array} = \{yyy, yyn, yny, ynn, nyy, nyn\}$$

Note that the word "or" will denote set union. The event that "the primary system *and* the first backup is operable" is given by the set $A \cap B$, the intersection of set A with set B. *The intersection of two sets consists of all sample points that are in both sets.* That is, it is the set of points that they have in common. Here

$$A \cap B = \text{primary } and \text{ first backup operable} = \{yyy, yyn\}$$

Note that the word "and" will denote the set intersection. The event that "the primary system or the first backup is operable but the second backup is inoperable" is given by $(A \cap B) \cap C'$, where C' denotes the complement of set C. *The complement of a set consists of the sample points in the sample space that are not in the given set.* Thus

$$(A \cup B) \cap C' = \begin{array}{c} \text{primary } or \text{ first backup operable} \\ but \text{ second backup inoperable} \end{array} = \{yyn, ynn, nyn\}$$

Note that the word "but" is also translated as a set intersection; the word "not" translates as a set complement.

Let us pause briefly to consider a basic difference between the sample space

$$S_1 = \{(\text{brown, blue}), (\text{blue, brown}), (\text{blue, blue}), (\text{brown, brown})\}$$

of Example 1.1.3 and

$$S_2 = \{yyy, yyn, yny, ynn, nyy, nyn, nny, nnn\}$$

of Example 1.2.1. Since each parent is just as likely to contribute a gene for brown eyes as for blue eyes, the sample points of S_1 are equally likely. This allows us to use the classical method to find the probability that a child born to a couple heterozygous for eye color will be brown-eyed. If we denote this event by A, then we can conclude that

$$P[A] = P[\{(\text{brown, blue}), (\text{blue, brown}), (\text{brown, brown})\}]$$
$$= \frac{n(A)}{n(S)} = \frac{3}{4}$$

However, it is not correct to assume that the sample points of S_2 are equally likely. This would be true if and only if each of the three computer systems is just as likely to fail as to be operable at launch time. Our technology is much better than that! The primary question to be answered is "What is the probability that at least one system will be operable at the time of the launch?" That is, what is

$$P[\{yyy, yyn, yny, ynn, nyy, nyn, nny\}]?$$

As will be shown later, this question can be answered. However, since the sample points are not equally likely, it cannot be answered using the classical method.

Not all trees are symmetric as is that pictured in Fig. 1.2. In some settings, paths end at different stages of the game. Example 1.2.3 illustrates an experiment of this sort.

Example 1.2.3. Consider a production process that is known to produce defective parts at the rate of one per hundred. The process is monitored by testing randomly selected parts during the production process. Suppose that as soon as a defective part is found, the process will be stopped and all machine settings will be checked. We are interested in studying the number of parts that are tested in order to obtain the first defective part. In the tree of Fig. 1.3, c represents that the sampling continues

First part sampled	Second part sampled	Third part sampled	Fourth part sampled
↓	↓	↓	↓

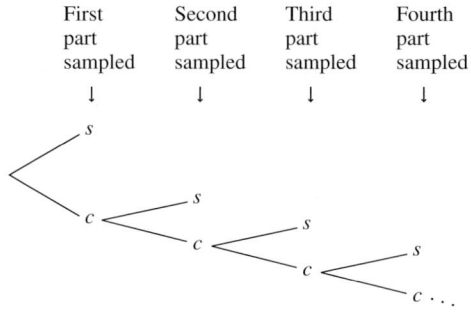

Figure 1.3
Sampling a production line for defective parts.

and s represents that production is stopped. Notice that as soon as a defective item is found, the process ends and the path also ends. For this reason, some paths are much shorter than others. Notice also that theoretically this tree continues indefinitely. The sample space generated by the tree is

$$S = \{s, cs, ccs, cccs, ccccs, \ldots\}$$

Since defective parts occur with probability .01, it should be evident that the paths of this tree are not equally likely.

Mutually Exclusive Events

Occasionally interest centers on two or more events that cannot occur at the same time. That is, the occurrence of one event precludes the occurrence of the other. Such events are said to be *mutually exclusive*.

Example 1.2.4. Consider the sample space

$$S = \{yyy, yyn, yny, ynn, nyy, nyn, nny, nnn\}$$

of Example 1.2.1. The events

$$A_1: \text{primary system operable} = \{yyy, yyn, yny, ynn\}$$

$$A_2: \text{primary system inoperable} = \{nyy, nyn, nny, nnn\}$$

are mutually exclusive. It is impossible for the primary system to be both operable and inoperable at the same time. Mathematically, A_1 and A_2 have no sample points in common. That is, $A_1 \cap A_2 = \varnothing$.

Example 1.2.4 suggests the mathematical definition of the term "mutually exclusive events."

Definition 1.2.5 (Mutually exclusive events). Two events A_1 and A_2 are mutually exclusive if and only if $A_1 \cap A_2 = \varnothing$. Events $A_1, A_2, A_3, \ldots$ are mutually exclusive if and only if $A_i \cup A_j = \varnothing$ for $i \neq j$.

CHAPTER SUMMARY

In this chapter we discussed how to interpret probabilities. We also presented three methods for assigning probabilities to events. These are called the personal, relative frequency, and classical approaches. We also introduced and defined important terms that you should know. These are

Sample space   Mutually exclusive events
Sample point
Event
Impossible event
Certain event

EXERCISES

Section 1.1

1. One environmental hazard recently identified is overexposure to airborne asbestos. In a sample of 10 public buildings over 20 years old, three were found to be insulated with materials that produced an excess number of airborne asbestos bodies. What is the approximate probability that another building of this type will have this problem? What method are you using to assign this probability?

2. A sample of 75 bridges in a given state is selected, and the bridges chosen are inspected for structural weaknesses. If 30 of the bridges sampled are found to have serious problems, what is the estimated probability that the next bridge sampled in the state will have serious structural problems? What method for assigning probabilities are you using to obtain this estimate?

3. Hemophilia is a sex-linked hereditary blood defect of males characterized by delayed clotting of the blood which makes it difficult to control bleeding, even in the case of a minor injury. When a woman is a carrier of classical hemophilia, there is a 50% chance that a male child will inherit the disease. If a carrier gives birth to two sons, what is the probability that both boys will have the disease? What approach to probability are you using to answer this question?

4. A foundry produces brake pads for use in Ford motor cars. A particular lot of 50 such pads contains 2 that have burrs (or rough spots) that were missed in the grinding process. If one part is selected at random from the lot to be installed in your car, what is the probability that it will have a burr? Is this a relative frequency approximation or a classical probability?

Section 1.2

5. Fission occurs when the nucleus of an atom captures a subatomic particle called a neutron and splits into two lighter nuclei. This causes energy to be released. At the same time other neutrons are emitted, two or three on the average. If at least one of these is captured by another fissionable nucleus, then a chain reaction is possible.

   (*a*) Consider a reaction in which three neutrons are emitted initially. Let *c* denote that a given neutron is captured by another nucleus; let *n* denote that the neutron is not captured by another nucleus. Construct a tree denoting the possible behavior for these three neutrons.

   (*b*) List the sample points generated by the tree.

   (*c*) List the sample points that constitute each of these events:
   A_1: a chain reaction is possible
   A_2: all three neutrons are captured
   A_3: a chain reaction is not possible

   (*d*) Are A_1 and A_2 mutually exclusive?
   Are A_1 and A_3 mutually exclusive?
   Are A_2 and A_3 mutually exclusive?
   Are A_1, A_2, and A_3 mutually exclusive?

   (*e*) The probability that a neutron will be captured depends on its neutron energy and is not the same for each neutron. Under these circumstances, is it correct to say that the probability that all three neutrons will be captured is 1/8 because this can occur in only one way and there are eight paths through the tree of part (*a*)? Explain your answer.

Figure 1.4
50% of the shells fall in the inner ellipse.

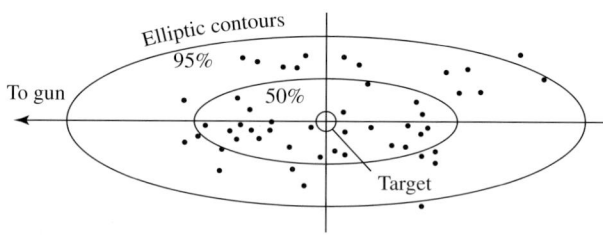

6. In ballistics studies conducted during World War II it was found that, in ground-to-ground firing, artillery shells tended to fall in an elliptical pattern such as that of Fig. 1.4. The probability that a shell would fall in the inner ellipse is .50; the probability that it would fall in the outer ellipse is .95. ("Statistics and Probability Applied to Problems of Antiaircraft Fire in World War II," E. S. Pearson, *Statistics: A Guide to the Unknown,* Holden-Day, San Francisco, 1972, pp. 407–415.)

    (*a*) A firing is considered to be a success (*s*) if the shell falls within the inner ellipse; otherwise it is failure (*f*). Construct a tree to represent the firing of three shells in succession.

    (*b*) List the sample points generated by the tree.

    (*c*) Let A_1 denote the event that the first firing is successful, A_2 the event that the second firing is successful, and A_3 the event that the third firing is successful. List the sample points that make up each of these three events. Are the events mutually exclusive? Explain from both a practical and a mathematical point of view.

    (*d*) Describe the event A_1' verbally, and then list the sample points that make up this event.

    (*e*) Describe the event $A_1 \cap A_2' \cap A_3'$ verbally, and list the sample points that make up this event.

    (*f*) Explain why classical probability can be used to find the probability of the event described in part (*e*), and find this probability.

7. A home computer is tied to a mainframe computer via a telephone modem. The home computer will dial repeatedly until contact is made. Once contact has been established, the dialing process will, of course, end. Let *c* denote the fact that contact is made on a particular attempt and *n* denote that contact is not made.

    (*a*) Construct a tree diagram to represent the dialing process.

    (*b*) Are the paths through the tree equally likely?

    (*c*) List the sample points generated by the tree. Can this list ever be completed?

    (*d*) List the sample points that constitute event *A*: contact is made in at most four attempts.

    (*e*) Give an example of two events that are not impossible but that are mutually exclusive.

8. A missile battery can fire five missiles in rapid succession. As soon as the target is hit, firing will cease. Let *h* denote a hit and *m* a miss.

    (*a*) Draw the tree to represent the possible firing of these missiles at a single incoming target.

    (*b*) Is there any difference in the tree drawn here and that illustrated in Example 1.2.3? Explain.

(*c*) List the sample points generated by the tree.
(*d*) List the sample points that constitute the events
A_1: exactly two shots are fired
A_2: at most two shots are fired
Are these events mutually exclusive? Explain.

REVIEW EXERCISES

9. An electrical control panel has three toggle switches labeled I, II, and III, each of which can be either on (*O*) or off (*F*).
   (*a*) Construct a tree to represent the possible configurations for these three switches.
   (*b*) List the elements of the sample space generated by the tree.
   (*c*) List the sample points that constitute the events
   A: at least one switch is on
   B: switch I is on
   C: no switch is on
   D: four switches are on
   (*d*) Are events *A* and *B* mutually exclusive? Are events *A* and *C* mutually exclusive? Are events *A* and *D* mutually exclusive?
   (*e*) What is the name given to an event such as *D*?
   (*f*) If at any given time each switch is just as likely to be on as off, what is the probability that no switch is on?

10. Two items are randomly selected one at a time from an assembly line and classed as to whether they are of superior quality (+), average quality (0), or inferior quality (−).
   (*a*) Construct a tree for this two-stage experiment.
   (*b*) List the elements of the sample space generated by the tree.
   (*c*) List the sample points that constitute the events
   A: the first item selected is of inferior quality
   B: the quality of each of the items is the same
   C: the quality of the first item exceeds that of the second
   (*d*) Are the events *A* and *B* mutually exclusive? Are the events *A* and *C* mutually exclusive?
   (*e*) Give a brief verbal description of these events:
   $A' \cap B$ $A' \cap B'$
   $A \cap B'$ $A \cap C' \cap B$
   (*f*) It is known that 90% of the items produced are of average quality, 1% are of superior quality, and the rest are of inferior quality. It is argued that since the classification experiment can proceed in nine ways with only one of these resulting in two items of average quality, the probability of obtaining two such items is 1/9. Criticize this argument.

11. An experiment consists of selecting a digit from among the digits 0 to 9 in such a way that each digit has the same chance of being selected as any other. We name the digit selected *A*. These lines of code are then executed:

   IF $A < 2$ THEN $B = 12$; ELSE $B = 17$;
   IF $B = 12$ THEN $C = A - 1$; ELSE $C = 0$;

(a) Construct a tree to illustrate the ways in which values can be assigned to the variables $A, B,$ and C.
(b) Find the sample space generated by the tree.
(c) Are the 10 possible outcomes for this experiment equally likely?
(d) Find the probability that A is an even number.
(e) Find the probability that C is negative.
(f) Find the probability that $C = 0$.
(g) Find the probability that $C \leq 1$.

CHAPTER 2

Some Probability Laws

In Chap. 1 we considered how to interpret probabilities. In this chapter we consider some laws that govern their behavior. The laws that we shall present are those that will have a direct application to problem solving. These laws will be stated and illustrated numerically. Their derivations are not hard, and most of them are left as exercises.

2.1 AXIOMS OF PROBABILITY

You have probably seen the development of a mathematical system in your study of high school geometry. In developing any mathematical system, one begins by stating a few basic definitions and axioms that underlie the system. The definitions are the technical terms of the system; axioms are statements that are assumed to be true and therefore require no proof. Usually one starts with as few axioms as possible and then uses these axioms and the technical definitions to develop whatever theorems follow logically. Some technical terms such as sample space, sample point, event, and mutually exclusive events have already been introduced. One can develop a useful system of theorems pertaining to probability with the aid of these definitions and three axioms, called the axioms of probability.

> **Axioms of probability.**
> 1. Let S denote a sample space for an experiment:
> $$P[S] = 1$$
> 2. $P[A] \geq 0$ for every event A.
> 3. Let $A_1, A_2, A_3, \ldots$ be a finite or an infinite collection of mutually exclusive events. Then $P[A_1 \cup A_2 \cup A_3 \cdots] = P[A_1] + P[A_2] + P[A_3] + \cdots$.

Axiom 1 states a fact that most people regard as obvious; namely, the probability assigned to the certain event S is 1. Axiom 2 ensures that probabilities can never be negative. Axiom 3 guarantees that when one deals with mutually exclusive events, the probability that at least one of the events will occur can be found by adding the individual probabilities. An important consequence of this axiom is that it gives us the ability to find the probability of an event when the sample points in the same space for the experiment are not equally likely. Example 2.1.1 illustrates this point.

Example 2.1.1. The distribution of blood types in the United States is roughly 41% type A, 9% type B, 4% type AB, and 46% type O. An individual is brought into an emergency room and is to be blood-typed. What is the probability that the type will be A, B, or AB?

The sample space for this experiment is

$$S = \{A, B, AB, O\}$$

The sample points are not equally likely, so the classical approach to probability is not applicable. That is, we cannot say that since there are four blood types and three of them are A, B, or AB the probability of obtaining one of these types is ¾. Let A_1, A_2, and A_3 denote the events that the patient has type A, B, and AB blood, respectively. The events A_1, A_2, and A_3 are mutually exclusive because one cannot have two different blood types at the same time. We are looking for $P[A_1 \cup A_2 \cup A_3]$. By axiom 3,

$$
\begin{aligned}
P[A_1 \cup A_2 \cup A_3] &= P[A_1] + P[A_2] + P[A_3] \\
&= .41 + .09 + .04 \\
&= .54
\end{aligned}
$$

An immediate consequence of these axioms is the fact that the probability assigned to the impossible event is 0, as you should suspect. The derivation of this result is outlined in Exercise 12.

Theorem 2.1.1. $P[\varnothing] = 0$.

Another consequence of the axioms is that the probability that an event will *not* occur is equal to 1 minus the probability that it will occur. For example, if the probability of a successful space shuttle mission is .99, then the probability that it will not be successful is $1 - .99 = .01$. This idea is stated in Theorem 2.1.2. Its derivation is outlined in Exercise 12.

Theorem 2.1.2. $P[A'] = 1 - P[A]$.

The General Addition Rule

We have seen how to handle questions concerning the probability of one or another event occurring if those events are mutually exclusive. We now develop a more general rule that will allow us to find the probability that at least one of two events will occur when the events are not necessarily mutually exclusive. This rule is suggested by considering the Venn diagram of Fig. 2.1. Assume that the shaded

Figure 2.1
$A_1 \cap A_2 \neq \varnothing$.

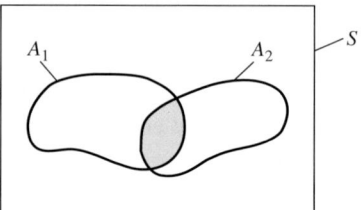

region in the diagram, $A_1 \cap A_2$, is not empty so that A_1 and A_2 are not mutually exclusive. If we claim that

$$P[A_1 \cup A_2] = P[A_1] + P[A_2]$$

we have committed an obvious error. Since $A_1 \cap A_2$ is contained in A_1 and $A_1 \cap A_2$ is contained in A_2, $P[A_1 \cap A_2]$ has been included twice in our calculation. To correct this error, we subtract $P[A_1 \cap A_2]$ from the right-hand side of the equation to obtain the general addition rule:

General addition rule

$$P[A_1 \cup A_2] = P[A_1] + P[A_2] - P[A_1 \cap A_2]$$

This rule can be derived from the axioms of probability and the theorems that we have already developed. Its proof is outlined in Exercise 12. The key word that signals its use is the word "or."

Example 2.1.2. Components of a propulsion system can be arranged in series. However, this arrangement has a serious drawback; if one component fails, the system fails. This is obviously a risky arrangement for space travel! Consider a system in which the main engine has a backup. These engines are designed to operate independently in that the success or failure of one has no effect on the other. The engine component is operable if one *or* the other of these two engines is operable. Such a system is said to have the engine component in parallel. Assume that each engine is 90% reliable. That is, each functions correctly with probability .9. As we shall show later, it is then reasonable to assume that both engines operate correctly with probability .81. Find the probability that the engine component is operable. Let A_1: the main engine is operable, and A_2: the backup engine is operable. We are given that $P[A_1] = P[A_2] = .9$ and that $P[A_1 \cap A_2] = .81$. We want to find $P[A_1 \cup A_2]$. By the addition rule

$$P[A_1 \cup A_2] = P[A_1] + P[A_2] - P[A_1 \cap A_2]$$
$$= .9 + .9 - .81 = .99$$

The addition rule links the operations of union and intersection. If $P[A_1 \cap A_2]$ is known, the addition rule can be used to find $P[A_1 \cup A_2]$. Similarly, if $P[A_1 \cup A_2]$ is known, we can use the rule to find $P[A_1 \cap A_2]$. Venn diagrams are helpful when using this rule.

Example 2.1.3. A chemist analyzes seawater samples for two heavy metals: lead and mercury. Past experience indicates that 38% of the samples taken from near the mouth of a river on which numerous industrial plants are located contain toxic levels of lead or mercury: 32% contain toxic levels of lead and 16% contain toxic levels of mercury. What is the probability that a randomly selected sample will contain toxic levels of lead only? Let A_1 denote the event that the sample contains toxic levels of lead, and let A_2 denote that the sample contains toxic levels of mercury. We are given that $P[A_1] = .32, P[A_2] = .16$, and $P[A_1 \cup A_2] = .38$. By the addition rule

$$P[A_1 \cup A_2] = P[A_1] + P[A_2] - P[A_1 \cap A_2]$$

or
$$.38 = .32 + .16 - P[A_1 \cap A_2]$$

Solving this equation, we obtain $P[A_1 \cap A_2] = .10$. This is indicated in Fig. 2.2(*a*). Since $P[A_1] = .32$ and $A_1 \cap A_2$ is contained in A_1, the probability associated with the shaded region in Fig. 2.2(*b*) is .22. Similarly, since $A_1 \cap A_2$ is contained in A_2, a probability of .06 is associated with the shaded region of Fig. 2.2(*c*). Finally, since

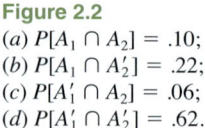

Figure 2.2
(a) $P[A_1 \cap A_2] = .10$;
(b) $P[A_1 \cap A_2'] = .22$;
(c) $P[A_1' \cap A_2] = .06$;
(d) $P[A_1' \cap A_2'] = .62$.

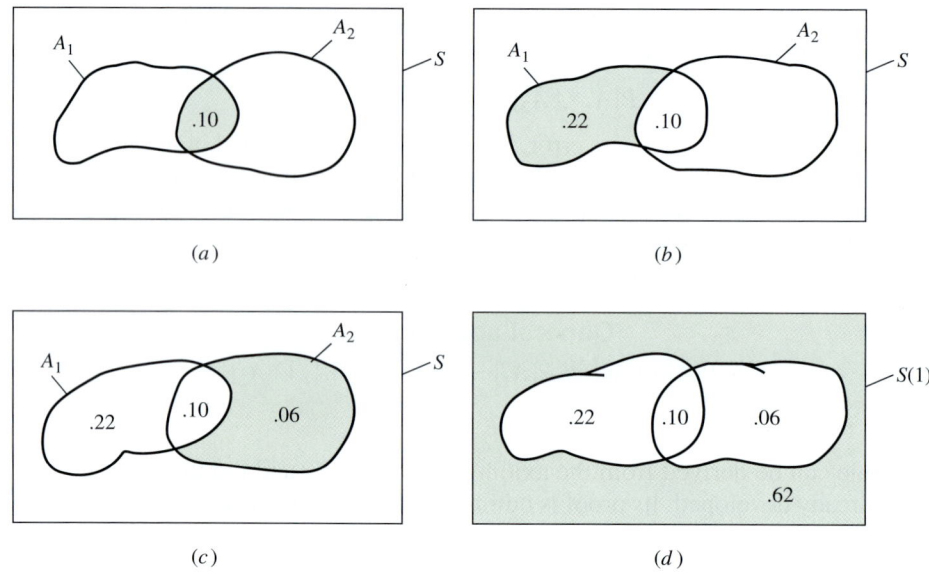

$P[S] = 1$, the probability assigned to the shaded area in Fig. 2.2(d) is .62. We are asked to find the probability that the sample will contain only lead. That is, we want to find $P[A_1 \cap A_2']$. This probability, .22, can be read from Fig. 2.2(b).

Notice that if the percentages reported in problems such as these are based on population data, then the probabilities calculated by use of the general addition rule are exact. However, if the percentages reported are based on samples drawn from a larger population, then the probabilities computed are relative frequency probabilities. They are *approximations* to the true probability of the occurrence of the event in question. Since most percentages reported in the literature are based on samples, most of them are properly viewed as being relative frequency probabilities. We use the word "probability" with the understanding that the probabilities given and computed by using the theorems in this chapter are, in most cases, only approximations.

2.2 CONDITIONAL PROBABILITY

In this section we introduce the notion of conditional probability. The name itself is indicative of what is to be done. We wish to determine the probability that some event A_2 will occur, "conditional on" the assumption that some other event A_1 has occurred. The key words to look for in identifying a conditional question are "if" and "given that." We use the notation $P[A_2 | A_1]$ to denote the conditional probability of event A_2 occurring given that event A_1 has occurred. A simple example will suggest the way to define this probability.

Example 2.2.1. In trying to determine the sex of a child a pregnancy test called "starch gel electrophoresis" is used. This test may reveal the presence of a protein zone called the pregnancy zone. This zone is present in 43% of all pregnant women. Furthermore, it is known that 51% of all children born are male. Seventeen percent of all children born are male and the pregnancy zone is present. The Venn diagram for these data is shown in Fig. 2.3. Let A_1 denote the event that the pregnancy zone is present, and A_2 that the child is male. We know that, for a randomly selected pregnant woman, $P[A_1] = .43, P[A_2] = .51, P[A_1 \cap A_2] = .17$. If asked, "What is the probability

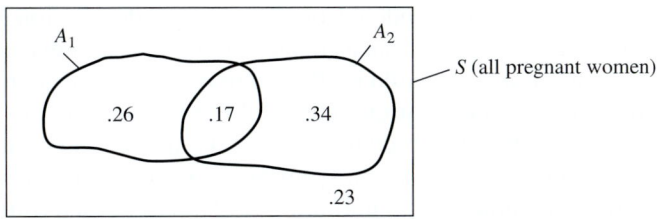

Figure 2.3
Partition of S.

that the child is male?" the answer is .51. Suppose we are *given* the information that the pregnancy zone is present and asked, "What is the probability that the child is male?" We now have information that was not available originally. What effect, if any, does this new information have on our belief that the child is male? That is, what is $P[A_2|A_1]$? Once we know that the pregnancy zone is present, our sample space no longer includes all pregnant women; it consists only of the 43% with this characteristic. Of these, $.17/.43 \doteq .395$ have male children. Logic implies that

$$P[\text{male} | \text{zone present}] = P[A_2|A_1] = .395$$

Receipt of the information that the pregnancy zone is present reduces from .51 to .395 the probability that the child is male.

To formalize the reasoning used in the previous example, note that $P[A_2|A_1]$ is found by forming a ratio whose denominator is $P[A_1]$, the probability that the *given* event will occur. The numerator is $P[A_1 \cap A_2]$, the probability that *both* the given event and the event in question will occur. That is, we define the conditional probability as follows:

> **Definition 2.2.1 (Conditional probability).** Let A_1 and A_2 be events such that $P[A_1] \neq 0$. The conditional probability of A_2 given A_1, denoted by $P[A_2|A_1]$, is defined by
>
> $$P[A_2|A_1] = \frac{P[A_1 \cap A_2]}{P[A_1]}$$

Sometimes receipt of the information that event A_1 has occurred has no effect on the probability assigned to event A_2. That is,

$$P[A_2|A_1] = P[A_2]$$

When this happens, A_1 and A_2 have a special relationship to one another. The nature of this relationship will be explored in the next section. In the meantime don't be surprised if you find that a particular conditional probability does not differ from the original probability assigned to the event!

2.3 INDEPENDENCE AND THE MULTIPLICATION RULE

We have used the word "independent" informally in several previous examples. Webster's dictionary defines independent objects as objects acting "irrespective of each other." Thus two events are independent if one may occur irrespective of the other. That is, the occurrence or nonoccurrence of one does not alter the likelihood of occurrence or nonoccurrence of the other. In some cases it is reasonable to assume that two events are independent from the physical description of the events themselves. For example, suppose that a couple heterozygous for eye color has two children. Since the eye color of a child is affected only by the genetic makeup of the parents and not by the eye color of the other child, it is reasonable to assume that the events A_1: the first child has brown eyes, and A_2: the second child has brown eyes, are independent. However, in most instances the issue is not clear-cut. In these cases we need a mathematical definition of the term to determine without a doubt whether two events are, in fact, independent.

To see how to characterize independence, let us consider a simple experiment that consists of rolling a single fair die once and then tossing a fair coin once. Let the first member of each ordered pair denote the number appearing on the die and the second, the face showing on the coin (H = heads, T = tails). A sample space for this experiment is

$$S = \{(1, H), (1, T), (2, H), (2, T), (3, H), (3, T),$$
$$(4, H), (4, T), (5, H), (5, T), (6, H), (6, T)\}$$

Since the die and the coin are considered to be fair, these 12 outcomes are equally likely. Consider these events:

> A: the die shows one or two
>
> B: the coin shows heads
>
> $A \cap B$: the die shows one or two and the coin shows heads

Since knowing the result of the die roll gives us no additional information on how the coin will land, it is reasonable to assume that the events A and B are independent. Using classical probability, we easily see that

$$P[A] = P[\{(1, H), (1, T), (2, H), (2, T)\}] = 4/12 = 1/3$$
$$P[B] = P[\{(1, H), (2, H), (3, H), (4, H), (5, H), (6, H)\}]$$
$$= 6/12 = 1/2$$
$$P[A \cap B] = P[\{(1, H), (2, H)\}] = 2/12 = 1/6$$

More importantly, it is easy to see that for these physically independent events

$$P[A \cap B] = P[A] \cdot P[B]$$

Consider now an experiment that consists of drawing two coins in succession from a box containing a nickel (N), a dime (D), and a quarter (Q). The first coin is not replaced before the second is drawn. A sample space for this experiment is

$$S = \{(N, D), (N, Q), (D, N), (D, Q), (Q, N), (Q, D)\}$$

These outcomes are equally likely. Consider these events:

> A: the first coin is a dime
>
> B: the second coin is a dime

Since we do not replace the first coin before the second draw, it is evident that if event A occurs, event B cannot occur. That is, knowledge that event A has occurred does give us information on whether or not event B will occur! These events are not independent. Using classical probability, we easily see that

$$P[A] = P[\{(D, N), (D, Q)\}] = 2/6$$
$$P[B] = P[\{(N, D), (Q, D)\}] = 2/6$$
$$P[A \cap B] = P[\varnothing] = 0$$

More importantly, it is easy to see that for these events that are not independent

$$P[A \cap B] \neq P[A]P[B]$$

Thus we have noticed that when A and B are clearly independent, $P[A \cap B] = P[A]P[B]$; when they are clearly dependent, $P[A \cap B] \neq P[A]P[B]$. This is not coincidental. It is natural to use this mathematical characterization as our technical definition of the term "independent events."

> **Definition 2.3.1 (Independent events).** Events A_1 and A_2 are independent if and only if
>
> $$P[A_1 \cap A_2] = P[A_1]P[A_2]$$

This definition is useful in two ways. If exact probabilities are available, then it serves as a test for independence. However, since most probabilities encountered in scientific studies are approximations, it is most useful as a way to find the probability that two events will occur when the events are clearly independent. Example 2.3.1 illustrates its use as a test for independence.

Example 2.3.1. Consider the experiment of drawing a card from a well-shuffled deck of 52 cards. Let

$$A_1: \text{a spade is drawn}$$

$$A_2: \text{an honor } (10, J, Q, K, A) \text{ is drawn}$$

Classical probability is used to see that $P[A_1] = 13/52$ and $P[A_2] = 20/52$. The probability that a spade and an honor, $P[A_1 \cap A_2]$, is drawn is 5/52. Notice that these probabilities are exact. They are not approximations based on observations of card draws. Are the events A_1 and A_2 independent? To decide, note that

$$P[A_1]P[A_2] = (13/52)(20/52) = 5/52$$

and $\quad P[A_1 \cap A_2] = 5/52$

Since $P[A_1 \cap A_2] = P[A_1]P[A_2]$, we can conclude that these events are independent.

In Chap. 12 a test for independence will be developed that can be used when working with real data rather than with classical probabilities. Its derivation is based on the definition of independent events just discussed.

Example 2.3.2 illustrates the use of Definition 2.3.1 in finding the probability that two events will occur simultaneously when the events are clearly independent.

Example 2.3.2. In Example 1.1.3, we found that the probability that a couple heterozygous for eye color will parent a brown-eyed child is 3/4 for each child. Genetic studies indicate that the eye color of one child is independent of that of the other. Thus if the couple has two children, then the probability that both will be brown-eyed is

$$P\left[\begin{array}{c} \text{first} \\ \text{brown} \end{array} \text{ and } \begin{array}{c} \text{second} \\ \text{brown} \end{array}\right] = P\left[\begin{array}{c} \text{first} \\ \text{brown} \end{array}\right] P\left[\begin{array}{c} \text{second} \\ \text{brown} \end{array}\right]$$

$$= \frac{3}{4} \cdot \frac{3}{4}$$

$$= \frac{9}{16}$$

Definition 2.3.1 defines independence for *any* events A_1 and A_2. If at least one of the events A_1 or A_2 occurs with *nonzero* probability, then an appealing characterization of independence can be obtained. To see how this is done, assume that $P[A_1] \neq 0$. By Definition 2.3.1, A_1 and A_2 are independent if and only if

$$P[A_1 \cap A_2] = P[A_1]P[A_2]$$

Dividing by $P[A_1]$, we can conclude that A_1 and A_2 are independent if and only if

$$\frac{P[A_1 \cap A_2]}{P[A_1]} = P[A_2|A_1] = P[A_2]$$

A similar argument holds if $P[A_2] \neq 0$. We have thus derived the result given in Theorem 2.3.1.

> **Theorem 2.3.1.** Let A_1 and A_2 be events such that at least one of $P[A_1]$ or $P[A_2]$ is nonzero. A_1 and A_2 are independent if and only if
>
> $$P[A_2|A_1] = P[A_2] \qquad \text{if } P[A_1] \neq 0 \qquad \text{and}$$
> $$P[A_1|A_2] = P[A_1] \qquad \text{if } P[A_2] \neq 0$$

Since most events of real interest do occur with nonzero probability, Theorem 2.3.1 is used as a test for independence. To understand the logic behind the theorem, let us reconsider the data of Example 2.3.1.

> **Example 2.3.3.** Consider the events A_1, a spade is drawn, and A_2, an honor is drawn. We know that $P[A_1] = 13/52$, $P[A_2] = 20/52$, and $P[A_1 \cap A_2] = 5/52$. Suppose we are asked, "What is the probability that a randomly selected card is an honor?" Our answer is 20/52. Suppose we are now told that the card is a spade and are asked, "What is the probability that the card is an honor?" That is, "What is $P[A_2|A_1]$?" If A_1 and A_2 are independent, the new information is irrelevant and our answer should not change. That is, $P[A_2|A_1] = P[A_2]$. Otherwise our answer should change, and $P[A_2|A_1] \neq P[A_2]$. In this setting, is $P[A_2|A_1] = P[A_2]$? To answer this question, note that
>
> $$P[A_2|A_1] = \frac{P[A_1 \cap A_2]}{P[A_1]} = \frac{5/52}{13/52} = 5/13$$
>
> and $\qquad P[A_2] = 20/52 = 5/13$

Since these probabilities are the same, we conclude via Theorem 2.3.1 that A_1 and A_2 are independent.

Occasionally we must deal with more than two events. Again, the question arises, "When are these events considered independent?" Definition 2.3.2 answers this question by extending our previous definition to include more than two events.

> **Definition 2.3.2.** Let $C = \{A_i: i = 1, 2, \ldots, n\}$ be a finite collection of events. These events are independent if and only if, given any subcollection $A_{(1)}, A_{(2)}, \ldots, A_{(m)}$ of elements of C,
>
> $$P[A_{(1)} \cap A_{(2)} \cap \cdots \cap A_{(m)}] = P[A_{(1)}]P[A_{(2)}] \cdots P[A_{(m)}]$$

Although this definition can be used to test a collection of events for independence, its main purpose is to provide a way to find the probability that a series of events that are assumed to be independent will occur. To illustrate, we reconsider a problem encountered in Chap. 1 (Example 1.2.1).

Example 2.3.4. During a space shot, the primary computer system is backed up by two secondary systems. They operate independently of one another, and each is 90% reliable. What is the probability that all three systems will be operable at the time of the launch? Let

$$A_1: \text{the main system is operable}$$
$$A_2: \text{the first backup is operable}$$
$$A_3: \text{the second backup is operable}$$

We are given that $P[A_1] = P[A_2] = P[A_3] = .9$. We want $P[A_1 \cap A_2 \cap A_3]$. Since these events are assumed to be independent,

$$P[A_1 \cap A_2 \cap A_3] = P[A_1]P[A_2]P[A_3]$$
$$= (.9)(.9)(.9)$$
$$= .729$$

Definition 2.3.2 must be used with care. In particular, one must be certain that it is reasonable to assume that events are independent before it is applied to compute the probability that a series of events will occur. The danger of erroneously assumed independence is illustrated in Example 2.3.5.

Example 2.3.5. An Atomic Energy Commission Study, WASH 1400, reported the probability of a nuclear accident such as that which occurred at Three Mile Island in March 1978 to be one in 10 million. Yet the accident did occur. According to Mark Stephens, "The methodology of WASH 1400 made use of event trees—sequences of actions that would be necessary for accidents to take place. These event trees did not assume any interrelation between events—that they might be caused by the same error in judgment or as part of the same mistaken action. The statisticians who assigned probabilities in the writing of WASH 1400 said, for example, that there was a one-in-a-thousand risk of one of the auxiliary feed-water control valves—the twelves—being closed. And if there is a one-in-a-thousand chance of one valve being closed, the chances of both valves being closed is one-thousandth of that, or a million to one. But both of the twelves were closed by the same man on March 26—and one had never been closed without the other." The events A_1: the first valve is closed, and A_2: the second valve is closed were not independent. However, they were treated as such when calculating the probability of an accident. This, among other things, led to an underestimate of the accident potential (from *Three Mile Island* by Mark Stephens, Random House, 1980).

The Multiplication Rule

There is one further point to be made before we conclude this section. We can find $P[A_1 \cap A_2]$ if the events are assumed to be independent. Furthermore, if the proper information is given, the general addition rule can be used to find this probability. Is there any other way to find the probability of the simultaneous occurrence of two events if the events are not independent? The answer is yes, and the method is easy to derive. We know that

$$P[A_2|A_1] = \frac{P[A_1 \cap A_2]}{P[A_1]} \qquad P[A_1] \neq 0$$

regardless of whether the events are independent. Multiplying each side of this equation by $P[A_1]$, we obtain the following formula, called the *multiplication rule:*

Multiplication rule

$$P[A_1 \cap A_2] = P[A_2|A_1]P[A_1]$$

The use of this rule is illustrated in Example 2.3.6.

Example 2.3.6.   Recent research indicates that approximately 49% of all infections involve anaerobic bacteria. Furthermore, 70% of all anaerobic infections are polymicrobic; that is, they involve more than one anaerobe. What is the probability that a given infection involves anaerobic bacteria *and* is polymicrobic? Let A_1 denote the event that the infection is anaerobic, and A_2 that it is polymicrobic. We are given that $P[A_1] = .49$ and that $P[A_2|A_1] = .70$. We want to find $P[A_1 \cap A_2]$. By the multiplication rule,

$$P[A_1 \cap A_2] = P[A_2|A_1]P[A_1]$$
$$= (.70)(.49)$$
$$= .343$$

2.4   BAYES' THEOREM

The topic of this section is the theorem formulated by the Reverend Thomas Bayes (1761). It deals with conditional probability. Bayes' theorem is used to find $P[A|B]$ when the available information is not immediately compatible with that required to apply the definition of conditional probability directly.

Example 2.4.1 is a typical problem calling for the use of Bayes' theorem. You will find applying Bayes' rule quite natural without having seen a formal statement of the theorem!

Example 2.4.1.   Assume that 40% of all interstate highway accidents involve excessive speed on the part of at least one of the drivers (event E) and that 30% involve alcohol use by at least one driver (event A). If alcohol is involved there is a 60% chance that excessive speed is also involved; otherwise, this probability is only 10%. An accident involves speeding. What is the probability that alcohol is involved? We are given these probabilities:

$$P[E] = .40 \quad P[A] = .30 \quad P[E|A] = .60$$
$$P[E'] = .60 \quad P[A'] = .70 \quad P[E|A'] = .10$$

We are being asked to find $P[A|E]$. Since this is a conditional question, it is natural to turn to the definition of conditional probability for a solution. In this case,

$$P[A|E] = \frac{P[E \cap A]}{P[E]}$$

Unfortunately, neither of the probabilities needed for the solution is immediately available. However, each can be obtained easily. By the multiplication rule,

$$P[E \cap A] = P[E|A]P[A]$$

Note that if excessive speed was involved, alcohol use either was or was not also involved. Hence event E can be subdivided into two mutually exclusive events as follows:

$$E = (E \cap A) \cup (E \cap A')$$

Thus $$P[E] = P[E \cap A] + P[E \cap A']$$

An expression has already been found for the first probability on the right; the multiplication rule can be applied to the second probability to see that

$$P[E \cap A'] = P[E|A']P[A']$$

Substitution now yields

$$P[A|E] = \frac{P[E \cap A]}{P[E]}$$

$$= \frac{P[E|A]P[A]}{P[E|A]P[A] + P[E|A']P[A']}$$

Note the pattern in this solution. In the numerator the conditional expression is the reverse of that in the original question; in the denominator, the conditional expressions run through all of the alternatives to the event in question, in this case A and A'. The numerical solution can now be obtained by substitution as follows:

$$P[A|E] = \frac{P[E|A]P[A]}{P[E|A]P[A] + P[E|A']P[A']}$$

$$= \frac{(.60)(.30)}{(.60)(.30) + (.10)(.70)}$$

$$= .72$$

If excessive speed was involved in an accident, there is a 72% chance that alcohol was also involved.

In the previous example, there were two mutually exclusive events, A and A', whose union is S. Bayes' theorem can also be applied when S is subdivided into more than two mutually exclusive events. We state the theorem in this more general setting.

> **Theorem 2.4.1 (Bayes' theorem).** Let $A_1, A_2, A_3, \ldots, A_n$ be a collection of mutually exclusive events whose union is S. Let B be an event such that $P[B] \neq 0$. Then for any of the events $A_j, j = 1, 2, 3, \ldots, n,$
>
> $$P[A_j|B] = \frac{P[B|A_j]P[A_j]}{\sum_{i=1}^{n} P[B|A_i]P[A_i]}$$

To see that Bayes' theorem could have been used directly to answer the question posed in Example 2.4.1, note that events A and A' are mutually exclusive events whose union is S and that event E occurs with nonzero probability. Hence we can make the following identifications:

$$A_1 = A \qquad A_2 = A' \qquad B = E$$

By applying Bayes' theorem directly we obtain

$$P[A_1|B] = \frac{P[B|A_1]P[A_1]}{P[B|A_1]P[A_1] + P[B|A_2]P[A_2]}$$

or
$$P[A|E] = \frac{P[E|A]P[A]}{P[E|A]P[A] + P[E|A']P[A']}$$

A quick comparison will show that this is the same as the solution derived in Example 2.4.1 using the multiplication rule.

The next example illustrates the use of Bayes' theorem in a setting in which the sample space is subdivided into four mutually exclusive events rather than two.

Example 2.4.2. The blood type distribution in the United States is type A, 41%; type B, 9%; type AB, 4%; and type O, 46%. It is estimated that during World War II, 4% of inductees with type O blood were typed as having type A; 88% of those with type A were correctly typed; 4% with type B blood were typed as A; and 10% with type AB were typed as A. A soldier was wounded and brought to surgery. He was typed as having type A blood. What is the probability that this is his true blood type? Let

$$A_1: \text{he has type A blood}$$
$$A_2: \text{he has type B blood}$$
$$A_3: \text{he has type AB blood}$$
$$A_4: \text{he has type O blood}$$
$$B: \text{he is typed as type A}$$

Note that the events A_1, A_2, A_3, A_4 are mutually exclusive, and their union is S because each individual can have only one blood type and all possible blood types have been listed. We are being asked to find $P[A_1|B]$. We are given that

$$P[A_1] = .41 \quad P[B|A_1] = .88$$
$$P[A_2] = .09 \quad P[B|A_2] = .04$$
$$P[A_3] = .04 \quad P[B|A_3] = .10$$
$$P[A_4] = .46 \quad P[B|A_4] = .04$$

Substitution into the expression given by Bayes' theorem yields

$$P[A_1|B] = \frac{(.88)(.41)}{(.88)(.41) + (.04)(.09) + (.10)(.04) + (.04)(.46)}$$
$$\doteq .93$$

If a person was typed as having type A blood, there was approximately a 93% chance that his true type was in fact type A.

CHAPTER SUMMARY

In this chapter we presented some of the laws that govern the behavior of probabilities. We began with the axioms, and from those we were able to derive the remaining laws. In particular, we derived the addition rule, which deals with the probability of the union of two events; the multiplication rule, which deals with the probability of the intersection of two events; and Bayes' theorem, which deals with conditional probability. We introduced and defined important terms that you should know. These are:

Conditional probability   Independent events

Care must be taken when using the concept of independence. In an applied problem, be sure that it is reasonable to assume that events A and B are independent before finding the probability of their joint occurrence via the definition $P[A \cap B] = P[A]P[B]$.

EXERCISES

Section 2.1

1. The probability that a wildcat well will produce oil is 1/13. What is the probability that it will not be productive?

2. The theft of precious metals from companies in the United States was and is a serious problem. The estimated probability that such a theft will involve a particular metal is given below: (Based on data reported in "Materials Theft," *Materials Engineering,* February 1982, pp. 27–31.)

tin: 1/35	platinum: 1/35	nickel: 1/35
steel: 11/35	gold: 5/35	zinc: 1/35
copper: 8/35	aluminum: 2/35	silver: 4/35
titanium: 1/35		

(Note that these events are assumed to be mutually exclusive.)
   (*a*) What is the probability that a theft of precious metal will involve gold, silver, or platinum?
   (*b*) What is the probability that a theft will not involve steel?

3. Assuming the blood type distribution to be A: 41%, B: 9%, AB: 4%, O: 46%, what is the probability that the blood of a randomly selected individual will contain the A antigen? That it will contain the B antigen? That it will contain neither the A nor the B antigen?

4. Assume that the engine component of a spacecraft consists of two engines in parallel. If the main engine is 95% reliable, the backup is 80% reliable, and the engine component as a whole is 99% reliable, what is the probability that both engines will be operable? Use a Venn diagram to find the probability that the main engine will fail but the backup will be operable. Find the probability that the backup engine will fail but the main engine will be operable. What is the probability that the engine component will fail?

5. When an individual is exposed to radiation, death may ensue. Factors affecting the outcome are the size of the dose, the length and intensity of the exposure, and the biological makeup of the individual. The term LD_{50} is used to denote the dose that is usually lethal for 50% of the individuals exposed to it. Assume that in a nuclear accident 30% of the workers are exposed to the LD_{50} and die; 40% of the workers die; and 68% are exposed to the LD_{50} or die. What is the probability that a randomly selected worker is exposed to the LD_{50}? Use a Venn diagram to find the probability that a randomly selected worker is exposed to the LD_{50} but does not die. Find the probability that a randomly selected worker is not exposed to the LD_{50} but dies.

6. When a computer goes down, there is a 75% chance that it is due to an overload and a 15% chance that it is due to a software problem. There is an 85% chance that it is due to an overload or a software problem. What is the probability that both of these problems are at fault? What is the probability that there is a software problem but no overload?

7. Due to the recent energy crisis in California, rolling blackouts were necessary and more might be necessary in the future. Assume that there is a 60% chance that the temperature will exceed 85° F on any given day in July in a particular area. Assume that there is a 30% chance that a rolling blackout will be needed in that area. There is a 20% chance that both events will occur. Find the probability that the temperature will exceed 85° F on a given July day but that no rolling blackout will be needed on that day.

8. Experience shows that 25% of all complaints about home telephone lines involve static on the line. Fifty percent involve line deterioration. Thirty-five percent involve only line deterioration. What is the probability that a randomly selected complaint will involve both problems? Will involve neither problem?

9. Assume that in a particular military exercise involving two units, Red and Blue, there is a 60% chance that the Red unit will successfully meet its objectives and a 70% chance that the Blue unit will do so. There is an 18% chance that only the Red unit will be successful. What is the probability that both units will meet their objectives? What is the probability that one or the other but not both of the units will be successful?

10. It has been found that 80% of all accidents at foundries involve human error and 40% involve equipment malfunction. Thirty-five percent involve both problems. An accident at a foundry is investigated. What is the probability that human error alone was involved?

11. Assume that 1% of all tires of a particular brand are defective due to a problem with a supplier of an important chemical component of the tire. Assume that .5% of this brand of tire will eventually fail due to sidewall blowouts. Also, 1.4% of this brand of tire experience at least one of these problems. What is the probability that in a future accident involving these tires, a blowout will occur but there will be no problem found with the chemical composition of the tire?

12. (a) Derive Theorem 2.1.1.
    Hint: Note that $S = S \cup \varnothing$ and that S and $\varnothing$ are mutually exclusive. Apply axioms 3 and 1.
    (b) Derive Theorem 2.1.2.
    Hint: Note that $S = A \cup A'$ and that A and A' are mutually exclusive. Apply axioms 3 and 1.
    (c) Let A be a subset of B. Show that $P[A] \leq P[B]$.
    Hint: $B = A \cup (A' \cap B)$. Apply axioms 3 and 2.
    (d) Show that the probability of any event A is at most 1.
    Hint: $A \subseteq S$. Apply Exercise 12C and axiom 1.
    (e) Let A_1 and A_2 be mutually exclusive. By axiom 3, $P[A_1 \cup A_2] = P[A_1] + P[A_2]$. Show that the general addition rule yields the same result.

Section 2.2

13. Use the data of Exercise 5 to answer these questions.
    (a) What is the probability that a randomly selected worker will die given that he is exposed to the lethal dose of radiation?
    (b) What is the probability that a randomly selected worker will not die given that he is exposed to the lethal dose of radiation?
    (c) What theorem allows you to determine the answer to (b) from knowledge of the answer to (a)?
    (d) What is the probability that a randomly selected worker will die given that he is not exposed to the lethal dose?
    (e) Is $P[\text{die}] = P[\text{die}|\text{exposed to lethal dose}]$? Did you expect these to be the same? Explain.

14. Use the data of Exercise 4 to answer these questions.
    (a) What is the probability that in an engine system such as that described the backup engine will function given that the main engine fails?
    (b) Is $P[\text{backup functions}] = P[\text{backup functions}|\text{main fails}]$? Did you expect these to be the same? Explain.

15. In a study of waters near power plants and other industrial plants that release wastewater into the water system it was found that 5% showed signs of chemical and thermal pollution, 40% showed signs of chemical pollution, and 35% showed evidence of thermal pollution. Assume that the results of the study accurately reflect the general situation. What is the probability that a stream that shows some thermal pollution will also show signs of chemical pollution? What is the probability that a stream showing chemical pollution will not show signs of thermal pollution?

16. A random digit generator on an electronic calculator is activated twice to simulate a random two-digit number. Theoretically, each digit from 0 to 9 is just as likely to appear on a given trial as any other digit.
    (a) How many random two-digit numbers are possible?
    (b) How many of these numbers begin with the digit 2?
    (c) How many of these numbers end with the digit 9?
    (d) How many of these numbers begin with the digit 2 and end with the digit 9?
    (e) What is the probability that a randomly formed number ends with 9 given that it begins with a 2. Did you anticipate this result?

17. In studying the causes of power failures, these data have been gathered.

    5% are due to transformer damage
    80% are due to line damage
    1% involve both problems

    Based on these percentages, approximate the probability that a given power failure involves
    (a) line damage given that there is transformer damage
    (b) transformer damage given that there is line damage
    (c) transformer damage but not line damage
    (d) transformer damage given that there is no line damage
    (e) transformer damage or line damage

Section 2.3

18. Let A_1 and A_2 be events such that $P[A_1] = .5$, $P[A_2] = .7$. What must $P[A_1 \cap A_2]$ equal for A_1 and A_2 to be independent?

19. Let A_1 and A_2 be events such that $P[A_1] = .6$, $P[A_2] = .4$, and $P[A_1 \cup A_2] = .8$. Are A_1 and A_2 independent?

20. Consider your answer to Exercise 14(b). Are the events A_1: the backup engine functions, and A_2: the main engine fails independent?

21. Studies in population genetics indicate that 39% of the available genes for determining the Rh blood factor are negative. Rh negative blood occurs if and only if the individual has two negative genes. One gene is inherited independently from each parent. What is the probability that a randomly selected individual will have Rh negative blood?

22. An individual's blood group (A, B, AB, O) is independent of the Rh classification. Find the probability that a randomly selected individual will have AB negative blood. *Hint:* See Example 2.1.1 and Exercise 21.

23. The use of plant appearance in prospecting for ore deposits is called geobotanical prospecting. One indicator of copper is a small mint with a mauve-colored flower. Suppose that, for a given region, there is a 30% chance that the soil has a high copper content and a 23% chance that the mint will be present there. If the copper content is high, there is a 70% chance that the mint will be present.

(*a*) Find the probability that the copper content will be high and the mint will be present.

(*b*) Find the probability that the copper content will be high given that the mint is present.

24. The most common water pollutants are organic. Since most organic materials are broken down by bacteria that require oxygen, an excess of organic matter may result in a depletion of available oxygen. In turn this can be harmful to other organisms living in the water. The demand for oxygen by the bacteria is called the biological oxygen demand (BOD). A study of streams located near an industrial complex revealed that 35% have a high BOD, 10% show high acidity, and 40% of streams with high acidity have a high BOD. Find the probability that a randomly selected stream will exhibit both characteristics.

25. A study of major flash floods that occurred over the last 15 years indicates that the probability that a flash flood warning will be issued is .5 and that the probability of dam failure during the flood is .33. The probability of dam failure given that a warning is issued is .17. Find the probability that a flash flood warning will be issued and a dam failure will occur. (Based on data reported in *McGraw-Hill Yearbook of Science and Technology,* 1980, pp. 185–186.)

26. The ability to observe and recall details is important in science. Unfortunately, the power of suggestion can distort memory. A study of recall is conducted as follows: Subjects are shown a film in which a car is moving along a country road. There is no barn in the film. The subjects are then asked a series of questions concerning the film. Half the subjects are asked, "How fast was the car moving when it passed the barn?" The other half is not asked the question. Later each subject is asked, "Is there a barn in the film?" Of those asked the first question concerning the barn, 17% answer "yes"; only 3% of the others answer "yes." What is the probability that a randomly selected participant in this study claims to have seen the nonexistent barn? Is claiming to see the barn independent of being asked the first question about the barn? *Hint:*

$$P[\text{yes}] = P[\text{yes and asked about barn}] + P[\text{yes and not asked about barn}]$$

(Based on a study reported in *McGraw-Hill Yearbook of Science and Technology,* 1981, pp. 249–251.)

27. The probability that a unit of blood was donated by a paid donor is .67. If the donor was paid, the probability of contracting serum hepatitis from the unit is .0144. If the donor was not paid, this probability is .0012. A patient receives a unit of blood. What is the probability of the patient's contracting serum hepatitis from this source?

28. Show that the impossible event is independent of every other event.

29. Consider the percentages given in Exercise 7. Find the probability of a rolling blackout occurring on a day on which the temperature exceeds 85° F. If the probabilities given are assumed to be exact, is the event that a rolling blackout occurs independent of the event that the temperature exceed 85° F? Explain based on the probability that you just computed.

30. Assume that there is a 50% chance of hard drive damage if a power line to which a computer is connected is hit during an electrical storm. There is a 5% chance that an electrical storm will occur on any given summer day in a given area. If there is a .1% chance that the line will be hit during a storm, what is the probability that the line will be hit and there will be hard drive damage during the next electrical storm in this area?

31. A foundry is producing cast iron parts to be used in the automatic transmissions of trucks. There are two crucial dimensions to the part, A and B. Assume that if the part meets specifications on dimension A then there is a 98% chance that it will also meet specifications on dimension B. There is a 95% chance that it will meet specifications on dimension A and a 97% chance that it will meet specifications on dimension B. A part is randomly selected and inspected. What is the probability that it will meet specifications on both dimensions?

32. Let A_1 and A_2 be mutually exclusive events such that $P[A_1]P[A_2] > 0$. Show that these events are not independent.

33. Let A_1 and A_2 be independent events such that $P[A_1]P[A_2] > 0$. Show that these events are not mutually exclusive.

Section 2.4

34. Use the data of Example 2.4.2 to find the probability that an inductee who was typed as having type A blood actually had type B blood.

35. A test has been developed to detect a particular type of arthritis in individuals over 50 years old. From a national survey it is known that approximately 10% of the individuals in this age group suffer from this form of arthritis. The proposed test was given to individuals with confirmed arthritic disease, and a correct test result was obtained in 85% of the cases. When the test was administered to individuals of the same age group who were known to be free of the disease, 4% were reported to have the disease. What is the probability that an individual has this disease given that the test indicates its presence?

36. It is reported that 50% of all computer chips produced are defective. Inspection ensures that only 5% of the chips legally marketed are defective. Unfortunately, some chips are stolen before inspection. If 1% of all chips on the market are stolen, find the probability that a given chip is stolen given that it is defective.

37. As society becomes dependent on computers, data must be communicated via public communication networks such as satellites, microwave systems, and telephones. When a message is received, it must be authenticated. This is done by using a secret enciphering key. Even though the key is secret, there is always the possibility that it will fall into the wrong hands, thus allowing an unauthentic message to appear to be authentic. Assume that 95% of all messages received are authentic. Furthermore, assume that only .1% of all unauthentic messages are sent using the correct key and that all authentic messages are sent using the correct key. Find the probability that a message is authentic given that the correct key is used.

REVIEW EXERCISES

38. A survey of engineering firms reveals that 80% have their own mainframe computer (M), 10% anticipate purchasing a mainframe computer in the near future (B), and 5% have a mainframe computer and anticipate buying another in the near future. Find the probability that a randomly selected firm:

(a) has a mainframe computer or anticipates purchasing one in the near future

(b) does not have a mainframe computer and does not anticipate purchasing one in the near future

(c) anticipates purchasing a mainframe computer given that it does not currently have one

(d) has a mainframe computer given that it anticipates purchasing one in the near future

39. In a simulation program, three random two-digit numbers will be generated independently of one another. These numbers assume the values 00, 01, 02, . . . , 99 with equal probability.

(a) What is the probability that a given number will be less than 50?

(b) What is the probability that each of the three numbers generated will be less than 50?

40. A power network involves three substations A, B, and C. Overloads at any of these substations might result in a blackout of the entire network. Past history has shown that if substation A alone experiences an overload, then there is a 1% chance of a network blackout. For stations B and C alone these percentages are 2% and 3%, respectively. Overloads at two or more substations simultaneously result in a blackout 5% of the time. During a heat wave there is a 60% chance that substation A alone will experience an overload. For stations B and C these percentages are 20 and 15%, respectively. There is a 5% chance of an overload at two or more substations simultaneously. During a particular heat wave a blackout due to an overload occurred. Find the probability that the overload occurred at substation A alone; substation B alone; substation C alone; two or more substations simultaneously.

41. A computer center has three printers, A, B, and C, which print at different speeds. Programs are routed to the first available printer. The probability that a program is routed to printers A, B, and C are .6, .3, and .1, respectively. Occasionally a printer will jam and destroy a printout. The probability that printers A, B, and C will jam are .01, .05, and .04, respectively. Your program is destroyed when a printer jams. What is the probability that printer A is involved? Printer B is involved? Printer C is involved?

42. A chemical engineer is in charge of a particular process at an oil refinery. Past experience indicates that 10% of all shutdowns are due to equipment failure *alone*, 5% are due to a combination of equipment failure and operator error, and 40% involve operator error. A shutdown occurs. Find the probability that

(a) equipment failure or operator error is involved

(b) operator error alone is involved

(c) neither operator error nor equipment failure is involved

(d) operator error is involved given that equipment failure occurs

(e) operator error is involved given that equipment failure does not occur

43. Assume that the probability that the air brakes on large trucks will fail on a particularly long downgrade is .001. Assume also that the emergency brakes on such trucks can stop a truck on this downgrade with probability .8. These braking systems operate independently of one another. Find the probability that

(a) the air brakes fail but the emergency brakes can stop the truck

(b) the air brakes fail and the emergency brakes cannot stop the truck

(c) the emergency brakes cannot stop the truck given that the air brakes fail

44. Consider the problem of Example 1.2.3. Assume that sampling is independent and that at each stage the probability of obtaining a defective part when the process is working correctly is .01. If the process is working correctly, what is the probability that the first defective part will be obtained on the fourth sample? On or before the fourth sample?

CHAPTER 3

Discrete Distributions

In the sciences one often deals with "variables." Webster's dictionary defines a variable as a "quantity that may assume any one of a set of values." In statistics we deal with *random variables*—variables whose observed value is determined by chance. Many of the examples presented in previous chapters involved random variables even though the term was not used at the time. Random variables usually fall into one of two categories; they are either discrete or continuous. We begin by learning to recognize discrete random variables. The remainder of the chapter is devoted to the study of random variables of this type.

3.1 RANDOM VARIABLES

We begin by considering three examples, each of which involves a random variable. Random variables will be denoted by uppercase letters and their observed numerical values by lowercase letters.

Example 3.1.1. Consider the random variable X, the number of brown-eyed children born to a couple heterozygous for eye color. If the couple is assumed to have two children, a priori, before the fact, the variable X can assume any one of the values 0, 1, or 2. The variable is random in that brown eyes depend on the chance inheritance of a dominant gene at conception. If for a particular couple there are two brown-eyed children, we write $x = 2$.

Example 3.1.2. The basic premise underlying the field of immunology is that an animal is immunized by injection of a suitable antigen. In one study malignant plasmacytoma cells are exposed to lymphocytes carrying a specific antigen. It is hoped that these cells will fuse, because the fused cells retain the ability to grow continuously and also to retain the antibody characteristics of the antigen fused. In this way the animal is quickly immunized. Cells are exposed to the lymphocytes one at a time in the presence of polyethylene glycol, a fusion-promoting agent. It is known that the probability that such a cell will fuse is 1/2. Let Y denote the number of cells exposed to obtain the first fusion. The variable Y is random; a priori it can assume any value in the set $\{1, 2, 3, \ldots\}$. Recall from your study of calculus that a set such as this that consists of an infinite collection of isolated points is called a *countably infinite set*.

Example 3.1.3. In Example 1.1.2 we considered the variable T, the time at which the peak demand for electricity occurs per day. This variable is random, since its value is affected by such chance factors as time of the year, humidity, and temperature. It can conceivably assume any value in the 24-hour time span from 12 midnight one day to 12 midnight the next day.

It is easy to distinguish a discrete random variable from one that is not discrete. Just ask the question, "What are the possible values for the variable?" If the answer is a finite set or a countably infinite set, then the random variable is discrete; otherwise it is not. This idea leads to the following definition:

Definition 3.1.1 (Discrete random variable). A random variable is discrete if it can assume at most a finite or a countably infinite number of possible values.

The random variable X, the number of brown-eyed children in a two-child family, is discrete. Its set of possible values is the finite set $\{0, 1, 2\}$. The set $\{1, 2, 3, \ldots\}$ of possible values for Y, the number of cells exposed to obtain the first fusion of Example 3.1.2, is countably infinite. Thus Y is also a discrete random variable. The random variable T, the time of the peak demand for electricity at a power plant, is different from the others. Time is measured continuously, and T can conceivably assume any value in the interval $[0, 24)$, where 0 denotes 12 midnight one day and 24 denotes 12 midnight the next. This set of real numbers is neither finite nor countably infinite. Any time that you ask yourself the question, "What are the possible values for the random variable?" and are forced to admit that the set of possibilities includes some interval or continuous span of real numbers, then the random variable being studied is not discrete.

3.2 DISCRETE PROBABILITY DENSITIES

When dealing with a random variable, it is not enough just to determine what values are possible. We also need to determine what is probable. We must be able to predict in some sense the values that the variable is likely to assume at any time. Since the behavior of a random variable is governed by chance, these predictions must be made in the face of a great deal of uncertainty. The best that can be done is to describe the behavior of the random variable in terms of probabilities. Two functions are used to accomplish this. We shall refer to these as the *density function* and the *cumulative distribution function*. The former is known by a variety of names in the discrete case, some of the most commonly encountered ones being the probability function, the probability mass function, and the probability density function. In the discrete case, the density is denoted by either $p(x)$ or $f(x)$; in the continuous case it is almost always denoted by $f(x)$. For consistency we shall use $f(x)$ for the density in both cases. We begin by defining the density function for discrete random variables.

Definition 3.2.1 (Discrete density). Let X be a discrete random variable. The function f given by

$$f(x) = P[X = x]$$

for x real is called the density function for X.

There are several facts to note concerning the density in the discrete case. First, f is defined on the entire real line, and for any given real number x, $f(x)$ is the probability that the random variable X assumes the value x. For example, $f(2)$ is the probability that the random variable X assumes the numerical value of 2. Second, since $f(x)$ is a probability, $f(x) \geq 0$ regardless of the value of x. Third, if we sum f over all values of X that occur with nonzero probability, the sum must be 1. The following two conditions are necessary and sufficient conditions for a function f to be a discrete density. That is, if a function satisfies both of these conditions then it can be viewed as representing the density for some discrete random variable; if it fails to satisfy both then it cannot be the density for any discrete random variable:

> **Necessary and Sufficient Conditions for a Function to be a Discrete Density**
>
> **1.** $f(x) \geq 0$
>
> **2.** $\displaystyle\sum_{\text{all } x} f(x) = 1$

The next example illustrates these ideas.

Example 3.2.1. Two fair dice are rolled, and the obtained numbers are observed. Consider the random variable Y which is defined as the product of these two numbers. Table 3.1 shows the possible values of Y. We can summarize the probability structure for Y in a density table like Table 3.2.

Note that even though a discrete density is defined on the entire real line, it is only necessary to specify the density for those values y for when $f(y) \neq 0$. For instance, in Table 3.2, it is understood that $f(y) = 0$ for all other real numbers.

Some people may prefer to use graphical methods (like a bar chart) to present the density function as shown in Fig. 3.1.

Table 3.1

		First die					
		1	2	3	4	5	6
	1	1	2	3	4	5	6
	2	2	4	6	8	10	12
Second die	3	3	6	9	12	15	18
	4	4	8	12	16	20	24
	5	5	10	15	20	25	30
	6	6	12	18	24	30	36

Table 3.2

y	1	2	3	4	5	6	8	9	10	12	15	16	18	20	24	25	30	36
$P[Y = y] = f(y)$	1/36	2/36	2/36	3/36	2/36	4/36	2/36	1/36	2/36	4/36	2/36	1/36	2/36	2/36	2/36	1/36	2/36	1/36

Figure 3.1
Density function of the
random variable in
Example 3.2.1.

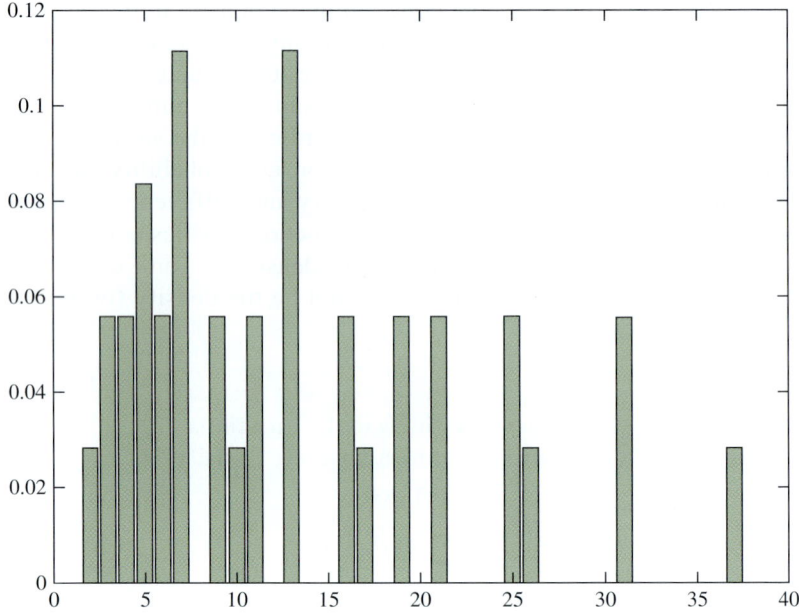

Cumulative Distribution

The second function used to compute probabilities is the cumulative distribution function F. Most of the statistical tables used in the material that follows are tables of the cumulative distribution function for some pertinent random variable.

The word "cumulative" suggests the role of this function. It sums or accumulates the probabilities found by means of the density. This function is defined as follows:

> **Definition 3.2.2 (Cumulative distribution—discrete).** Let X be a discrete random variable with density f. The cumulative distribution function for X, denoted by F, is defined by
>
> $$F(x) = P[X \le x] \qquad \text{for } x \text{ real}$$

Consider a specific real number x_0. To find $P[X \le x_0] = F(x_0)$, we sum the density f over all values of X that occur with nonzero probability that are less than or equal to x_0. That is, computationally,

$$F(x_0) = \sum_{x \le x_0} f(x)$$

This idea is illustrated in Example 3.2.2.

Example 3.2.2.   The density for Y, the product of the face-up numbers of two dice, and its cumulative distribution are shown in Table 3.3. Notice that

$$F(1) = P[Y \le 1] = P[Y = 1] = 1/36$$
$$F(2) = P[Y \le 2] = P[Y = 1] + P[Y = 2] = 1/36 + 2/36$$

Table 3.3

y	1	2	3	4	5	6	8	9	10	12	15	16	18	20	24	25	30	36
$P[Y \leq y] = F(y)$	1/36	3/36	5/36	8/36	10/36	14/36	16/36	17/36	19/36	23/36	25/36	26/36	28/36	30/36	32/36	33/36	35/36	36/36

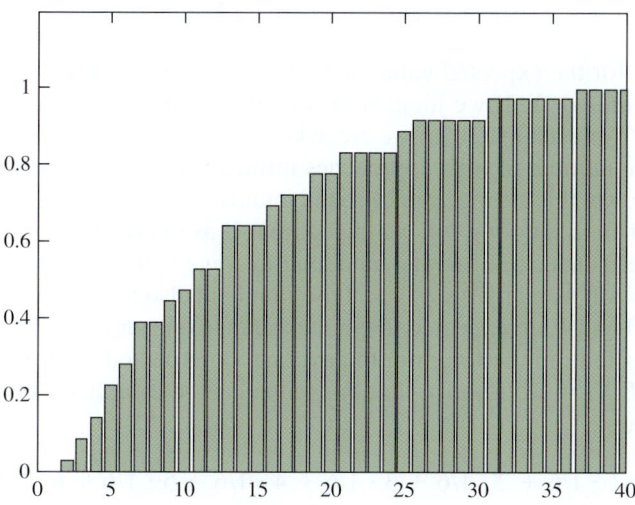

Figure 3.2
Cumulative distribution of the random variable in Example 3.2.2.

$$F(3) = P[Y \leq 3] = P[Y = 1] + P[Y = 2] + P[Y = 3]$$
$$= 1/36 + 2/36 + 2/36$$

$$\cdots$$

$$F(36) = P[Y \leq 36] = 36/36 = 1$$

For discrete random variables that can assume only a finite number of possible values, the last entry in the bottom row of the cumulative table will always be 1.

We can also present the cumulative distribution by graphical method as shown in Fig. 3.2.

3.3   EXPECTATION AND DISTRIBUTION PARAMETERS

The density function of a random variable completely describes the behavior of the variable. However, associated with any random variable are constants, or "parameters," that are descriptive. Knowledge of the numerical values of these parameters gives the researcher quick insight into the nature of the variables. We consider three such parameters: the mean μ, the variance σ^2, and the standard deviation σ. If the exact density of the random variable is known, then the numerical value of each parameter can be found from mathematical considerations. That is the topic of this section. If the only thing available to the researcher is a set of observations on the random variable (a data set), then the values of these parameters cannot be found exactly. They must be approximated by using statistical techniques. That is the topic of much of the remainder of this text.

To understand the reasoning behind most statistical methods, it is necessary to become familiar with one general concept, namely, the idea of *mathematical expectation* or *expected value*. This concept is used in defining many statistical parameters and provides the logical basis for most of the methods of statistical inference presented later in this text.

A simple example will illustrate the basic idea of expectation. Consider the roll of a single fair die, and let X denote the number that is obtained. The possible values for X are $1, 2, 3, 4, 5, 6$, and since the die is fair, the probability associated with each value is 1/6. The density for X is given by

$$f(x) = 1/6 \qquad x = 1, 2, 3, 4, 5, 6$$

When we ask for the expected value of X, we are asking for the *long-run theoretical average value of X*. If we imagine rolling the die over and over and recording the value of X for each roll, then we are asking for the theoretical average value of the rolls as the number of rolls approaches infinity. Since the density for X is symmetric and known, this average can be found intuitively. Notice that since $P[X = 1] = P[X = 6] = 1/6$, in the long run we expect to roll as many 1's as 6's. These values should counterbalance one another, and their average value is $(6 + 1)/2 = 3.5$. We also expect to roll as many 2's as 4's; these numbers also average to 3.5. Likewise, the numbers 3 and 4 are expected to counterbalance one another; they average 3.5. Logic dictates that, in the long run the average or expected value of X is 3.5. We write this as $E[X] = 3.5$. Notice that this value can be calculated from the density for X as follows:

$$E[X] = 1 \cdot 1/6 + 2 \cdot 1/6 + 3 \cdot 1/6 + 4 \cdot 1/6 + 5 \cdot 1/6 + 6 \cdot 1/6 = 3.5$$

or

$$E[X] = \sum_{\text{all } x} (\text{value of } x)(\text{probability})$$

Of course, the characteristic that makes finding this expectation easy is the symmetry of the density. Can we develop a definition of expectation that will work for nonsymmetric densities and that will apply not only to X, but also to random variables that are functions of X? The answer is "yes," and the desired definition is given in Definition 3.3.1. Let us point out that in most problems interest centers first on $E[X]$. However, expectations for functions of X such as X^2, $(X - c)^2$, where c is a constant and e^{tX} are especially useful in statistical theory. For this reason, the definition of expected value is given in general terms. We now define what we mean by the expected value of some function of X which we denote by $H(X)$.

Definition 3.3.1 (Expected value). Let X be a discrete random variable with density f. Let $H(X)$ be a random variable. The expected value of $H(X)$, denoted by $E[H(X)]$, is given by

$$E[H(X)] = \sum_{\text{all } x} H(x)f(x)$$

provided $\sum_{\text{all } x} |H(x)| f(x)$ is finite. Summation is over all values of X that occur with nonzero probability.

Note that in the special case in which $H(X) = X$, we obtain the expected value of X from this definition. Thus we see that

Expected Value of X

$$E[X] = \sum_{\text{all } x} x f(x)$$

One other thing to note concerning this definition is the fact the restriction that $\sum_{\text{all } x} |H(x)| f(x)$ exists is not particularly restrictive in practice. If the set of possible values for X is finite, it *will* be satisfied; if the set of possible values for X is countably infinite, it will *usually* be satisfied. However, it is possible to concoct a density f and a function $H(X)$ for which the series $\sum_{\text{all } x} |H(x)| f(x)$ does not converge. (See Exercise 22.) In this case we say that the expected value of the random variable $H(X)$ does not exist. An example will illustrate the use of Definition 3.3.1. Please realize that the density has been greatly oversimplified for purposes of illustration!

Example 3.3.1.   A drug is used to maintain a steady heart rate in patients who have suffered a mild heart attack. Let X denote the number of heartbeats per minute obtained per patient. Consider the hypothetical density given in Table 3.4. What is the average heart rate obtained by all patients receiving this drug? That is, what is $E[X]$? By Definition 3.3.1,

$$E[X] = \sum_{\text{all } x} H(x)f(x)$$

$$= \sum_{\text{all } x} x f(x)$$

$$= 40(.01) + 60(.04) + 68(.05) + \cdots + 100(.01)$$

$$= 70$$

Since the number of possible values for X is finite, $\sum_{\text{all } x} |x| f(x)$ exists. Thus we can say that the *average* heart rate obtained by patients using this drug is 70 heartbeats per minute. Intuitively, we should have expected this result. Notice the symmetry of the density. In the long run we would expect as many patients with heart rates of 100 as with heart rates of 40; as many with a rate of 60 as with a rate of 80. Similarly, the rates of 68 and 72 occur with the same frequency. Each of these pairs averages to 70, the value obtained by the remaining 80% of the patients. Common sense points to 70 as the expected value for X.

When used in a statistical context, the expected value of a random variable X is referred to as its *mean* and is denoted by μ or μ_X. That is, the terms *expected value* and *mean* are interchangeable, as are the symbols $E[X]$ and μ. The mean can be thought of as a measure of the "center of location" in the sense that it indicates where the "center" of the density lies. For this reason, the mean is often referred to as a "location" parameter. To emphasize these points, let us summarize the preceding discussion.

Table 3.4

x	40	60	68	70	72	80	100
$f(x)$	.01	.04	.05	.80	.05	.04	.01

Notes on the Expected Value of a Random Variable X

1. The expected value of a random variable is its theoretical average value. It is denoted by $E[X]$ and can be calculated from knowledge of the density for X.
2. In a statistical setting, the average value of X is called its *mean value*. Hence the terms *average value, mean value,* and *expected value* are interchangeable.
3. The mean value of X is denoted by the Greek symbol μ (*mu*). Hence the symbols μ and $E[X]$ are interchangeable.
4. The mean or expected value of X is one measure of the location of the center of the X values. For this reason, μ is called a "location" parameter.

There are three rules for handling expected values that are useful in justifying statistical procedures in later chapters. These rules hold for both continuous and discrete random variables. The rules are stated and illustrated here.

> **Theorem 3.3.1 (Rules for expectation).** Let X and Y be random variables and let c be any real number.
> 1. $E[c] = c$ (The expected value of any constant is that constant.)
> 2. $E[cX] = cE[X]$ (Constants can be factored from expectations.)
> 3. $E[X + Y] = E[X] + E[Y]$ (The expected value of a sum is equal to the sum of the expected values.)

Example 3.3.2. Let X and Y be random variables with $E[X] = 7$ and $E[Y] = -5$. Then

$$
\begin{aligned}
E[4X - 2Y + 6] &= E[4X] + E[-2Y] + E[6] &\quad \text{Rule 3} \\
&= 4E[X] + (-2)E[Y] + E[6] &\quad \text{Rule 2} \\
&= 4E[X] - 2E[Y] + 6 &\quad \text{Rule 1} \\
&= 4(7) - 2(-5) + 6 \\
&= 44
\end{aligned}
$$

Variance and Standard Deviation

Knowledge of the mean of a random variable is important, but this knowledge *alone* can be misleading. The next example should show you the problem.

Example 3.3.3. Suppose that we wish to compare a new drug to that of Example 3.3.1. Let X denote the number of heartbeats per minute obtained using the old drug and Y the number per minute obtained with the new drug. The hypothetical density of each of these variables is given in Table 3.5. Since each of the densities is symmetric, inspection shows that $\mu_X = \mu_Y = 70$. Each drug produces *on the average* the same number of heartbeats per minute. However, there is obviously a drastic difference between the two drugs that is not being detected by the mean. The old drug produces fairly consistent reactions in patients, with 90% differing from the mean by at most 2; very few (2%) have an extreme reaction to the drug. However, the new drug produces highly diverse responses. Only 10% of the patients have heart rates within 2 units of the mean, whereas 80% show an extreme reaction. If we examined only the mean, we would conclude that the two drugs had identical effects—but nothing could be further from the truth!

Table 3.5

x	40	60	68	70	72	80	100
$f(x)$	.01	.04	.05	.80	.05	.04	.01

y	40	60	68	70	72	80	100
$f(y)$	.40	.05	.04	.02	.04	.05	.40

It is obvious from Example 3.3.3 that something is not being measured by the mean. That something is *variability*. We must find a parameter that reflects consistency or the lack of it. We want the measure to assume a large positive value if the random variable fluctuates in the sense that it often assumes values far from its mean; the measure should assume a small positive value if the values of X tend to cluster closely about the mean. There are several ways to define such a measure. The most widely used is the *variance*.

Definition 3.3.2 (Variance). Let X be a random variable with mean μ. The variance of X, denoted by Var X, or σ^2, is given by

$$\text{Var } X = \sigma^2 = E[(X - \mu)^2]$$

Note that the variance measures variability by considering $X - \mu$, the difference between the variable and its mean. The difference is squared so that negative values will not cancel positive ones in the process of finding the expected value. When expressed in the form $E[(X - \mu)^2]$, it is easy to see that σ^2 has the properties that we want. When the variable X often assumes values far from μ, σ^2 will be a large positive number; when the values of X tend to fall close to μ, σ^2 will assume a small positive value. Figure 3.3 illustrates the idea.

Usually, the definition of σ^2 is not used to compute the variance. Rather, we use an alternative form which is given in the following theorem.

Theorem 3.3.2 (Computational formula for σ^2)

$$\sigma^2 = \text{Var } X = E[X^2] - (E[X])^2$$

Proof. By definition

$$\text{Var } X = E[(X - \mu)^2]$$
$$= E[X^2 - 2\mu X + \mu^2]$$

(*a*)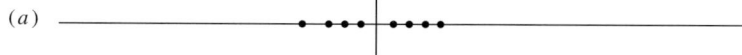

(*b*)

Figure 3.3
(*a*) A distribution with a small variance. Most of the data points, denoted by dots, lie fairly close to the average value, μ. Hence most of the differences, $x - \mu$, will be small; (*b*) a distribution with a large variance. Many of the data points lie far from the average value, μ.

Using the rules of expectation, Theorem 3.3.1, we obtain

$$\text{Var } X = E[X^2] - 2\mu E[X] + \mu^2$$

Since the symbols μ and $E[X]$ are interchangeable,

$$\text{Var } X = E[X^2] - 2(E[X])^2 + (E[X])^2$$
$$= E[X^2] - (E[X])^2$$

We illustrate the theorem by computing the variance of each of the random variables of Example 3.3.4.

Example 3.3.4.   To find σ_X^2 and σ_Y^2 for the variables of Example 3.3.3, we first use Table 3.5 to find $E[X^2]$ and $E[Y^2]$. We know that $E[X] = E[Y] = 70$.

$$E[X^2] = \sum_{\text{all } x} x^2 f(x)$$
$$= (40^2)(.01) + (60^2)(.04) + \cdots + (100^2)(.01)$$
$$= 4926.4$$

$$E[Y^2] = \sum_{\text{all } y} y^2 f(y)$$
$$= (40^2)(.40) + (60^2)(.05) + \cdots + (100^2)(.40)$$
$$= 5630.32$$

By Theorem 3.3.2,

$$\text{Var } X = E[X^2] - (E[X])^2$$
$$= 4926.4 - 70^2 = 26.4$$
$$\text{Var } Y = E[Y^2] - (E[Y])^2$$
$$= 5630.32 - 70^2 = 730.32$$

As expected, Var Y > Var X. Even though the drugs produce the same mean number of heartbeats per minute, they do not behave in the same way. The new drug is not as consistent in its effect as the old.

Note that the variance of a random variable reported alone is not very informative. Is a variance of 26.4 large or small? Only when this value is compared to the variance of a similar variable does it take on meaning. Hence variances are used often for comparative purposes to choose between two variables that otherwise appear to be identical. Also, note that the variance of a random variable is essentially a pure number whose associated units are often physically meaningless. When this occurs, the unit can be omitted. For example, the unit associated with the variance of Example 3.3.4 is a "squared heartbeat." This makes little sense, so in this case variance can be reported with no unit attached. To overcome this problem, a second measure of variability is employed. This measure is the nonnegative square root of the variance, and it is called the *standard deviation*. It has the advantage of having associated with it the same units as the original data.

Definition 3.3.3 (Standard deviation). Let X be a random variable with variance σ^2. The standard deviation of X, denoted by σ, is given by

$$\sigma = \sqrt{\text{Var } X} = \sqrt{\sigma^2}$$

Example 3.3.5. The standard deviations of variables X and Y of Example 3.3.4 are, respectively,

$$\sigma_X = \sqrt{\text{Var } X} = \sqrt{26.4} = 5.14 \text{ heartbeats per minute}$$

$$\sigma_Y = \sqrt{\text{Var } Y} = \sqrt{730.32} = 27.02 \text{ heartbeats per minute}$$

To emphasize these points we present a brief summary of the important aspects of the standard deviation of a random variable X.

Properties of standard deviation

1. The standard deviation of X is defined as the nonnegative square root of its variance.
2. The standard deviation is denoted by σ, and the variance of X is denoted by σ^2.
3. A large standard deviation implies that the random variable X is rather inconsistent and somewhat hard to predict; a small standard deviation is an indication of consistency and stability.
4. Standard deviation is always reported in physical measurement units that match the original data. Variance is often unitless.

Just as there are three rules for expectation that help in simplifying complex expressions, so are there three rules for variance. These rules parallel those for expectation. Rules 1 and 2 can be proved by using the rules for expectation (see Exercise 20). The proof of rule 3 must be deferred until the notion of "independent random variables" has been formalized.

> **Theorem 3.3.3 (Rules for variance).** Let X and Y be random variables and c any real number. Then
>
> 1. Var $c = 0$
> 2. Var $cX = c^2$ Var X
> 3. If X and Y are independent, then Var$(X + Y) = $ Var $X + $ Var Y
>
> (Two variables are independent if knowledge of the value assumed by one gives no clue to the value assumed by the other.)

Example 3.3.6. Let X and Y be independent with $\sigma_X^2 = 9$ and $\sigma_Y^2 = 3$. Then

$$\begin{aligned}
\text{Var}[4X - 2Y + 6] &= \text{Var}[4X] + \text{Var}[-2Y] + \text{Var } 6 \qquad \text{Rule 3} \\
&= 16 \text{ Var } X + 4 \text{ Var } Y + \text{Var } 6 \qquad \text{Rule 2} \\
&= 16 \text{ Var } X + 4 \text{ Var } Y + 0 \qquad \text{Rule 1} \\
&= 16(9) + 4(3) = 156
\end{aligned}$$

In this section we discussed three *theoretical* parameters associated with a random variable X. We showed not only how to determine their numerical values from knowledge of the density, but also how to interpret them physically. Keep these things in mind, for they play a major role in the study of statistical methods for analyzing experimental data.

3.4 BINOMIAL DISTRIBUTION

We next study the *binomial* distribution. Once again, you have already seen some binomial random variables even though they were not labeled as such at the time. The theoretical basis for working with this distribution is the binomial theorem presented in most beginning algebra courses. The statement of this theorem is as follows:

Binomial theorem

For any two real numbers a and b and any positive integer n,

$$(a + b)^n = \sum_{k=0}^{n} \binom{n}{k} a^k b^{n-k}$$

where $\binom{n}{k}$ is given by $\dfrac{n!}{k!(n-k)!}$.

To recognize a situation that involves a binomial random variable, you must be familiar with the assumptions that underlie this distribution, which are as follows:

Binomial properties

1. The experiment consists of a *fixed* number, n, of trials, which are called Bernoulli trials, that result in either a "success" (s) or a "failure" (f).

2. The trials are identical and independent, and therefore the probability of success, p, remains the same from trial to trial.

3. The random variable X denotes the number of successes obtained in the n trials.

Once we realize that the binomial model is appropriate from the physical description of the experiment, we shall want to describe the behavior of the binomial random variable involved. To do so, we need to consider the density for the random variable. To get an idea of the general form for the binomial density, let us consider the case in which $n = 3$. The sample space for such an experiment is

$$S = \{fff,\ sff,\ fsf,\ ffs,\ ssf,\ sfs,\ fss,\ sss\}$$

Since the trials are independent, the probability assigned to each sample point is found by multiplying. For example, the probabilities assigned to the sample points *fff* and *sff* are $(1 - p)(1 - p)(1 - p) = (1 - p)^3$ and $p(1 - p)(1 - p) = p(1 - p)^2$, respectively. The random variable X assumes the value 0 only if the experiment results in the outcome *fff*. That is,

$$P[X = 0] = (1 - p)^3$$

However, X assumes the value 1 if the experiment results in any one of the outcomes *sff, fsf,* or *ffs*. Thus

$$P[X = 1] = 3 \cdot p(1 - p)^2$$

Similarly,

$$P[X = 2] = 3 \cdot p^2(1 - p)$$

and

$$P[X = 3] = p^3$$

It is evident that for $x = 0, 1, 2, 3$

$$P[X = x] = c(x)p^x(1-p)^{3-x}$$

where $c(x)$ denotes the number of sample points that correspond to x successes. Such a sample point is expressed as a permutation of three letters, with x of these being s's and the rest, $3 - x$, of these being f's. Using the formula for the number of permutations of indistinguishable objects, we see that

$$c(x) = \frac{3!}{x!(3-x)!} = \binom{3}{x}$$

Thus the density for this binomial random variable is given by

$$f(x) = \binom{3}{x}p^x(1-p)^{3-x} \qquad x = 0, 1, 2, 3$$

To generalize this idea to n trials, we replace 3 by n to obtain the expression

$$f(x) = \binom{n}{x}p^x(1-p)^{n-x} \qquad x = 0, 1, 2, \ldots, n$$

This suggests the formal definition of the binomial distribution.

Definition 3.4.1 (Binomial distribution). A random variable X has a binomial distribution with parameters n and p if its density is given by

$$f(x) = \binom{n}{x}p^x(1-p)^{n-x} \qquad x = 0, 1, 2, \ldots, n$$
$$0 < p < 1$$

where n is a positive integer.

To see that the function given in this definition is a density, note that it is non-negative. Furthermore, by applying the binomial theorem with $k = x$, $a = p$, and $b = 1 - p$ it can be seen that

$$\sum_{x=0}^{n} \binom{n}{x}p^x(1-p)^{n-x} = [p + (1-p)]^n = 1$$

as desired.

Example 3.4.1. Recent studies of German air traffic controllers have shown that it is difficult to maintain accuracy when working for long periods of time on data display screens. A surprising aspect of the study is that the ability to detect spots on a radar screen decreases as their appearance becomes too rare. The probability of correctly identifying a signal is approximately .9 when 100 signals arrive per 30-minute period. This probability drops to .5 when only 10 signals arrive at random over a 30-minute period. The hypothesis is that unstimulated minds tend to wander. Let X denote the number of signals correctly identified in a 30-minute time span in which 10 signals arrive. This experiment consists of a series of $n = 10$ independent and identical Bernoulli trials with "success" being the correct identification of a signal. The probability of success is $p = 1/2$. Since X denotes the number of successes in a fixed number of trials, X is binomial. Its density is found by letting $n = 10$ and $p = 1/2$ in the expression for f given in Definition 3.4.1. That is,

$$f(x) = \binom{n}{x} p^x (1-p)^{n-x} \qquad x = 0, 1, 2, \ldots, n$$

or

$$f(x) = \binom{10}{x} (1/2)^x (1/2)^{10-x} \qquad x = 0, 1, 2, \ldots, 10$$

The next theorem summarizes other theoretical properties of the binomial distribution. Its proof is left as an exercise (Exercise 43).

Theorem 3.4.1. Let X be a binomial random variable with parameters n and p.

1. $E[X] = \mu = np$
2. $\text{Var } X = \sigma^2 = npq$

Example 3.4.2. The random variable X, the number of radar signals properly identified in a 30-minute period, is a binomial random variable with parameters $n = 10$ and $p = 1/2$. Its mean is $\mu = np = 10(1/2) = 5$, and its variance is $\sigma^2 = npq = 10(1/2)(1/2) = 10/4$.

In statistical studies we shall usually be interested in computing the probability that the random variable assumes certain values. This probability can be computed from the density function, f, or from the cumulative distribution function, F. Since the binomial distribution comes into play in such a wide variety of physical applications, tables of the cumulative distribution function for selected values of n and p have been compiled. Table I of App. A is one such table. That is, Table I gives the values of

$$F(t) = \sum_{x=0}^{[t]} \binom{n}{x} p^x (1-p)^{n-x}$$

for selected values of n and p, where $[t]$ represents the greatest integer less than or equal to t. Its use is illustrated in the following example.

Example 3.4.3 Let X denote the number of radar signals properly identified in a 30-minute time period in which 10 signals are received. Assuming that X is binomial with $n = 10$ and $p = 1/2$, find the probability that at most seven signals will be identified correctly. This probability can be found by summing the density from $x = 0$ to $x = 7$. That is,

$$P[X \le 7] = \sum_{x=0}^{7} \binom{10}{x} (1/2)^x (1/2)^{10-x}$$

Evaluating this probability directly entails a large amount of arithmetic. However, its value can be read from Table I of App. A. We first look at the group of values labeled $n = 10$. The desired probability of .9453 is found in the column labeled .5 and the row labeled 7. That is,

$$P[X \le 7] = F(7) = .9453$$

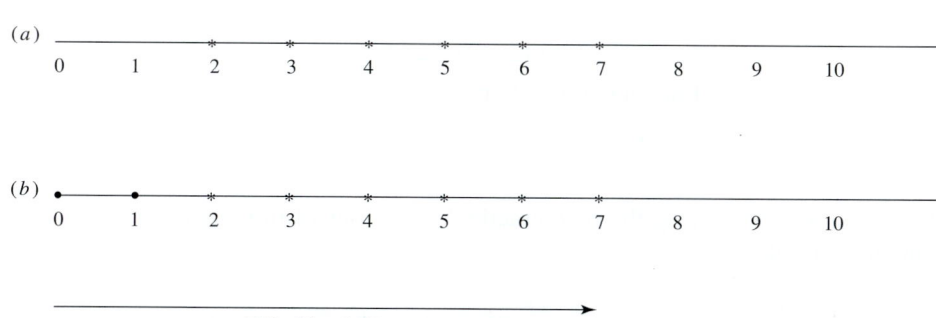

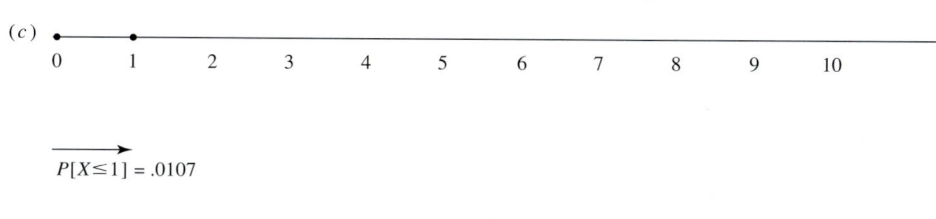

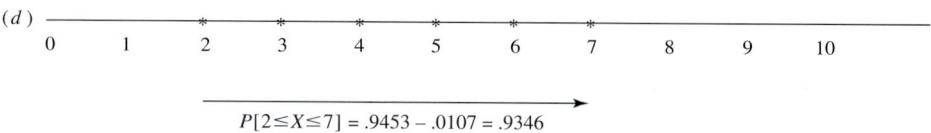

Figure 3.4
(a) The probability that X lies between 2 and 7 inclusive is the probability associated with the starred points; (b) P[X ≤ 7] = .9453 includes the probability associated with 0 and 1; (c) the probability associated with the unwanted points 0 and 1 is .0107; (d) the desired probability is found by subtraction.

Other probabilities can be found. For example, find $P[2 \leq X \leq 7]$. Figure 3.4 suggests how this is done. Notice that in Fig. 3.4 we want the probability associated with points that are starred. To determine the desired probability, we first find the number 7 in Table I of App. A. Since the table is cumulative, the probability given, .9453, is the probability that X is at most 7. This probability includes the probability that $X = 0$ or $X = 1$. Since we did not want to include those values, $P[X \leq 1] = F(1) = .0107$ must be subtracted from .9453. Thus

$$P[2 \leq X \leq 7] = P[X \leq 7] - P[X < 2]$$
$$= P[X \leq 7] - P[X \leq 1]$$
$$= F(7) - F(1)$$
$$= .9453 - .0107$$
$$= .9346$$

Later in the text we shall show ways of approximating binomial probabilities when the values of n and p are such that no appropriate binomial table is available.

3.5   POISSON DISTRIBUTION

The next discrete family to be considered is the family of *Poisson* random variables, named after the French mathematician Simeon Denis Poisson (1781–1840). The Maclaurin series expansion for the function e^z studied in beginning calculus courses provides the theoretical basis for this distribution. This series is given by

> **Maclaurin series**
>
> For z a real number,
>
> $$e^z = 1 + z + z^2/2! + z^3/3! + z^4/4! + \cdots$$

We begin by considering the mathematical properties of this important family of random variables.

> **Definition 3.5.1 (Poisson distribution).** A random variable X is said to have a Poisson distribution with parameter k if its density f is given by
>
> $$f(x) = \frac{e^{-k}k^x}{x!} \qquad \begin{array}{l} x = 0, 1, 2, \ldots \\ k > 0 \end{array}$$

The function f given in this definition is nonnegative. To see that it sums to 1, note that

$$\sum_{x=0}^{\infty} \frac{e^{-k}k^x}{x!} = e^{-k}(1 + k + k^2/2! + k^3/3! + \cdots)$$

The series on the right is the Maclaurin series for e^k. Thus

$$\sum_{x=0}^{\infty} \frac{e^{-k}k^x}{x!} = e^{-k}e^k = e^0 = 1$$

as desired.

> **Theorem 3.5.1.** Let X be a Poisson random variable with parameter k.
>
> 1. $E[X] = k$
> 2. $\text{Var } X = k$

Poisson random variables usually arise in connection with what are called *Poisson processes*. Poisson processes involve observing discrete events in a continuous "interval" of time, length, or space. We use the word "interval" in describing the general Poisson process with the understanding that we may not be dealing with an interval in the usual mathematical sense. For example, we might observe the number of white blood cells in a drop of blood. The discrete event of interest is the observation of a white cell, whereas the continuous "interval" involved is a drop of blood. We might observe the number of times radioactive gases are emitted from a nuclear power plant during a 3-month period. The discrete event of concern is the emission of radioactive gases. The continuous interval consists of a period of 3 months. The variable of interest in a Poisson process is X, the number of occurrences of the event in an interval of length s units. Although the derivation is a bit tricky, it can be shown using differential equations that X is a Poisson random variable with parameter $k = \lambda s$, where λ is a positive number that characterizes the underlying Poisson process. To understand the physical significance of the constant λ, note that by Definition 3.5.1 the density for X is given by

$$f(x) = \frac{e^{-\lambda s}(\lambda s)^x}{x!} \qquad x = 0, 1, 2, 3, \ldots$$

By Theorem 3.5.1 the expected value of X is λs. That is, the average number of occurrences of the event of interest in an interval of s units is λs. Thus the average number of occurrences of the event in 1 unit of time, length, area, or space is $\lambda s/s = \lambda$. That is, physically, the *parameter λ of a Poisson process represents the average number of occurrences of the event in question per measurement unit.*

The following steps are used in the solution of an applied Poisson problem:

Steps in Solving a Poisson Problem

1. Determine the basic unit of measurement being used.
2. Determine the average number of occurrences of the event per unit. This number is denoted by λ.
3. Determine the length or size of the observation period. This number is denoted by s.
4. The random variable X, the number of occurrences of the event in the interval of size s follows a Poisson distribution with parameter $k = \lambda s$.

These steps are illustrated in Example 3.5.1.

Example 3.5.1. The white blood cell count of a healthy individual can average as low as 6000 per cubic millimeter of blood. To detect a white-cell deficiency, a .001 cubic millimeter drop of blood is taken and the number of white cells X is found. How many white cells are expected in a healthy individual? If at most two are found, is there evidence of a white cell deficiency?

This experiment can be viewed as involving a Poisson process. The discrete event of interest is the occurrence of a white cell; the continuous interval is a drop of blood.

Let the measurement unit be a cubic millimeter; then $s = .001$ and λ, the average number of occurrences of the event per unit, is 6000. Thus X is a Poisson random variable with parameter $\lambda s = 6000(.001) = 6$. By Theorem 3.5.1, $E[X] = \lambda s = 6$. In a healthy individual we would expect, on the average, to see six white cells. How rare is it to see at most two? That is, what is $P[X \leq 2]$? From Definition 3.5.1,

$$P[X \leq 2] = \sum_{x=0}^{2} f(x) = \sum_{x=0}^{2} \frac{e^{-6}6^x}{x!}$$
$$= \frac{e^{-6}6^0}{0!} + \frac{e^{-6}6^1}{1!} + \frac{e^{-6}6^2}{2!}$$

Evaluating this type of expression directly does entail some arithmetic.

Once again, because of the wide appeal of the Poisson model, the values of the cumulative distribution function for selected values of the parameter $k = \lambda s$ are tabulated. Table II of App. A is one such table. The desired probability of .062 is found by looking under the column labeled $k = 6$ in the row labeled 2. Is there evidence of a white-cell deficiency? There are no rules that say at what point probabilities are considered to be small. To answer this question, a value judgment must be made. If you consider .062 to be small, then the natural conclusion is that the individual does have a white-cell deficiency.

3.6 SIMULATING A DISCRETE DISTRIBUTION

In designing operating systems of various types, one often needs to simulate the system before it is built. Simulation is usually done with the aid of a computer. However, the idea behind simulation can be illustrated by using a random digit

Table 3.6

	Column	Random digits		
Row		**(1)**	**(2)**	**(3)**
1		10480	15011	01536
2		22368	46573	25595
3		24130	48360	22527
4		42167	93093	06243
5		37570	39975	81837
6		77921	06907	11008
7		99562	72905	56420
8		96301	91977	05463
9		89579	14342	63661
10		85485	36857	43342

table. A portion of such a table is given in Table III of App. A. Its use is illustrated in the following example.

Example 3.6.1.   Table 3.6 presents a portion of the random digit table in the appendix. Let us read a sequence of random two-digit numbers from this table. To do so, we must get a random start. This can be done by writing the integers 1 through 14 on slips of paper, placing the slips in a bowl, stirring, and drawing one slip at random from the bowl. The number selected identifies the column in which our starting number is located. In a similar way, we can select the row in which the starting number is located. Suppose that this process results in the selection of column 2 and row 5. This identifies the random starting point as 39975.

Since we want two-digit numbers, we need only read the first two digits of this number. Thus our first random number is 39. Since a random digit table is constructed in such a way that the digit appearing at each position in the table is just as likely to be one digit as any other, the table can be read in any way. Let us agree to read down the second column so that the next four two-digit numbers are 06, 72, 91, and 14.

The next example illustrates the use of a random digit table in a simple simulation experiment.

Example 3.6.2.   Suppose that at a particular airport planes arrive at an average rate of one per minute and depart at the same average rate. We are interested in simulating the behavior of the random variable Z, the number of planes on the ground at a given time. We will simulate Z for five consecutive one-minute periods. Note that for each of these periods the random variables X, the number of arrivals, and Y, the number of departures, are both Poisson variables with parameter $k = 1$. The density for X and Y is obtained from Table II of App. A and is shown below:

$$P[X = 0] = P[Y = 0] = .368$$
$$P[X = 1] = P[Y = 1] = .368$$
$$P[X = 2] = P[Y = 2] = .184$$
$$P[X = 3] = P[Y = 3] = .061$$
$$P[X = 4] = P[Y = 4] = .015$$
$$P[X = 5] = P[Y = 5] = .003$$
$$P[X = 6] = P[Y = 6] = .001$$
$$P[X > 6] = P[Y > 6] \doteq 0$$

Table 3.7

Random number	Number of arrivals (x)	Number of departures (y)	$P[X = x] = P[Y = y]$
000–367	0	0	.368
368–735	1	1	.368
736–919	2	2	.184
920–980	3	3	.061
981–995	4	4	.015
996–998	5	5	.003
999	6	6	.001

Table 3.8

Time span, min	Random 3-digit number	Number of arrivals (x)	Number of departures (y)	Number on ground at end of time period (z)
1	015	0		100
	255		0	100
2	225	0		
	062		0	100
3	818	2		
	110		0	102
4	564	1		
	054		0	103
5	636	1		
	433		1	103

There are 1000 possible three-digit numbers. We divide them into seven categories to reflect the above probabilities. This division is shown in Table 3.7. To perform the simulation, we read a total of 10 random three-digit numbers using the procedure demonstrated in Example 3.6.1. Assume that at the beginning of the simulation there are 100 planes on the ground and that our random starting point is the number 01536 found in line 1 and column 3 of Table 3.6. The first number read corresponds to the arrivals during the first minute of observation, the second to the departures during this time span, and so forth. The results of the simulation are shown in Table 3.8. If this simulation were continued over a long period of time, we could begin to answer such questions as: "On the average, how many planes are on the ground at a given time?" and "How much variability is there in the number of planes on the ground?"

CHAPTER SUMMARY

In this chapter we introduced the concept of a random variable and showed you how to distinguish a discrete random variable from one that is not discrete. We studied two functions, the density function and the cumulative distribution function, that are used to compute probabilities. The density gives the probability that X assumes a specific value x; the cumulative distribution gives the probability that X assumes a value less than or equal to x. The concept of expected value was introduced and used to define three important parameters, the mean (μ), the variance (σ^2), and the standard deviation (σ). The mean is a measure of the center of location of the distribution; the variance and standard deviation measure the variability of the random variable about its mean. Special discrete distributions that find extensive use in all areas of application were presented. These are the binomial, Bernoulli, uniform, and Poisson distributions. We also discussed briefly how to simulate a discrete distribution. We introduced and defined terms that you should know. These are:

Random variable	Variance
Discrete random variable	Standard deviation
Discrete density	Bernoulli trial
Cumulative distribution	Sampling with replacement
Expected value	Sampling without replacement
Mean	

EXERCISES

Section 3.1

In each of the following, identify the variable as discrete or not discrete.

1. *T:* the turnaround time for a computer job (the time it takes to run the program and receive the results).
2. *M:* the number of meteorites hitting a satellite per day.
3. *N:* the number of neutrons expelled per thermal neutron absorbed in fission of uranium-235.
4. Neutrons emitted as a result of fission are either prompt neutrons or delayed neutrons. Prompt neutrons account for about 99% of all neutrons emitted and are released within 10^{-14} s of the instant of fission. Delayed neutrons are emitted over a period of several hours. Let *D* denote the time at which a delayed neutron is emitted in a fission reaction.
5. Electrical resistance is the opposition offered by electrical conductors to the flow of current. The unit of resistance is the ohm. For example, a 2½-inch electric bell will usually have a resistance somewhere between 1.5 and 3 ohms. Let *O* denote the actual resistance of a randomly selected bell of this type.
6. The number of power failures per month in the Tennessee Valley power network.

Section 3.2

7. Grafting, the uniting of the stem of one plant with the stem or root of another, is widely used commercially to grow the stem of one variety that produces fine fruit on the root system of another variety with a hardy root system. Most Florida sweet oranges grow on trees grafted to the root of a sour orange variety. The density for *X,* the number of grafts that fail in a series of five trials, is given by Table 3.9.
    (*a*) Find $f(5)$.
    (*b*) Find the table for *F*.
    (*c*) Use *F* to find the probability that at most three grafts fail; that at least two grafts fail.
    (*d*) Use *F* to verify that the probability of exactly three failures is .03.
8. In blasting soft rock such as limestone, the holes bored to hold the explosives are drilled with a Kelly bar. This drill is designed so that the explosives can be packed into the hole before the drill is removed. This is necessary since in soft rock the hole often collapses as the drill is removed. The bits for these drills must be changed fairly often. Let *X* denote the number of holes that can be drilled per bit. The density for *X* is given in Table 3.10.
    (*a*) Find $f(8)$.
    (*b*) Find the table for *F*.

Table 3.9

x	0	1	2	3	4	5
$f(x)$	.7	.2	.05	.03	.01	?

Table 3.10

x	1	2	3	4	5	6	7	8
$f(x)$	.02	.03	.05	.2	.4	.2	.07	?

(c) Use F to find the probability that a randomly selected bit can be used to drill between three and five holes inclusive.

(d) Find $P[X \leq 4]$ and $P[X < 4]$. Are these probabilities the same?

(e) Find $F(-3)$ and $F(10)$. *Hint:* Express these in terms of the probabilities that they represent and their values will become obvious.

9. Consider Example 1.2.1. Let X denote the number of computer systems operable at the time of the launch. Assume that the probability that each system is operable is .9.

(a) Use the tree of Fig. 1.2 to find the density table.

(b) There is a pattern to the probabilities in the density table. In particular,

$$f(x) = k(x)(.9)^x(.1)^{3-x}$$

where $k(x)$ gives the number of paths through the tree yielding a particular value for X. Verify that $k(x) = \binom{n}{x}$ for $x = 0, 1, 2, 3$

(c) Find the table for F.

(d) Use F to find the probability that at least one system is operable at launch time.

(e) Use F to find the probability that at most one system is operable at the time of the launch.

10. It is known that the probability of being able to log on to a computer from a remote terminal at any given time is .7. Let X denote the number of attempts that must be made to gain access to the computer.

(a) Find the first four terms of the density table.

(b) Find a closed-form expression for $f(x)$.

(c) Find $P[X = 6]$.

(d) Find a closed-form expression for $F(x)$.

(e) Use F to find the probability that at most four attempts must be made to gain access to the computer.

(f) Use F to find the probability that at least five attempts must be made to gain access to the computer.

11. Knitting machines at a factory making elastic use a laser to detect broken threads. When a thread breaks, the machine must be stopped and the broken thread must be found and repaired by a technician. Assume that the density for X, the number of times per day that a specific machine is stopped, is given by

$$f(x) = \left(\frac{16}{31}\right)\left(\frac{1}{2}\right)^x \qquad x = 0, 1, 2, 3, 4$$

Table 3.11

x	0	1	2	3	4	5	6
$F(x)$	.05	.15	.35	.65	.85	.95	1.0

(a) Find the density table for X, and verify that the sum of the probabilities given in the table is 1.

(b) If $x < 0$, what is the numerical value of $F(x)$?

(c) If $x > 4$, what is the numerical value of $F(x)$?

12. Past experience shows that over time the rivets in bridge supports can become dangerously loose. Assume that X, the number of loose rivets found per 10 feet beam on bridges over 20 years old, has the cumulative distribution shown in Table 3.11.

(a) Find the density table for X.

(b) Verify that $f(x) = \dfrac{6 - 2|x - 3|}{20}$ $x = 1, 2, 3, 4, 5$

$$f(x) = \dfrac{4 - |x - 3|}{20} \qquad x = 0 \text{ or } 6$$

13. Explain why the cumulative distribution function for a discrete random variable can never decrease in value.

Section 3.3

14. In an experiment to graft Florida sweet orange trees to the root of a sour orange variety, a series of five trials is conducted. Let X denote the number of grafts that fail. The density for X is given in Table 3.9.

(a) Find $E[X]$.

(b) Find μ_X.

(c) Find $E[X^2]$.

(d) Find Var X.

(e) Find σ_X^2.

(f) Find the standard deviation for X.

(g) What physical unit is associated with σ_X?

15. The density for X, the number of holes that can be drilled per bit while drilling into limestone is given in Table 3.10.

(a) Find $E[X]$ and $E[X^2]$.

(b) Find Var X and σ_X.

(c) What physical unit is associated with σ_X?

16. Use the density derived in Exercise 9 to find the expected value and variance for X, the number of computer systems operable at the time of the launch. Can you express $E[X]$ and Var X in terms of n, the number of systems available, and p, the probability that a given system will be operable?

17. The probability p of being able to log on to a computer from a remote terminal at any given time is .7. Let X denote the number of attempts that must be made to gain access to the computer. Find $E[X]$. Can you express $E[X]$ in terms of p? *Hint:* The series $\sum_{x=1}^{\infty} x(.7)(.3)^{x-1} = E[X]$ is not geometric. To find $E[X]$, expand this series and the series $.3E[X]$. Subtract the two to form the series $.7E[X]$. Evaluate this *geometric* series, and solve for $E[X]$.

18. The probability that a cell will fuse in the presence of polyethylene glycol is 1/2. Let Y denote the number of cells exposed to antigen-carrying lymphocytes to obtain the first fusion. Use the method of Exercise 17 to find $E[Y]$.

19. Let X be a discrete random variable with density f. Let c be any real number. Show that

(a) $E[c] = c$. *Hint:* Remember that constants can be factored from summations and that $\sum_{\text{all } x} f(x) = 1$.

(b) $E[cX] = cE[X]$.

20. Use the rules for expectation to verify that $\text{Var } c = 0$ and $\text{Var } cX = c^2 \text{Var } X$ for any real number c. *Hint:* $\text{Var } c = E[c^2] - (E[c])^2$.

21. Let X and Y be independent random variables with $E[X] = 3$, $E[X^2] = 25$, $E[Y] = 10$ and $E[Y^2] = 164$.

(a) Find $E[3X + Y - 8]$.

(b) Find $E[2X - 3Y + 7]$.

(c) Find $\text{Var } X$.

(d) Find σ_X.

(e) Find $\text{Var } Y$.

(f) Find σ_Y.

(g) Find $\text{Var}[3X + Y - 8]$.

(h) Find $\text{Var}[2X - 3Y + 7]$.

(i) Find $E[(X - 3)/4]$ and $\text{Var}[(X - 3)/4]$.

(j) Find $E[(Y - 10)/8]$ and $\text{Var}[(Y - 10)/8]$.

(k) The results of parts (i) and (j) are not coincidental. Can you generalize and verify the conjecture suggested by these two exercises?

22. Consider the function f defined by

$$f(x) = (1/2)2^{-|x|} \qquad x = \pm 1, \pm 2, \pm 3, \pm 4, \ldots$$

(a) Verify that this is the density for a discrete random variable X. *Hint:* Expand the series $\sum_{\text{all } x} f(x)$ for a few terms. A recognizable series will develop!

(b) Let $g(X) = (-1)^{|X|-1} [2^{|X|}/(2|X| - 1)]$. Show that $\sum_{\text{all } x} g(x)f(x) < \infty$. *Hint:* Expand the series for a few terms. You will obtain an alternating series that can be shown to converge.

(c) Show that $\sum_{\text{all } x} |g(x)| f(x)$ does not converge. This will show that $E[g(X)]$ does not exist. *Hint:* Expand the series for a few terms. You will obtain a series that is term by term larger than the diverging harmonic type series $(1/3)\sum_{x=1}^{\infty} 1/x$.

23. *(An application to sort algorithms.)* In studying various sort algorithms in computer science, it is of interest to compare their efficiency by estimating the average number of interchanges needed to sort random arrays of various sizes. It is also of interest to compare these estimated averages to the "ideal" average, where by "ideal" we mean the expected minimum number of interchanges needed to sort the array. In this exercise you will derive this ideal average. (American Mathematical Association of Two-Year Colleges, "A Note on the Minimum Number of Interchanges Needed to Sort a Random Array," with T. McMillan, I. Liss, and J. Milton, Fall 1990.)

(a) Consider a random array of length n. When the positions of exactly two elements of the array are exchanged, we say that an "interchange" has taken place. Let X_n denote the minimum number of interchanges necessary to sort an array of size n. Note that

$$X_n = X_{n-1} + I$$

where $I = 0$ if the last element of the array is in the correct position and $I = 1$ otherwise. Argue that $P[I = 0] = 1/n$ and $P[I = 1] = 1 - (1/n)$.

(b) Show that

$$E[I] = 1 - \frac{1}{n}$$

(c) Argue that

$$E[X_n] = E[X_{n-1}] + 1 - \frac{1}{n}$$

$$E[X_{n-1}] = E[X_{n-2}] + 1 - \frac{1}{n-1}$$

$$E[X_{n-2}] = E[X_{n-3}] + 1 - \frac{1}{n-2}$$

$$\vdots$$

$$E[X_3] = E[X_2] + 1 - \frac{1}{3}$$

$$E[X_2] = E[X_1] + 1 - \frac{1}{2}$$

$$E[X_1] = 0$$

(d) Use a recursive argument to show that

$$E[X_n] = (n-1) - \sum_{i=2}^{n} \frac{1}{i}$$

(e) Illustrate the expression given in part (d) by finding $E[X_5]$.

(f) Elementary calculus can be used to approximate $E[X_n]$ by noting that

$$\sum_{i=2}^{n} \frac{1}{i} \doteq \int_{1.5}^{n+.5} \frac{1}{t} \, dt$$

Use this idea to approximate $E[X_5]$ and to compare the result to the exact solution found in part (e).

(g) A random digit generator is used to generate sets of 100 different three-digit numbers lying between 0 and 1. What is the ideal average number of interchanges needed to sort such an array?

Section 3.4

24. Let X be binomial with parameters $n = 15$ and $p = .2$.
   (a) Find the expression for the density for X.
   (b) Find $E[X]$ and Var X.
   (c) Find $E[X]$, $E[X^2]$, and Var X using the moment generating function, thus verifying your answer to part (c) of this exercise.
   (d) Find $P[X \le 1]$ by evaluating the density directly. Compare your answer to that given in Table I of App. A.
   (e) Draw dot diagrams similar to that of Fig. 3.4 to illustrate each of these probabilities, and find the probabilities using Table I of App. A.

$P[X \le 5]$	$P[X \ge 3]$
$P[X < 5]$	$F(9)$
$P[2 \le X \le 7]$	$F(20)$
$P[2 \le X < 7]$	$P[X = 10]$

25. Albino rats used to study the hormonal regulation of a metabolic pathway are injected with a drug that inhibits body synthesis of protein. The probability that a rat will die from the drug before the experiment is over is .2. If 10 animals are treated with the drug, how many are expected to die before the experiment ends? What is the probability that at least eight will survive? Would you be surprised if at least five died during the course of the experiment? Explain, based on the probability of this occurring.

26. Consider Example 1.2.1. The random variable X is the number of computer systems operable at the time of a space launch. The systems are assumed to operate independently. Each is operable with probability .9.

    (*a*) Argue that X is binomial and find its density. Compare your answer to that obtained in Exercise 9(*b*).

    (*b*) Find $E[X]$ and Var X.

27. In humans, geneticists have identified two sex chromosomes, R and Y. Every individual has an R chromosome, and the presence of a Y chromosome distinguishes the individual as male. Thus the two sexes are characterized as RR (female) and RY (male). Color blindness is caused by a recessive allele on the R chromosome, which we denote by r. The Y chromosome has no bearing on color blindness. Thus relative to color blindness, there are three genotypes for females and two for males:

Female	Male
RR (normal)	RY (normal)
Rr (carrier)	rY (color-blind)
rr (color-blind)	

A child inherits one sex chromosome randomly from each parent.

    (*a*) A carrier of color blindness parents a child with a normal male. Construct a tree to represent the possible genotypes for the child. Use the tree to find the probability that a given child will be a color-blind male.

    (*b*) If the couple has five children, what is the expected number of color-blind males? What is the probability that three or more will be color-blind males?

28. In scanning electron microscopy photography, a specimen is placed in a vacuum chamber and scanned by an electron beam. Secondary electrons emitted from the specimen are collected by a detector, and an image is displayed on a cathode-ray tube. This image is photographed. In the past a 4- $\times$ 5-inch camera has been used. It is thought that a 35-millimeter (mm) camera can obtain the same clarity. This type of camera is faster and more economical than the 4- $\times$ 5-inch variety.

    (*a*) Photographs of 15 specimens are made using each camera system. These unmarked photographs are judged for clarity by an impartial judge. The judge is asked to select the better of the two photographs from each pair. Let X denote the number selected taken by a 35-mm camera. If there is really no difference in clarity and the judge is randomly selecting photographs, what is the expected value of X?

    (*b*) Would you be surprised if the judge selected 12 or more photographs taken by the 35-mm camera? Explain, based on the probability involved.

    (*c*) If $X \geq 12$, do you think that there is reason to suspect that the judge is not selecting the photographs at random?

29. It has been found that 80% of all printers used on home computers operate correctly at the time of installation. The rest require some adjustment. A particular dealer sells 10 units during a given month.
   (*a*) Find the probability that at least nine of the printers operate correctly upon installation.
   (*b*) Consider 5 months in which 10 units are sold per month. What is the probability that at least 9 units operate correctly in each of the 5 months?

30. It is possible for a computer to pick up an erroneous signal that does not show up as an error on the screen. The error is called a silent paging error. A particular terminal is defective, and when using the system word processor, it introduces a silent paging error with probability .1. The word processor is used 20 times during a given week.
   (*a*) Find the probability that no silent paging errors occur.
   (*b*) Find the probability that at least one such error occurs.
   (*c*) Would it be unusual for more than four such errors to occur? Explain, based on the probability involved.

31. Assume that each time a metal detector at an airport signals, there is a 25% chance that the cause is change in the passenger's pocket. During a given hour, 15 passengers are stopped because of a signal from the metal detector.
   (*a*) Find the probability that at least 3 persons will have been stopped due to change in their pockets.
   (*b*) If 15 passengers are stopped by the detector, would it be unusual for none of these to have been stopped due to change in the pocket? Explain based on the probability of this occurring.

32. (*Point binomial or Bernoulli distribution.*) Assume that an experiment is conducted and that the outcome is considered to be either a success or a failure. Let p denote the probability of success. Define X to be 1 if the experiment is a success and 0 if it is a failure. X is said to have a *point binomial* or a *Bernoulli* distribution with parameter p.
   (*a*) Argue that X is a binomial random variable with $n = 1$.
   (*b*) Find the density for X.
   (*c*) Find the mean and variance for X.
   (*d*) In DNA replication errors can occur that are chemically induced. Some of these errors are "silent" in that they do not lead to an observable mutation. Growing bacteria are exposed to a chemical that has probability .14 of inducing an observable error. Let X be 1 if an observable mutation results, and let X be 0 otherwise. Find $E[X]$.

33. A binomial random variable has mean 5 and variance 4. Find the values of n and p that characterize the distribution of this random variable.

Section 3.5

34. Let X be a Poisson random variable with parameter $k = 10$.
   (*a*) Find $E[X]$.
   (*b*) Find Var X.
   (*c*) Find σ_X.
   (*d*) Find the expression for the density for X.
   (*e*) Find $P[X \le 4]$.
   (*f*) Find $P[X < 4]$.
   (*g*) Find $P[X = 4]$.
   (*h*) Find $P[X \ge 4]$.
   (*i*) Find $P[4 \le X \le 9]$.

35. A particular nuclear plant releases a detectable amount of radioactive gases twice a month on the average. Find the probability that there will be at most four such emissions during a month. What is the expected number of emissions during a 3-month period? If, in fact, 12 or more emissions are detected during a 3-month period, do you think that there is a reason to suspect the reported average figure of twice a month? Explain, on the basis of the probability involved.

36. Geophysicists determine the age of a zircon by counting the number of uranium fission tracks on a polished surface. A particular zircon is of such an age that the average number of tracks per square centimeter is five. What is the probability that a 2-centimeter-square sample of this zircon will reveal at most three tracks, thus leading to an underestimation of the age of the material?

37. California is hit by approximately 500 earthquakes that are large enough to be felt every year. However, those of destructive magnitude occur on the average once every year. Find the probability that California will experience at least one earthquake of this magnitude during a 6-month period. Would it be unusual to have 3 or more earthquakes of destructive magnitude in a 6-month period? Explain, based on the probability of this occurring.

38. Load-bearing structures in underground mines are often required to carry additional loads while mining operations are in progress. As the structures adjust to this new weight, small-scale displacements take place that result in the release of seismic and acoustic energy, called *rock noise*. This energy can be detected using special geophysical equipment. Assume that in a particular mine the average number of rock noises recorded during normal activity is 3 per hour. Would you consider it unusual if more than 10 were detected in a 2-hour period? Explain, based on the probability involved.

39. A burr is a thin ridge or rough area that occurs when shaping a metal part. These must be removed by hand or by means of some newer method such as water jets, thermal energy, or electrochemical processing before the part can be used. Assume that a part used in automatic transmissions typically averages two burrs each. What is the probability that the total number of burrs found on seven randomly selected parts will be at most four?

40. Cast iron is an alloy composed primarily of iron together with smaller amounts of other elements, including carbon, silicon, sulfur, and phosphorus. The carbon occurs as graphite, which is soft, or iron carbide, which is very hard and brittle. The type of cast iron produced is determined by the amount and distribution of carbon in the iron. Five types of cast iron are identifiable. These are gray, compacted graphite, ductile, malleable, and white. In malleable cast iron the carbon is present as discrete graphite particles. Assume that in a particular casting these particles average 20 per square inch. Would it be unusual to see a 1/4-inch-square area of this casting with fewer than two graphite particles? Explain, based on the probability involved.

41. A Poisson random variable is such that it assumes the values 0 and 1 with equal probability. Find the value of the Poisson parameter k for this variable.

42. If the sensitivity of a motion-activated light is set correctly, the average number of times that it will be activated per week by squirrels and other small woods animals is .5. What is the average number of times that you would expect the light to be activated by these animals in a two-week period? If this occurred at least 5 times during a two-week period, would you suspect that the sensitivity needed to be adjusted? Explain based on the probability involved.

43. *Escherichia coli,* a bacterium often found in the human digestive tract, can mutate from being streptomycin sensitive to being streptomycin resistant, which can cause the individual involved to become resistant to the antibiotic streptomycin. Assume that there is an average of two streptomycin-resistant bacteria on cultures drawn from a particular patient. Each culture has an area of 80 square centimeters. What is the probability that a one-square-centimeter random sample from a single culture will contain at least one resistant bacterium? What is the probability that at least one will be found in 5 randomly selected one-square-centimeter samples? (Assume that the 5 samples are independent of one another.)

Section 3.6

44. An engine contains 5 seals that operate independently. If 3 or more seals fail, then the engine will fail. It is thought that when the temperature drops below $0°$ F each seal has a 10% chance of failure. Let X denote the number of seals that fail so that X is binomial with $n = 5$ and $p = .10$. Simulate the performance of 10 such engines under $0°$ conditions. Use the 10 simulations to estimate the average number of seals that will fail per engine by averaging your 10 values of X. Compare your estimate to the theoretical mean of .5. In your simulation, how many of the 10 engines would have failed?

45. Use Table II of App. A to simulate the arrival and departure of planes to the airport described in Example 3.6.2 for 10 more 1-minute periods. Based on these data, approximate the average number of planes on the ground at a given time by finding the arithmetic average of the values of Z simulated in the experiment.

46. Consider the random variable X, the number of runs conducted to produce an unacceptable lot when coating steel tubes (see Exercise 25). X is geometric with $p = .05$. Divide the 100 possible two-digit numbers into two categories, with numbers 00–04 denoting the production of an unacceptable lot and the remaining numbers denoting the production of an acceptable lot. Simulate the experiment of producing lots until an unacceptable one is obtained 10 times. Record the value obtained for X in each simulation. Based on these data, approximate the average value of X. Does your approximate value lie close to the theoretical mean value of 20? If not, run the simulation 10 more times. Is the arithmetic average of your observed values for X closer to 20 this time?

REVIEW EXERCISES

47. A large microprocessor chip contains multiple copies of circuits. If a circuit fails, the chip knows it and knows how to select the proper logic to repair itself. The average number of defects per chip is 300. What is the probability that 10 or fewer defects will be found in a randomly selected region that comprises 5% of the total surface area? What is the probability that more than 10 defects will be found?

48. When a program is submitted to the computer in a time-sharing system, it is processed on a space-available basis. Past experience shows that a program submitted to one such system is accepted for processing within 1 minute with probability .25. Assume that during the course of a day five programs are submitted with enough time between submissions to ensure independence. Let X denote the number of programs accepted for processing within 1 minute.
(*a*) Find $E[X]$ and Var X.
(*b*) Find the probability that none of these programs will be accepted for processing within 1 minute.

(c) Five programs are submitted on each of two consecutive days. What is the probability that no programs will be accepted for processing within 1 minute during this two-day period?

49. A new type of brake lining is being studied. It is thought that the lining will last for at least 70,000 miles on 90% of the cars in which it is used. Laboratory trials are conducted to simulate the driving experience of 100 cars in which this lining is used. Let X denote the number of cars whose brakes must be relined before the 70,000-mile mark.
    (a) What is the distribution of X? What is $E[X]$?
    (b) What distribution can be used to approximate probabilities for X?
    (c) Suppose that we agree that the 90% figure is too high if 17 or more of the 100 cars require a relinement prior to the 70,000-mile mark. What is the probability that we will come to this conclusion by chance even though the 90% figure is correct?

50. An automobile repair shop has 10 rebuilt transmissions in stock. Three are not in correct working order and have an internal defect that will cause trouble within the first 1000 miles of operation. Four of these transmissions are randomly selected and installed in customers' cars. Find the probability that no defective transmissions are installed. Find the probability that exactly one defective transmission is installed.

51. A computer terminal can pick up an erroneous signal from the keyboard that does not show up on the screen. This creates a silent error that is difficult to detect. Assume that for a particular keyboard the probability that this will occur per entry is 1/1000. In 12,000 entries find the probability that no silent errors occur. Find the probability of at least one silent error.

52. It is thought that 1 of every 10 cars on the road has a speedometer that is miscalibrated to the extent that it reads at least 5 miles per hour low. During the course of a day 15 drivers are stopped and charged with exceeding the speed limit by at least 5 miles per hour. Would you be surprised to find that at least 5 of the cars involved have miscalibrated speedometers? Explain, based on the probability of observing a result this unusual by chance.

53. Let

$$f(x) = \frac{x^2}{14} \qquad x = 1, 2, 3$$

(a) Show that f is the density for a discrete random variable.
(b) Find $E[X]$ and $E[X^2]$ from the definition of these terms.
(c) Find Var X and σ.

CHAPTER 4

Continuous Distributions

In Chap. 3 we learned to distinguish a discrete random variable from one that is not discrete. In this chapter we consider a large class of nondiscrete random variables. In particular, we consider random variables that are called *continuous*. We first study the general properties of variables of the continuous type and then present some important families of continuous random variables.

4.1 CONTINUOUS DENSITIES

In Chap. 3 we considered the random variable T, the time of the peak demand for electricity at a particular power plant. We agreed that this random variable is not discrete since, "a priori"—before the fact—we cannot limit the set of possible values for T to some finite or countably infinite collection of times. Time is measured continuously, and T can conceivably assume any value in the time interval $[0, 24)$, where 0 denotes 12 midnight one day and 24 denotes 12 midnight the next day. Furthermore, if we ask *before* the day begins, What is the probability that the peak demand will occur exactly 12.013 278 650 931 271? the answer is 0. It is virtually impossible for the peak load to occur at this split second in time, not the slightest bit earlier or later. These two properties, possible values occurring as intervals and the a priori probability of assuming any specific value being 0, are the characteristics that identify a random variable as being continuous. This leads us to our next definition.

> **Definition 4.1.1 (Continuous random variable).** A random variable is continuous if it can assume any value in some interval or intervals of real numbers and the probability that it assumes any specific value is 0.

Note that the statement that the probability that a continuous random variable assumes any specific value is 0 is essential to the definition. Discrete variables have no such restriction. For this reason, we calculate probabilities in the continuous case differently than we do in the discrete case. In the discrete case we defined a function f, called the density, which enabled us to compute probabilities associated with the random variable X. This function is given by

$$f(x) = P[X = x] \qquad x \text{ real}$$

This definition cannot be used in the continuous case because $P[X = x]$ is always 0. However, we do need a function that will enable us to compute probabilities associated with a continuous random variable. Such a function is also called a density.

Definition 4.1.2 (Continuous density). Let X be a continuous random variable. A function f such that

1. $f(x) \geq 0$ for x real

2. $\displaystyle\int_{-\infty}^{\infty} f(x)\,dx = 1$

3. $\displaystyle P[a \leq X \leq b] = \int_{a}^{b} f(x)\,dx$ for a and b real

is called a density for X.

Although this definition may look arbitrary at first glance, it is not. Note that, as in the discrete case, f is defined over the entire real line and is nonnegative. Recall from elementary calculus that integration is the natural extension of summation in the sense that the integral is the limit of a sequence of Riemann sums. In the discrete case we require that $\sum_{\text{all } x} f(x) = 1$. The natural extension of this requirement to the continuous case is that the density integrate to 1. Therefore the necessary and sufficient conditions for a function to be a density for a continuous random variable are as follows:

**Necessary and Sufficient Conditions
for a Function to be a Continuous Density**

1. $f(x) \geq 0$

2. $\displaystyle\int_{-\infty}^{\infty} f(x)\,dx = 1$

In the discrete case we find the probability that X assumes a value in some set A by summing $f(x)$ over all values of x in A. That is,

$$P[X \in A] = \sum_{x \in A} f(x).$$

In the continuous case we shall be interested in finding the probability that X assumes values in some interval $[a, b]$. Replacing A by $[a, b]$ and substituting integration for summation in the previous expression suggest property 3 of Definition 4.1.2. That is,

$$P[a \leq X \leq b] = \int_{a}^{b} f(x)\,dx$$

It is evident that the term "density" in the continuous case is just an extension of the ideas presented in the discrete case, with summation being replaced by integration. This is an important notion, as it will allow us to define the concept of expected value in the continuous case quite naturally.

Example 4.1.1. The lead concentration in gasoline currently ranges from .1 to .5 grams per liter. What is the probability that the lead concentration in a randomly selected liter of gasoline will lie between .2 and .3 grams inclusive? To answer this question, we need a density, f, for the random variable X, the number of grams of lead per liter of gasoline. Consider the function

$$f(x) = \begin{cases} 12.5x - 1.25 & .1 \leq x \leq .5 \\ 0 & \text{elsewhere} \end{cases}$$

Figure 4.1
Graph of

$$f(x) = \begin{cases} 12.5x - 1.25 & .1 \le x \le .5 \\ 0 & \text{elsewhere} \end{cases}$$

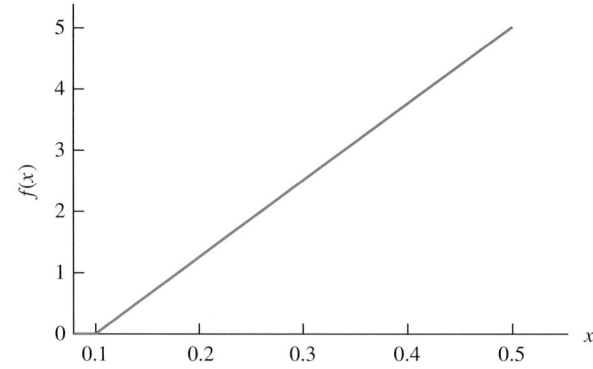

The graph of f is shown in Fig. 4.1. The function is nonnegative. Furthermore,

$$\int_{-\infty}^{\infty} f(x)\,dx = \int_{.1}^{.5} (12.5x - 1.25)\,dx$$

$$= \left[\frac{12.5x^2}{2} - 1.25x \right]_{.1}^{.5}$$

$$= \left[\frac{12.5(.5)^2}{2} - 1.25(.5) \right] - \left[\frac{12.5(.1)^2}{2} - 1.25(.1) \right]$$

$$= .9375 - (-.0625) = 1$$

Thus f satisfies properties 1 and 2 of Definition 4.1.2. Property 3 allows us to use f to find the desired probability. In particular,

$$P[.2 \le X \le .3] = \int_{.2}^{.3} f(x)\,dx$$

$$= \int_{.2}^{.3} (12.5x - 1.25)\,dx$$

$$= \left[\frac{12.5x^2}{2} - 1.25x \right]_{.2}^{.3}$$

$$= \left[\frac{12.5(.3)^2}{2} - 1.25(.3) \right] - \left[\frac{12.5(.2)^2}{2} - 1.25(.2) \right]$$

$$= .1875$$

There are several important points to be made concerning the density in the continuous case. First, we shall follow the convention of defining f only over intervals for which $f(x)$ may be nonzero. For values of x not explicitly mentioned, $f(x)$ is assumed to be 0. In Example 4.1.1 we could have written f as

$$f(x) = 12.5x - 1.25 \qquad .1 \le x \le .5$$

with the understanding that $f(x) = 0$ elsewhere. Second, since the integral of a nonnegative function can be thought of as an area, properties 2 and 3 of Definition 4.1.2 can be expressed in terms of areas. In particular, property 2 requires that *the total area under the graph of f be 1*. Property 3 implies that the probability that the variable assumes a value between two points a and b is *the area under the graph of f between $x = a$ and $x = b$*. These ideas as they apply to Example 4.1.1 are demonstrated in Figs. 4.2(*a*) and (*b*), respectively. Third, since $P[X = a] = P[X = b] = 0$ in the continuous case,

$$P[a \le X \le b] = P[a \le X < b] = P[a < X \le b] = P[a < X < b].$$

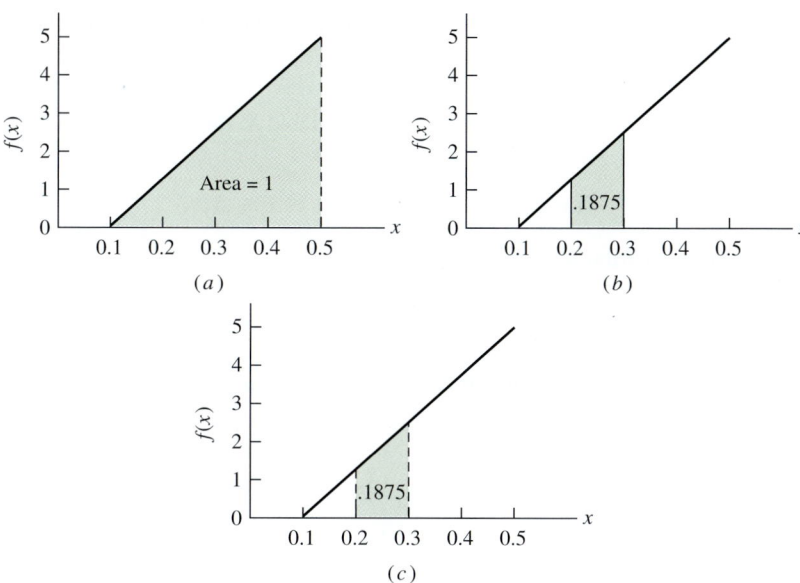

Figure 4.2
(a) $\int_{-\infty}^{\infty} f(x)\,dx = 1$ implies that the total area under the graph of f is 1; (b) $P[.2 \leq X \leq .3] = \int_{.2}^{.3} (12.5x - 1.25)\,dx = .1875$ implies that the area under the graph of f between $x = .2$ and $x = .3$ is .1875; (c) $P[.2 < X < .3] = P[.2 \leq X \leq .3] = .1875$.

In Example 4.1.1 the probability that the lead concentration in a liter of gasoline lies between .2 and .3 gram inclusive, $P[.2 \leq X \leq .3]$, is the same as $P[.2 < X < .3]$, the probability that it lies strictly between .2 and .3 gram. See Fig. 4.2(c). Fourth, properties 1 and 2 of Definition 4.1.2 are necessary and sufficient conditions for a function to be a density for a continuous random variable X. However, the density chosen for X cannot be just any function satisfying these conditions. It should be a function that assigns reasonable probabilities to events via property 3 of Definition 4.1.2. Whether or not the function f given in Example 4.1.1 satisfies this criteria is debatable. It was chosen for illustrative purposes only. Finding an appropriate density is not always easy.

Cumulative Distribution

The idea of a cumulative distribution function in the continuous case is useful. It is defined exactly as in the discrete case although found by using integration rather than summation.

> **Definition 4.1.3 (Cumulative distribution—continuous).** Let X be continuous with density f. The cumulative distribution function for X, denoted by F, is defined by
>
> $$F(x) = P[X \leq x] \qquad x \text{ real}$$

To find $F(x)$ for a specific real number x, we integrate the density over all real numbers that are less than or equal to x.

> **Computing F Continuous Case**
>
> $$P[X \leq x] = F(x) = \int_{-\infty}^{x} f(t)\,dt \qquad x \text{ real}$$

Graphically, this probability corresponds to the area under the graph of the density to the left of and including the point x.

Example 4.1.2. The density for the random variable X, the lead content in a liter of gasoline, is

$$f(x) = 12.5x - 1.25 \qquad .1 \leq x \leq .5$$

The cumulative distribution function for X is

$$P[X \leq x] = F(x) = \int_{-\infty}^{x} f(t)\,dt$$

For $x < .1$ this integral has value 0 since for these values of x, $f(t)$ is itself 0. For $.1 \leq x \leq .5$,

$$F(x) = \int_{-\infty}^{x} f(t)\,dt = \int_{.1}^{x} (12.5t - 1.25)\,dt$$

$$= \left[\frac{12.5t^2}{2} - 1.25t \right]_{.1}^{x}$$

$$= 6.25x^2 - 1.25x + .0625$$

For $x > .5$ the integral has value 1 since for these values of x we have integrated the density over its entire set of possible values. Summarizing, F is given by

$$F(x) = \begin{cases} 0 & x < .1 \\ 6.25x^2 - 1.25x + .0625 & .1 \leq x \leq .5 \\ 1 & x > .5 \end{cases}$$

What is the probability that the lead concentration in a randomly selected liter of gasoline will lie between .2 and .3 gram per liter? To answer this question, we rewrite it in terms of the cumulative distribution

$$P[.2 \leq X \leq .3] = P[X \leq .3] - P[X < .2]$$
$$= P[X \leq .3] - P[X \leq .2] \qquad (X \text{ is continuous})$$
$$= F(.3) - F(.2)$$

By substitution,

$$F(.3) = 6.25(.3)^2 - 1.25(.3) + .0625 = .2500$$
$$F(.2) = 6.25(.2)^2 - 1.25(.2) + .0625 = .0625$$

Thus

$$P[.2 \leq X \leq .3] = F(.3) - F(.2)$$
$$= .2500 - .0625 = .1875$$

Note that this agrees with the result obtained in Example 4.1.1 using direct integration. Note also that $F(.3)$ gives the area to the left of .3 shown in Fig. 4.3(a); $F(.2)$ gives the area to the left of .2 shown in Fig. 4.3(b). When we form the difference $F(.3) - F(.2)$, we naturally obtain the area between .2 and .3 given in Fig. 4.3(c).

Recall that in the discrete case, the cumulative distribution, F, was obtained from the density by addition; if F was available, f could be obtained by subtraction, the operation that reverses addition. The same sort of thing happens in the continuous case. We obtain the cumulative distribution from the density by integrating f;

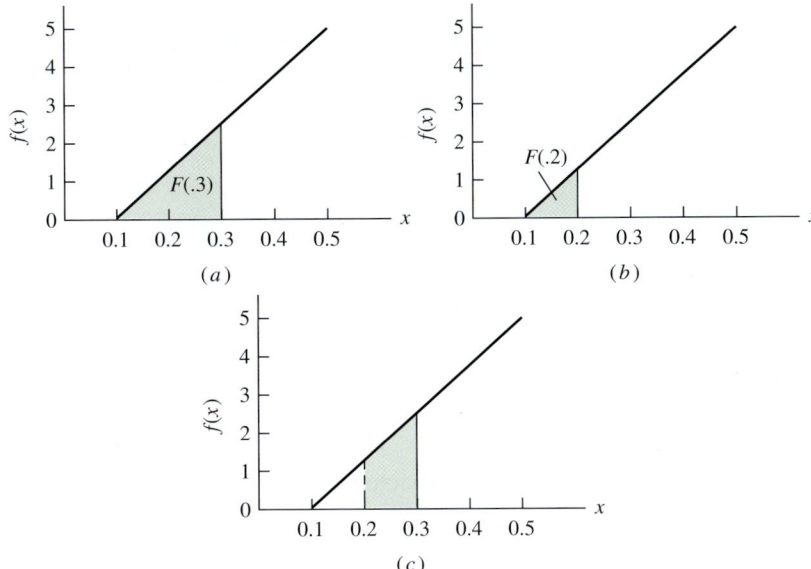

Figure 4.3
(a) $F(.3) = P[X \le .3]$;
(b) $F(.2) = P[X \le .2]$;
(c) $F(.3) - F(.2) = P[.2 \le X \le .3]$.

if F is available, we can retrieve f by reversing the integration operation via differentiation. That is, in the continuous case,

> **Obtaining f from F in the Continuous Case**
> $$f(x) = F'(x)$$

Example 4.1.3. In Example 4.1.2, we derived the cumulative distribution

$$F(x) = 6.25x^2 - 1.25x + .0625 \qquad .1 \le x \le .5$$

Note that

$$F'(x) = 12.5x - 1.25 \qquad .1 \le x \le .5$$

This is, as expected, the expression for the density for X that was given in Example 4.1.2.

Uniform Distribution

Perhaps the simplest continuous distribution with which to work is the *uniform* distribution. This distribution of events occur with equal or uniform probability. Since it is easy and instructive to develop the properties of this family of random variables directly from the definition, we leave the derivations to you.

4.2   EXPECTATION AND DISTRIBUTION PARAMETERS

In this section we define the term *expected value for continuous random variables*. We also discuss how to use the definition to find the mean and the variance of a variable of the continuous type. As you will see, the definition parallels that given in the discrete case, with the summation operation being replaced by integration.

> **Definition 4.2.1 (Expected value).** Let X be a continuous random variable with density f. Let $H(X)$ be a random variable. The expected value of $H(X)$, denoted by $E[H(X)]$, is given by
>
> $$E[H(X)] = \int_{-\infty}^{\infty} H(x)f(x)\,dx$$
>
> provided
>
> $$\int_{-\infty}^{\infty} |H(x)|\,f(x)\,dx$$
>
> is finite.

As in the discrete case, the mean or expected value of X is a special case of the above definition.

> **Expected Value of X**
>
> $$E[X] = \int_{-\infty}^{\infty} xf(x)\,dx$$

We illustrate the use of this definition by finding the mean and variance of the random variable X of Example 4.1.1. Recall that, by Theorem 3.3.2, the variance for X can be found via the computational shortcut

$$\sigma^2 = \text{Var}\,(X) = E[X^2] - (E[X])^2$$

Example 4.2.1. The density for X, the lead concentration in gasoline in grams per liter, is given by

$$f(x) = 12.5x - 1.25 \qquad .1 \le x \le .5$$

The mean or expected value of X is

$$
\begin{aligned}
\mu = E[X] &= \int_{-\infty}^{\infty} x f(x)\,dx \\
&= \int_{.1}^{.5} x(12.5x - 1.25)\,dx \\
&= \left[\frac{12.5x^3}{3} - \frac{1.25x^2}{2} \right]_{.1}^{.5} \\
&= \left[\frac{(12.5)(.5)^3}{3} - \frac{1.25(.5)^2}{2} \right] - \left[\frac{12.5(.1)^3}{3} - \frac{1.25(.1)^2}{2} \right] \\
&\doteq .3667 \text{ g/liter}
\end{aligned}
$$

Since integration is over an interval of finite length

$$\int_{-\infty}^{\infty} |x|\,f(x)\,dx$$

exists. We can conclude that, on the average, a liter of gasoline contains approximately .3667 g of lead. How much variability is there from liter to liter?

To answer this question, we find $E[X^2]$ and apply Theorem 3.3.2 to find the variance of X:

$$E[X^2] = \int_{-\infty}^{\infty} x^2 f(x)\,dx$$

$$= \int_{.1}^{.5} x^2(12.5x - 1.25)\,dx$$

$$= \left[\frac{12.5x^4}{4} - \frac{1.25x^3}{3} \right]_{.1}^{.5} \doteq .1433$$

By Theorem 3.3.2,

$$\text{Var } X = E[X^2] - (E[X])^2 \doteq .1433 - (.3667)^2 \doteq .00883$$

The standard deviation of X is

$$\sigma = \sqrt{\text{Var } X} = \sqrt{.00883} = .09396 \text{ g/liter}$$

It should be pointed out that there is a nice geometric interpretation of the mean in the case of a continuous random variable. Imagine cutting out of a piece of thin rigid metal the region bounded by the graph of f and the x axis, and attempting to balance this region on a knife-edge held parallel to the vertical axis. The point at which the region would balance, if such a point exists, is the mean of X. Thus, μ_X is a "location" parameter in that it indicates the position of the center of the density along the x axis. The variance can also be interpreted pictorially. In the continuous case variance is a "shape" parameter in the sense that a random variable with small variance will have a compact density; one with a large variance will have a density that is rather spread out or flat.

4.3 NORMAL DISTRIBUTION

The normal distribution is a distribution that underlies many of the statistical methods used in data analysis. It was first described in 1733 by De Moivre as being the limiting form of the binomial density as the number of trials becomes infinite. This discovery did not get much attention, and the distribution was "discovered" again by both Laplace and Gauss a half-century later. Both men dealt with problems of astronomy, and each derived the normal distribution as a distribution that seemingly described the behavior of errors in astronomical measurements. The distribution is often referred to as the "gaussian" distribution.

Definition 4.3.1 (Normal distribution). A random variable X with density

$$f(x) = \frac{1}{\sqrt{2\pi}\,\sigma} e^{-(1/2)[(x-\mu)/\sigma]^2} \qquad -\infty < x < \infty$$

$$-\infty < \mu < \infty$$

$$\sigma > 0$$

is said to have a normal distribution with parameters μ and σ.

One implication of this definition is that

$$\int_{-\infty}^{\infty} \frac{1}{\sqrt{2\pi}\,\sigma}\, e^{-(1/2)[(x-\mu)/\sigma]^2} dx = 1$$

To verify this requires a transformation to polar coordinates. This technique is beyond the mathematical level assumed here. A detailed proof can be found in [49]. Note that Definition 4.3.1 states only that μ is a real number and that σ is positive. As you might suspect from the notation used, the parameters that appear in the equation for the density for a normal random variable are, in fact, its mean and its standard deviation.

Theorem 4.3.1. Let X be a normal random variable with parameters μ and σ. Then μ is the mean of X and σ is its standard deviation.

The graph of the density of a normal random variable is a symmetric, bell-shaped curve centered at its mean. The points of inflection occur at $\mu \pm \sigma$.

Example 4.3.1. One of the major contributors to air pollution is hydrocarbons emitted from the exhaust system of automobiles. Let X denote the number of grams of hydrocarbons emitted by an automobile per mile. Assume that X is normally distributed with a mean of 1 gram and a standard deviation of .25 gram. The density for X is given by

$$f(x) = \frac{1}{\sqrt{2\pi}(.25)}\, e^{-(1/2)[(x-1)/.25]^2}$$

The graph of this density is a symmetric, bell-shaped curve centered at $\mu = 1$ with inflection points at $\mu \pm \sigma$, or $1 \pm .25$. A sketch of the density is given in Fig. 4.4.

One point must be made. Theoretically speaking, a normal random variable must be able to assume any value whatsoever. This is clearly unrealistic here. It is impossible for an automobile to emit a negative amount of hydrocarbons. When we say that X is normally distributed, we mean that over the range of physically reasonable values of X, the given normal curve yields acceptable probabilities. With this understanding, we can at least approximate, for example, the probability that a

Figure 4.4
Graph of the density for a normal random variable with mean 1 and standard deviation .25.

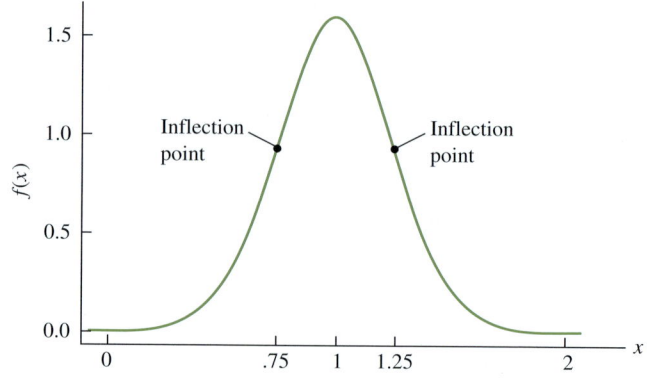

randomly selected automobile will emit between .9 and 1.54 grams of hydrocarbons by finding the area under the graph of *f* between these two points.

Standard Normal Distribution

There are infinitely many normal random variables each of which is uniquely characterized by the two parameters μ and σ. To calculate probabilities associated with a specific normal curve requires that one integrate the normal density over a particular interval. However, the normal density is not integrable in closed form. To find areas under the normal curve requires the use of numerical integration techniques. A simple algebraic transformation is employed to overcome this problem. By means of this transformation, called the *standardization procedure,* any question about any normal random variable can be transformed to an equivalent question concerning a normal random variable with mean 0 and standard deviation 1. This particular normal random variable is denoted by Z and is called the *standard normal* variable.

> **Theorem 4.3.2 (Standardization theorem).** Let *X* be normal with mean μ and standard deviation σ. The variable $(X - \mu)/\sigma$ is standard normal.

You have already verified that the transformation yields a random variable with mean 0 and standard deviation 1 (see Chap. 3, Exercise 21). To prove that the transformed variable is normal requires the use of moment generating function techniques which are covered in this book.

The cumulative distribution function for the standard normal random variable is given in Table V of App. A. The use of the standardization theorem and this table is illustrated in the following example.

Example 4.3.2. Let *X* denote the number of grams of hydrocarbons emitted by an automobile per mile. Assuming that *X* is normal with $\mu = 1$ gram and $\sigma = .25$ gram, find the probability that a randomly selected automobile will emit between .9 and 1.54 grams of hydrocarbons per mile. The desired probability is shown in Fig. 4.5. To find $P[.9 \leq X \leq 1.54]$, we first standardize by subtracting the mean of 1 and dividing by the standard deviation of .25 across the inequality. That is,

$$P[.9 \leq X \leq 1.54] = P[(.9 - 1)/.25 \leq (X - 1)/.25 \leq (1.54 - 1)/.25]$$

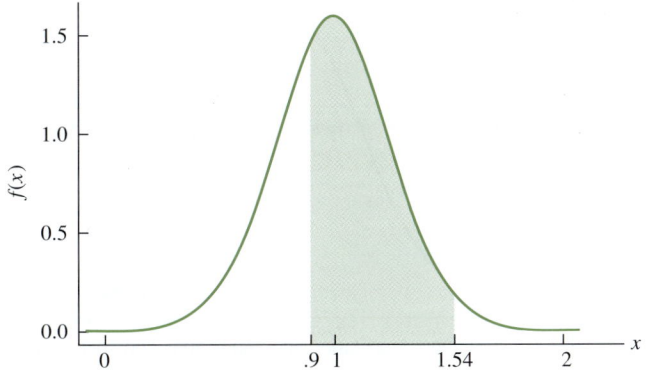

Figure 4.5
Shaded area = $P[.9 \leq X \leq 1.54]$.

The random variable $(X - 1)/.25$ is now Z. Therefore the problem is to find $P[-.4 \leq Z \leq 2.16]$ from Table V. We first express the desired probability in terms of the cumulative distribution as follows:

$$P[-.4 \leq Z \leq 2.16] = P[Z \leq 2.16] - P[Z < -.4]$$
$$= P[Z \leq 2.16] - P[Z \leq -.4] \qquad (Z \text{ is continuous})$$
$$= F(2.16) - F(-.4)$$

$F(2.16)$ is found by locating the first two digits (2.1) in the column headed z; since the third digit is 6, the desired probability of .9846 is found in the row labeled 2.1 and the column labeled .06. Similarly, $F(-.4)$ or .3446 is found in the row labeled -0.4 and the column labeled .00. We now see that the probability that a randomly selected automobile will emit between .9 and 1.54 grams of hydrocarbons per mile is

$$P[.9 \leq X \leq 1.54] = P[-.4 \leq Z \leq 2.16]$$
$$= F(2.16) - F(-.4)$$
$$= .9846 - .3446 = .64$$

Interpreting this probability as a percentage, we can say that 64% of the automobiles in operation emit between .9 and 1.54 grams of hydrocarbons per mile driven.

We shall have occasion to read Table V in reverse. That is, given a particular probability r we shall need to find the point with r of the area to its right. This point is denoted by z_r. Thus, notationally, z_r denotes that point associated with a standard normal random variable such that

$$P[Z \geq z_r] = r$$

To see how this need arises, consider Example 4.3.3.

Example 4.3.3.   Let X denote the amount of radiation that can be absorbed by an individual before death ensues. Assume that X is normal with a mean of 500 roentgens and a standard deviation of 150 roentgens. Above what dosage level will only 5% of those exposed survive? Here we are asked to find the point x_0 shown in Fig. 4.6. In terms of probabilities, we want to find the point x_0 such that

$$P[X \geq x_0] = .05$$

Figure 4.6
$P[X \geq x_0] = .05$.

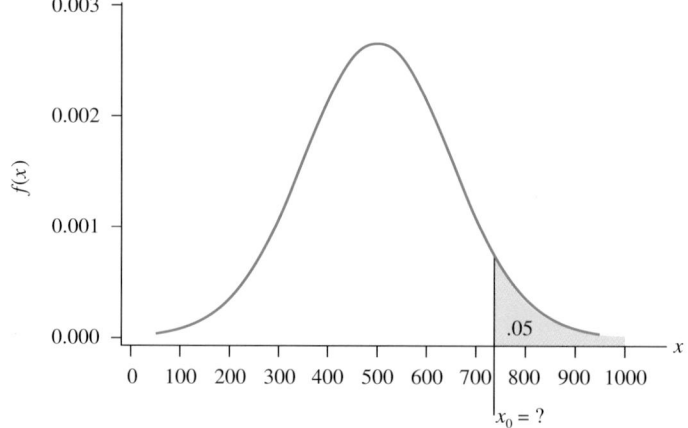

Standardizing gives

$$P[X \geq x_0] = P\left[\frac{X - 500}{150} \geq \frac{x_0 - 500}{150}\right]$$

$$= P\left[Z \geq \frac{x_0 - 500}{150}\right] = .05$$

Thus $(x_0 - 500)/150$ is the point on the standard normal curve with 5% of the area under the curve to its right and 95% to its left. That is, $(x_0 - 500)/150$ is the point $z_{.05}$. From Table V the numerical value of this point is approximately 1.645 (we have interpolated). Equating these, we get

$$\frac{x_0 - 500}{150} = 1.645$$

Solving this equation for x_0 gives the desired dosage level:

$$x_0 = 150(1.645) + 500 = 746.75 \text{ roentgens}$$

4.4   NORMAL PROBABILITY RULE AND CHEBYSHEV'S INEQUALITY

It is sometimes useful to have a quick way of determining which values of a random variable are common and which are considered to be rare. In the case of a normally distributed random variable, a rule of thumb, called the *normal probability rule,* can be developed easily. This rule is given in Theorem 4.4.1.

> **Theorem 4.4.1 (Normal probability rule).** Let X be normally distributed with parameters μ and σ. Then
>
> $$P[-\sigma < X - \mu < \sigma] \doteq .68$$
> $$P[-2\sigma < X - \mu < 2\sigma] \doteq .95$$
> $$P[-3\sigma < X - \mu < 3\sigma] \doteq .997$$

Proof. Note that division by σ yields

$$P[-\sigma < X - \mu < \sigma] = P\left[-1 < \frac{X - \mu}{\sigma} < 1\right]$$

By Theorem 4.3.2, $(X - \mu)/\sigma$ follows the standard normal distribution. From Table V of App. A,

$$P[-1 < Z < 1] = .8413 - .1587 = .6826$$

This probability can be rounded to .68. The other results given in the theorem are proved similarly.

The normal probability rule can be expressed in terms of percentages. In particular, it implies that in repeated sampling from a normal distribution approximately 68% of the observed values of X should lie within 1 standard deviation of its mean; 95% should lie within two standard deviations, and 99.7% within 3 standard deviations of the mean. Thus an observed value that falls farther than 3 standard deviations from μ is indeed rare, since such values occur with probability .003. This rule will be used later to obtain a quick estimate of the standard deviation of a normally distributed random variable.

Figure 4.7 illustrates the normal probability rule as it applies to the standard normal distribution. Recall that for this distribution $\sigma = 1, 2\sigma = 2,$ and $3\sigma = 3$.

Figure 4.7
(*a*) The probability that a normally distributed random variable will lie within one standard deviation of its mean is approximately .68 or 68%. (*b*) The probability that a normally distributed random variable will lie within two standard deviations of its mean is approximately .95 or 95%. (*c*) The probability that a normally distributed random variable will lie within three standard deviations of its mean is approximately .997 or 99.7%.

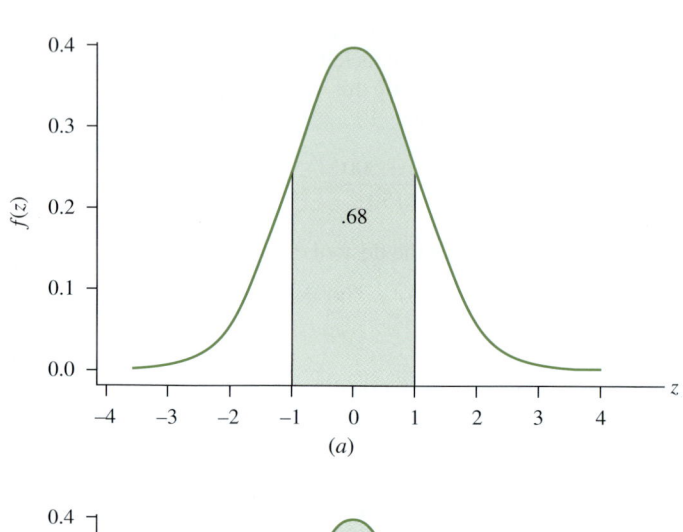

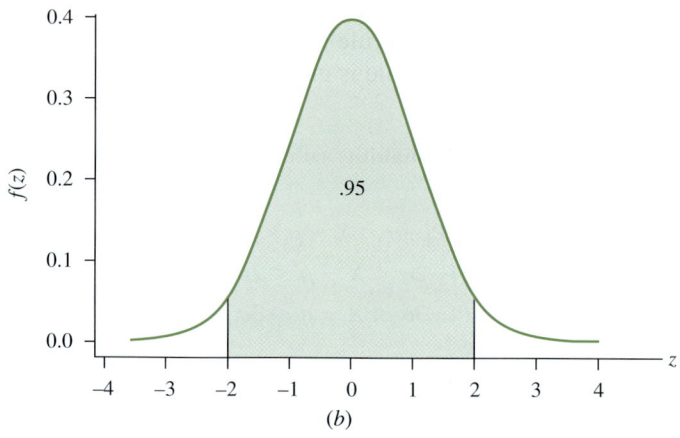

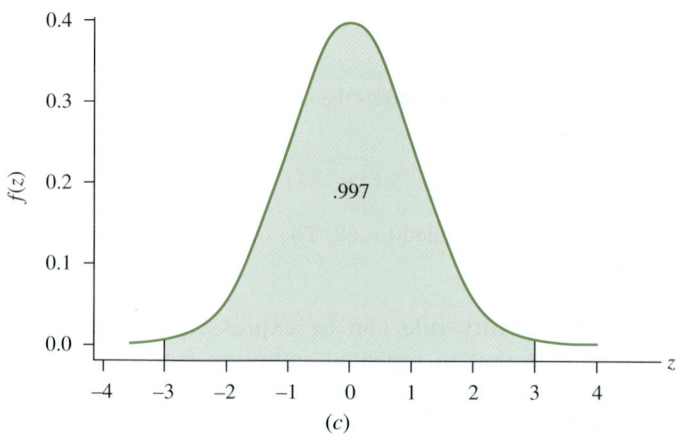

Chebyshev's Inequality

A second rule of thumb that can be used to gauge the rarity of observed values of a random variable is *Chebyshev's inequality*. This inequality was derived by the Russian probabilist P. L. Chebyshev (Tchebysheff, 1821–1894). The inequality differs from the normal probability rule in that it does *not* require that the random variable involved be normally distributed. Although we shall prove the theorem in the continuous setting, continuity is *not* required. The inequality holds for any random variable.

> **Theorem 4.4.2 (Chebyshev's inequality).** Let X be a random variable with mean μ and standard deviation σ. Then for any positive number k,
>
> $$P[\,|X - \mu| < k\sigma\,] \geq 1 - \frac{1}{k^2}$$

See Appendix C for the proof of this theorem.

Some examples will clarify the difference between Theorems 4.4.1 and 4.4.2.

Example 4.4.1. The viscosity of a fluid can be measured roughly by dropping a small ball into a calibrated tube containing the fluid and observing X, the time that it takes for the ball to drop a measured distance. Assume that this random variable is normally distributed with a mean of 20 s and a standard deviation of .5 s. By the normal probability rule, approximately 95% of the observed values of X will lie within 1 s (2 standard deviations) of the mean. That is, X will fall between 19 and 21 s with probability .95. Since Chebyshev's inequality applies to any random variable, it is appropriate here. This inequality guarantees that X will fall between 19 and 21 s (within $k = 2$ standard deviations of its mean) with probability *at least* $1 - 1/k^2 = .75$. Note that when the random variable in question is normally distributed, the normal probability rule yields a stronger statement than does Chebyshev's inequality.

Example 4.4.2. The safety record of an industrial plant is measured in terms of M, the total staffing-hours worked without a serious accident. Past experience indicates that M has a mean of 2 million with a standard deviation of .1 million. A serious accident has just occurred. Would it be unusual for the next serious accident to occur within the next 1.6 million staffing-hours? To answer this question, we must assess $P[M \leq 1.6]$. Since we have no reason to assume that M is normally distributed, the normal probability rule is inappropriate here. However, we know from Chebyshev's inequality with $k = 4$ that

$$P[1.6 < M < 2.4] \geq 1 - (1/16) = .9375$$

This implies that

$$P[M \leq 1.6] + P[M \geq 2.4] \leq .0625$$

Since it is possible for M to exceed 2.4, we can safely say that

$$P[M \leq 1.6] < .0625$$

No stronger statement can be made without some knowledge of the shape of the density of M. However, if it is known that the density is symmetric, then we can go one step further and state that

$$P[M \leq 1.6] \leq .0625/2 = .03125$$

4.5 NORMAL APPROXIMATION TO THE BINOMIAL DISTRIBUTION

The binomial tables given in this text or in any other text are necessarily limited in scope due to the fact that n can vary from 1 to infinity and p can assume any value between 0 and 1. It is impossible to table every combination of n and p. Due to the advances in computer and calculator technology, it is now possible to find exact binomial probabilities for any combination of n and p. Prior to this time, the normal curve was used to give good approximations of binomial probabilities. The technique introduced in this section is still useful in situations in which the needed technology tools are not readily available. To see how such approximations were suggested, we consider four binomial random variables each with probability of success .4 but with differing values for n. The densities for these variables, obtained from Table I of App. A, together with a sketch for each, are given in Fig. 4.8(a) to (d).

The point to note from these diagrams is made in Fig. 4.8(d). Namely, it is not hard to imagine a smooth bell curve that closely fits the block diagram shown. This suggests that binomial probabilities represented by one or more blocks in the diagram can be approximated reasonably well by a carefully selected area under an appropriately chosen normal curve. Which of the infinitely many normal curves is appropriate? Common sense indicates that the normal variable selected should have the same mean and variance as the binomial variable that it approximates. Theorem 4.5.1 summarizes these ideas.

> **Theorem 4.5.1 (Normal approximation to the binomial distribution).** Let X be binomial with parameters n and p. For large n, X is approximately normal with mean np and variance $np(1 - p)$.

The proof of this theorem is based on the Central Limit Theorem, which will be considered in Chap. 6. Admittedly, Theorem 4.5.1 is a bit vague in the sense that the word "large" is not well defined. In the strictest mathematical sense, "large" means as n approaches infinity. For most practical purposes the approximation is acceptable for values of n and p such that either $p \leq .5$ and $np > 5$ or $p > .5$ and $n(1 - p) > 5$.

Example 4.5.1. A study is performed to investigate the connection between maternal smoking during pregnancy and birth defects in children. Of the mothers studied, 40% smoke and 60% do not. When the babies were born, 20 were found to have some sort of birth defect. Let X denote the number of children whose mother smoked while pregnant. If there is no relationship between maternal smoking and birth defects, then X is binomial with $n = 20$ and $p = .4$. What is the probability that 12 or more of the affected children had mothers who smoked?

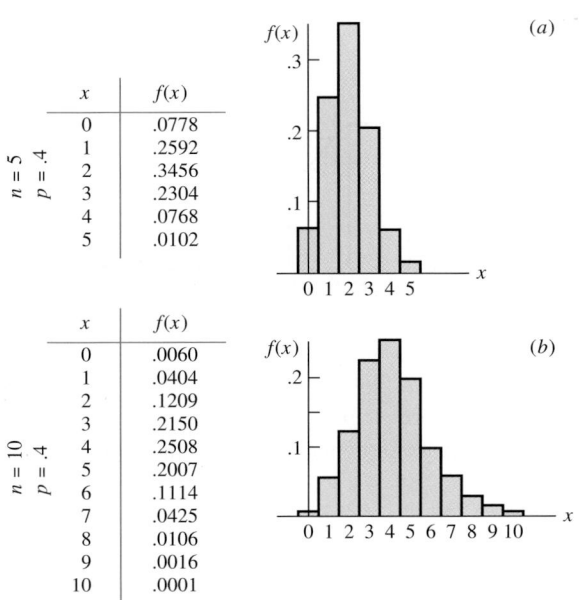

Figure 4.8
Density for X binomial:
(a) $n = 5, p = .4$; (b) $n = 10$,
$p = .4$; (c) $n = 15, p = .4$;
(d) $n = 20, p = .4$.

$n = 5$, $p = .4$

x	$f(x)$
0	.0778
1	.2592
2	.3456
3	.2304
4	.0768
5	.0102

$n = 10$, $p = .4$

x	$f(x)$
0	.0060
1	.0404
2	.1209
3	.2150
4	.2508
5	.2007
6	.1114
7	.0425
8	.0106
9	.0016
10	.0001

$n = 15$, $p = .4$

x	$f(x)$	x	$f(x)$
0	.0005	8	.1181
1	.0047	9	.0612
2	.0219	10	.0245
3	.0634	11	.0074
4	.1268	12	.0016
5	.1859	13	.0003
6	.2066	14	~ 0
7	.1771	15	~ 0

$n = 20$, $p = .4$

x	$f(x)$	x	$f(x)$
0	~ 0	11	.0710
1	.0005	12	.0355
2	.0031	13	.0145
3	.0124	14	.0049
4	.0350	15	.0013
5	.0746	16	.0003
6	.1244	17	~ 0
7	.1659	18	~ 0
8	.1797	19	~ 0
9	.1597	20	~ 0
10	.1172		

To answer this question, we need to find $P[X \geq 12]$ under the assumption that X is binomial with $n = 20$ and $p = .4$. This probability, .0565, can be found from Table I of App. A. Note that since $p = .4 \leq .5$ and $np = 20(.4) = 8 > 5$, the normal approximation should give a result quite close to .0565. We shall approximate probabilities associated with X using a normal random variable Y with mean $np = 20(.4) = 8$ and standard deviation $\sqrt{np(1-p)} = \sqrt{20(.4)(.6)} = \sqrt{4.8}$.

The exact probability of .0565 is given by the sum of the areas of the blocks centered at 12, 13, 14, 15, 16, 17, 18, 19, and 20, as shown in Fig. 4.9. The

Figure 4.9
$P[X \geq 12]$ = area of shaded
blocks $\doteq$ area under curve
beyond 11.5.

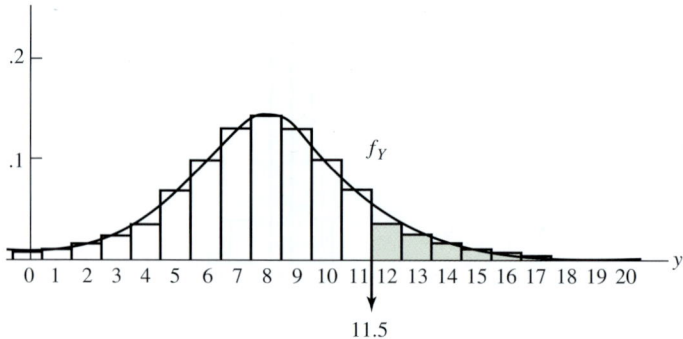

approximate probability is given by the area under the normal curve shown above 11.5. That is,

$$P[X \geq 12] \doteq P[Y \geq 11.5]$$

The number .5 is called the *half-unit correction* for continuity. It is subtracted from 12 in the approximation because otherwise half the area of the block centered at 12 will be inadvertently ignored, leading to an unnecessary error in the calculation. From this point on the calculation is routine:

$$P[X \geq 12] \doteq P[Y \geq 11.5]$$

$$= P\left[\frac{Y - 8}{\sqrt{4.8}} \geq \frac{11.5 - 8}{\sqrt{4.8}}\right]$$

$$= P[Z \geq 1.59]$$

$$= 1 - .9441 = .0559$$

Note that even with n as small as 20, the approximated value of .0559 compares quite favorably with the exact value of .0565. In practice, of course, one would not approximate a probability that could be found directly from a binomial table. This was done here only for comparative purposes.

4.6   SIMULATING A CONTINUOUS DISTRIBUTION

In Sec. 3.6 we showed how to simulate a discrete distribution using a random digit table. The table also can be used to simulate a continuous distribution. The idea is as follows:

1. We find the cumulative distribution function F for the random variable and its inverse.
2. We select a random two- (or three-) digit number from Table III of App. A and interpret this number as a probability, that is, as a number between 0 and 1.
3. We evaluate F^{-1} at this randomly selected point to obtain a randomly generated value for the random variable X.

This procedure is illustrated in Example 4.6.1.

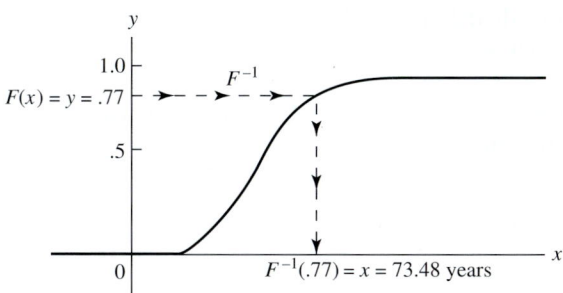

Figure 4.10
$F(x) = y = .77$ if and only if
$F^{-1}(.77) = x = 73.48$ years.

Example 4.6.1. Consider the random variable X, the time to failure of a computer chip. Assume that X has a Weibull distribution with parameters $\alpha = .02$ and $\beta = 1$. The density for X is

$$f(x) = .02e^{-.02x} \qquad x > 0$$

and its cumulative distribution is

$$y = F(x) = 1 - e^{-.02x}$$

The inverse of F is found by solving this equation for x as follows:

$$y = 1 - e^{-.02x}$$
$$e^{-.02x} = 1 - y$$
$$-.02x = \ln(1 - y)$$
$$x = \frac{-\ln(1 - y)}{.02}$$

To simulate an observation on X, we select a random two-digit number from Table III of App. A. Suppose the number selected is 77, which is interpreted as the probability $y = .77$. For this value of y our simulated observation on X is

$$x = \frac{-\ln(1 - .77)}{.02} = 73.48 \text{ years}$$

This procedure can be repeated to generate as many random values for X as desired. Figure 4.10 illustrates this procedure graphically.

CHAPTER SUMMARY

In this chapter we considered the general properties underlying random variables of the continuous type. These are random variables that assume their values in intervals of real numbers rather than at isolated points. The density function was introduced as a means of computing probabilities. These densities are defined in such a way that probabilities correspond to areas. The ideas of expected value were defined by replacing the summation operation, used in the discrete case, with integration. We studied the normal distribution and showed how to use this distribution to approximate binomial and Poisson probabilities. The log-normal, uniform, and Cauchy distributions were introduced as exercises. We saw how to simulate continuous distributions. We introduced and defined important terms that you should know. These are:

Continuous random variable   Continuous density
Continuous distribution function
Half-unit correction
Standard normal

EXERCISES

Section 4.1

1. Consider the function

$$f(x) = kx \qquad 2 \le x \le 4$$

   (a) Find the value of k that makes this a density for a continuous random variable.
   (b) Find $P[2.5 \le X \le 3]$.
   (c) Find $P[X = 2.5]$.
   (d) Find $P[2.5 < X \le 3]$.

2. Consider the areas shown in Fig. 4.11. In each case, state what probability is being depicted. What is the relationship between the areas depicted in Figs. 4.11(a) and (b)? Between those in Figs. 4.11(d) and (e)?

3. Let X denote the length in minutes of a long-distance telephone conversation. Assume that the density for X is given by

$$f(x) = (1/10)e^{-x/10} \qquad x > 0$$

   (a) Verify that f is a density for a continuous random variable.
   (b) Assuming that f adequately describes the behavior of the random variable X, find the probability that a randomly selected call will last at most 7 minutes; at least 7 minutes; exactly 7 minutes.
   (c) Would it be unusual for a call to last between 1 and 2 minutes? Explain, based on the probability of this occurring.
   (d) Sketch the graph of f and indicate in the sketch the area corresponding to each of the probabilities found in part (b).

4. Some plastics in scrapped cars can be stripped out and broken down to recover the chemical components. The greatest success has been in processing the flexible polyurethane cushioning found in these cars. Let X denote the amount of this material, in pounds, found per car. Assume that the density for X is given by

$$f(x) = \frac{1}{\ln 2} \frac{1}{x} \qquad 25 \le x \le 50$$

   (a) Verify that f is a density for a continuous random variable.
   (b) Use f to find the probability that a randomly selected auto will contain between 30 and 40 pounds of polyurethane cushioning.
   (c) Sketch the graph of f, and indicate in the sketch the area corresponding to the probability found in part (b).

5. (*Continuous uniform distribution.*) A random variable X is said to be uniformly distributed over an interval (a, b) if its density is given by

Figure 4.11

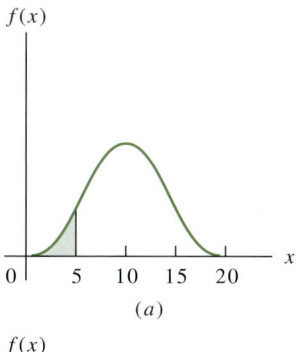

(a)

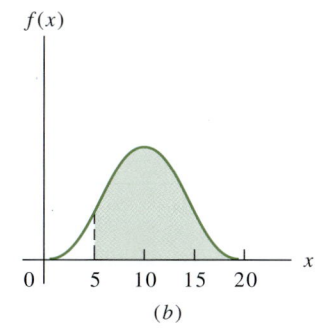

(b)

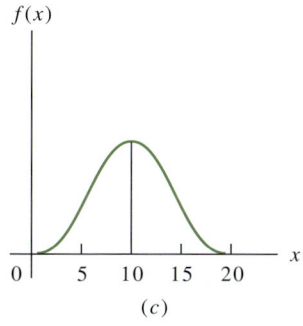

(c)

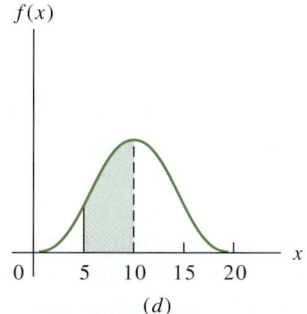

(d)

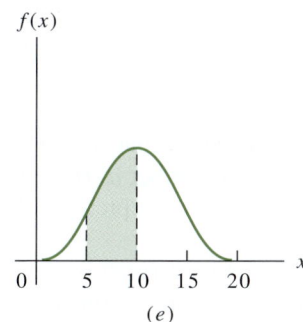

(e)

$$f(x) = \frac{1}{b - a} \qquad a < x < b$$

(a) Show that this is a density for a continuous random variable.

(b) Sketch the graph of the uniform density.

(c) Shade the area in the graph of part (b) that represents $P[X \leq (a + b)/2]$.

(d) Find the probability pictured in part (c).

(e) Let (c, d) and (e, f) be subintervals of (a, b) of equal length. What is the relationship between $P[c \leq X \leq d]$ and $P[e \leq X \leq f]$? Generalize the idea suggested by this example, thus justifying the name "uniform" distribution.

6. If a pair of coils were placed around a homing pigeon and a magnetic field was applied that reverses the earth's field, it is thought that the bird would become disoriented. Under these circumstances it is just as likely to fly in one direction as in any other. Let θ denote the direction in radians of the bird's initial flight. See Fig. 4.12. θ is uniformly distributed over the interval $[0, 2\pi]$.

(a) Find the density for θ.

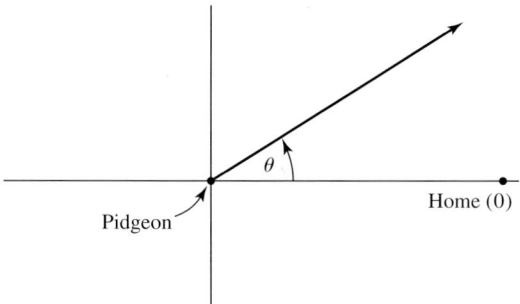

(b) Sketch the graph of the density. The uniform distribution is sometimes called the "rectangular" distribution. Do you see why?

(c) Shade the area corresponding to the probability that a bird will orient within $\pi/4$ radians of home, and find this area using plane geometry.

(d) Find the probability that a bird will orient within $\pi/4$ radians of home by integrating the density over the appropriate region(s), and compare your answer to that obtained in part (c).

(e) If 10 birds are released independently and at least seven orient within $\pi/4$ radians of home, would you suspect that perhaps the coils are not disorienting the birds to the extent expected? Explain, based on the probability of this occurring.

7. Use Definition 4.1.2 to show that for a continuous random variable X, $P[X = a] = 0$ for every real number a. *Hint:* Write $P[X = a]$ as $P[a \leq X \leq a]$.

8. Express each of the probabilities depicted in Fig. 4.15 in terms of the cumulative distribution function F.

9. Consider the random variable of Exercise 1.
    (a) Find the cumulative distribution function F.
    (b) Use F to find $P[2.5 \leq X \leq 3]$, and compare your answer to that obtained previously.
    (c) Find $F'(x)$, and verify that your result is the density given in Exercise 1.

10. *(Uniform distribution.)* Find the general expression for the cumulative distribution function for a random variable X that is uniformly distributed over the interval (a, b). See Exercise 5.

11. *(Uniform distribution.)* Consider the random variable of Exercise 6.
    (a) Use Exercise 10 to find the cumulative distribution function F.
    (b) Find $F'(x)$, and verify that your result is, as expected, the uniform density over the interval $[0, 2\pi]$.

12. Find the cumulative distribution function for the random variable of Exercise 3. Use F to find $P[1 \leq X \leq 2]$, and compare your answer to that obtained previously.

13. Find the cumulative distribution function for the random variable of Exercise 4. Use F to find $P[30 \leq X \leq 40]$, and compare your answer to that obtained previously.

14. In parts (a) and (b) proposed cumulative distributions are given. In each case, find the "density" that would be associated with each, and decide whether it really does define a valid continuous density. If it does not, explain what property fails.

(*a*) Consider the function F defined by

$$F(x) = \begin{cases} 0 & x < -1 \\ x + 1 & -1 \le x \le 0 \\ 1 & x > 0 \end{cases}$$

(*b*) Consider the function defined by

$$F(x) = \begin{cases} 0 & x \le 0 \\ x^2 & 0 < x \le 1/2 \\ (1/2)x & 1/2 < x \le 1 \\ 1 & x > 1 \end{cases}$$

Section 4.2

15. Consider the random variable X with density

$$f(x) = (1/6)x \qquad 2 \le x \le 4$$

(*a*) Find $E[X]$.
(*b*) Find $E[X^2]$.
(*c*) Find σ^2 and σ.

16. Let X denote the amount in pounds of polyurethane cushioning found in a car. (See Exercise 4.) The density for X is given by

$$f(x) = \frac{1}{\ln 2}\frac{1}{x} \qquad 25 \le x \le 50$$

Find the mean, variance, and standard deviation for X.

17. (*Uniform distribution.*) The density for a random variable X distributed uniformly over (a, b) is

$$f(x) = \frac{1}{b - a} \qquad a < x < b$$

Use Definition 4.2.1 to show that

$$E[X] = \frac{a + b}{2} \qquad \text{and} \qquad \text{Var } X = \frac{(b - a)^2}{12}$$

18. (*Uniform distribution.*) Let θ denote the direction in radians of the flight of a bird whose sense of direction has been disoriented as described in Exercise 6. Assume that θ is uniformly distributed over the interval $[0, 2\pi]$. Use the results of Exercise 18 to find the mean, variance, and standard deviation of θ.

19. Figure 4.13 gives the graphs of the densities of four continuous random variables whose means do exist. In each case, approximate the value of μ_X from the graph.

20. Consider the two densities given in Fig. 4.14. What is μ_X? What is μ_Y? Which random variable has the larger variance?

21. *(Cauchy distribution.)* A random variable X with density

$$f(x) = \frac{1}{\pi} \frac{a}{a^2 + (x-b)^2} \qquad \begin{array}{c} -\infty < x < \infty \\ -\infty < b < \infty \\ a > 0 \end{array}$$

is said to have a Cauchy distribution with parameters a and b. This distribution is interesting in that it provides an example of a continuous random variable whose mean does not exist. Let $a = 1$ and $b = 0$ to obtain a special case of the Cauchy distribution with density

Figure 4.13

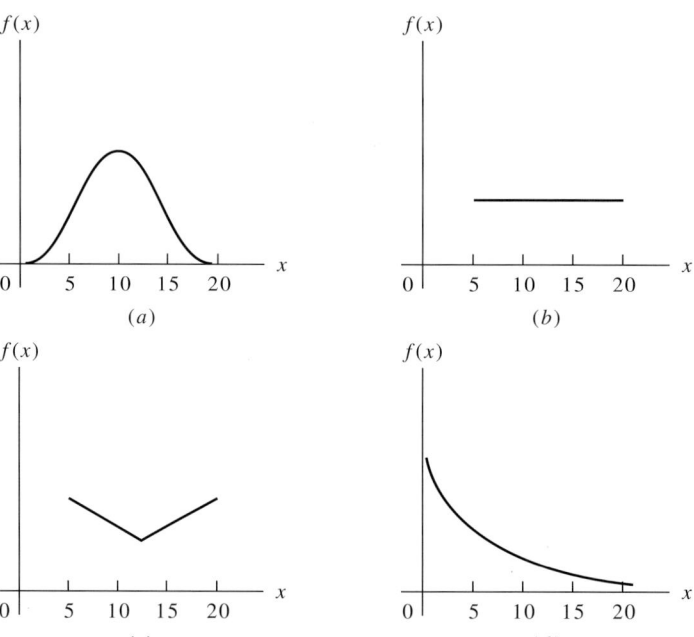

Figure 4.14

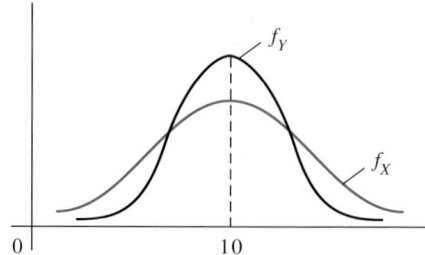

$$f(x) = \frac{1}{\pi} \frac{1}{1 + x^2} \qquad -\infty < x < \infty$$

Show that $\int_{-\infty}^{\infty} |x| f(x) \, dx$ does not exist, thus showing that $E[X]$ does not exist. *Hint:* Write

$$\int_{-\infty}^{\infty} |x| \frac{1}{\pi} \frac{1}{1 + x^2} \, dx = \frac{1}{\pi} \int_{-\infty}^{0} \frac{-x}{1 + x^2} \, dx + \frac{1}{\pi} \int_{0}^{\infty} \frac{x}{1 + x^2} \, dx$$

and recall that $\int (du/u) = \ln |u|$.

22. Let X denote the amount of time in hours that a battery on a solar calculator will operate adequately between exposures to light sufficient to recharge the battery. Assume that the density for X is given by

$$f(x) = (50/6)x^{-3} \qquad 2 < x < 10$$

   (a) Verify that this is a valid continuous density.
   (b) Find the expression for the cumulative distribution function for X, and use it to find the probability that a randomly selected solar battery will last at most 4 hours before needing to be recharged.
   (c) Find the average time that a battery will last before needing to be recharged.
   (d) Find $E[X^2]$, and use this to find the variance of X.

23. Assume that the increase in demand for electric power in millions of kilowatt hours over the next 2 years in a particular area is a random variable whose density is given by

$$f(x) = (1/64)x^3 \qquad 0 < x < 4$$

   (a) Verify that this is a valid density.
   (b) Find the expression for the cumulative distribution for X, and use it to find the probability that the demand will be at most 2 million kilowatt hours.
   (c) If the area only has the capacity to generate an additional 3 million kilowatt hours, what is the probability that demand will exceed supply?
   (d) Find the average increase in demand.

Section 4.3

24. Use Table V of App. A to find each of the following:
   (a) $P[Z \leq 1.57]$.   (b) $P[Z < 1.57]$.
   (c) $P[Z = 1.57]$.   (d) $P[Z > 1.57]$.
   (e) $P[-1.25 \leq Z \leq 1.75]$.   (f) $z_{.10}$.
   (g) $z_{.90}$.
   (h) The point z such that $P[-z \leq Z \leq z] = .95$.
   (i) The point z such that $P[-z \leq Z \leq z] = .90$.

25. The bulk density of soil is defined as the mass of dry solids per unit bulk volume. A high bulk density implies a compact soil with few pores. Bulk density is an important factor in influencing root development, seedling emergence, and aeration. Let X denote the bulk density of Pima clay loam. Studies show that X is normally distributed with $\mu = 1.5$ and $\sigma = .2$ g/cm^3.

(a) What is the density for *X*? Sketch a graph of the density function. Indicate on this graph the probability that *X* lies between 1.1 and 1.9. Find this probability.

(b) Find the probability that a randomly selected sample of Pima clay loam will have bulk density less than .9 g/cm^3.

(c) Would you be surprised if a randomly selected sample of this type of soil has a bulk density in excess of 2.0 g/cm^3? Explain, based on the probability of this occurring.

(d) What point has the property that only 10% of the soil samples have bulk density this high or higher?

26. Most galaxies take the form of a flattened disc, with the major part of the light coming from this very thin fundamental plane. The degree of flattening differs from galaxy to galaxy. In the Milky Way Galaxy most gases are concentrated near the center of the fundamental plane. Let *X* denote the perpendicular distance from this center to a gaseous mass. *X* is normally distributed with mean 0 and standard deviation 100 parsecs. (A parsec is equal to approximately 19.2 trillion miles.)

(a) Sketch a graph of the density for *X*. Indicate on this graph the probability that a gaseous mass is located within 200 parsecs of the center of the fundamental plane. Find this probability.

(b) Approximately what percentage of the gaseous masses are located more than 250 parsecs from the center of the plane?

(c) What distance has the property that 20% of the gaseous masses are at least this far from the fundamental plane?

27. Among diabetics, the fasting blood glucose level *X* may be assumed to be approximately normally distributed with mean 106 milligrams per 100 milliliters and standard deviation 8 milligrams per 100 milliliters.

(a) Sketch a graph of the density for *X*. Indicate on this graph the probability that a randomly selected diabetic will have a blood glucose level between 90 and 122 mg/100 ml. Find this probability.

(b) Find $P[X \leq 120 \text{ mg/100 ml}]$.

(c) Find the point that has the property that 25% of all diabetics have a fasting glucose level of this value or lower.

(d) If a randomly selected diabetic is found to have fasting blood glucose level in excess of 130, do you think there is cause for concern? Explain, based on the probability of this occurring naturally.

28. Let *X* denote the time in hours needed to locate and correct a problem in the software that governs the timing of traffic lights in the downtown area of a large city. Assume that *X* is normally distributed with mean 10 hours and variance 9.

(a) Find the probability that the next problem will require at most 15 hours to find and correct.

(b) The fastest 5% of repairs take at most how many hours to complete?

29. Assume that during seasons of normal rainfall the water level in feet at a particular lake follows a normal distribution with mean of 1876 feet and standard deviation of 6 inches.

(a) During such a season, would it be unusual to observe a water level of at most 1875 feet? Explain based on the probability of this occurring.

(b) Suppose that the water will crest the spillway if the level exceeds 1878 feet. What is the probability that this will occur during a season of normal rainfall?

30. *(Log-normal distribution.)* The log-normal distribution is the distribution of a random variable whose natural logarithm follows a normal distribution. Thus if X is a normal random variable, then $Y = e^X$ follows a log-normal distribution. Complete the argument below, thus deriving the density for a log-normal random variable.

Let X be normal with mean μ and variance σ^2. Let G denote the cumulative distribution function for $Y = e^X$, and let F denote the cumulative distribution function for X.

(a) Show that $G(y) = F(\ln y)$.

(b) Show that $G'(y) = F'(\ln y)/y$.

(c) Show that the density for Y is given by

$$g(y) = \frac{1}{\sqrt{2\pi}\sigma y} \exp\left[-\frac{1}{2}\frac{(\ln y - \mu)^2}{\sigma^2}\right] \quad \begin{array}{l} -\infty < \mu < \infty \\ \sigma > 0 \\ y > 0 \end{array}$$

Note that μ and σ are the mean and standard deviation of the underlying normal distribution; they are not the mean and standard deviation of Y itself.

31. Let Y denote the diameter in millimeters of Styrofoam pellets used in packing. Assume that Y has a log-normal distribution with parameters $\mu = .8$ and $\sigma = .1$.

(a) Find the probability that a randomly selected pellet has a diameter that exceeds 2.7 millimeters.

(b) Between what two values will Y fall with probability approximately .95?

Section 4.4

32. Verify the normal probability rule.

33. The number of Btu's of petroleum and petroleum products used per person in the United States in 1975 was normally distributed with mean 153 million Btu's and standard deviation 25 million Btu's. Approximately what percentage of the population used between 128 and 178 million Btu's during that year? Approximately what percentage of the population used in excess of 228 million Btu's?

34. Reconsider Exercises 26(a), 27(a), and 28(a) in light of the normal probability rule.

35. For a normal random variable, $P[|X - \mu| < 3\sigma] \doteq .997$. What value is assigned to this probability via Chebyshev's inequality? Are the results consistent? Which rule gives a stronger statement in the case of a normal variable?

36. Animals have an excellent spatial memory. In an experiment to confirm this statement, an eight-armed maze such as that shown in Fig. 4.15 is used. At the beginning of a test, one pellet of food is placed at the end of each arm. A hungry animal is placed at the center of the maze and is allowed to choose freely from among the arms. The optimal strategy is to run to the end of each arm exactly once. This requires that the animal remember where it has been. Let X denote the number of correct arms (arms still containing food) selected among its first eight choices. Studies indicate that $\mu = 7.9$.

(a) Is X normally distributed?

Figure 4.15
An eight-armed maze.

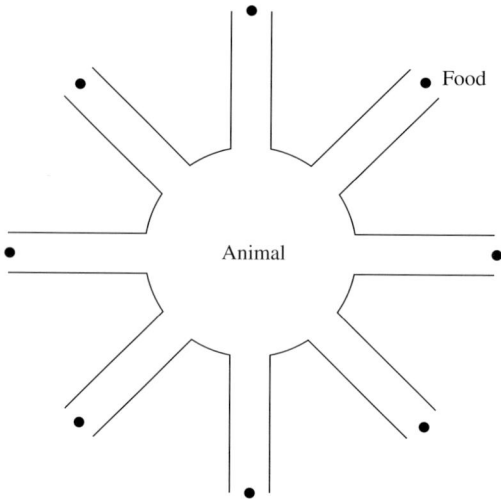

(b) State and interpret Chebyshev's inequality in the context of this problem for $k = .5, 1, 2,$ and 3. At what point does the inequality begin to give us some practical information?

Section 4.5

37. Let X be binomial with $n = 20$ and $p = .3$. Use the normal approximation to approximate each of the following. Compare your results with the values obtained from Table I of App. A.
  (a) $P[X \leq 3]$.
  (b) $P[3 \leq X \leq 6]$.
  (c) $P[X \geq 4]$.
  (d) $P[X = 4]$.

38. Although errors are likely when taking measurements from photographic images, these errors are often very small. For sharp images with negligible distortion, errors in measuring distances are often no larger than .0004 inch. Assume that the probability of a serious measurement error is .05. A series of 150 independent measurements are made. Let X denote the number of serious errors made.
  (a) In finding the probability of making at least one serious error, is the normal approximation appropriate? If so, approximate the probability using this method.
  (b) Approximate the probability that at most three serious errors will be made.

39. A chemical reaction is run in which the usual yield is 70%. A new process has been devised that should improve the yield. Proponents of the new process claim that it produces better yields than the old process more than 90% of the time. The new process is tested 60 times. Let X denote the number of trials in which the yield exceeds 70%.
  (a) If the probability of an increased yield is .9, is the normal approximation appropriate?
  (b) If $p = .9$, what is $E[X]$?
  (c) If $p > .9$ as claimed, then, on the average, more than 54 of every 60 trials will result in an increased yield. Let us agree to accept the claim if X is at

least 59. What is the probability that we will accept the claim if p is really only .9?

(d) What is the probability that we shall not accept the claim ($X \leq 58$) if it is true, and p is really .95?

40. Opponents of a nuclear power project claim that the majority of those living near a proposed site are opposed to the project. To justify this statement, a random sample of 75 residents is selected and their opinions are sought. Let X denote the number opposed to the project.

(a) If the probability that an individual is opposed to the project is .5, is the normal approximation appropriate?

(b) If $p = .5$, what is $E[X]$?

(c) If $p > .5$ as claimed, then, on the average, more than 37.5 of every 75 individuals are opposed to the project. Let us agree to accept the claim if X is at least 46. What is the probability that we shall accept the claim if p is really only .5?

(d) What is the probability that we shall not accept the claim ($X \leq 45$) even though it is true and p is really .7?

41. *(Normal approximation to the Poisson distribution.)* Let X be Poisson with parameter λs. Then for large values of λs, X is approximately normal with mean λs and variance λs. (The proof of this theorem is also based on the Central Limit Theorem and will be considered in Chap. 7.) Let X be a Poisson random variable with parameter $\lambda s = 15$. Find $P[X \leq 12]$ from Table II of App. A. Approximate this probability using a normal curve. Be sure to employ the half-unit correction factor.

42. The average number of jets either arriving at or departing from O'Hare Airport is one every 40 seconds. What is the approximate probability that at least 75 such flights will occur during a randomly selected hour? What is the probability that fewer than 100 such flights will take place in an hour?

Section 4.6

43. Use Table III of App. A to generate nine more observations on the random variable X, the time to failure of a computer chip. (See Example 4.6.1.) Based on these data, approximate the average time to failure by finding the arithmetic average of the values of X simulated in the experiment. Does this value agree well with the theoretical mean value of 50 years?

44. Simulate 20 observations on the random variable X, the time to failure of the signal lights on an automobile. (See Exercise 60.) Approximate the average time to failure for these lights based on the simulated data. Does this value agree well with the theoretical mean value for X?

45. A satellite has malfunctioned and is expected to reenter the earth's atmosphere sometime during a 4-hour period. Let X denote the time of reentry. Assume that X is uniformly distributed over the interval $[0, 4]$. Simulate 20 observations on X. (See Exercise 17.)

REVIEW EXERCISES

46. Let X be a continuous random variable with density

$$f(x) = cx^2 \qquad -3 \leq x \leq 3$$

(a) Assuming that $f(x) = 0$ elsewhere, find the value of c that makes this a density.
(b) Find $E[X]$ and $E[X^2]$ from the definitions of these terms.
(c) Find Var X and σ.
(d) Find $P[X \le 2]$; $P[-1 \le X \le 2]$; $P[X > 1]$ by direct integration.
(e) Find the closed-form expression for the cumulative distribution function F.
(f) Use F to find each of the probabilities of part (d), and compare your answers to those obtained earlier.

47. Find $\int_0^\infty z^{10} e^{-z} \, dz$.

48. A computer firm introduces a new home computer. Past experience shows that the random variable X, the time of peak demand measured in months after its introduction, follows a gamma distribution with variance 36.
(a) If the expected value of X is 18 months, find α and β.
(b) Find $P[X \le 7.01]$; $P[X \ge 26]$; $P[13.7 \le X \le 31.5]$.

49. Let X denote the lag time in a printing queue at a particular computer center. That is, X denotes the difference between the time that a program is placed in the queue and the time at which printing begins. Assume that X is normally distributed with mean 15 minutes and variance 25.
(a) Find the expression for the density for X.
(b) Find the probability that a program will reach the printer within 3 minutes of arriving in the queue.
(c) Would it be unusual for a program to stay in the queue between 10 and 20 minutes? Explain, based on the approximate probability of this occurring. You do not have to use the Z table to answer this question!
(d) Would you be surprised if it took longer than 30 minutes for the program to reach the printer? Explain, based on the probability of this occurring.

50. A computer center maintains a telephone consulting service to troubleshoot for its users. The service is available from 9 a.m. to 5 p.m. each working day. Past experience shows that the random variable X, the number of calls received per day, follows a Poisson distribution with $\lambda = 50$. For a given day, find the probability that the first call of the day will be received by 9:15 a.m.; after 3 p.m.; between 9:30 a.m. and 10 a.m.

51. Let X denote the time required to upgrade a computer system in hours. Assume that the density for X is given by

$$f(x) = k \exp(-2x) \qquad 0 < x < \infty$$

(a) Find the numerical value of k that makes this a valid density.
(b) Find the probability that it will take at most 1 hour to upgrade a given system.
(c) Find the average time required to upgrade a system.
(d) Find the standard deviation in the time required for the upgrade.

52. Past evidence shows that when a customer complains of an out-of-order phone there is an 8% chance that the problem is with the inside wiring. During a 1-month period, 100 complaints are lodged. Assume that there have been no wide-scale problems that could be expected to affect many phones at once, and that, for this reason, these failures are considered to be independent. Find the expected number of failures due to a problem with the inside wiring. Find the probability that at least 10 failures are due to a problem with the inside wiring. Would it be unusual if at most 5 were due to problems with the inside wiring? Explain, based on the probability of this occurring.

53. The cumulative distribution function for a continuous random variable X is defined by

$$F(x) = \begin{cases} 0 & x < 0 \\ \dfrac{x^3 + x^2}{2} & 0 \le x \le 1 \\ 1 & x > 1 \end{cases}$$

Find the density for X.

54. An electronic counter records the number of vehicles exiting the interstate at a particular point. Assume that the average number of vehicles leaving in a 5-minute period is 10. Approximate the probability that between 100 and 120 vehicles inclusive will exit at this point in a 1-hour period.

CHAPTER 5

Descriptive Statistics

Thus far we have considered random variables from a theoretical point of view. We have studied two functions, the density and the cumulative distribution function, that enable us to predict the behavior of the variable in a probabilistic sense. We have also considered three parameters that characterize or describe a random variable, namely, μ, σ^2, and σ. In practice, the exact distribution of a random variable is seldom known. Rather, we must determine a reasonable form for the density and appropriate values for the distribution parameters from a data set. In this chapter we consider some simple graphical and analytic methods for doing so.

5.1 RANDOM SAMPLING

We begin by considering a typical problem that calls for a statistical solution. Suppose that we wish to study the performance of the lithium batteries used in a particular model of pocket calculator. The purpose of our study is to determine the mean effective life span of these batteries so that we can place a limited warranty on them in the future. Since this type of battery has not been used in this model before, no one can tell us the distribution of the random variable, X, the life span of a battery. We must attempt to discover its distribution for ourselves. This is inherently a statistical problem. What characteristics identify it as such? Simply the following:

Characteristics of a Statistical Problem

1. Associated with the problem is a large group of objects about which inferences are to be made. This group of objects is called the *population.*
2. There is at least one random variable whose behavior is to be studied relative to the population.
3. The population is too large to study in its entirety, or techniques used in the study are destructive in nature. In either case we must draw conclusions about the population based on observing only a portion or "sample" of objects drawn from the population.

In our example the population is large and hypothetical in the sense that it consists of all lithium batteries used in this model calculator in the past, present, and future. Since we cannot observe the life span of batteries not yet produced, the population obviously cannot be studied in its entirety! Furthermore, to determine the life span of a battery, it must be used until it fails. That is, the method of study destroys the object being studied. For these reasons, we must devise methods for approximating the characteristics of the life span of a lithium battery based on observing only a sample of these batteries.

To draw inferences about a population using statistical methods, the sample drawn should be "random." To understand what we mean by this term, let us return to our example. Here we have a large population that consists of all lithium batteries produced for a certain model of pocket calculator. Associated with the population is a random variable X. We do not know the form of its density, nor do we know its mean or variance. We want to select a subset of n batteries from the population "at random." That is, we want to select n batteries for study in such a way that the selection of one battery neither ensures nor precludes the selection of any other. In this way the selection of one battery is independent of the selection of any other. This collection of objects can be thought of as a "random sample."

Note that, prior to the actual selection of the batteries to be studied, X_i ($i = 1, 2, 3, \ldots, n$), the life span of the ith battery selected is a random variable. It has the same distribution as X, the life span of batteries in the population. Furthermore, these random variables are independent in the sense that the value assumed by one has no effect on the value assumed by any of the others. The random variables $X_1, X_2, X_3, \ldots, X_n$ and can be thought of as a "random sample."

Once we have actually selected n batteries for study and have observed the life span of each battery, we shall have available n numbers, $x_1, x_2, x_3, \ldots, x_n$. These numbers are the observed values of the random variables $X_1, X_2, X_3, \ldots, X_n$ and can be thought of as a "random sample."

As you can see, the term "random sample" is used in three different but closely related ways in applied statistics. It may refer to the *objects* selected for study, to the *random variables* associated with the objects to be selected, or to the *numerical values* assumed by those variables. It is usually clear from the context of the discussion which is intended. These ideas are illustrated in Fig. 5.1.

Even though the term "random sample" is used in these three ways, the formal definition of the term is mathematical in nature. When we use the term in stating theoretical results, we mean the following:

> **Definition 5.1.1 (Random sample).** A random sample of size n from the distribution of X is a collection of n independent random variables, each with the same distribution as X.

The theorems and definitions presented later use the term "random sample" in the sense just described. When objects are selected from a finite population, this type of sample results only when sampling is done with replacement. That is, an object is drawn, observed, and placed back in the population for possible reselection. This ensures that $X_1, X_2, X_3, \ldots, X_n$ are indeed independent and identically distributed. Usually, sampling from a finite population is done without replacement. This means that the random variables $X_1, X_2, X_3, \ldots, X_n$ are not independent. However, if the sample is small relative to the population itself, then removal of a few items does not drastically alter the composition of the population. A generally accepted guideline is that for all practical purposes we may assume independence whenever the sample constitutes at most 5% of the population. If this is not true, then the techniques used to estimate parameters must be altered to take this into account. We shall be assuming that for all practical purposes $X_1, X_2, X_3, \ldots, X_n$ are independent in the discussions that follow.

Once a random sample has been drawn, we commonly use the data gathered to evaluate pertinent *statistics*. What is a statistic? Roughly speaking, a statistic is a

Figure 5.1

The objects selected generate *n numbers* $x_1, x_2, x_3, \ldots, x_n$, which are the observed values of the random variables $X_1, X_2, X_3, \ldots, X_n$.

A statistician has a population about which to draw inferences

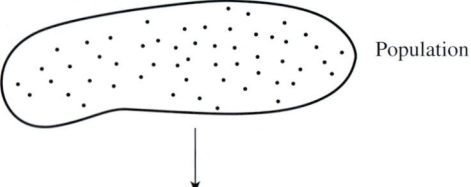

Population

Prior to the selection of the objects for study, interest centers on the n independent and identically distributed *random variables* $X_1, X_2, X_3, \ldots, X_n$

A set of *n objects* is selected from the population for study

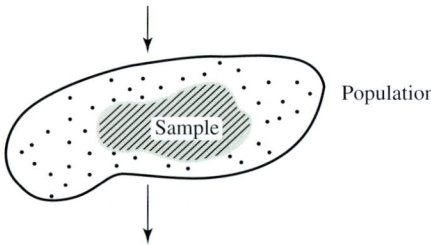

Population

The objects selected generate *n numbers* $x_1, x_2, x_3, \ldots, x_n$, which are the observed values of the random variables $X_1, X_2, X_3, \ldots, X_n$

random variable whose numerical value can be determined from a random sample. That is, a statistic is a random variable that is a function of the elements of a random sample $X_1, X_2, X_3, \ldots, X_n$. Typical statistics of interest to statisticians are $\sum_{i=1}^{n} X_i$, $\sum_{i=1}^{n} X_i^2$, $\sum_{i=1}^{n} X_i/n$, $\max_i\{X_i\}$, and $\min_i\{X_i\}$. These ideas are illustrated in Example 5.1.1.

Example 5.1.1. Consider the random variable X, the number of times per hour that a television signal is interrupted by random interference. Assume that this random variable has a Poisson distribution with unknown mean μ and unknown variance σ^2. To approximate the value of each of these parameters, we intend to observe the signal for ten randomly selected nonoverlapping one-hour periods over a week's time. Let X_i $(i = 1, 2, 3, \ldots, 10)$ denote the number of interruptions that occur during the ith observation period. The random variables $X_1, X_2, X_3, \ldots, X_{10}$ constitute a random sample of size 10 from a Poisson distribution with unknown mean μ and unknown variance σ^2. When the experiment is conducted, these data result:

$$
\begin{array}{lllll}
x_1 = 1 & x_3 = 0 & x_5 = 1 & x_7 = 0 & x_9 = 3 \\
x_2 = 0 & x_4 = 2 & x_6 = 1 & x_8 = 0 & x_{10} = 0
\end{array}
$$

The observed values of the statistics $\sum X_i$, $\sum X_i^2$, $\sum X_i/n$, $\max_i\{X_i\}$, and $\min_i\{X_i\}$ based on this sample are 8, 16, .8, 3, and 0, respectively. Note that the random variable $X_1 - \mu$ is *not* a statistic. Since μ is unknown, we cannot determine its numerical value from a random sample.

5.2 PICTURING THE DISTRIBUTION

When studying a random variable X, one important question to be answered is, "To which family of random variables does X belong?" That is, we need to determine whether X is binomial, Poisson, normal, exponential, or belongs to some other family of variables. In the discrete case it is often possible to determine the appropriate family from the physical description of the experiment. The only job left for the statistician is to approximate the values of the parameters that characterize the distribution. Continuous random variables are more difficult to handle. To determine the family to which such a variable belongs, we must get an idea of the *shape* of its density. For example, if the density appears to be flat, then it is reasonable to suspect that X is uniformly distributed; if it is bell-shaped, then X may be normally distributed.

If the distribution appears to be nonsymmetric with a long tail to the left or the right, then it is called *skewed* left or skewed right, respectively. Distributions such as the exponential, chi-squared, and gamma distributions exhibit this property.

Stem-and-Leaf Diagram

Here we consider some graphical methods for studying the distribution of a continuous random variable. The first method entails constructing what is called a *stem-and-leaf* diagram. This method was first introduced by John Tukey in 1977 [50].

A stem-and-leaf diagram consists of a series of horizontal rows of numbers. Each row is labeled via a number called its stem; the other numbers in the rows are called *leaves*. There are no rigid rules as to how to construct such a diagram. Basically these steps are followed:

Constructing a Stem-and-Leaf Diagram

1. Choose some convenient numbers to serve as stems. The stems are usually the first one or two digits of the numbers in the data set.
2. Label the rows via the stems selected.
3. Reproduce the data set graphically by recording the digit following the stem as a leaf.
4. Turn the graph on its side to get an idea of the shape of the distribution.

These ideas are illustrated in Example 5.2.1.

Example 5.2.1. To study the random variable X, the life span in hours of the lithium battery in a particular model of pocket calculator, we obtain a random sample of 50 batteries and determine the life span of each we obtain. These data result:

4285	564	1278	205	3920
2066	604	209	602	1379
2584	14	349	3770	99
1009	4152	478	726	510
318	737	3032	3894	582
1429	852	1461	2662	308
981	1560	701	497	3367
1402	1786	1406	35	99
1137	520	261	2778	373
414	396	83	1379	454

To construct a stem-and-leaf diagram for these data, we first choose numbers to serve as "stems." It is often convenient to use the first digit of a number as its stem. If a three-digit number such as 318 is expressed as a four-digit number (0318) by including a leading zero, then this data set entails the use of the five stems 0, 1, 2, 3, 4. We shall use the second digit of a number as its "leaf." The diagram is constructed by listing the stems as a vertical column as shown in Fig. 5.2(a). The first observation, 4285, has a stem of 4 and a leaf of 2. It is represented in the diagram as shown in Fig. 5.2(b). The entire data set, recorded in the order in which the observations appear, is shown in Fig. 5.2(c).

Is it reasonable to assume that X is normally distributed? To answer this question, turn the stem-and-leaf diagram on its side and look for the bell-shape characteristic of a normal density. This bell shape is not present, leading us to suspect that X is *not* a member of the family of normal random variables.

Notice that, in the above example, the first stem has a very large number of leaves. This often occurs when data sets are large or when there is not much variability in the data. In this case it is usually constructive to create what is called a *double* stem-and-leaf diagram. This is done by using each stem twice. We plot the low leaves of 0, 1, 2, 3, 4 on the first stem and the high leaves of 5, 6, 7, 8, 9 on the second. The double stem-and-leaf diagram for the data of Example 5.2.1 is shown in Fig. 5.3. This diagram was produced by MINITAB. This diagram shows even more clearly than that of Fig. 5.2 that the distribution from which this sample was drawn is probably not normal. In fact, it resembles a distribution that is exponential. We know now that a reasonable density for X assumes the general form

$$f(x) = (1/\beta) \exp(-1/\beta) \qquad x > 0 \qquad \beta > 0$$

It is now the job of the researcher to estimate the numerical value of β so that probabilities can be estimated in the future via the exponential density.

Figure 5.2
(a) The integers 0, 1, 2, 3, 4 form the stems for a stem-and-leaf diagram; (b) the number 4285 has a stem of 4 and a leaf of 2; (c) complete stem-and-leaf diagram for the sample of battery life spans of Example 5.2.1.

```
0 |          0 |          0 | 39456078532347202674005530 34
1 |          1 |          1 | 04415724433
2 |          2 |          2 | 0567
3 |          3 |          3 | 07893
4 |          4 | 2        4 | 21
(a)         (b)                   (c)
```

Figure 5.3
A double stem-and-leaf diagram with leaves in order.

```
Stem-and-leaf of hours      N  = 50
Leaf Unit = 1.0

   17      0 00000222333334444
  (11)     0 55556677789
   22      1 012334444
   13      1 57
   11      2 0
   10      2 567
    7      3 03
    5      3 789
    2      4 12
```

Histograms and Ogives

The stem-and-leaf diagram provides a quick look at a data set. It is a useful way to get an idea of the shape of a distribution when the data set is moderate in size. It has the advantage of preserving, to some extent, the ability to read the actual data values from the diagram. However, the technique does not work well when data sets are large. In this case, we turn to a technique that has been used for many years and that is often seen in data displays in journals, newspapers, corporate reports, and other presentations. This plot, called a *histogram,* is a vertical or horizontal bar graph. The bars or categories are defined in such a way that each observation belongs to one and only one category. We make the width of each bar the same so that the area of the bar is proportional to the number of observations in the respective category. This allows for easy visual comparisons of category frequencies and percentages. It also allows us to get an idea of the family of random variables to which the variable under study belongs by observing the shape of the histogram.

There are many ways to select category boundaries. Statistical packages each use their own algorithm for doing so, and these may differ from package to package. If several different packages are used to plot a given data set via its default technique, then the histograms can vary slightly in terms of number of categories chosen and category boundary values. They will all give the same general impression of shape.

We present here an algorithm for selecting the number of categories and category boundaries. This algorithm will guarantee that each data point falls into exactly one category, that categories are the same width, and that no data point can assume a boundary value. Some computer packages allow the user to select the number of categories or to specify boundary values. If so, then this algorithm can be used to control the construction of the histogram if desired.

Rules for Breaking Data into Categories

1. Decide on the number of categories wanted. The number chosen depends on the number of observations available. Table 5.1 gives suggested guidelines for the number of categories to be used as a function of sample size. It is based on Sturges' rule, a formula developed by H. A. Sturges in 1926.
2. Locate the largest observation and the smallest observation.

Table 5.1
Suggested number of categories to be used in subdividing numeric data as a function of sample size

Sample size	Number of categories
Fewer than 16	Not enough data
16–31	5
32–63	6
64–127	7
128–255	8
256–511	9
512–1023	10
1024–2047	11
2048–4095	12
4096–8190	13

3. Find the difference between the largest and the smallest observations. Subtract in the order of the largest minus the smallest. This difference is called the *range* of the data.

4. Find the minimum length required to cover this range by dividing the range by the number of categories desired. This length is the minimum length required to cover the range if the lower boundary for the first category is taken to be the smallest data point. However, to ensure that no data point falls on a boundary, we shall define boundaries in such a way that they involve one more decimal place than the data. Hence we shall start the first category slightly *below* the first data point. By doing this, the minimum category length required to cover the range is not long enough to trap the largest data point in the last category. For this reason, the actual length used must be a little longer than minimum.

5. The actual category length to be used is found by rounding the minimum length *up* to the same number of decimal places as the data itself. If the minimum length by chance already has the same number of decimal places as the data, we shall round up 1 unit. For example, if we have data reported to one decimal place accuracy and the minimum length required to cover the range is found to be 1.7, we bump this up to 1.8 to obtain the actual category length to be used.

6. The lower boundary for the first category lies 1/2 unit below the smallest observation. Table 5.2 gives units and half units for various types of data sets.

7. The remaining category boundaries are found by adding the category length to the preceding boundary value.

> **Example 5.2.2.**   Consider the data of Example 5.2.1. The data set has 50 observations. From Table 5.1 we see that the suggested number of categories to be used is 6. Now we locate the largest data point (4285) and the smallest (14). These are used to find the range, that is, the length of the interval containing all the data points. In this case the data are covered by an interval of length $4285 - 14 = 4271$ units. To find the minimum length required for each category, we divide this number by the number of categories desired. Here the minimum category length is $4271/6 \doteq 711.83$ units. To find the actual category length to be used in splitting the data, we round up the minimum length to the same number of decimal places as the data. Here the data are reported in whole numbers. Thus we round up the minimum length, 711.83, to the nearest whole number, 712. The categories actually used will be of length 712. The first category starts 1/2 unit below the smallest observation. From Table 5.2 we see that 1/2 unit is .5 in the case of integer data. That is, the lower boundary for the first category is $14 - .5 = 13.5$. The remaining category boundaries are found by successively adding the category length (712) to the preceding boundary until all

Table 5.2
Units and half units for data reported to the stated degree of accuracy

Data reported to nearest	Unit	1/2 unit
Whole number	1	.5
Tenth (1 decimal place)	.1	.05
Hundredth (2 decimal places)	.01	.005
Thousandth (3 decimal places)	.001	.000 5
Ten thousandth (4 decimal places)	.000 1	.0000 5

data points are covered. In this way we obtain the following six finite categories for the battery lives:

13.5 to 725.5	2149.5 to 2861.5
725.5 to 1437.5	2861.5 to 3573.5
1437.5 to 2149.5	3573.5 to 4285.5

Note that since the boundaries have one more decimal place than the data, no data point can fall on a boundary; each data point must fall into exactly one category. The data can be summarized now in table form by recording the number (frequency) and the percentage (relative frequency) of the observations in each category, as shown in Table 5.3. From this table we can construct a histogram of the data. If the frequency per category is plotted along the vertical axis, the resulting bar graph is called a *frequency histogram;* if the vertical axis is used to plot the relative frequency per category, then the diagram is called a *relative frequency histogram.* Both plots provide a visual display of the data that conveys an idea of the shape of the density of the random variable X under study. The relative frequency histogram for the data of Example 5.2.1 is shown in Fig. 5.4. Since the histogram does not exhibit a bell shape, we see once again that these data do not support an assumption of normality. In fact, the distribution suggested by the data is the exponential distribution. In this case it is now the job of the researcher to estimate β, the parameter that describes

Table 5.3

Category	Boundaries	Frequency	Relative frequency
1	13.5 to 725.5	24	24/50 = 48%
2	725.5 to 1437.5	12	12/50 = 24%
3	1437.5 to 2149.5	4	4/50 = 8%
4	2149.5 to 2861.5	3	3/50 = 6%
5	2861.5 to 3573.5	2	2/50 = 4%
6	3573.5 to 4285.5	5	5/50 = 10%

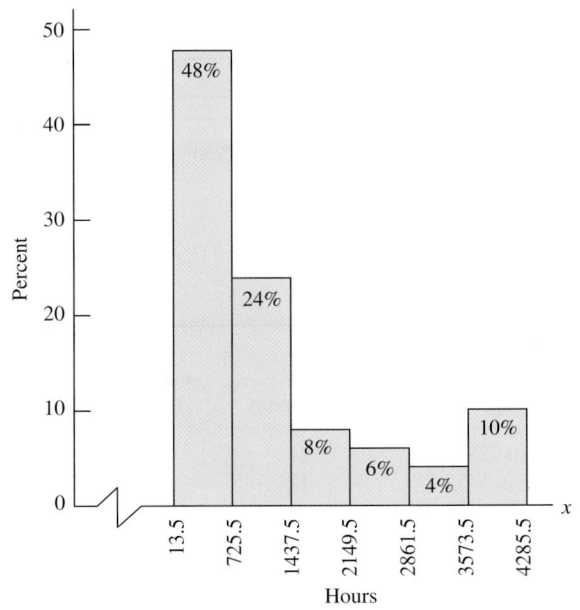

Figure 5.4
Relative frequency histogram for the sample of battery life spans of Example 5.2.1.

this distribution. By so doing, we are able to estimate the density for *X*. This estimated density can then be used to approximate probabilities in the future.

Figure 5.5 shows the histogram produced by MINITAB's default settings. Notice that more categories and different boundaries are chosen by the computer algorithm than is the case with the textbook procedure. We still get the same impression of a distribution that is skewed to the right.

Cumulative Distribution Plots (Ogives)

In addition to the frequency distribution among categories, it is of interest to consider the cumulative frequency distribution of the observations. The cumulative frequency distribution is found by determining for each category the number and percentage of observations falling in or below that category. The cumulative distribution of the data of Example 5.2.1 is shown in Table 5.4.

When the random variable under study is continuous, the cumulative distribution can be used to construct a graph that approximates its cumulative distribution function *F*. The graph is a line graph obtained by plotting the upper boundary of each category on the horizontal axis against the relative cumulative frequency. This type of graph is called a *relative cumulative frequency ogive*. The ogive for the data of Example 5.2.1 is shown in Fig. 5.6. From the ogive we can answer questions

Figure 5.5
Histogram produced via
MINITAB default settings.

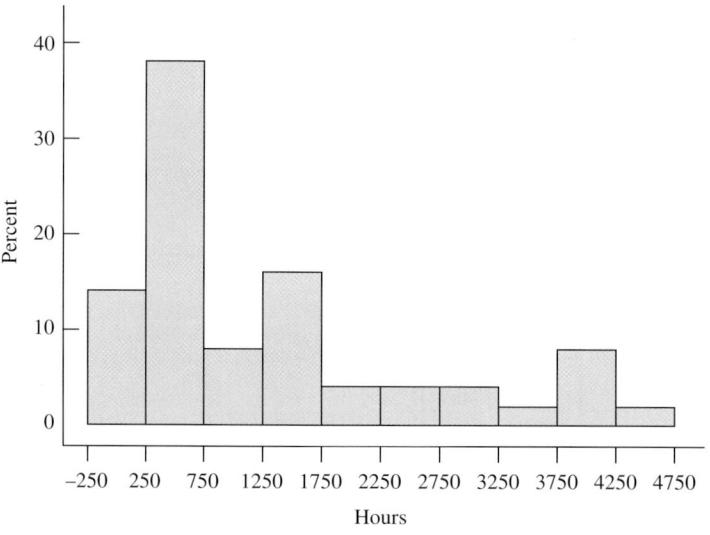

Table 5.4

Category	Boundaries	Frequency	Cumulative frequency	Relative cumulative frequency
1	13.5 to 725.5	24	24	24/50 = 48%
2	725.5 to 1437.5	12	36	36/50 = 72%
3	1437.5 to 2149.5	4	40	40/50 = 80%
4	2149.5 to 2861.5	3	43	43/50 = 86%
5	2861.5 to 3573.5	2	45	45/50 = 90%
6	3573.5 to 4285.5	5	50	50/50 = 100%

such as, "Approximately what percentage of batteries fail during the first 1500 hours of operation?" and "What time represents the midway point in the sense that half the batteries fail on or before this time?"

The first question can be answered graphically by locating 1500 on the horizontal axis, projecting a vertical line up to the ogive, and then projecting a horizontal line over to the vertical axis, as shown in Fig. 5.7. The desired percentage is seen to be approximately 72%. The second question is answered by locating .5 on the vertical axis and reversing the process. The answer is seen to be a little over 725 hours. (See Fig. 5.7.)

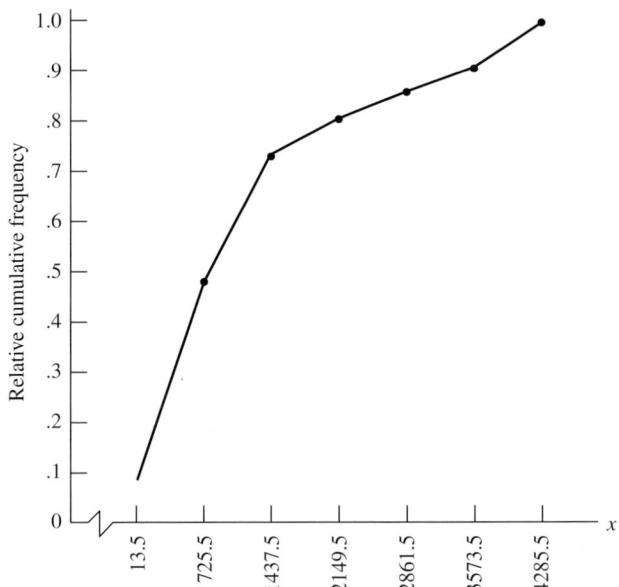

Figure 5.6
Relative cumulative frequency ogive for the sample of battery life spans of Example 5.2.1.

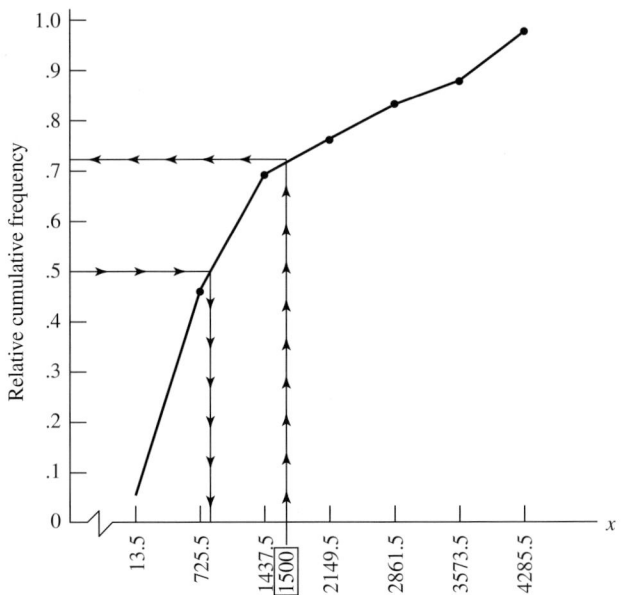

Figure 5.7
Projective method of approximating probabilities using a relative cumulative frequency ogive.

5.3 SAMPLE STATISTICS

We have seen that the behavior of a random variable X is determined by its density. We have also seen that the parameters μ, the theoretical average value of the random variable, and σ^2, its variability about the mean, are helpful in describing X. In the last section we considered some graphical methods for getting an idea of the shape of the density. In this section we consider some statistics that allow us to summarize a data set analytically. Since it is hoped that the data set reflects the population as a whole, these statistics also give us some idea of the values of the parameters that characterize X over the population under study. In particular, we consider two measures of location or central tendency in a data set, the *sample mean* and the *sample median*. We also consider three measures of variability within the data set, the *sample variance*, the *sample standard deviation*, and the *sample range*. The word "sample" is used to emphasize the fact that the data sets presented are based on experiments involving only a small portion of objects that constitute the population being studied. That is, they represent a random sample from the distribution of X.

Location Statistics

The mean or theoretical average value of X is our primary measure of the center of location of X. The primary measure of the center of location of a data set is its arithmetic average. Since we view a data set as a set of observations on X, the arithmetic average for a particular set of observations is just the observed value of the *statistic* $\sum_{i=1}^{n} X_i/n$. This statistic, called the *sample mean*, is defined formally in the next definition.

> **Definition 5.3.1 (Sample mean).** Let $X_1, X_2, X_3, \ldots, X_n$ be a random sample from the distribution of X. The statistic $\sum_{i=1}^{n} X_i/n$ is called the sample mean and is denoted by $\overline{X}$.

Note that μ_X and $\overline{X}$ are *not* the same. The parameter μ_X is the theoretical average value for X over the entire population; $\overline{X}$ is a statistic which, when evaluated over a particular random sample, gives the average value of X *for that sample*. It is hoped, of course, that the observed value of $\overline{X}$ is close to μ_X. In reporting sample means, we shall usually retain one more decimal place than that of the data. Rounding will be used rather than truncation.

Example 5.3.1. A random sample of size 9 yields the following observations on the random variable X, the coal consumption in millions of tons by electric utilities for a given year:

406	395	400	450	390	410	415	401	408

The observed value of the sample mean for these data is

$$\bar{x} = \sum_{i=1}^{n} x_i/n = (406 + 395 + 400 + \cdots + 408)/9$$

$$= 3675/9 \doteq 408.3 \text{ million tons}$$

The average value for X for this sample is 408.3 million tons. What is the average number of tons of coal used by electric utilities across the country in this particular year? That is, What is μ_X? Unfortunately, this question cannot be answered with certainty from this sample. However, the sample leads us to believe that μ_X lies close to 408.3 million tons. Admittedly, the word "close" is a bit vague. In Chap. 7 we shall consider a method for determining how close μ_X is likely to be to 408.3 million tons.

A second measure of the center of location of a random variable X is its *median*. The median of a random variable is its 50th percentile (see Exercise 12). That is, the median for X is that number M such that

$$P[X < M] \le .50 \qquad \text{and} \qquad P[X \le M] \ge .50$$

If X is continuous, then its median is the "halfway point" in the sense that an observation on X is just as likely to fall below M as it is to fall above it. We define the median for a sample with this in mind.

> **Definition 5.3.2.** Let $x_1, x_2, \ldots, x_n$ be a sample of observations arranged in order from the smallest to the largest. The sample median is the middle observation if n is odd. It is the average of the two middle observations if n is even. We shall denote the median of a sample by $\tilde{x}$.

If n is small, it is easy to spot the middle of a data set. However, if n is large, it is useful to have a formula that pinpoints the location of the middle observation or observations. The formula is given below, and its use is illustrated in Example 5.3.2.

$$\text{Median location} = \frac{n + 1}{2}$$

Example 5.3.2. The nine observations on X, the coal consumption in millions of tons by electric utilities for a given year, arranged in order, are

| 390 | 395 | 400 | 401 | 406 | 408 | 410 | 415 | 450 |

The median location is $\dfrac{n + 1}{2} = \dfrac{9 + 1}{2} = 5$. The median is the fifth data point in the ordered list. In this case, $\tilde{x} = 406$. This observation is the middle value in our ordered list. Note that this is the median for this data set. It gives us a *rough* idea of the median coal consumption across the country during the year.

Measures of Variability

Recall that we are usually concerned not only with the mean of a random variable, but also with its variance. The variance of a random variable, given by

$$\sigma^2 = E[(X - \mu)^2]$$

measures the variability of X about the population mean. We want to develop an analogous measure of variability within a sample. To do so, we parallel the logic used in defining σ^2. We do not know the value of the population mean, but we shall have available an observed value for the sample mean. We cannot observe the differences $(X - \mu)^2$ for all members of the population, but we can observe the difference $(X_i - \overline{X})^2$ for each element X_i of the random sample. Since σ^2 is an expectation, a theoretical average value, logic dictates that we replace this operation

Figure 5.8
(a) The statistic $\Sigma_{i=1}^{n}(X_i - \bar{X})^2/n$ tends to underestimate σ^2. On the average, it will produce estimates that are a bit too small. It is not an unbiased estimator for σ^2; (b) the statistic $\Sigma_{i=1}^{n}(X_i - \bar{X})^2/(n-1)$ is unbiased for σ^2. On the average, it will produce estimates that are centered at σ^2.

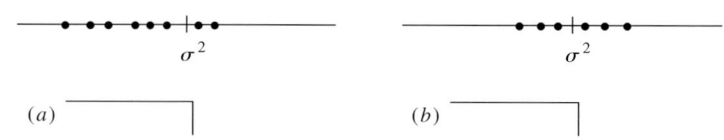

by an arithmetic average of sample values. That is, the natural measure of variability within a sample that parallels our definition of variability within the population is

$$\sum_{i=1}^{n} \frac{(X_i - \bar{X})^2}{n}$$

This method of measuring variability within a sample is acceptable. In fact, many electronic calculators with built-in statistical capability utilize this formula to compute the variance of a sample. In most cases we shall be using the variability in the sample to approximate σ^2. However, it can be shown that this statistic tends, on the average, to underestimate σ^2. To improve the situation, we divide $\Sigma_{i=1}^{n}(X_i - \bar{X})^2$ by $n-1$ rather than by n. In this way we obtain a statistic that is unbiased for σ^2. The term "unbiased" is a technical term. It is defined formally in Sec. 6.1. Basically, it means "centered at the right spot." Since the sample variance is used to estimate σ^2, in this case "the right spot" is σ^2. Successive estimates for σ^2 based on the formula $[\Sigma_{i=1}^{n}(X_i - \bar{X})^2]/(n-1)$ should be centered at σ^2. Figure 5.8 illustrates the expected behavior of the two statistics just discussed. So that the statistic used to estimate σ^2 will be unbiased for σ^2, we choose to define the variance of a sample as given in Definition 5.3.3. The definition of the term "sample standard deviation" follows logically.

Definition 5.3.3 (Sample variance and sample standard deviation). Let $X_1, X_2, X_3, \ldots, X_n$ be a random sample of size n from the distribution of X. Then the statistic

$$S^2 = \sum_{i=1}^{n} \frac{(X_i - \bar{X})^2}{n-1}$$

is called the *sample variance*. Furthermore, the statistic $S = \sqrt{S^2}$ is called the *sample standard deviation.*

Recall that when we computed the value of σ^2 in Chap. 3, the actual definition of the term "variance" was seldom used; a computational formula was developed that was arithmetically easier to handle than the definition. The same is true here. When S^2 is evaluated from a sample, Definition 5.3.3 is not commonly used. Rather, we use a computational formula.

Theorem 5.3.1 (A computational formula for S^2). Let $X_1, X_2, X_3, \ldots, X_n$ be a random sample of size n from the distribution of X. The sample variance is given by

$$S^2 = \frac{n\sum_{i=1}^{n} X_i^2 - \left(\sum_{i=1}^{n} X_i\right)^2}{n(n-1)}$$

The above formula was convenient before the advent of calculators with built-in statistical capabilities and statistical computer packages. Since most calculators will find s for you by simply entering the data in a statistical mode, this formula is not often needed. We present it here because you might encounter it in some other setting and wonder about its validity. You are encouraged to use whatever computing aids you have available to find $\bar{x}$, s^2, and s. However, the use of the formula is illustrated in Example 5.3.3. In reporting s^2, we shall usually retain two more decimal places than that of the data; s will be reported to one more decimal place. Rounding will be used.

Example 5.3.3. These data constitute a sample of observations on X, the coal consumption in millions of tons by electric utilities for a given year:

390	400	406	410	450	395	401	408	415

To compute the sample variance, we must evaluate the statistics $\sum_{i=1}^{n} X_i$ and $\sum_{i=1}^{n} X_i^2$ for this sample. The observed values are

$$\sum_{i=1}^{9} x_i = 3675 \qquad \sum_{i=1}^{9} x_i^2 = 1{,}503{,}051$$

The observed value of S^2 is

$$S^2 = \frac{9 \sum_{i=1}^{9} x_i^2 - \left(\sum_{i=1}^{9} x_i \right)^2}{9(8)} = \frac{9(1{,}503{,}051) - (3675)^2}{9(8)} \doteq 303.25$$

Remember that variance is usually considered to be unitless because the physical unit attached to it is often meaningless. The observed value of S is

$$s = \sqrt{s^2} = \sqrt{303.25} \doteq 17.4 \text{ million tons}$$

Notice that the physical measurement unit associated with s matches that of the original data and that 17.4 million tons is the standard deviation for this sample. It is not the standard deviation in coal consumption for all electric utilities across the country for the given year. However, it does indicate that σ probably has a value close to 17.4 million tons.

The last sample statistic to be considered is the *sample range*. This statistic was used in categorizing data in Sec. 5.2.

Definition 5.3.4 (Sample range). The sample range is defined to be the difference between the largest and smallest observations with subtraction in the order largest minus smallest.

The sample range for the data of Example 5.3.3 is $450 - 390 = 60$ million tons.

One word of caution is in order. We have assumed that the data set presented in this section represents a random sample drawn from a larger population because this is the situation most often encountered in practice. Occasionally you will encounter a data set that is *not* a sample. Rather, it represents an observation on X for *every* member of the population. If this is the case, then the population mean is just the arithmetic average of these observations; that is, $\mu = \bar{x}$. Furthermore, the population variance is given by

$$\boxed{\begin{array}{c} \textbf{Population Variance} \\[4pt] \sigma^2 = \sum_{i=1}^{n} \frac{(x_i - \bar{x})^2}{n} \end{array}}$$

Be careful! Be sure that you understand the nature of your data set before you begin to summarize its properties.

5.4 BOXPLOTS

In summarizing data, it is useful to report all the statistics considered in Sec. 5.3. This is especially true if the data set contains a value that is unusually large or unusually small. A value that appears to be atypical in that it seems to be far removed from the bulk of the data is called an *outlier* or a "wild" number. It is important to be able to detect such numbers and to understand the effect that they have on the usual sample statistics.

Outliers arise for two reasons: (1) They are legitimate observations whose values are simply unusually large or unusually small, or (2) they are the result of an error in measurement, poor experimental technique, or a mistake in recording or entering the data. In the first case it is suggested that the presence of the outlier be reported and that sample statistics be reported both with and without the outlier. In the second case the data point can be corrected if possible or else dropped from the data set.

Of the statistics presented thus far the sample mean, the variance, the standard deviation, and the range are adversely affected by the presence of an outlier; however, the sample median is not so affected. Thus in the presence of an outlier the sample median may be preferable to the sample mean measure of location. We say that the median is *resistant* to outliers.

Sometimes outliers are so obvious that their presence can be detected by inspection. However, it is useful to have an analytical and graphical technique for identifying values that are truly unusual. One such technique is the *boxplot*. Its construction is based on the interquartile range, a measure of variability that is resistant to outliers. The sample interquartile range, iqr, represents the length of the interval that contains roughly the middle 50% of the data. If the iqr is small, then much of the data lies close to the center of the distribution; if it is large, the data tend to be widely dispersed. These steps are used to calculate the iqr.

Finding the Sample Interquartile Range

1. Find the median location $(n + 1)/2$, where n is the sample size.
2. Truncate the median location by rounding it *down* to the nearest whole number.
3. Find the quartile location q by

$$q = \frac{\text{truncated median location} + 1}{2}$$

4. Find q_1 by counting up from the smallest data point to location q. If q is an integer, then q_1 is the data point in position q. If q is not an integer, then q_1 is the average of the data points in positions $q - .5$ and $q + .5$. Approximately 25% of the data will fall on or below q_1.

5. Find q_3 by counting down from the largest data point to position q as in part 4. Approximately 75% of the data will fall on or below q_3.

6. Define iqr by iqr $= q_3 - q_1$.

Example 5.4.1. A study of the type of sediment found at two different deep-sea drilling sites is conducted. The random variable of interest is the percentage by volume of cement found in core samples. By cement we mean dissolved and reprecipitated carbonate material. The following data are obtained:

Site I, % cement				Site II, % cement		
10	21	12	12	1	10	14
20	13	24	36	9	21	19
31	18	17	16	15	17	13
37	16	32	13	25	22	20
14	49	25	19	24	12	23
13	32	27		15	20	18

The double stem-and-leaf diagram for the data of site I is shown in Fig. 5.9. The sample is size $n = 23$. The median location is $(n + 1)/2 = 12$. The quartile location is $q = (12 + 1)/2 = 6.5$. To find q_1, we use the stem-and-leaf diagram to locate the sixth and seventh data points, counting from the smaller numbers up. These values are 13 and 14, respectively. Hence $q_1 = (13 + 14)/2 = 13.5$. To find q_3, we find the sixth and seventh data points counting from the higher numbers down. These points are 31 and 27, respectively, yielding $q_3 = (31 + 27)/2 = 29$. The sample interquartile range is $q_3 - q_1 = 29 - 13.5 = 15.5$. For site II you can verify that $q_1 = 13$ and $q_3 = 21$.

A word of caution is in order. All computer software and statistical calculators calculate the median as we have done. However, different algorithms are sometimes used to find the quartiles; some will agree with our values, but others will not. All produce good estimates of the population quartiles. For example, if the TI83 calculator is used to find q_1 and q_3 for the data of Example 5.4.1, site I, it reports $q_1 = 13$ and $q_3 = 31$. These values differ slightly from those that we found previously. That calculator's answers will agree with ours for the data of site II. MINITAB reports $q_1 = 13$ and $q_3 = 31$ for site I and thus agrees with the TI83 calculator. However, it yields $q_1 = 12.75$ and $q_3 = 21.25$ for the quartiles of site II. These do not agree with our estimates or those of the TI83. Just be aware that different technologies can yield slightly different quartiles and therefore will produce slightly different boxplots when applied to the same set of data.

Once the interquartile range has been found, it can be used to construct a boxplot. The *boxplot* is a graphical representation of a data set that gives a visual impression of location, spread, and the degree and direction of skewness. For an approximately bell-shaped distribution the *boxplot* also allows us to identify outliers. It is especially useful when we want to compare two or more data sets.

```
1 | 0433223
1 | 86769
2 | 014
2 | 57
3 | 122
3 | 76
4 |
4 | 9
```

Figure 5.9
Double stem-and-leaf diagram for the percentage by volume of cement in core samples taken at deep-sea drilling site I.

Constructing a Boxplot

1. A horizontal or vertical reference scale is constructed.
2. Find the sample median, q_1, q_3, and iqr.
3. Find two points f_1 and f_3, called *inner fences*, by

$$f_1 = q_1 - 1.5(\text{iqr})$$
$$f_3 = q_3 + 1.5(\text{iqr})$$

These points will be used to identify outliers.
They are *not* a visible part of the boxplot.

4. Find two points a_1 and a_3, called *adjacent values*. The point a_1 is the data point that is closest to f_1 without lying below f_1 in value. The point a_3 is the data point that is closest to f_3 without lying above f_3 in value.
5. Find two points F_1 and F_3, called *outer fences*, by

$$F_1 = q_1 - 2(1.5)(\text{iqr})$$
$$F_3 = q_3 + 2(1.5)(\text{iqr})$$

These fences, as with inner fences, are not visible on the boxplot.

6. Locate the points found thus far on the horizontal or vertical scale. Their relative positions are shown in Fig. 5.10(a).
7. Construct a box with ends at q_1 and q_3 with an interior line drawn at the median, as shown in Fig. 5.10(b).
8. Indicate adjacent values by x, and connect them to the box with dashed lines. Locate any data points falling between the inner and outer fences, and denote these by open circles. These points are considered to be mild outliers. Indicate data points that fall beyond the outer fences with asterisks. These points are considered to be extreme outliers [see Fig. 5.10(c)].

The location of the midline of the box is an indication of the shape of the distribution. If the line is badly off center, then we know that the distribution is skewed in the direction of the longer end of the box.

Before we illustrate this technique, the notion of fences needs to be clarified. It can be shown that when sampling from a normal distribution, only about 7 values in every 1000 fall beyond the inner fences. You are asked to verify this result in Exercises 26 and 27. Since these values are very unusual, they are deemed to be outliers. Outliers must be treated with care since, as you have already seen, their presence can have a dramatic impact on $\bar{x}$, s^2, and s, the usual measures of location

Figure 5.10
(a) Relative positions of median ($\tilde{x}$), quartiles (q_1 and q_3), adjacent values (a_1 and a_3), inner fences (f_1 and f_3), and outer fences (F_1 and F_3);
(b) a box is drawn with ends at q_1 and q_3 and interior line at $\tilde{x}$;
(c) adjacent values are indicated by x. Mild outliers are indicated by open circles; extreme outliers are given by asterisks.

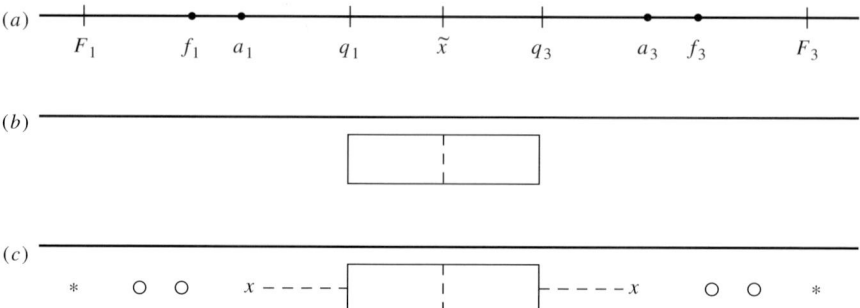

and variation. When an outlier is found, we should consider its source. Is it a legitimate data point whose value is simply unusually large or small? Is it a misrecorded value? Is it the result of some error or accident in experimentation? In the last two instances the point can be deleted from the data set and the analysis completed on the remaining data. In the first case we suggest that the presence of the outlier be made known and that statistics be reported both with and without the outlier. In this way the decision of whether or not to include the outlier in future analyses can be made by the researcher who is the subject matter expert.

Example 5.4.2. A study of posttraumatic amnesia after a closed head injury is conducted. One variable studied is the length of hospitalization in days. The stem-and-leaf diagram for the data is shown in Fig. 5.11. (Based on information found in Jerry Mysia et al., "Prospective Assessment of Posttraumatic Amnesia: A Comparison of GOAT and the OGMS," *Journal of Head Trauma Rehabilitation,* March 1990, pp. 65–77.) For these data the median location is $(n + 1)/2 = 11$ and the median is 40 days. Quartile location is $q = $ (truncated median location $+ 1)/2 = 6$. The points q_1 and q_3 are 32 and 47, respectively. The interquartile range is iqr $= q_3 - q_1 = 15$. The inner fences are

$$f_1 = q_1 - 1.5(\text{iqr}) \qquad f_3 = q_3 + 1.5(\text{iqr})$$
$$= 32 - 22.5 \qquad\qquad = 47 + 22.5$$
$$= 9.5 \qquad\qquad\quad = 69.5$$

The adjacent values are $a_1 = 12$ and $a_3 = 61$. The outer fences are

$$F_1 = q_1 - 2(1.5)(\text{iqr}) \qquad F_3 = q_3 + 2(1.5)(\text{iqr})$$
$$= 32 - 45 \qquad\qquad\quad = 47 + 45$$
$$= -13 \qquad\qquad\qquad = 92$$

The data set contains two points, 8 and 89, that qualify as mild outliers. The point 108 qualifies as an extreme outlier. Notice that since F_1 is negative, it is physically impossible to see an extreme outlier on the lower end of the scale. The boxplot is shown in Fig. 5.12. Notice that the midline of the box is near its center, indicating a nearly symmetric distribution. Are the outliers real observations that must be taken into account, or are they the result of errors in data collection? In this case it would be easy to check patient records to find the answer, and this should be done before proceeding with any further analysis of the data.

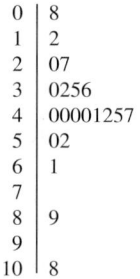

```
 0 | 8
 1 | 2
 2 | 07
 3 | 0256
 4 | 00001257
 5 | 02
 6 | 1
 7 |
 8 | 9
 9 |
10 | 8
```

Figure 5.11
Stem-and-leaf diagram for the data of Example 5.4.2. Data represent length of hospitalization in days of posttraumatic amnesia patients ($n = 21$).

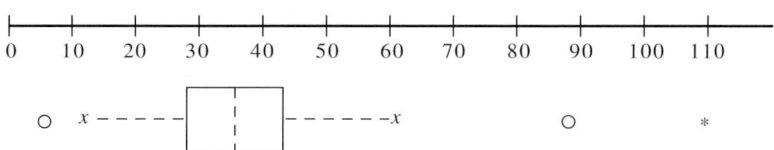

Figure 5.12
Boxplot for the data of Example 5.4.2.

As with any other statistical technique, the method given here for detecting outliers must be used with care. Since the location of the fences is chosen to detect unusual values when sampling from a normal distribution, this fact must be kept in mind when interpreting the boxplot. If the data set is large enough so that a histogram or a stem-and-leaf plot exhibits the bell characteristic of a normal curve, then legitimate data points that are flagged as outliers are unusual enough to warrant investigation. If the data set is small or appears to be drawn from a distribution that is not normal, then no real conclusions concerning outliers can be drawn. For example, the exponential distribution is far from symmetric and by nature has a long tail. In this case it is quite likely that the technique demonstrated in this section would flag the largest data point as an outlier. In fact, the point might not be unusual at all. David Hoaglin and John Tukey [50] have a nice discussion of the use of boxplots and outliers for distributions that are not normal.

CHAPTER SUMMARY

This chapter is a link between the study of probability in its own right and the use of probability in the study of applied statistics. We began by defining exactly what we mean by the term "random sample." In particular, we noted that the term is used in three ways. It can denote the objects sampled, the random variables associated with those objects, or the numerical values assumed by these random variables. We noted also that in this text we are assuming that either sampling is from an infinite population, sampling is done with replacement from a finite population, or sampling without replacement from a finite population is done in such a way that the sample constitutes at most 5% of the population. This ensures that it is reasonable to assume that the random variables $X_1, X_2, \ldots, X_n$ are, for all practical purposes, independent. We introduced three graphical methods for picturing the distribution of a data set. These methods, the stem-and-leaf chart, histograms, and boxplots, help to determine the type of random variable with which we are dealing. That is, they help us get an idea of the shape of the density f associated with the random variable. The relative cumulative frequency ogive was introduced as a means of approximating the cumulative distribution function, F, of a continuous random variable. We introduced some summary statistics that serve two purposes. They describe the data set at hand, and they help approximate the value of corresponding parameters associated with the population from which the sample was drawn. We introduced and defined important terms that you should know. These are:

Population	Sample mean
Percentile	Random sample
Median	Quartile
Statistic	Sample median
Decile	Stem and leaf
Sample variance	Interquartile range
Frequency histogram	Sample standard deviation
Relative frequency histogram	Sample range
Relative cumulative frequency ogive	Outlier
Inner fences	Mild outliers
Outer fences	Extreme outliers
Adjacent values	Boxplots
Resistant statistic	

EXERCISES

Section 5.1

In Exercises 1 through 5 a problem is described. In each case, decide whether a statistical study is appropriate. If so, explain why you think this is the case and identify the population(s) of interest.

1. A bridge is to be built across a deep canyon. An engineer is interested in determining the distribution of the random variable X, the maximum wind speed per day at the site, so that the bridge can be designed to withstand potential stresses that will be placed upon it from this source.

2. A botanist thinks that indoleacetic acid is effective in stimulating the formation of roots in cuttings from lemon trees. In an experiment to verify this contention two groups of cuttings are to be used. One group is to be treated with a dilute solution of indoleacetic acid; the other is given only water. Later a comparison of the root systems of the two groups will be made.

3. An architectural firm is to sublet a contract for a wiring project. Seven electrical contractors are available for the job. We want to determine the average estimated cost of the job and the average projected time required to complete the job for these seven contractors.

4. A computer system has a number of remote terminals attached to it. To decide whether or not to increase this number, it is necessary to study the random variable X, the length of time expended per session by users of the terminals currently in place.

5. Prior to changing from the traditional 8-hour-a-day, 5-day-a-week work schedule to a 10-hour-a-day, 4-day-a-week schedule, the opinion of the 50,000 workers who would be affected is to be sought.

6. Air quality is of concern to everyone. It is judged by the number of micrograms of particulate present per cubic meter of air. Assume that this variable is normally distributed with unknown mean and unknown variance. Monitoring stations sample air by sucking it through a thin fiberglass sheet that collects the fine particles suspended in the air. In a particular locality this is done for five randomly selected 24-hour periods each month. Thus each month a random sample of size $n = 5$ from a normal distribution is available.

   (*a*) Consider the random variable X_1, the particulate level for the first 24-hour period studied during a given month. What is the distribution of this random variable?

   (*b*) For a given month, these readings result:

   $$x_1 = 45 \qquad x_2 = 50 \qquad x_3 = 62 \qquad x_4 = 57 \qquad x_5 = 70$$

   For these data, evaluate the statistics $\Sigma X_i, \Sigma X_i^2, \Sigma X_i/n, \max_i\{X_i\}, \min_i\{X_i\}$.

   (*c*) Is the random variable $X_5 - \mu$ a statistic? Is the random variable $(X_5 - \mu)/\sigma$ a statistic? Explain.

Section 5.2

7. A data set containing 70 observations, each reported to one decimal place, is to be split into seven categories. The largest observation is 75.1, and the smallest is 16.3.

   (*a*) These data are covered by an interval of what length?

   (*b*) Using the method outlined in this section, each category will be of what length?

(c) What is the lower boundary for the first category?

(d) What are the boundaries for each of the seven categories?

8. Acute exposure to cadmium produces respiratory distress and kidney and liver damage, and may even result in death. For this reason, the level of airborne cadmium dust and cadmium oxide fume in the air is monitored. This level is measured in milligrams cadmium per cubic meter of air. A sample of 35 readings yields the following data:

.044	.030	.052	.044	.046
.020	.066	.052	.049	.030
.040	.045	.039	.039	.039
.057	.050	.056	.061	.042
.055	.037	.062	.062	.070
.061	.061	.058	.053	.060
.047	.051	.054	.042	.051

(a) Construct a stem-and-leaf diagram for these data. Use the numbers $02, 03$, $04, 05, 06$, and 07 as stems.

(b) Would you be surprised to hear someone claim that the random variable X, the cadmium level in the air, is normally distributed? Explain.

(c) Use the method outlined in this section to break these data into six categories. (Here a unit is .001 and a half unit is .0005.)

(d) Construct a frequency table and a relative frequency histogram for these data. Does the histogram exhibit the bell-shape characteristic of a normal density?

(e) Construct a cumulative frequency table and a relative cumulative frequency ogive for these data. Use the ogive to approximate that point above which 50% of the readings should fall.

9. Let X denote the time in minutes that a vehicle must wait to get through a traffic light at a busy intersection. The following data are obtained from a random sample of 36 vehicles:

.2	.5	.7	1.1	1.2	1.2	1.3	1.4	1.4	1.4
1.5	1.5	1.6	1.6	1.7	1.9	2.0	2.1	2.1	2.2
2.3	2.5	2.6	2.9	2.8	3.0	3.1	3.0	3.7	3.7
4.0	4.1	4.5	5.1	5.8	1.4				

(a) Construct a double stem-and-leaf diagram for these data.

(b) Do the data suggest that the distribution of X is skewed? If so, what is the direction of the skew?

10. Liquid products were first obtained from coal in England during the 1700s. Lamp oil was produced from coal in the United States as early as 1850, but the domestic coal chemicals industry did not develop until World War I. A modern coal-for-recovery system uses a battery of coke ovens to produce liquid products from the coal feed. These observations are obtained on the random variable X, the number of gallons of liquid product obtained per ton of coal feed:

7.6	8.2	7.1	10.0	6.5	9.6
6.1	6.2	7.6	6.2	9.5	6.7
7.4	9.5	9.2	8.0	8.5	9.3
8.8	9.6	9.7	6.8	7.1	7.7
8.7	7.8	8.7	8.2	8.2	7.4
9.0	8.8	7.3	7.9	7.1	7.9
7.6	6.7	8.1	6.2	5.3	7.4
7.7	9.1	7.9	8.7	8.4	8.1

(a) Construct a stem-and-leaf diagram for these data. Use the numbers $5, 6, 7$, $8, 9, 10$ as stems.

(b) Is the assumption that X is normally distributed justifiable? Explain.

(c) Use the method outlined in this section to break these data into six categories.

(d) Construct a frequency table and a relative frequency histogram for these data. Does the histogram exhibit the bell-shape characteristic of a normal density?

(e) Construct a cumulative frequency table and a relative cumulative frequency ogive for these data. Use the ogive to approximate the probability that a randomly selected ton of coal will yield less than 7 gallons of liquid product.

11. Some efforts are currently being made to make textile fibers out of peat fibers. This would provide a source of cheap feedstock for the textile and paper industries. One variable being studied is X, the percentage ash content of a particular variety of peat moss. Assume that a random sample of 50 mosses yields these observations:

.5	1.8	4.0	1.0	2.0
1.1	1.6	2.3	3.5	2.2
2.0	3.8	3.0	2.3	1.8
3.6	2.4	.8	3.4	1.4
1.9	2.3	1.2	1.9	2.3
2.6	3.1	2.5	1.7	5.0
1.3	3.0	2.7	1.2	1.5
3.2	2.4	2.5	1.9	3.1
2.4	2.8	2.7	4.5	2.1
1.5	.7	3.7	1.8	1.7

(a) Construct a stem-and-leaf diagram for these data. Use the numbers $0, 1, 2, 3, 4, 5$ as stems.

(b) Is there any reason to suspect that X is not normally distributed? Explain.

(c) Use the method outlined in this section to break these data into six categories.

(d) Construct a frequency table and a relative frequency histogram for these data. Does the histogram suggest that X might not be normally distributed? If so, what distribution might be appropriate?

(e) Construct a cumulative frequency table and a relative cumulative frequency ogive for these data. Use the ogive to approximate the probability that a randomly selected specimen of this variety of moss will have an ash content that exceeds 2%.

12. (Percentiles.) Let X be a random variable. The point $p_{k/100}$ $(k = 1, 2, 3, \ldots, 100)$ such that

$$P[X < p_{k/100}] \leq k/100 \qquad \text{and} \qquad P[X \leq p_{k/100}] \geq k/100$$

is called the kth percentile for X. For example, let X be binomial with $n = 20$ and $p = .5$. The 25th percentile for X is the point $p_{25/100} = 8$ since, from Table I of App. A, we see that

$$P[X < 8] = .1316 \leq .25 \qquad \text{and} \qquad P[X \leq 8] = .2517 \geq .25$$

(a) Let X be binomial with $n = 20$ and $p = .5$. Find the 60th percentile for X.

(b) Let X be Poisson with $\lambda s = 10$. Find the 30th percentile for X.

(c) Argue that in the case of a continuous random variable the kth percentile is that point such that $P[X \leq p_{k/100}] = k/100$.

(d) Let X be exponentially distributed with $\beta = 1$. Show that the 20th percentile for X is $-\ln .80$. Hint: Find the point p such that

$$\int_0^p e^{-x}\, dx = .20$$

13. *(Quartiles.)* The 25th, 50th, 75th, and 100th percentiles for X are called its first, second, third, and fourth *quartiles,* respectively.
    (a) State the definition of the first quartile in terms of probabilities.
    (b) Let X be binomial with $n = 20$ and $p = .5$. Find the first quartile for X.
    (c) Let X be exponentially distributed with $\beta = 1$. Find the first quartile for X.

14. *(Deciles.)* The 10th, 20th, 30th, 40th, 50th, 60th, 70th, 80th, 90th, and 100th percentiles for X are called its *deciles.*
    (a) State the definition of the 4th decile for X in terms of probabilities.
    (b) Let X be Poisson with $\lambda s = 10$. Find the 6th decile for X.
    (c) Let X be exponentially distributed with $\beta = 1$. Find the third decile for X.

15. The percentiles, quartiles, and deciles for a continuous random variable can be approximated from a relative cumulative frequency ogive using the projective method. For instance, in Fig. 5.5 we approximated the 50th percentile for X, the life span of a lithium battery, to be a little over 725 hours.
    (a) Approximate the first quartile for X, the cadmium level in the air, using the data of Exercise 8.
    (b) Approximate the fourth decile for X, the number of gallons of liquid product obtained per ton of coal fuel, using the data of Exercise 10.
    (c) Approximate the 50th percentile for X, the percentage ash content for a particular variety of moss, using the data of Exercise 11.

16. In running computer programs on a time-sharing basis, the costs vary from session to session. These observations are obtained on the random variable X, the cost per session to the user:

$1.08	.84	1.41	.99	.82
.89	.38	1.05	1.19	.65
1.09	1.03	.81	.55	.71
1.89	.47	.59	1.22	1.27
1.02	1.09	1.02	.86	1.23
1.23	.85	1.02	1.25	.80

Construct a relative cumulative frequency ogive for these data. Use the ogive to approximate the 50th percentile; the first quartile; the third quartile.

Section 5.3

17. Consider these data sets:

I				II			
1	3	2		1	2	4	1
2	5	4		2	5	2	5
4	3	3		1	5	5	3

    (a) Find the sample mean and sample median for each data set.
    (b) Find the sample range for each data set.
    (c) Find the sample variance and sample standard deviation for each data set.
    (d) Would you be surprised to hear someone claim that these data were drawn from the same population? Explain. *Hint:* Consider the shape of the distribution as well as the observed values of the sample statistics.

18. The observed values of the statistics $\sum_{i=1}^{50} X_i$ and $\sum_{i=1}^{50} X_i^2$ for the data of Example 5.2.1 are $\sum_{i=1}^{50} x_i = 63{,}707$ and $\sum_{i=1}^{50} x_i^2 = 154{,}924{,}261$.
    (a) Would you be surprised to hear someone claim that the mean lifespan of the lithium batteries used in this model calculator is 1270 hours? Explain.
    (b) Find the sample variance and sample standard deviation for these data.

19. Use the data of Example 5.1.1 to approximate the mean and variance of the random variable X, the number of times per hour that a television signal is interrupted by random interference.

20. Use the data of Exercise 8 to approximate the mean, variance, and standard deviation of the random variable X, the level of airborne cadmium dust and cadmium oxide fumes. Assume that these approximations are fairly accurate. Between what two values would you expect approximately 95% of the readings to fall? Explain.

21. Use the data of Exercise 10 to approximate the mean, variance, and standard deviation of the random variable X, the number of gallons of liquid product obtained per ton of coal feed.

22. Use the data of Exercise 11 to approximate the mean, variance, and standard deviation of the random variable X, the percentage ash content of a particular variety of peat moss.

23. Consider the data of Exercise 9.
  (*a*) Find the mean and median for these data.
  (*b*) Find the standard deviation and variance for these data.
  (*c*) What physical measurement unit is associated with each of the statistics in parts (*a*) and (*b*)?

24. There have been many improvements made in lighting in the last 10 years. One new bulb, the Philips' Earth Light, uses a compact screw-in fluorescent bulb with an electronic ballast incorporated in its base. It is thought to last 10 to 13 times longer than household bulbs used in the past. These data are obtained on the life span of a sample of these new bulbs (time is in thousands of hours):

9.1	10.1	9.0	11.4
10.5	9.5	12.0	9.1
12.2	13.1	10.0	9.3
9.0	9.6	11.1	9.1
13.3	10.7	9.1	9.0
9.0	11.0	9.2	11.6

(Based on information found in "Lighting Comes of Age with New Technology," *Research and Development*, November 1992, pp. 30–31.)
  (*a*) Construct a stem-and-leaf diagram for these data, and suggest a distribution from which these data might have been drawn.
  (*b*) Based on these data, approximate the value of μ, the average life span of these bulbs.
  (*c*) Approximate the median life span of these bulbs, and explain exactly what this value means.
  (*d*) Find the sample variance and sample standard deviation for these data.
  (*e*) Criticize the following statement:
    "Based on the normal probability rule, it is estimated that approximately 95% of all bulbs have a life span between 7,530 and 13,050 hours."
  (*f*) Based on Chebyshev's inequality, what can be said about the proportion of bulbs whose life span is expected to fall between 7,530 and 13,050 hours?

25. (*Approximating σ via the range.*) The range can play an important role in the design of statistical studies. To obtain a prespecified degree of accuracy when estimating population parameters, an adequate sized sample must be drawn. Most formulas used to determine sample size require knowledge of σ, the population standard deviation. Often the researcher will not have an estimate of σ available but will have an idea of the expected range of his or her data. In Sec. 4.4 we saw that when sampling from a normal distribution,

$$P[-2\sigma < X - \mu < 2\sigma] \doteq .95$$

If X is not normally distributed, then Chebyshev's inequality can be applied to conclude that

$$P[-3\sigma < X - \mu < 3\sigma] \geq .89$$

That is, X always lies within at most 3 standard deviations of its mean with high probability. From this it can be concluded that the estimated range covers an interval of roughly 4σ for normally distributed random variables and 6σ otherwise. In the normal case an estimate of σ can be obtained by solving the equation

$$4\sigma \doteq \text{estimated range}$$

for σ. Thus we see that

$$\sigma \doteq (\text{estimated range})/4$$

when X is normally distributed. If X is not normally distributed, then

$$\sigma \doteq (\text{estimated range})/6$$

These data are obtained on the random variable X, the cpu time in seconds required to run a program using a statistical package:

6.2	5.8	4.6	4.9	7.1	5.2
8.1	.2	3.4	4.5	8.0	7.9
6.1	5.6	5.5	3.1	6.8	4.6
3.8	2.6	4.5	4.6	7.7	3.8
4.1	6.1	4.1	4.4	5.2	1.5

(a) Construct a stem-and-leaf diagram for these data. Is the assumption justified that X is normally distributed?
(b) Approximate σ via the sample standard deviation s.
(c) Find the sample range for these data, and use it to approximate σ. Compare your result to that obtained in part (b).

Section 5.4

26. Consider the standard normal distribution.
   (a) Use the Z table to verify that q_1 is approximately $-.67$ and q_3 is approximately $.67$.
   (b) Find the interquartile range for Z, and explain what this means.
   (c) Verify that the inner fences for Z are $f_1 = -2.68$ and $f_3 = 2.68$.
   (d) Verify that the probability that a standard normal random variable will fall beyond the inner fences is approximately $.007$.
   (e) Find the outer fences for Z.
   (f) Find the probability that a standard normal random variable will fall beyond the outer fences.

27. Let X be normally distributed with mean μ and variance σ^2.
   (a) Verify that $q_3 = \mu + .67\sigma$ and that $q_1 = \mu - .67\sigma$.
   (b) Find the interquartile range for X.
   (c) Verify that the inner fences for X are $f_1 = \mu - 2.68\sigma$ and $f_3 = \mu + 2.68\sigma$.
   (d) Verify that the probability that X will fall beyond the inner fences is approximately $.007$.

28. Temperature differences between the warm upper surface of the ocean and the colder deeper levels can be utilized to convert thermal energy to mechanical energy. This mechanical energy can in turn be used to produce electrical power using a vapor turbine. Let X denote the difference in temperature between the surface of the water and the water at a depth of 1 kilometer. Measurements are taken at 15 randomly selected sites in the Gulf of Mexico. These data result in the following temperatures:

22.5	23.8	23.2	22.8	10.1*
23.5	24.0	23.2	24.2	24.3
23.3	23.4	23.0	23.5	22.8

(a) Construct a double stem-and-leaf diagram for these data.

(b) Find the sample mean, sample median, and sample standard deviation for these data.

(c) Note that the starred observation in the data set is very different from the others. It is a potential outlier. Construct a boxplot for these data to verify that the value 10.1 does, in fact, qualify as an outlier.

(d) To see the effect of this outlier, drop it from the data set and calculate the sample mean, median, and standard deviation for the remaining 14 observations. Which measure is least affected by the presence of the outlier? Do you see why it is desirable to report both the mean and median of a data set?

29. Most homes utilize a variety of electronic equipment and appliances. For this reason, both suppliers and consumers of these products have become interested in product reliability. One aspect of reliability is the ability of the appliance to withstand power surges. In a study of this phenomena the following data are obtained on the strength of a surge in kilovolts required to damage or upset the appliance (based on figures found in "The Effects of Surges on Electronic Appliances," Stephen B. Smith and Ronald B. Standler, *IEEE Power Engineering Review,* July 1992, p. 50):

Clocks

1.1	3.5	3.2	4.0
3.6	1.5	2.3	3.0
4.7	4.0	4.9	2.9
2.6	2.5	2.7	4.2
2.4	3.7	3.8	6.0
5.0	1.8	5.6	5.1
3.9	3.8	3.7	3.5

Television receivers

2.0	5.0	4.5	5.4
4.2	4.3	5.1	5.6
5.2	4.7	4.4	5.8
5.2	7.8	5.4	4.9
4.6	4.6	5.0	4.8
5.3	5.2	5.3	5.9

dc power supplies

4.2	4.4	5.1	4.1	6.1
4.5	5.9	3.9	5.0	5.8
5.0	4.8	4.3	5.4	
4.7	5.1	5.2	5.6	
4.9	4.7	4.6	4.8	

(a) Sketch a double stem-and-leaf diagram for the clock data. Based on this diagram, would you be surprised to hear a claim that these data are drawn from an exponential distribution? Explain.

(b) Use the boxplot technique to check for outliers in the clock data. Based on your results, which measure of location, the sample mean or the sample median, is probably the better measure of the location of the bulk of the data for these data?

(c) Sketch a stem-and-leaf diagram for the television data. Use the stem 4 five times and the stem 5 five times. Based on this diagram, does there appear to be at least one outlier in the data set?

(d) Use the boxplot technique on the television data to test the suspicious points. Do you think that they are truly outliers? If so, are they mild outliers or extreme outliers? Which measure of variability, the sample variance or the iqr, is probably a better measure of the variability of the bulk of the data?

(e) Sketch a double stem-and-leaf diagram for the dc power supply data. These data contain an outlier due to a misplaced decimal point. Do you see it? Calculate the mean for the data using the bad data point as written. Now correct the data point and recalculate the sample mean. In light of this, explain what it means to say that $\bar{x}$ is not resistant to outliers.

REVIEW EXERCISES

30. Bricks are produced in lots of size 1000. Before shipping a lot, a sample of 25 bricks is selected and inspected for quality. Two random variables are of interest. These are X, the number of chips per brick, and Y, the hardness of the brick. Assume that hardness is measured on a continuous scale from 1 to 10 with larger numbers indicating a harder brick:

		x					y		
2	5	0	1	2	3.2	6.3	6.4	6.7	7.3
0	3	0	0	2	7.1	5.4	4.6	5.8	9.1
1	1	0	1	3	7.7	6.1	8.1	5.9	6.2
2	1	1	7	4	6.0	6.8	7.2	6.3	8.2
0	2	3	5	1	5.1	4.2	6.9	4.5	5.0

(a) What is the name of the family of random variables to which X belongs?

(b) Approximate the mean, variance, standard deviation, and median of X based on these data.

(c) Construct a stem-and-leaf diagram for the hardness measurements. Based on this diagram, would it be unrealistic to assume that Y is approximately normally distributed?

(d) Approximate the mean, variance, standard deviation, and median of Y.

31. In an attempt to study the problem of failure in field-installed computer equipment, data is collected on fifty field trips made to repair equipment. The random variables studied are X, the time in hours required to locate and rectify the problem, and Y, the cause of the failure. We define Y by

$$Y = \begin{cases} 1 & \text{if the failure is due to a faulty microprocessor chip} \\ 0 & \text{otherwise} \end{cases}$$

These data are obtained:

			x							y				
1.52	1.83	2.25	4.73	2.89	1.49	1.34	0	0	0	0	0	0	1
2.15	2.66	2.79	1.35	1.54	4.59	4.27	0	0	0	0	0	0	0
3.91	2.76	3.03	3.52	5.97	1.45		0	0	0	0	0	0	
3.07	2.18	1.38	2.04	1.49	1.11		1	0	0	0	0	0	
1.24	4.84	2.82	3.16	4.58	3.28		0	0	0	0	0	0	
1.30	3.01	1.20	3.42	1.86	3.49		0	0	0	0	0	0	
3.93	2.56	2.63	5.60	4.60	5.34		0	0	0	0	0	0	
1.62	2.82	4.88	2.04	1.62	.24		0	0	1	0	0	0	

(*a*) Construct a relative frequency histogram for the data on the time required to locate and rectify the problem. Use six categories. Based on this histogram, would you be surprised to hear someone claim that X is approximately normally distributed? Explain.

(*b*) Approximate the mean, variance, and standard deviation for X.

(*c*) Construct a relative cumulative frequency ogive. Use this ogive to approximate the median for X. Approximately what percentage of problems can be located and rectified in 1.5 hours or less?

(*d*) Let p denote the probability that the failure is due to a faulty microprocessor chip. Assume that even though p is unknown its value is the same for each chip. Theoretically, Y follows a point binomial distribution with parameter p. What is the theoretical mean for Y? Approximate this mean based on these data. If asked to approximate the probability that a future failure is due to the failure of a microprocessor chip, what would you say?

(*e*) What is the theoretical variance for Y? Use your answer to part (*d*) to approximate the variance of Y. Use the sample variance to approximate σ_Y^2. Did you get the same result? Which answer is unbiased for σ_Y^2?

(*f*) Use the technique of Exercise 25 to estimate σ_X. Compare your answer to that of part (*b*).

(*g*) Construct a boxplot for the data on x.

32. Most people are familiar with sparklers burned to celebrate New Year's Day and the Fourth of July. Two random variables are of interest. These are X, the length of the chemical coating that covers the tip of the sparkler, and Y, the burn time of the sparkler in seconds. These data are obtained on these random variables (based on data gathered in 1993–1994 by students at Radford University and Virginia Polytechnic Institute and State University):

x (in)	y (s)	x (in)	y (s)
4.5	29	4.3	22
3.6	26	3.5	21
4.0	25	4.5	30
3.7	25	4.6	22
4.0	27	4.6	25
3.7	27	3.9	20
4.0	28	3.5	13
4.0	25	3.8	19
3.8	25	3.9	23
4.0	28	3.6	25
3.8	24	3.6	27
4.1	15	3.6	18
3.9	22	3.3	11
4.1	25	3.7	24
3.9	24	3.7	23
4.2	26	4.3	26
3.8	24	3.9	27

(*a*) Construct a stem-and-leaf diagram for the burn time data. Use each stem 5 times so that each stem will involve two leaves.

(*b*) Construct a boxplot for the burn time data. Are any data points flagged as outliers?

(*c*) Take a good look at the shape of the distribution as indicated by the stem-and-leaf diagram. Does the distribution appear to be skewed? If so, what is the direction of the skew? If we assume in this case that all data points are legitimate and not due to poor technique or recording errors, then from what family of random variables might these data have been drawn? Do you think that the "outliers" should be treated as such? Explain.

(d) Construct a stem-and-leaf diagram for the length data. Again, use each stem 5 times. Does the normality assumption appear fairly reasonable here?

(e) Construct a boxplot for the length data, and comment on any outliers that might be identified.

33. Let X denote the gasoline mileage obtained in tests on a newly designed SUV (sport utility vehicle). A sample of 21 simulated test runs yields these data:

15	16	17	18	17	20
16	17	18	20	18	
18	19	19	17	21	
17	19	18	17	22	

(a) Construct a stem-and-leaf diagram for these data. Do the data suggest that X is normally distributed?

(b) Calculate the mean and median for this sample.

(c) Calculate the standard deviation and variance for this sample.

(d) Find the values of q_1 and q_3 and the iqr for the sample. Compare these values to those obtained via a TI83 calculator or any other technology tool that you have at your disposal.

34. In designing airplanes and airplane seats it is important to consider such variables as height and weight of passengers. A random sample of 100 adult male passengers yielded these weights:

212.8	256.3	278.1	298.3
213.7	257.0	278.2	298.4
214.2	258.6	279.1	299.3
217.7	259.1	279.6	300.8
219.8	259.2	279.9	300.9
220.0	261.6	283.0	301.1
224.5	262.5	283.1	301.7
225.3	265.2	283.6	302.5
227.8	267.0	284.9	304.8
230.8	267.9	285.0	306.6
232.7	268.1	286.0	306.8
233.8	268.3	286.3	310.5
236.1	269.1	286.6	310.6
237.7	269.5	286.6	310.9
239.7	270.1	286.8	310.9
241.0	271.2	286.9	312.4
243.3	271.8	289.3	313.8
244.7	272.7	290.4	316.0
246.1	273.3	291.0	316.9
249.8	274.8	291.2	320.2
250.9	275.1	293.8	321.5
252.0	275.2	296.1	332.2
252.7	275.8	296.1	335.7
254.9	276.8	297.1	342.4
255.4	277.6	298.2	353.6

(a) Calculate $\bar{x}$ and s.

(b) Use whatever technology tools you have available to construct a histogram for these data.

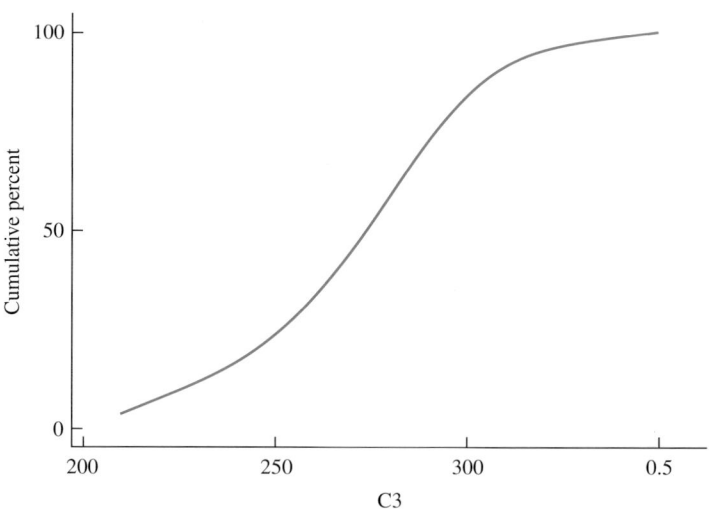

(c) It is thought that adult male weight is normally distributed with $\mu = 273$ pounds and $\sigma = 30$ pounds. Do your findings tend to support this notion?

(d) Figure 5.13 shows the ogive, the graph of the relative cumulative frequency distribution, for these data. Use it to estimate q_1, q_3, and the median.

(e) Use the textbook method or any other technology tool to estimate q_1, q_3, and the median, and compare these values to your graphical estimates.

35. A study of the lights used at railroad-highway grade crossings is conducted. The purpose of the study is to compare two types of lamps. These are 25-watt lamps to which a low warming voltage is applied during the off portion of the flashing cycle and 25-watt standard lamps. The data obtained in the study are found on the website. Variables are observation number, type of lamp with $1 =$ warmed lamp and $2 =$ standard lamp, and life span in thousands of hours.

(a) Plot a histogram for each type of lamp, and discuss the shape of the distribution from which each sample was drawn.

(b) Find the mean, median, standard deviation, and variance for each sample. Compare the values of these statistics. Do the samples seem similar in any way?

(c) Construct a boxplot for each sample, and note any outliers that are identified.

(d) If outliers are found, delete them and recompute the statistics requested in part(b) to see the effect that these outliers have on each statistic.

36. It is known that power surges or line spikes can damage sensitive electronic equipment. A study of these surges is conducted. The purpose of the study is to ascertain whether or not there are differences in the frequency of these surges among the seven days of the week. Date for the study is found on the website. Variables are observation number; day, with $m =$ Monday, $t =$ Tuesday, $w =$ Wednesday, $th =$ Thursday, $f =$ Friday, $s =$ Saturday, and $sn =$ Sunday; and number of spikes per day.

(a) Obtain descriptive statistics on the number of spikes per day for each day of the week. Discuss any differences among days that appear to exist.

(b) Construct boxplots for each day, and use the boxplots for a visual comparison of days.

CHAPTER 6

Estimation

In Chap. 5, we found that once the family to which a random variable belongs is determined, the problem of approximating or *estimating* the numerical value of pertinent parameters remains. Even though we were able to define sample statistics that allow us to estimate the mean, variance, and standard deviation of a random variable in a logical manner, we were unable to assess their effectiveness. In this chapter we consider the mathematical properties of these statistics. We also present a brief introduction to the theory of estimation. The ideas developed here will be used extensively throughout the remainder of the text.

6.1   POINT ESTIMATION

In an estimation problem there is at least one parameter θ whose value is to be approximated on the basis of a sample. The approximation is done by using an appropriate statistic. A statistic used to approximate or estimate a population parameter θ is called a *point estimator* for θ and is denoted by $\hat{\theta}$ (the symbol is called a "hat"); the numerical value assumed by this statistic when evaluated for a given sample is called a *point estimate* for θ. For example, in estimating the mean coal consumption by electric utilities for a given year (see Example 5.3.1), the statistic $\bar{X}$ was used. Thus $\bar{X}$ is a point estimator for μ and we write $\hat{\mu} = \bar{X}$. In Example 5.3.1 we evaluated this statistic for a particular sample and obtained the value 408.3 million tons. This number is called a *point estimate* for μ. Note that there is a difference in the terms "estimator" and "estimate." The estimator is the statistic used to generate the estimate; it is a random variable. An estimate is a number.

   Once a logical point estimator for a parameter θ has been developed, the natural question to ask is, "How good is this estimator?" Obviously, we want the estimator to generate estimates that can be expected to be close in value to θ. This can be expected to occur if the estimator $\hat{\theta}$ possesses two properties.

Desirable Properties of a Point Estimator

1. $\hat{\theta}$ to be *unbiased* for θ.
2. $\hat{\theta}$ to have a small variance for large sample sizes.

The word "unbiased" was explained graphically in Chap. 5. Basically, it means "centered at the right spot," where the right spot is the parameter being estimated. The term "unbiased" is a technical term. To be able to prove analytically that an estimator $\hat{\theta}$ is an unbiased estimator for a parameter θ, we need a formal definition for the term. This definition is given here.

> **Definition 6.1.1 (Unbiased).** An estimator $\hat{\theta}$ is an unbiased estimator for a parameter θ if and only if $E[\hat{\theta}] = \theta$.

Recall that $\hat{\theta}$ is a statistic; therefore it is also a random variable and, as such, has a mean, or expected, value. To say that $\hat{\theta}$ is unbiased for θ implies that the mean of the estimator $\hat{\theta}$ is equal to the parameter θ that it is estimating. Thus an estimator $\hat{\mu}$ is an unbiased estimator for μ if and only if $E[\hat{\mu}] = \mu$; an estimator $\hat{\sigma}^2$ is unbiased for σ^2 if and only if $E[\hat{\sigma}^2] = \sigma^2$; an estimator $\hat{\sigma}$ is unbiased for σ if and only if $E[\hat{\sigma}] = \sigma$. Let us reexamine the estimators $\bar{X}$, S^2, and S developed in Chap. 5 in light of this new definition.

> **Theorem 6.1.1.** Let $X_1, X_2, X_3, \ldots, X_n$ be a random sample of size n from a distribution with mean μ. The sample mean, $\bar{X}$, is an unbiased estimator for μ.

Proof. By Definition 5.3.1,

$$E[\bar{X}] = E[1/n(X_1 + X_2 + X_3 + \cdots + X_n)]$$

By the Rules for Expectation (Theorem 3.3.1),

$$E[\bar{X}] = 1/n(E[X_1] + E[X_2] + E[X_3] + \cdots + E[X_n])$$

Since $X_1, X_2, X_3, \ldots, X_n$ constitutes a random sample from a distribution with mean μ, each of these random variables has mean μ. Therefore

$$E[\bar{X}] = 1/n\bigg(\underbrace{\mu + \mu + \mu + \cdots + \mu}_{n \text{ terms}}\bigg) = 1/n(n\mu) = \mu$$

and the proof is complete.

It is important to realize that since $\hat{\theta}$ is a statistic, in repeated sampling the estimates generated will vary from sample to sample. To say that $\hat{\theta}$ is unbiased for θ implies that these estimates vary about θ; it also implies that the *average* value of these estimates can be expected to lie reasonably close to θ. For example, since $\bar{X}$ is unbiased for μ, for k repetitions of an experiment the observed sample means $\bar{x}_1, \bar{x}_2, \bar{x}_3, \ldots, \bar{x}_k$ will vary about μ and the *average* value of these k estimates should lie reasonably close to μ.

Example 6.1.1. Consider the experiment of rolling a single fair die. Let X denote the number obtained. X is discrete, with density given by

$$f(x) = 1/6 \qquad x = 1, 2, 3, 4, 5, 6$$

The average value of X is

$$\mu = E[X] = \Sigma\, x f(x) = 3.5$$

Now consider tossing a single die 30 times and recording the average toss, $\bar{x}$. If this process is repeated many times the $\bar{x}$ values will vary from sample to sample.

Figure 6.1
Plot of experimental x values of Example 6.1.1.

Since $\bar{X}$ is an unbiased estimator for the true average value, μ, the observed $\bar{x}$ values are expected to vary around the value 3.5. This experiment was conducted in class 56 times. The results were as follows:

3.43	3.33	3.60	2.97	3.50	4.20	3.07	3.33	3.86	3.80	3.00
3.30	3.40	4.00	3.83	3.43	3.90	3.07	3.47	3.23	3.20	3.76
3.50	3.70	4.13	3.97	3.42	3.57	3.47	3.80	3.53	3.20	3.13
3.63	3.33	3.33	3.56	3.47	4.33	3.53	3.33	3.47	3.73	3.90
3.32	4.21	3.63	3.67	3.53	3.43	3.40	3.53	3.63	3.42	3.67
3.57										

Figure 6.1 shows a dot plot of these data. Notice that, as expected, the $\bar{x}$ values vary and the value 3.5 is close to the center of the data points. The average of the 56 $\bar{x}$ values is 3.548, a little higher than the ideal theoretical value of 3.5.

It is equally important to understand what the term "unbiased" does *not* imply. It does not imply that any *one* estimate will be close in value to the parameter being estimated. In reference to Example 5.3.1, the estimated mean coal consumption by electric utilities was $\hat{\mu} = \bar{x} \doteq 408.3$ million tons. This estimate is unbiased in the sense that it was generated by means of the unbiased estimator $\bar{X}$. This *alone* does not guarantee that the actual mean coal consumption by electric utilities across the country is anywhere close to 408.3 million tons. This is unfortunate. Usually, statistical studies are not repeated over and over so that the estimates obtained can be averaged. In general, only one sample is drawn; one estimate is obtained. To have some assurance that this estimate is close in value to θ, the parameter being estimated, ideally the estimator used not only should be unbiased, but also it should have a small variance for large sample sizes. In this way, even though the estimated values fluctuate about θ, the variability is small. Each estimate produced can be expected to be fairly close in value to θ. Theorem 6.1.2 shows that $\bar{X}$ has this property.

> **Theorem 6.1.2.** Let $\bar{X}$ be the sample mean based on a random sample of size n from a distribution with mean μ and variance σ^2. Then
>
> $$\text{Var } \bar{X} = \frac{\sigma^2}{n}$$

The proof of this theorem is based on the Rules for Variance (Theorem 3.3.3) and is similar to that of Theorem 6.1.1. Note that since σ^2 is constant, as the sample size n increases, the variance of $\bar{X}$, σ^2/n, decreases and can be made as small as we wish by choosing n sufficiently large. This implies that a sample mean based on a large sample can be expected to lie reasonably close to μ; one based on a small sample may vary widely from the actual population mean. This points out the advantages of working with a large sample and the danger of placing too much emphasis on conclusions drawn from small samples. Keep in mind that many of the examples and exercises presented in this text are based on small samples. This is done for illustrative purposes only. We do *not* mean to imply that samples this small are common in research.

Since the standard deviation of any random variable is the square root of its variance, the standard deviation of the sample mean is the square root of the variance of $\bar{X}$. Thus the standard deviation of $\bar{X}$ is $\sqrt{\sigma^2/n} = \sigma/\sqrt{n}$. This standard deviation plays a vital role in the development of techniques used in making inferences on the true value of μ based on information concerning the observed value of $\bar{x}$. The name given to this special standard deviation is *standard error of the mean.*

> **Definition 6.1.2 (Standard error of the mean).** Let $\bar{X}$ denote the sample mean based on a sample of size n drawn from a distribution with standard deviation σ. The standard deviation of $\bar{X}$ is given by $\sigma/\sqrt{n}$ and is called the standard error of the mean.

In Chap. 5 we defined the sample variance S^2 by dividing $\sum_{i=1}^{n}(X_i - \bar{X})^2$ by $n - 1$. This was done so that the resulting estimator would be unbiased for σ^2. This result is stated formally in Theorem 6.1.3. The proof of this theorem is found in Appendix C.

> **Theorem 6.1.3.** Let S^2 be the sample variance based on a random sample of size n from a distribution with mean μ and variance σ^2. S^2 is an unbiased estimator for σ^2.

It should be noted that even though S^2 is an unbiased estimator for σ^2, it can be shown that S is not unbiased for σ (see Exercise 8). This emphasizes the fact that unbiasedness is desirable in an estimator but not essential.

6.2 FUNCTIONS OF RANDOM VARIABLES—DISTRIBUTION OF $\bar{X}$

There is one drawback to point estimation. It yields a single value for the unknown parameter θ. Is there any assurance that this estimate is even close in value to θ? The best answer is that in most cases the point estimators used are logical. To get an idea not only of the value of the parameter being estimated, but also of the accuracy of the estimate, researchers turn to the method of *interval estimation* or *confidence intervals.* An interval estimator is what the name implies. It is a random interval, an interval whose endpoints L_1 and L_2 are each statistics. It is used to determine a numerical interval based on a sample. It is hoped that the numerical interval obtained will contain the population parameter being estimated. By expanding from a point to an interval, we create a little room for error and in so doing gain the ability, based on probability theory, to report the confidence that we have in the estimate.

One of the more useful statistics that we have studied is $\bar{X}$, the sample mean. Since $\bar{X}$ is a statistic, it is also a random variable. It makes sense to ask, "What is the distribution of $\bar{X}$?" We have already seen that the center of location for $\bar{X}$ is μ, the mean of the population from which the sample is drawn. We have also seen that its variance is σ^2/n, the original population variance divided by the sample size. We have not yet mentioned the type of distribution possessed by the statistic. Does $\bar{X}$ follow some distribution such as the uniform, or normal distributions that we have already studied, or must we introduce a new distribution now?

In fact, one can prove that when sampling from a *normal* distribution the random variable $\bar{X}$ will itself be *normally* distributed.

> **Theorem 6.2.1 (Distribution of $\overline{X}$—normal population).** Let $X_1, X_2, \ldots, X_n$ be a random sample of size n from a normal distribution with mean μ and variance σ^2. Then $\overline{X}$ is normally distributed with mean μ and variance σ^2/n.

6.3   INTERVAL ESTIMATION AND THE CENTRAL LIMIT THEOREM

As mentioned previously, point estimation does not give us the ability to report the accuracy of our estimate. To do this, we must turn to the method of interval estimation. The statistics used to extend a point estimate for a parameter θ to an interval of values that should contain the true value of θ vary from parameter to parameter. However, the method for deriving these statistics is basically the same in each case. In this section we illustrate the method by deriving a "confidence interval" for the mean of a normal random variable when its variance is assumed to be known. In later chapters we apply the general technique illustrated here to find confidence intervals for other important parameters.

The term "confidence interval" is a technical term that we now define.

> **Definition 6.3.1 (Confidence interval).** A $100(1 - \alpha)\%$ confidence interval for a parameter θ is a random interval $[L_1, L_2]$ such that
> $$P[L_1 \leq \theta \leq L_2] \doteq 1 - \alpha$$
> regardless of the value of θ.

One general statement will guide in the construction of most of the confidence intervals presented in this text:

> To construct a $100(1 - \alpha)\%$ confidence interval for a parameter θ, we shall find a random variable whose expression involves θ and whose probability distribution is known at least approximately.

Confidence Interval on the Mean: Variance Known

To use this guideline to find a $100(1 - \alpha)\%$ confidence interval for the mean of a normal random variable whose variance is known, we must find a random variable whose expression involves μ and whose distribution is known. This is easy to do. Note that in Theorem 6.2.1, we showed that under the given conditions the sample mean, $\overline{X}$, is normally distributed with mean μ and variance σ^2/n. This implies that the random variable

$$\frac{\overline{X} - \mu}{\sigma/\sqrt{n}}$$

is *standard normal*. Note that this random variable involves the parameter μ and its distribution is known. We illustrate how this random variable can be used to generate a 95% confidence interval for μ. The technique used can be generalized easily to obtain any desired degree of confidence.

> **Example 6.3.1.** Acute myeloblastic leukemia is among the most deadly of cancers. Past experience indicates that the time in months that a patient survives after initial diagnosis of the disease is normally distributed with a mean of 13 months and a standard deviation of 3 months. A new treatment is being investigated which should

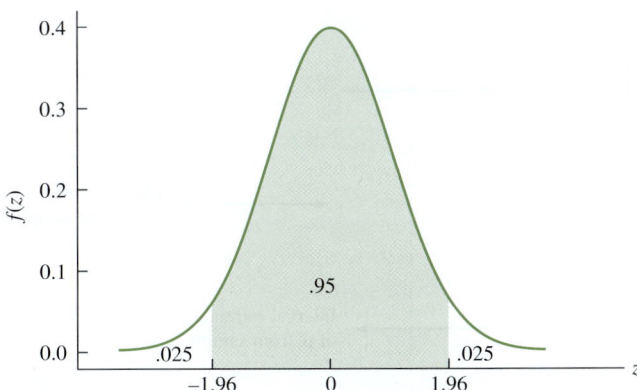

Figure 6.2
Partition of Z needed to obtain a 95% confidence interval for μ.

prolong the average survival time without affecting variability. Let $X_1, X_2, X_3, \ldots, X_n$ denote a random sample from the distribution of X, the survival time under the new treatment. We are assuming that X is normally distributed with $\sigma^2 = 9$ and μ unknown. We want to find statistics L_1 and L_2 so that $P[L_1 \leq \mu \leq L_2] \doteq .95$. To do so, consider the partition of the standard normal curve shown in Fig. 6.2. It can be seen that

$$P[-1.96 \leq Z \leq 1.96] = .95$$

In this case $Z = (\bar{X} - \mu)/(\sigma/\sqrt{n})$, and hence we may conclude that

$$P\left[-1.96 \leq \frac{\bar{X} - \mu}{\sigma/\sqrt{n}} \leq 1.96\right] = .95$$

To find L_1 and L_2, we algebraically isolate μ in the center of the preceding inequality as follows:

$$P[-1.96\sigma/\sqrt{n} \leq \bar{X} - \mu \leq 1.96\sigma/\sqrt{n}] = .95$$
$$P[-\bar{X} - 1.96\sigma/\sqrt{n} \leq -\mu \leq -\bar{X} + 1.96\sigma/\sqrt{n}] = .95$$
$$P[\bar{X} - 1.96\sigma/\sqrt{n} \leq \mu \leq \bar{X} + 1.96\sigma/\sqrt{n}] = .95$$

From this we see that the lower and upper bounds for a 95% confidence interval are

$$L_1 = \bar{X} - 1.96\sigma/\sqrt{n} \qquad L_2 = \bar{X} + 1.96\sigma/\sqrt{n}$$

These statistics have the property that in repeated sampling from the population, 95% of the numerical intervals generated are expected to contain μ; by chance, 5% will not. This idea is illustrated in Fig. 6.3.

Note that since we are assuming that σ^2 is known, the confidence bounds, $\bar{X} \pm 1.96\sigma/\sqrt{n}$, just derived are *statistics*. Given a particular set of observations on X, their numerical values can be determined easily.

Example 6.3.2. In Example 6.1.1 fifty-six samples, each of size 30, were generated. Each sample was obtained by tossing of a single fair die 30 times. The sample mean was found for each sample. For the single die experiment, it can be shown that $E[X^2] = 15.167$, and hence the variance of X is given by

$$\sigma^2 = \text{Var}(X) = E[X^2] - E[X]^2 = 15.167 - (3.5)^2 = 2.92$$

The standard deviation of X is $\sqrt{2.92} = 1.7088$. The standard error of the mean, $\sigma/\sqrt{30}$, has the value .3119. Thus the formula for a 95% confidence interval on μ in this case is

$$\bar{X} \pm 1.96(\sigma/\sqrt{n}) \qquad \text{or} \qquad \bar{X} \pm 1.96(.3119)$$

Figure 6.3
Of the intervals constructed by using $[L_1, L_2]$, 95% are expected to contain μ, the true but unknown population mean.

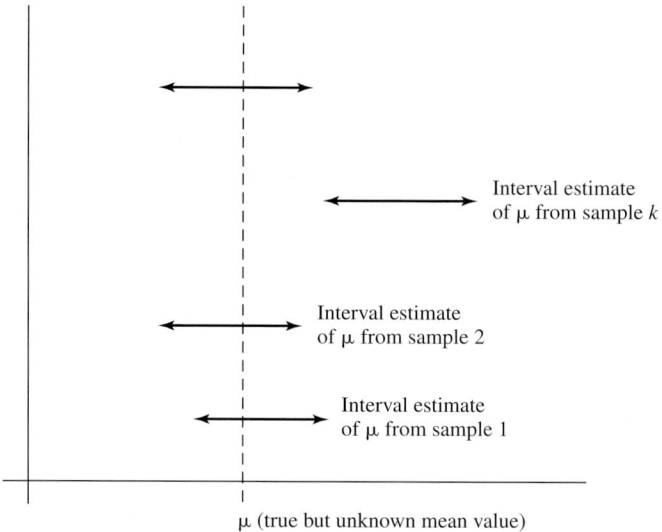

Figure 6.4
Results of the experiment of Example 6.3.2.

	xbar	lower	upper	result	caught		xbar	lower	upper	result	caught
1	3.43	2.81868	4.04132	1	trapped	29	3.47	2.85868	4.08132	1	trapped
2	3.33	2.71868	3.94132	1	trapped	30	3.80	3.18868	4.41132	1	trapped
3	3.60	2.98868	4.21132	1	trapped	31	3.53	2.91868	4.14132	1	trapped
4	2.97	2.35868	3.58132	1	trapped	32	3.20	2.58868	3.81132	1	trapped
5	3.50	2.88868	4.11132	1	trapped	33	3.13	2.51868	3.74132	1	trapped
6	4.20	3.58868	4.81132	0	missed	34	3.63	3.01868	4.24132	1	trapped
7	3.07	2.45868	3.68132	1	trapped	35	3.33	2.71868	3.94132	1	trapped
8	3.33	2.71868	3.94132	1	trapped	36	3.33	2.71868	3.94132	1	trapped
9	3.86	3.24868	4.47132	1	trapped	37	3.56	2.94868	4.17132	1	trapped
10	3.80	3.18868	4.41132	1	trapped	38	3.47	2.85868	4.08132	1	trapped
11	3.00	2.38868	3.61132	1	trapped	39	4.33	3.71868	4.94132	0	missed
12	3.30	2.68868	3.91132	1	trapped	40	3.53	2.91868	4.14132	1	trapped
13	3.40	2.78868	4.01132	1	trapped	41	3.33	2.71868	3.94132	1	trapped
14	4.00	3.38868	4.61132	1	trapped	42	3.47	2.85868	4.08132	1	trapped
15	3.83	3.21868	4.44132	1	trapped	43	3.73	3.11868	4.34132	1	trapped
16	3.43	2.81868	4.04132	1	trapped	44	3.90	3.28868	4.51132	1	trapped
17	3.90	3.28868	4.51132	1	trapped	45	3.32	2.70868	3.93132	1	trapped
18	3.07	2.45868	3.68132	1	trapped	46	4.21	3.59868	4.82132	0	missed
19	3.47	2.85868	4.08132	1	trapped	47	3.63	3.01868	4.24132	1	trapped
20	3.23	2.61868	3.84132	1	trapped	48	3.67	3.05868	4.28132	1	trapped
21	3.20	2.58868	3.81132	1	trapped	49	3.53	2.91868	4.14132	1	trapped
22	3.76	3.14868	4.37132	1	trapped	50	3.43	2.81868	4.04132	1	trapped
23	3.50	2.88868	4.11132	1	trapped	51	3.40	2.78868	4.01132	1	trapped
24	3.70	3.08868	4.31132	1	trapped	52	3.53	2.91868	4.14132	1	trapped
25	4.13	3.51868	4.74132	0	missed	53	3.63	3.01868	4.24132	1	trapped
26	3.97	3.35868	4.58132	1	trapped	54	3.42	2.80868	4.03132	1	trapped
27	3.42	2.80868	4.03132	1	trapped	55	3.67	3.05868	4.28132	1	trapped
28	3.57	2.95868	4.18132	1	trapped	56	3.57x	2.95868	4.18132	1	trapped

Each of the $\bar{x}$ values found in Example 6.1.1 is substituted into the above formula. We thus generate fifty six 95% confidence intervals on μ. Each is trying to trap the true mean value of 3.5. Some will succeed, and others will fail. Theoretically, 95% or about 53 will succeed and 5% or about 3 will fail. How well did the experiment work? Figure 6.4 gives the results of this exercise. The first column gives the value of $\bar{x}$, the second gives the lower 95% confidence limit, and the third shows the upper 95% confidence limit. The fourth column, result, is coded so that its value is 1 if the true mean of 3.5 falls between the lower and upper confidence limits. The last column states whether or not the interval in question actually trapped or missed the true mean. In this case, the results of our experiment agree extremely well with those predicted by theory even though X is not normal. You will soon see why.

Example 6.3.3. When the experiment of Example 6.3.1 is conducted, the following observations on X, the survival time under the new treatment, result:

8.0	13.6	13.2	13.6
12.5	14.2	14.9	14.5
13.4	8.6	11.5	16.0
14.2	19.0	17.9	17.0

Based on these data, $\hat{\mu} = \bar{x} = 13.88$ months. This point estimate is extended to a 95% confidence interval by evaluating the statistics L_1 and L_2. In particular,

$$L_1 = \bar{x} - 1.96\sigma/\sqrt{n} = 13.88 - 1.96(3/\sqrt{16})$$
$$= 13.88 - 1.47$$
$$= 12.41 \text{ months}$$
$$L_2 = \bar{x} + 1.96\sigma/\sqrt{n} = 13.88 + 1.47$$
$$= 15.35 \text{ months}$$

Based on these data, the interval estimate for μ is [12.41, 15.35]. Does the true mean survival time for patients receiving the new treatment really lie between 12.41 and 15.35 months? Unfortunately, there is no way of knowing. The interval [12.41, 15.35] is a 95% confidence interval. This means that the procedure used is expected to trap μ 95% of the time. We hope that the interval obtained from our particular sample does so.

To obtain the general formula for a $100(1 - \alpha)\%$ confidence interval on the mean of a normal random variable whose variance is known, we need only to partition the standard normal curve as shown in Fig. 6.5. The algebraic argument of Example 6.3.1 goes through exactly as presented with the point $x_{.025} = 1.96$ being replaced by $z_{\alpha/2}$. This change results in the general formula given in Theorem 6.3.1.

Theorem 6.3.1 [$100(1 - \alpha)\%$ Confidence interval on μ when σ^2 is known].
Let $X_1, X_2, X_3, \ldots, X_n$ be a random sample of size n from a normal distribution with mean μ and variance σ^2. A $100(1 - \alpha)\%$ confidence interval on μ is given by

$$\bar{X} \pm z_{\alpha/2}\sigma/\sqrt{n}$$

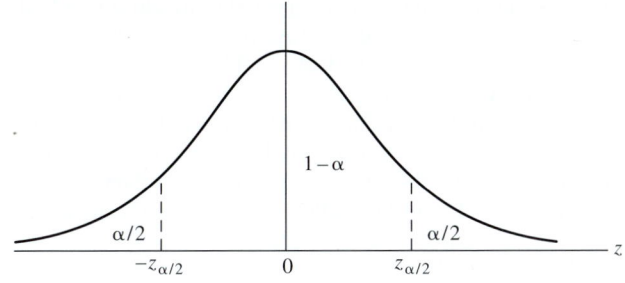

Figure 6.5
Partition of Z to obtain a $100(1 - \alpha)\%$ confidence interval for μ.

Let us point out that the preceding confidence interval is very idealistic. It is usable only in settings in which the population standard deviation, σ, is known. In practice, this is seldom the case. In most real life problems both μ and σ^2 must be estimated from available data. When this occurs, the previous confidence interval is *not* appropriate. In Sec. 8.2 we shall show how to overcome this problem. Meanwhile, view this interval as a prototype for confidence intervals in general. It is useful as an aid for understanding how confidence intervals are derived and interpreted.

There are several things to notice concerning the preceding formula. First, every confidence interval on μ is centered at $\bar{x}$, the unbiased point estimate for μ. Second, the length of the confidence interval is dependent on three factors. These are the desired confidence, the amount of variability in X, and the sample size (n). The desired confidence determines the value of the z point used. The higher the confidence desired, the larger this value becomes. When a random variable displays a high degree of variability, it is hard to predict its behavior. Thus the larger σ becomes, the longer the confidence interval must become. Sample size works in reverse. With all other factors held constant, as n increases, the length of the confidence interval decreases. We can say that the length of a confidence interval on μ is directly proportional to σ and to the confidence desired and inversely proportional to the sample size.

Central Limit Theorem

There is one further point to be made. Theorem 6.3.1 does require that the base variable X be normal. If this condition is not satisfied, then the confidence bounds given can be used as long as the sample is not too small. Empirical studies have shown that for samples as small as 25, the above bounds are usually satisfactory even though approximate. This is due to a remarkable theorem, first formulated in the early nineteenth century by Laplace and Gauss. This theorem, known as the *Central Limit Theorem*, gives the distribution of $\bar{X}$ when sampling from a distribution that is not necessarily normal.

Theorem 6.3.2 (Central Limit Theorem). Let $X_1, X_2, \ldots, X_n$ be a random sample of size n from a distribution with mean μ and variance σ^2. Then for large n, $\bar{X}$ is approximately normal with mean μ and variance σ^2/n. Furthermore, for large n, the random variable $(\bar{X} - \mu)/(\sigma/\sqrt{n})$ is approximately standard normal.

Example 6.3.4 illustrates the Central Limit Theorem graphically.

Example 6.3.4. Consider a single die toss. We have tossed a single die 30 times and have repeated the experiment 56 times to obtain 56 $\bar{x}$ values. According to the Central Limit Theorem, a histogram of these data is expected to exhibit an approximate bell shape. The center of the bell is expected to lie close to 3.5, the true value of μ; the variance of the data should be close in value to .0973, the true value of σ^2/n; and the standard deviation of the data should approximate well the true value of the standard error of the mean, .3119. Figure 6.6 shows the histogram for the data of Example 6.1.1. Notice that the bell shape is not perfect. There is a slight right skew due to the fact that there were a few relatively large $\bar{x}$ values obtained via the experimentation. The mean for these data is 3.548, a little higher than the true mean of 3.5; the sample variance is .0911, a little smaller than the theoretical value of .0973; the estimated value of the standard error of the mean based on these data is .3019, a little smaller than the theoretical value of .3119. As the size of the sample upon which

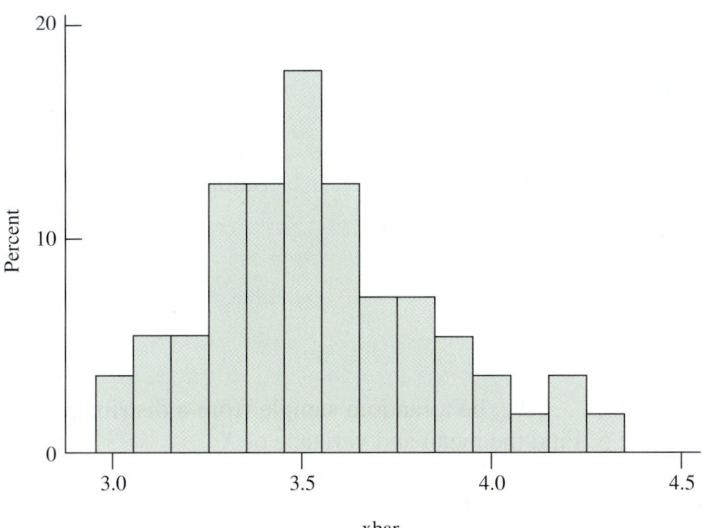

Figure 6.6
Histogram of the 56 $\bar{x}$ values given in Example 6.1.1.

each $\bar{x}$ value is based increases, the histogram is expected to exhibit a more pronounced bell shape and the estimates for the mean, variance, and standard deviation of $\bar{X}$ are expected to agree more closely with those predicted by theory.

Please note the differences between the Central Limit Theorem and Theorem 6.2.1. The former does not require that sampling be from a normal distribution, whereas normality is assumed in the latter; the former claims that $\bar{X}$ will be approximately normally distributed for large sample sizes, whereas the latter claims that $\bar{X}$ will be exactly normally distributed regardless of the sample size involved.

The Central Limit Theorem is important to us for two reasons. First, it allows us to make inferences on the mean of a distribution based on relatively large samples without having to be overly concerned as to whether or not we are sampling from a normal distribution. Second, it allows us to justify analytically the normal approximations to the binomial distribution.

CHAPTER SUMMARY

In this chapter we considered the ideas of point and interval estimation. We introduced unbiased estimators, whose mean value is equal to the parameter being estimated. We showed that $\bar{X}$ is unbiased for μ, that S^2 is unbiased for σ^2, but that S is not unbiased for σ.

We introduced the general concept of a $100(1 - \alpha)\%$ confidence interval on a parameter θ. This is a random interval, an interval of the form $[L_1, L_2]$, where L_1 and L_2 are statistics with the property that a priori θ will be trapped between L_1 and L_2 with probability $1 - \alpha$. We used information just developed on the distribution of $\bar{X}$ to develop specific formulas for constructing a $100(1 - \alpha)\%$ confidence interval on the mean of a normal distribution. Finally, we considered the Central Limit Theorem. This theorem concerns the approximate distribution of $\bar{X}$ when sampling from a nonnormal distribution. It allows us to make inferences on the mean of any distribution when relatively large samples are available. It also allows us to justify some of the approximation techniques presented earlier in the text.

We introduced and defined important terms that you should know. These are:

Point estimator   Point estimate
Unbiased   Weighted mean
Confidence interval or interval estimator   Interval estimate
Standard error of the mean   Sample standard error
Central Limit Theorem

EXERCISES

Section 6.1

1. Let $X_1, X_2, X_3, \ldots, X_{20}$ be a random sample from a distribution with mean 8 and variance 5. Find the mean and variance of $\bar{X}$.

2. Let $X_1, X_2, X_3, \ldots, X_{15}$ be a random sample from a Poisson distribution with parameter λs. Give an unbiased estimator for this parameter.

3. Let X denote the number of paint defects found in a square yard section of a car body painted by a robot. These data are obtained:

8	5	0	10
0	3	1	12
2	7	9	6

   Assume that X has a Poisson distribution with parameter λs.
   (a) Find an unbiased estimate for λs.
   (b) Find an unbiased estimate for the average number of flaws per square yard.
   (c) Find an unbiased estimate for the average number of flaws per square foot.

4. An interactive computer system is available at a large installation. Let X denote the number of requests for this system received per hour. Assume that X has a Poisson distribution with parameter λs. These data are obtained:

25	20	20
30	24	15
10	23	4

   (a) Find an unbiased estimate for λs.
   (b) Find an unbiased estimate for the average number of requests received per hour.
   (c) Find an unbiased estimate for the average number of requests received per quarter hour.

5. Let X_1, X_2, X_3, X_4, X_5 be a random sample from a binomial distribution with $n = 10$ and p unknown.
   (a) Show that $\bar{X}/10$ is an unbiased estimator for p.
   (b) Estimate p based on these data: $3, 4, 4, 5, 6$.

6. An experiment is conducted to study the effect of a power surge on data stored in a digital computer. A "word" is a sequence of 8 bits. Each bit is either "on" (activated) or "off" (not activated) at any given time. Twenty 8-bit words are stored, and a power surge is induced. Let X denote the number of bit reversals that result per word. Assume that X is binomially distributed with $n = 8$ and p, the probability of a bit reversal, unknown. These data result:

1	0	0	0
0	0	1	1
0	1	2	1
1	0	1	0
2	2	3	0

(a) Find an unbiased estimate for p.
(b) Based on the estimate for p just found, approximate the probability that in another 8-bit word a similar power surge will result in no bit reversals.
(c) A data line utilizes 64 bits. Based on the estimate for p just found, approximate the probability that at most one bit reversal will occur.

7. Stress tests are conducted on fiberglass rods used in communications networks. The random variable studied is X, the distance in inches from the anchored end of the rod to the crack location when the rod is subjected to extreme stress. Assume that X is uniformly distributed over the interval $(0, b)$. These data are obtained on 10 test rods:

10	7	11	12	8
8	9	10	9	13

(a) Find an unbiased estimate for the average distance from the anchored end of the rod to the crack.
(b) Find an unbiased estimate for the variance of X.
(c) Find an unbiased estimate for b.
(d) Find an estimate for σ, the standard deviation of X. Is this estimate unbiased?

8. Note that S is a statistic, and unless X is constant, its value will vary from sample to sample. Therefore Var $S > 0$. To show that S is not unbiased for σ, use proof by contradiction. That is, assume that $E[S] = \sigma$ and obtain a contradiction. *Hint:* Use Theorem 3.3.2.

9. *(Weighted means.)* Assume that one has k independent random samples of sizes $n_1, n_2, n_3, \ldots, n_k$ from the same distribution. These samples generate k unbiased estimators for the mean, namely, $\bar{X}_1, \bar{X}_2, \bar{X}_3, \ldots, \bar{X}_k$.
(a) Show that the arithmetic average of these estimators, $(\bar{X}_1 + \bar{X}_2 + \bar{X}_3 + \cdots + \bar{X}_k)/k$, is also unbiased for μ.
(b) Certain mineral elements required by plants are classed as macronutrients. Macronutrients are measured in terms of their percentage of the dry weight of the plant. Proportions of each element vary in different species and in the same species grown under differing conditions. One macronutrient is sulfur. In a study of winter cress, a member of the mustard family, these data, based on three independent random samples, are obtained:

$$\bar{x}_1 = .8 \qquad \bar{x}_2 = .95 \qquad \bar{x}_3 = .7$$
$$n_1 = 9 \qquad n_2 = 3 \qquad n_3 = 200$$

Use the result of part (a) to obtain an unbiased estimate for μ, the mean proportion of sulfur by dry weight in winter cress. By averaging the three values .8, .95, and .7 to obtain the estimate for μ, each sample is being given equal importance or "weight." Does this seem reasonable in this problem? Explain.
(c) To take sample sizes into account, a "weighted" mean is used. This estimator, $\hat{\mu}_W$, is given by

$$\hat{\mu}_W = \frac{n_1\bar{X}_1 + n_2\bar{X}_2 + \cdots + n_k\bar{X}_k}{n_1 + n_2 + \cdots + n_k}$$

Show that $\hat{\mu}_W$ is an unbiased estimator for μ.
(d) Use the data of part (b) to find the weighted estimate for the mean proportion of sulfur by dry weight in winter cress. Compare your answer to the estimate found in part (b).

10. Let X denote the number of heads obtained when a fair coin is tossed 4 times.
   (a) What is $E[X]$ and Var X?
   (b) Perform the experiment of tossing a fair coin 4 times and recording the number of heads obtained 10 times. You thus obtain a random sample of size 10 from a binomial distribution with $n = 4$ and $p = 1/2$.
   (c) Based on your 10 observations, estimate the mean and variance of X. Compare your answers to those of your classmates. Do the observed values of $\bar{X}$ fluctuate about the theoretical mean of 2? Do the observed values of S^2 fluctuate about the theoretical variance of 1?
   (d) Average the values of $\bar{X}$ that you have available. Is the average value close to 2? Average the values of S^2 that you have available. Is the average value of S^2 close to 1?

11. Let X denote the number of heads obtained when a fair coin is tossed 4 times. Perform this experiment 3 times, and record the value of X for each set of four tosses. In this way you obtain a single sample of size 3 from a binomial distribution with $n = 4$ and $p = 1/2$.
   (a) Find the numerical value of $\bar{X}$ for your sample.
   (b) Repeat the experiment 9 more times, recording the value of $\bar{X}$ each time.
   (c) What is $E[\bar{X}]$? Average your 10 values of $\bar{X}$. Is the average value close to the theoretical mean of 2?
   (d) What is Var $\bar{X}$? Find the value of S^2 for the 10 observations on $\bar{X}$. Does this value lie close to the theoretical value of 1/3?

12. Consider the experiment of rolling a pair of fair dice until a sum of 7 is obtained. Let X denote the number of trials needed to obtain a sum of 7.
   (a) Notice that X is discrete. What is the distribution of X?
   (b) What is the theoretical average value of X? That is, what is μ?
   (c) What is the theoretical variance of X? That is, what is σ^2?
   (d) Perform the experiment described 25 times, and thus obtain a sample of size $n = 25$ observations on X. Plot a stem-and-leaf diagram for your data. Does the distribution appear to be symmetric? Use your data to obtain unbiased estimates for μ and σ^2. Compare your answers to the true values of these parameters found in parts (b) and (c), respectively.
   (e) Consider the random variable $\bar{X}$, the average number of trials needed to roll a sum of 7 based on 25 trials. What is $E[\bar{X}]$? What is Var $\bar{X}$?
   (f) Pool the class observations on $\bar{X}$. Plot these values on a number line. Do they fluctuate about μ as expected? Find the average value of these observed $\bar{X}$ values. Is it close to μ as expected? Find the variance of the $\bar{X}$ values. Is this sample variance close in value to $\sigma^2/25$ as expected?

13. Ozone levels around Los Angeles have been measured as high as 220 parts per billion (ppb). Concentrations this high can cause the eyes to burn and are a hazard to both plant and animal life. These data were obtained on the ozone level in a forested area near Seattle, Washington (based on information found in "Twigs," *Americans Forests*, April 1990, p. 71):

160	176	160	180	167	164
165	163	162	168	173	179
170	196	185	163	162	163
172	162	167	161	169	178
161					

   (a) Construct a double stem-and-leaf diagram for these data. Do these data appear to be skewed? If so, in which direction?

(b) Construct a boxplot for these data, and identify the potential outlier that is flagged by this technique. Assume that the point in question is a legitimate data point. In this case, do you believe that it is truly an outlier or probably simply a natural consequence of the distribution involved? Explain.

(c) Use these data to estimate the mean and variance of the ozone level in this area.

14. In this exercise you will show that the most logical estimator for σ^2, namely, $\sum_{i=1}^{n}(X_i - \bar{X})^2/n$, is a biased estimator for σ^2 and tends to underestimate the true variance. Let $X_1, X_2, \ldots, X_n$ be a random sample of size n from a distribution with mean μ and variance σ^2.

(a) Show that $\sum_{i=1}^{n}(X_i - \bar{X})^2/n = (n-1)S^2/n$.

(b) Verify that $E[\sum_{i=1}^{n}(X_i - \bar{X})^2/n] \neq \sigma^2$, thus showing that this estimator is not an unbiased estimator for σ^2. Argue that it tends to underestimate σ^2.

(c) Consider the theoretical setting described in Exercise 12. Based on samples of size $n = 25$, what is $E[S^2]$? What is $E[\sum_{i=1}^{n}(X_i - \bar{X})^2/25]$?

Section 6.2

15. Let X denote the time required to do a computation using an algorithm written in programming language A, and let Y denote the time required to do the same calculation using an algorithm written in programming language B. Assume that X is normally distributed with mean 10 seconds and standard deviation 3 seconds and that Y is normally distributed with mean 9 seconds and standard deviation 4 seconds.

(a) What is the distribution of the random variable $X - Y$?

(b) Find the probability that a given calculation will run faster using A than when using B.

Section 6.3

16. As heat is added to a material its temperature rises. The heat capacity is a quantitative statement of the increase in temperature for a specified addition of heat. These data are obtained on X, the measured heat capacity of liquid ethylene glycol at constant pressure and 80° C. Measurements are in calories per gram degree Celsius:

.645	.654	.640	.627	.626
.649	.629	.631	.643	.633
.646	.630	.634	.631	.651
.659	.638	.645	.655	.624
.658	.658	.658	.647	.665

Past experience indicates that $\sigma = .01$.

(a) Evaluate $\bar{X}$ for these data, thereby obtaining an unbiased point estimate for μ.

(b) Assume that X is normally distributed. Find a 95% confidence interval for μ.

(c) Would you expect a 90% confidence interval for μ based on these data to be longer or shorter than the interval of part (b)? Explain. Verify your answer by finding a 90% confidence interval on μ. *Hint:* Begin by sketching a curve similar to that shown in Fig. 6.3 with $1 - \alpha = .90$ and $\alpha/2 = .05$.

(d) Would you expect a 99% confidence interval for μ based on these data to be longer or shorter than the interval of part (b)? Explain. Verify your answer by finding a 99% confidence interval on μ.

17. The late manifestation of an injury following exposure to a sufficient dose of radiation is common. These data are obtained on the variable X, the time in

days that elapses between the exposure to radiation and the appearance of peak erythema (skin redness):

16	12	14	16	13	9	15	7
20	19	11	14	9	13	11	3
8	21	16	16	12	16	14	20
7	14	18	14	18	13	11	16
18	16	11	13	14	16	15	15

(a) Even though the time at which the peak redness appears is recorded to the nearest day, time is actually a continuous random variable. Sketch a stem-and-leaf diagram for these data. Does the diagram lend support to the assumption that X is normally distributed?

(b) Evaluate $\bar{X}$ for these data.

(c) Assume that $\sigma = 4$ and find a 95% confidence interval on the mean time to the appearance of peak redness. Would you be surprised to hear a claim that $\mu = 17$ days? Explain, based on the confidence interval.

18. When fission occurs, many of the nuclear fragments formed have too many neutrons for stability. Some of these neutrons are expelled almost instantaneously. These observations are obtained on X, the number of neutrons released during fission of plutonium-239:

3	2	2	2	2	3	3	3
3	3	3	3	4	3	2	3
3	2	3	3	3	3	3	1
3	3	3	3	3	3	3	3
3	3	2	3	3	3	3	3

(a) Is X normally distributed? Explain.

(b) Estimate the mean number of neutrons expelled during fission of plutonium-239.

(c) Assume that $\sigma = .5$. Find a 99% confidence interval on μ. What theorem justifies the procedure you used to construct this interval?

(d) The reported value of μ is 3.0. Do these data refute this value? Explain.

19. (Central Limit Theorem.) Consider an infinite population with 25% of the elements having the value 1, 25% the value 2, 25% the value 3, and 25% the value 4. If X is the value of a randomly selected item, then X is a discrete random variable whose possible values are 1, 2, 3, and 4.

(a) Find the population mean μ and population variance σ^2 for the random variable X.

(b) List all 16 possible distinguishable samples of size 2, and for each calculate the value of the sample mean. Represent the value of the sample mean $\bar{X}$ using a probability histogram (use one bar for each of the possible values for $\bar{X}$). Note that although this is a very small sample, the distribution of $\bar{X}$ does not look like the population distribution and has the general shape of the normal distribution.

(c) Calculate the mean and variance of the distribution of $\bar{X}$ and show that, as expected, they are equal to μ and σ^2/n, respectively.

REVIEW EXERCISES

20. Studies have shown that the random variable X, the processing time required to do a multiplication on a new 3-D computer, is normally distributed with mean μ and standard deviation 2 microseconds. A random sample of 16 observations is to be taken.

(a) What is the distribution of $\bar{X}$?

(b) These data are obtained:

42.65	45.15	39.32	44.44
41.63	41.54	41.59	45.68
46.50	41.35	44.37	40.27
43.87	43.79	43.28	40.70

Based on these data, find an unbiased estimate for μ.

(c) Find a 95% confidence interval for μ. Would you be surprised to read that the average time required to process a multiplication on this system is 42.2 microseconds? Explain, based on the confidence interval.

21. Let X denote the unit price of a 3.5-inch floppy diskette. These observations are obtained from a random sample of 10 suppliers:

$3.83	3.54	3.44	3.89	3.65
3.70	3.59	3.37	4.04	3.93

(a) Find an unbiased estimate for the mean price of these diskettes.

(b) Find an unbiased estimate for the variance in the price of these diskettes.

(c) Find the sample standard deviation. Is this an unbiased estimate for σ?

22. *(Central Limit Theorem.)* In an attempt to approximate the proportion p of improperly sealed packages produced on an assembly line, a random sample of 100 packages is selected and inspected. Let

$$X_i = \begin{cases} 1 & \text{if the } i\text{th package selected is improperly sealed} \\ 0 & \text{otherwise} \end{cases}$$

(a) What is the distribution of X_i?

(b) Based on the Central Limit Theorem, what is the approximate distribution of $\bar{X}$?

(c) When the experiment is conducted, we observe five improperly sealed packages. Find a point estimate for the proportion of improperly sealed packages being produced on this assembly line.

23. *(Central Limit Theorem.)* In a study of the size of various computer systems the random variable X, the number of files stored, is considered. Past experience indicates that $\sigma = 5$. These data are obtained:

7	8	4	5	9	9
4	12	8	1	8	7
3	13	2	1	17	7
12	5	6	2	1	13
14	10	2	4	9	11
3	5	12	6	10	7

(a) Find an unbiased estimate for μ, the mean number of files per system.

(b) Based on the Central Limit Theorem, what is the approximate distribution of $\bar{X}$?

(c) Find an approximate 98% confidence interval on μ.

(d) In describing the size of such systems, an executive states that the average number of files exceeds 10. Does this statement surprise you? Explain.

24. Let X be normally distributed with mean 2 and variance 25.

(a) What is the distribution of the random variable $(X - 2)/5$?

(b) What is the distribution of the random variable $[(X - 2)/5]^2$?

(c) Let $X_1, X_2, X_3, \ldots, X_{10}$ represent a random sample from the distribution of X. What is the distribution of the random variable

$$\sum_{i=1}^{10} \left(\frac{X_i - 2}{5} \right)^2$$

25. *(Central Limit Theorem.)* Consider the experiment of tossing a fair die once. Let X denote the number that occurs. Theoretically, X follows a discrete uniform distribution.
    (a) Find the theoretical density, mean, and variance for X.
    (b) Now consider an experiment in which the die is tossed 20 times and the results averaged. By the Central Limit Theorem, what is the theoretical mean and variance for the random variable $\bar{X}$?
    (c) Perform the experiment of part (b) 25 times and record the value of $\bar{X}$ each time. (You will toss the die 500 times and obtain a data set that consists of 25 averages.) What shape should the stem-and-leaf diagram for these data assume? Explain. Construct a stem-and-leaf diagram for your data. Did the diagram take the shape that you expected?
    (d) Approximately what value would you expect to obtain if you averaged the data of part (c)? Average your 25 observations on $\bar{X}$. Did the result come out as expected?
    (e) Approximately what value would you expect to obtain if you found the sample variance for the data of part (c)? Explain. Find s^2 for your 25 observations on $\bar{X}$. Did the result come out as expected?
    (f) If you were to construct 95% confidence intervals on μ based on each of the values of $\bar{X}$ found in part (c), approximately how many of them would you expect to contain the true value of μ? From your data, can you find an example of a confidence interval that does contain μ? of a confidence interval that does not contain μ?

26. Assume that a single fair die is tossed 30 times. Let X denote the number obtained per toss. Suppose that $\bar{x}$ assumes the value 2.83 for these 30 tosses.
    (a) Find a 95% confidence interval for the mean value of X. Did the interval you constructed trap the true mean of 3.5?
    (b) If we construct a 90% confidence interval on μ, will the interval have a chance of trapping μ? Explain based on what you learned in part (a).
    (c) If we construct a 99% confidence interval on μ, will the interval have a chance of trapping the true mean? Explain.
    (d) Construct a 99% confidence interval on μ. Did this interval trap the mean?

CHAPTER 7

Inferences on the Mean and Variance of a Distribution

One of the modern-day "miracles" is the rise of Japan's industrial strength after World War II. Much of this success has been attributed to the work of the American statistician W. Edwards Deming. This man not only helped the Japanese to implement the methods of statistical process control, but he also developed and taught his system of total quality management (TQM). TQM is a management system that is based on Deming's 14 points. One of the aims of TQM is to reduce variability. That is, with respect to a process, the aim is to produce goods that not only satisfy some target average value, but that also do so consistently. To achieve this, random variation must be reduced at every stage of the production process from the procurement of raw material through the marketing and servicing of the finished product. For this reason, there is interest in both the mean (the target value) and the variance of the random variable involved. In this chapter we present some statistical techniques that can be used to draw conclusions about these population parameters based on information obtained from samples.

We have seen how to estimate both the mean and the variance of a distribution via point estimation. We have also seen how to generate a confidence interval for the mean of a normal distribution when its variance is assumed to be *known*. Unfortunately, in most statistical studies, the assumption that σ^2 is known is unrealistic. If it is necessary to estimate the mean of a distribution, then its variance is usually unknown also. In this chapter we turn our attention to the problem of making inferences on the mean and variance of a distribution when both of these parameters are assumed to be unknown. We begin by considering the construction of a confidence interval for σ^2.

7.1 INTERVAL ESTIMATION OF VARIABILITY

In Theorem 6.1.3 we showed that the statistic S^2 is an unbiased estimator for σ^2. To obtain a $100(1 - \alpha)\%$ confidence interval for σ^2, we need a random variable whose expression involves σ^2 and whose probability distribution is known. The following theorem provides the random variable needed to construct a confidence interval for σ^2. Its proof is found in App C.

> **Theorem 7.1.1 [Distribution of $(n - 1)S^2/\sigma^2$].** Let $X_1, X_2, X_3, \ldots, X_n$ be a random sample from a normal distribution with mean μ and variance σ^2. The random variable
>
> $$(n - 1)S^2/\sigma^2 = \sum_{i=1}^{n}(X_i - \bar{X})^2/\sigma^2$$
>
> has a chi-squared distribution with $n - 1$ degrees of freedom.

To use the random variable $(n - 1)S^2/\sigma^2$ to derive a $100(1 - \alpha)\%$ confidence interval on σ^2, we first partition the X_{n-1}^2 curve as shown in Fig. 7.1. Remember that in our notational convention, the subscript associated with a point denotes the area to the *right* of the point. If we partition the chi-squared curve so that $(1 - \alpha)\%$ of the area is in the center of the curve, then the missing α is divided in half. Thus the right-hand chi-squared point has $\alpha/2\%$ of the area to its right; it is denoted by $\chi_{\alpha/2}^2$. Since the chi-squared distribution is never negative, the left-hand chi-squared point is not the negative of the right-hand point as was the case in the Z-type confidence interval on the mean developed in Sec. 6.3. Rather, it is simply a point with $\alpha/2$ area to its left and $1 - \alpha/2$ area to its right; it is denoted by $\chi_{1-\alpha/2}^2$. To derive the confidence interval, we begin by giving a probability statement based on Fig. 7.1 that can be set equal to $1 - \alpha$. It is evident that

$$P[\chi_{1-\alpha/2}^2 \leq (n - 1)S^2/\sigma^2 \leq \chi_{\alpha/2}^2] = 1 - \alpha$$

To find the lower and upper bounds for the confidence interval, we isolate σ^2 in the center of the inequality by inverting each term and solving for σ^2.

$$P[1/\chi_{\alpha/2}^2 \leq \sigma^2/(n - 1)S^2 \leq 1/\chi_{1-\alpha/2}^2] = 1 - \alpha$$

or

$$P\left[\frac{(n - 1)S^2}{\chi_{\alpha/2}^2} \leq \sigma^2 \leq \frac{(n - 1)S^2}{\chi_{1-\alpha/2}^2}\right] = 1 - \alpha$$

The desired confidence bounds can be read from the latter inequality and are given in Theorem 7.1.2.

> **Theorem 7.1.2 [$100(1 - \alpha)\%$ confidence interval on σ^2].** Let $X_1, X_2, X_3, \ldots, X_n$ be a random sample of size n from a normal distribution with mean μ and variance σ^2. The lower and upper bounds, L_1 and L_2, respectively, for a $100(1 - \alpha)\%$ confidence interval on σ^2, are given by
>
> $$L_1 = (n - 1)S^2/\chi_{\alpha/2}^2 \qquad \text{and} \qquad L_2 = (n - 1)S^2/\chi_{1-\alpha/2}^2$$

Figure 7.1
Partition of the X_{n-1}^2 curve needed to derive a $100(1 - \alpha)\%$ confidence interval on σ^2.

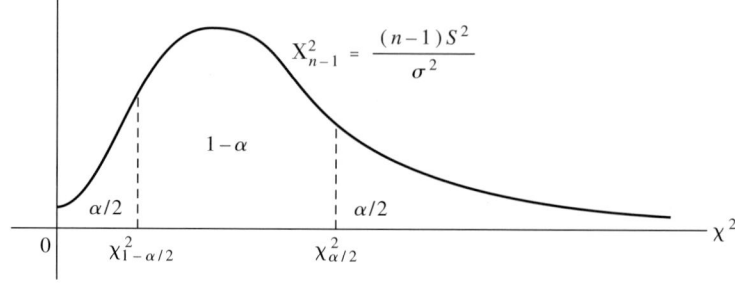

As one would suspect, to obtain the bounds for a $100(1 - \alpha)\%$ confidence interval on the standard deviation of a normal random variable, we take the nonnegative square root of the bounds given in Theorem 7.1.2.

Example 7.1.1.   In computing, "workload" is defined as a collection of processor and input-output (I/O) resource requests during a particular period of time. Workloads are compared via a measure called *relative I/O content*. The average commercial batch MVS installation provides the base for this measure and is given a relative I/O content rating of 1. Other installations are rated relative to this base. These observations on the relative I/O content for a large consulting firm over randomly selected 1-hour periods are obtained:

3.4	3.6	4.0	0.4	2.0
3.0	3.1	4.1	1.4	2.5
1.4	2.0	3.1	1.8	1.6
3.5	2.5	1.7	5.1	.7
4.2	1.5	3.0	3.9	3.0

Let us construct a 95% confidence interval on the standard deviation of the relative I/O content for this installation.

Firstly, we check the assumption about the population distribution. According to Theorem 7.1.2, we have to ensure that these data are a random sample from a normal distribution. The stem-and-leaf diagram for these data is shown in Fig. 7.2. This diagram does not suggest a serious departure from normality.

Next, we find the required value of χ^2. The partition of the χ^2_{24} curve needed to construct the confidence interval is shown in Fig. 7.3. From the Table IV of App. A, the required values are

$$\chi^2_{.975} = 12.4 \qquad \text{and} \qquad \chi^2_{.025} = 39.4$$

```
0 | 47
1 | 457486
2 | 0505
3 | 405611090
4 | 201
5 | 1
```

Figure 7.2
Stem-and-leaf diagram of the relative I/O content of the consulting firm of Example 7.1.1.

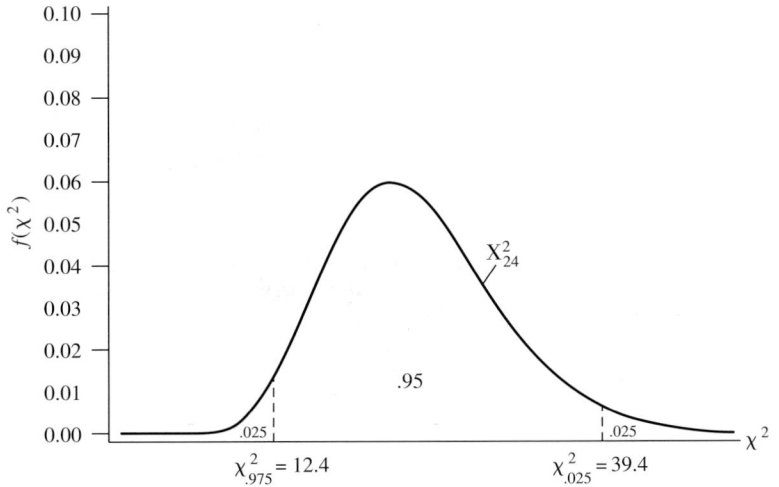

Figure 7.3
Partition of the X^2_{24} curve needed to construct a 95% confidence interval on the variance in relative I/O content of the consulting firm of Example 7.1.1.

The values of s and s^2 obtained via a statistical calculator are $s = 1.186381052$ and $s^2 = 1.4075$. Since the sample variance is reported to two more decimal places than the data, $s^2 = 1.408$. The sample standard is reported to one more decimal place than the data. Here $s = 1.19$.

Finally, we apply the formula in Theorem 7.1.2. The bounds for a 95% confidence interval on σ^2 are

$$L_1 = (n - 1)s^2/\chi^2_{.025} = 24(1.408)/39.4 = .858$$
$$L_2 = (n - 1)s^2/\chi^2_{.975} = 24(1.408)/12.4 = 2.725$$

The bounds for a 95% confidence interval on σ are

$$L_1 = \sqrt{.858} \doteq .926$$
$$L_2 = \sqrt{2.725} \doteq 1.65$$

Thus we can say that we are 95% confident that the true variance in the relative I/O content at this consulting firm lies between .858 and 2.725; we are 95% confident that the true standard deviation lies between .926 and 1.65.

7.2 ESTIMATING THE MEAN AND THE STUDENT-t DISTRIBUTION

Note that to obtain a point estimate for a population mean μ, it is not necessary to know the population variance; the sample mean $\bar{X}$ provides an unbiased estimator for μ regardless of the value of σ^2. However, the bounds for a $100(1 - \alpha)\%$ confidence interval on μ given in Sec. 6.3 are $\bar{X} \pm z_{\alpha/2}\sigma/\sqrt{n}$. It is assumed that even though the population mean is unknown, the population variance is known. Practically speaking, this assumption is not very realistic. In most instances when a statistical study is being conducted, it is being done for the first time; there is no way to know prior to the study either the mean or the variance of the population of interest. We consider in this section the more realistic problem of constructing a confidence interval on a population mean when the population variance is assumed to be *unknown*.

To derive a general formula for a $100(1 - \alpha)\%$ confidence interval on μ under these circumstances, it is natural to begin by considering the random variable used earlier, namely,

$$\frac{\bar{X} - \mu}{\sigma/\sqrt{n}}$$

There are two problems to overcome:

1. The value of σ is not known and must be estimated.
2. The distribution of the random variable obtained by replacing σ by an estimator is not known.

The first problem is easy to overcome. We shall use the sample standard deviation S as an estimator for σ. The second problem is a little more difficult to solve. When we replace σ by its estimator S, the random variable $(\bar{X} - \mu)/(S/\sqrt{n})$ results. It can be shown that the distribution of this random variable is no longer standard normal. Rather, when sampling from a normal distribution, it follows what is called a *Student-t*, or simply a T distribution. This distribution was first described by W. S. Gosset in 1908. He used the pen name "Student" because his employers, an Irish brewery, did not want their competitors to know that they were using statistical methods in their work. We pause briefly to consider this distribution.

The *T* Distribution

Definition 7.2.1 (*T* Distribution). Let Z be a standard normal random variable and let X_γ^2 be an independent chi-squared random variable with γ degrees of freedom. The random variable

$$T = \frac{Z}{\sqrt{X_\gamma^2/\gamma}}$$

is said to follow a *T* distribution with γ degrees of freedom.

This definition implies that to show that a random variable follows a *T* distribution, we must show that it can be written as a ratio of a standard normal random variable to the square root of an independent chi-squared random variable divided by its degrees of freedom.

We note here the characteristics of *T* distributions that will be useful in the work that follows:

Properties of the *T* Distribution

1. There are infinitely many *T* distributions, each identified by one parameter γ, called *degrees of freedom*. This parameter is always a positive integer. The notation T_γ denotes a *T* random variable with γ degrees of freedom.
2. Each *T* random variable is continuous. The density for a *T* random variable with γ degrees of freedom is given by

$$f(t) = \frac{\Gamma(\gamma+1)/2}{\Gamma(\gamma/2)\sqrt{\pi\gamma}} \left(1 + \frac{t^2}{\gamma}\right)^{-(\gamma+1)/2} \qquad -\infty < t < \infty$$

3. The graph of the density of a T_γ random variable is a symmetric bell-shaped curve centered at 0.
4. The parameter γ is a shape parameter in the sense that as its value increases, the variance of the random variable T_γ decreases. Thus as the value of γ increases, the bell-shaped curve associated with T_γ becomes more compact.
5. As the number of degrees of freedom increases, the bell-shaped curve associated with the T_γ random variable approaches the standard normal curve.

These ideas are illustrated in Fig. 7.4.

A partial summary of the cumulative distribution for selected values of γ is given in Table VI of App. A. The table is read just as the chi-squared table is read. That is, the degrees of freedom are listed as row headings, pertinent probabilities are listed as column headings, and the points associated with those probabilities are listed in the body of the table. We use our previous convention of denoting by t_r the point associated with the T_γ curve such that the area to the right of the point is r.

Figure 7.4
(a) Typical relationship
between two T curves with
$\gamma_1 > \gamma_2$; (b) typical
relationship between a T curve
and the standard normal curve.

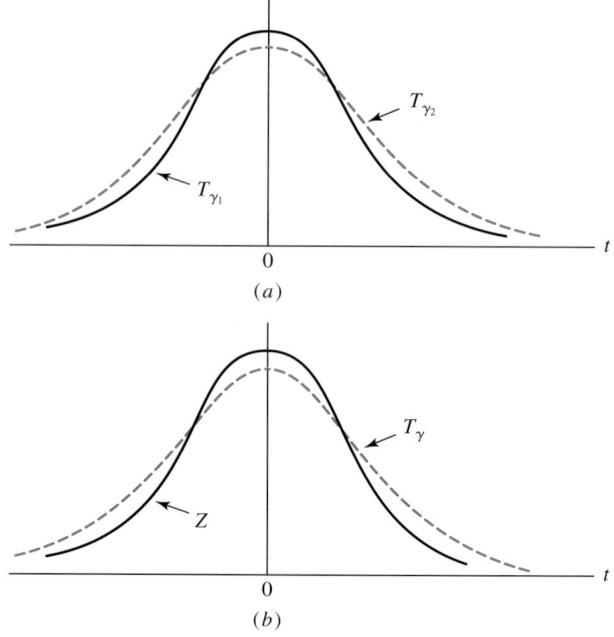

Example 7.2.1. Consider the random variable T_{10}.

1. From Table VI of App. A, $P[T_{10} \le 1.372] = F(1.372) = .90$. By our notational convention, $t_{.10} = 1.372$. [See Fig. 7.5(a).]

2. Due to the symmetry of the T curve, $t_{.90} = -t_{.10} = -1.372$.

3. The point t such that $P[-t \le T_{10} \le t] = .95$ is $t_{.025} = 2.228$. [See Fig. 7.5(b).]

The last row in Table VI of App. A is labeled ∞. The points listed in that row are actually points associated with the standard normal curve. Note that as γ increases, the values in each column of the table approach the value listed in the last row. This occurs because, for large values of γ, the graph of the density for the T_γ random variable, for all practical purposes, coincides with that of the standard normal or Z curve. For small values of γ there is quite a bit of difference between the two curves.

Let us now show that the random variable $(\bar{X} - \mu)/(S/\sqrt{n})$ follows a T distribution as claimed. The proof of this theorem depends on a result that is beyond the scope of this discussion mathematically. In particular, it can be shown that when sampling from a normal distribution, the sample mean $\bar{X}$ and the sample standard deviation S are independent. This result is not surprising. It says simply that knowledge of the center of location of a normal random variable does not contribute to knowledge of its variability. The next theorem provides the basis for the construction of a $100(1 - \alpha)\%$ confidence interval on μ when σ^2 is assumed to be unknown.

Theorem 7.2.1. Let $X_1, X_2, X_3, \ldots, X_n$ be a random sample from a normal distribution with mean μ and variance σ^2. The random variable

$$\frac{\bar{X} - \mu}{S/\sqrt{n}}$$

follows a T distribution with $n - 1$ degrees of freedom.

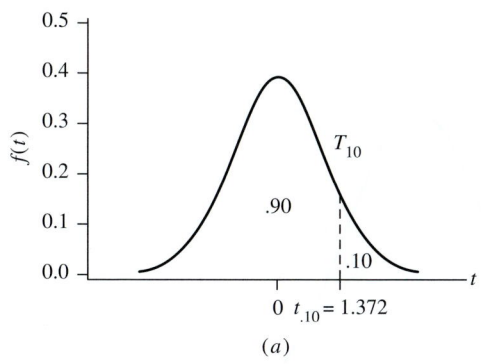

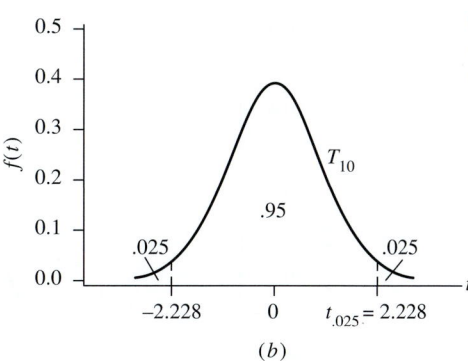

Figure 7.5
(a) $P[T_{10} \leq 1.372] = .90$;
(b) $P[-2.228 \leq T_{10} \leq 2.228] = .95$.

Confidence Interval on the Mean: Variance Estimated

It is now easy to determine the general form for a $100(1 - \alpha)\%$ confidence interval on μ when σ^2 is unknown. We need only note that the two random variables

$$Z = \frac{\bar{X} - \mu}{\sigma/\sqrt{n}} \quad \text{and} \quad T_\gamma = \frac{\bar{X} - \mu}{S/\sqrt{n}}$$

have the same algebraic structure. Thus the algebraic argument given in Sec. 6.3 will hold with σ being replaced by S and $z_{\alpha/2}$ being replaced by $t_{\alpha/2}$. These substitutions result in Theorem 7.2.2.

> **Theorem 7.2.2 [$100(1 - \alpha)\%$ Confidence interval on μ when σ^2 is unknown].** Let $X_1, X_2, X_3, \ldots, X_n$ be a random sample from a normal distribution with mean μ and variance σ^2. A $100(1 - \alpha)\%$ confidence interval on μ is given by
>
> $$\bar{X} \pm t_{\alpha/2}S/\sqrt{n}$$

Example 7.2.2 illustrates the use of this theorem.

Example 7.2.2. Sulfur dioxide and nitrogen oxide are both products of fossil fuel consumption. These compounds can be carried long distances and converted to acid before being deposited in the form of "acid rain." These data were obtained on the sulfur dioxide concentration (in micrograms per cubic meter) in a Bavarian forest thought to have been damaged by acid rain:

52.7	43.9	41.7	71.5	47.6	55.1
62.2	56.5	33.4	61.8	54.3	50.0
45.3	63.4	53.9	65.5	66.6	70.0
52.4	38.6	46.1	44.4	60.7	56.4

Given that the average concentration of sulfur dioxide concentration in undamaged areas of the country is 20 mg/m^3, is there evidence of an elevated concentration of this compound in the damaged forest?

We can answer this question by considering the confidence interval from the sample. Firstly, we find the required t values. The partition of the T_{23} curve needed to find a 95% confidence interval on the mean sulfur dioxide concentration in this forest is shown in Fig. 7.6. From Table VI of App. A, the required values are

$$t_{.975} = -t_{.025} = -2.069 \quad \text{and} \quad t_{.025} = 2.069$$

Figure 7.6
Partition of the T_{23} curve needed to construct a 95% confidence interval on the mean sulfur dioxide concentration in a Bavarian forest.

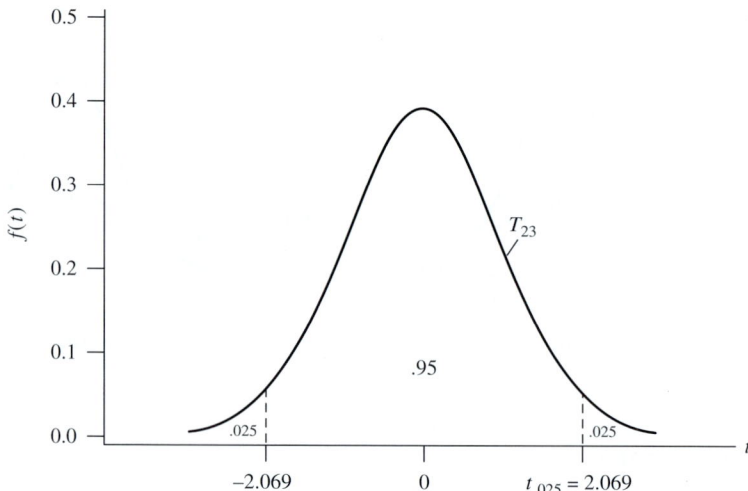

A statistical calculator yields these values:

$$\bar{x} = 53.91666667 \qquad s = 10.07371382 \qquad s^2 = 101.4797102$$

Our rounding guidelines yield

$$\bar{x} \doteq 53.92 \ \mu g/m^3 \qquad s \doteq 10.07 \ \mu g/m^3 \qquad s^2 \doteq 101.480$$

Applying the formula in Theorem 7.2.2, the confidence bounds for the interval are

$$\bar{x} \pm t_{\alpha/2} s / \sqrt{n} = 53.92 \pm 2.069(10.07)/\sqrt{24}$$

That is, we are 95% confident that the mean sulfur dioxide concentration in this forest lies in the interval [49.67, 58.17]. The average concentration of this compound in undamaged areas of the country is 20 mg/m³. Since this value is not included in the above interval, there is evidence of an elevated sulfur dioxide concentration in the damaged forest.

Several things should be pointed out. First, the number of degrees of freedom involved in finding a confidence interval on μ when σ^2 is unknown is $n - 1$, the sample size minus 1. For large samples this value may not be listed in Table VI of App. A. In this case the last row in the table (∞) is used to find points of interest. Thus in the case of large samples we are in effect estimating a desired t point via a z point. As mentioned earlier, this is appropriate, since for large samples the Z and T curves are virtually identical. Second, once again, a normality assumption has been made. We can check the validity of this assumption graphically using a stem-and-leaf diagram or a histogram. More precise methods for testing for normality are available. If there is reason to suspect that the variable under study has a distribution that is not normal and the sample size is small, then methods based on the T distribution may *not* be appropriate. Rather, some nonparametric technique should be employed. Some of these techniques are discussed in Sec. 7.7.

7.3 HYPOTHESIS TESTING

We have considered the basic ideas of estimation in some detail. Recall that in a typical estimation problem there is some population parameter, θ, whose value is to be approximated based on a sample. Usually, there is *no* preconceived notion concerning the actual value of this parameter. We are attempting simply to ascertain its value to the best of our ability. In contrast, when testing a hypothesis on θ, there is a preconceived notion concerning its value. This implies that two theories, or hypotheses, are, in fact, involved in any statistical study of this sort: the hypothesis being proposed by the experimenter and the negation of this hypothesis. The former, denoted by H_1, is called the *alternative* or *research hypothesis*; the latter is denoted by H_0 and is called the *null hypothesis.* The purpose of the experiment is to decide whether the evidence tends to refute the null hypothesis. These three guidelines help in deciding how to state H_0 and H_1:

Guidelines for Hypothesis Testing

1. When testing a hypothesis concerning the value of some parameter θ, the statement of equality will always be included in H_0. In this way H_0 pinpoints a specific numerical value that could be the actual value of θ. This value is called the *null value* and is denoted by θ_0.
2. Whatever is to be detected or supported is the alternative hypothesis.
3. Since our research hypothesis is H_1, it is hoped that the evidence leads us to reject H_0 and thereby to accept H_1.

An example will help to clarify these ideas.

Example 7.3.1. Highway engineers have found that many factors affect the performance of reflective highway signs. One is the proper alignment of the automobile's headlights. It is thought that more than 50% of the automobiles on the road have misaimed headlights. If this contention can be supported statistically, then a new tougher inspection program will be put into operation. Let p denote the proportion of automobiles in operation that have misaimed headlights. Since we wish to support the statement that $p > .5$, this contention is taken as the alternative or research hypothesis, H_1. The null hypothesis is automatically the negation of H_1, namely, $p \le .5$. Thus the two hypotheses are

$$H_0: p \le .5$$
$$H_1: p > .5$$

Note that the statement of equality appears in the null hypothesis. This pinpoints the value .5 as a possible value for p; that is, the "null value" for p is $p_0 = .5$. Note also that if H_0 is rejected, then our research hypothesis is accepted and the new inspection program will be implemented.

Once a sample has been selected and the data have been collected, a decision must be made. The decision will be either to reject H_0 or to fail to do so. The decision is made by observing the value of some statistic whose probability distribution is known *under the assumption that the null value is the true value of θ.* Such a statistic is called a *test statistic.* If the test statistic assumes a value that is rarely seen when $\theta = \theta_0$ and tends to lend credence to the alternative hypothesis, then we reject H_0 in favor of H_1; if the value observed is a commonly occurring one under the assumption that $\theta = \theta_0$, then we do not reject the null hypothesis. This means that at the end of any study we shall be forced into exactly one of the following situations:

Possible End Results for Any Test of a Hypothesis

1. We shall have rejected H_0 when it was true and shall have committed what is known as a *Type I error.*

2. We shall have made the correct decision of rejecting H_0 when the alternative, H_1, was true.

3. We shall have failed to reject H_0 when the alternative, H_1, was true. In this case we shall have committed what is known as a *Type II error.*

4. We shall have made the correct decision of failing to reject H_0 when H_0 was true.

Example 7.3.2.   In Example 7.3.1 we were testing

$$H_0: p \leq .5$$

$$H_1: p > .5 \qquad \text{(majority of automobiles in operation have misaimed headlights)}$$

If a Type I error is made, we shall have rejected H_0 when H_0 is true. Practically speaking, we shall have concluded that a majority of cars on the road have misaimed headlights when, in fact, this is not true. This error could lead to the implementation of an unnecessary inspection program. A Type II error occurs if we fail to reject H_0 when H_1 is true. In this case, the inspection program would not be implemented when, in fact, it is needed.

Note that regardless of what is done, an error is possible. Any time H_0 is rejected, a Type I error might occur; any time H_0 is not rejected, a Type II error might occur. There is no way to avoid this dilemma. The job of the statistician is to design methods for deciding whether or not to reject H_0 that keep the probabilities of making either error reasonably small.

Philosophically, there are two ways to determine whether or not to reject H_0. The first method, which we discuss in this section, is called *hypothesis testing*. This method has been used extensively in the past and is still used today. The second method, called *significance testing*, is becoming increasingly popular. It is discussed in the next section.

Hypothesis testing involves a procedure in which the values of the test statistic that lead to rejection of the null hypothesis are set before the experiment is conducted. These values constitute what is called the *critical*, or *rejection*, *region* for the test. The probability that the observed value of the test statistic will fall into this region by chance even though $\theta = \theta_0$ is called *alpha* (α), *the size of the test* or the *level of significance of the test.* If this occurs, a Type I error is committed. That is, in a hypothesis testing study, α is the probability of committing a Type I error. These ideas are summarized in Definition 7.3.1.

> **Definition 7.3.1 (Type I error and level of significance).** Consider a test of a hypothesis. A Type I error is an error that is made when the null hypothesis is rejected when, in fact, it is true. The probability of committing a Type I error is called the level of significance of the test and is denoted by the Greek letter alpha (α).

Example 7.3.3.   To test the hypothesis of Example 7.3.1,

$$H_0: p \leq .5$$

$$H_1: p > .5 \qquad \text{(majority of automobiles in operation have misaimed headlights)}$$

a random sample of 20 cars is selected and the headlights are tested. Let us design a test so that α, the probability of rejecting H_0 when p is equal to the null value of .5, is about .05. The test statistic that we shall use is X, the number of cars in the sample with misaimed headlights. If p is, in fact, equal to the null value, then X is binomial with $n = 20$, $p = .5$, and $E[X] = np = 10$. Thus if $p = .5$, then, on the average, 10 of every 20 cars tested will have misaimed headlights; if H_1 is true, this average value will be higher than 10. Logically, we should reject H_0 if the observed value of the test statistic X is somewhat larger than 10. Note from Table I of App. A that

$$P[X \geq 14|p = .5] = 1 - P[X < 14|p = .5]$$
$$= 1 - P[X \leq 13|p = .5]$$
$$= 1 - .9423$$
$$= .0577$$

Let us agree to reject H_0 in favor of H_1 if the observed value of the test statistic, X, is 14 or greater. In this way we have split the possible values of X into two sets: $C = \{14, 15, 16, 17, 18, 19, 20\}$ and $C' = \{0, 1, 2, \ldots, 13\}$. If the observed value of X lies in C, we reject H_0 and conclude that the majority of cars in operation have misaimed headlights. The set of values of the test statistic that leads to rejection of the null hypothesis C is the *critical*, or *rejection*, *region* for the test. We chose C so that the probability that the test statistic will fall into C by chance, even though $p = .5$, is .0577. That is, we designed the test so that the probability of committing a Type I error (α) is approximately .05 as desired.

There is one point to note. In the previous example we use the null value $p_0 = .5$ to determine the critical region for the test, even though the null hypothesis allows for values of p that are less than .5. It is safe to do this, since values of X that are too large to occur by chance when $p = .5$ are also certainly too large to occur by chance when $p < .5$. That is, any value of X that leads us to reject .5 as a reasonable value for p also leads us to reject any value less than .5. (See Exercise 29.)

It is possible that the observed value of the test statistic does not fall into the rejection region, even though H_0 is not true and should be rejected. If this occurs, a Type II error will be committed. The probability of this occurring is called *beta* (β). Definition 7.3.2 summarizes these ideas.

> **Definition 7.3.2 (Type II error and beta).** Consider a test of a hypothesis. A Type II error is an error that is made when the null hypothesis is not rejected when, in fact, the research theory is true. The probability of committing a Type II error is denoted by the Greek letter beta (β).

Beta is a little harder to handle than alpha, which can be dictated by the experimenter. For a particular test, β depends on the alternative. That is, β can be found only if a particular value of the alternative is specified. To illustrate, let us find β for the test designed in Example 7.3.3.

Example 7.3.4. The critical region for the test of Example 7.3.3 is $C = \{14, 15, 16, 17, 18, 19, 20\}$. Suppose that, unknown to the researcher, the true proportion of cars with misaimed headlights is .7. What is the probability that our test, as designed, is unable to detect this situation? To answer this question, we calculate β, the probability that H_0 will not be rejected given that $p = .7$. By definition

$$\beta = P[\text{Type II error}]$$
$$= P[\text{fail to reject } H_0|p = .7]$$
$$= P[X \text{ is not in the critical region}|p = .7]$$
$$= P[X \leq 13|p = .7] = .3920 \qquad (\text{Table I, App. A})$$

That is, for the test as designed there is not a very high probability that we shall be able to distinguish between $p = .5$ and $p = .7$. Beta is a function of the alternative in that if p is changed from .7 to .8, then β will change also. In this case

$$\beta = P[X \le 13 | p = .8] = .0867$$

Note that as the difference between the null value of .5 and the alternative value of p increases, β decreases.

There is one other important probability to consider. Put yourself in the position of a researcher who has put a great deal of time, effort, and money into designing and carrying out an experiment to gather evidence to support a research theory. We want the study designed in such a way that, if the research theory is true, there is a high probability that the study will show it to be true. That is, we want the probability of rejecting the null hypothesis when the research theory is true to be high. The probability of coming to this important correct decision is called the *power* of the test.

Definition 7.3.3 (Power). Consider a test of a hypothesis. The probability that the null hypothesis will be rejected when, in fact, the research theory is true is called the power of the test.

Power and beta are related. Notice that both of these probabilities are computed under the assumption that the research theory is true. In this case, we will either fail to reject the null hypothesis with probability β or we will reject the null hypothesis with probability power. Hence,

$$\beta + \text{power} = 1 \qquad \text{or} \qquad \text{power} = 1 - \beta$$

Example 7.3.5. In Example 7.3.3 we designed an experiment to test

$$H_0 : p \le .5$$
$$H_1 : p > .5$$

We discovered that if the research theory is true and $p = .7$, there is a 39.2% chance that we will not be able to detect this fact. That is, $\beta = .392$. The power of the test for detecting this alternative value of p is

$$\text{power} = 1 - \beta = 1 - .392 = .608$$

If it is important to detect the difference between a 50% rate of misaimed headlights and a 70% rate, then the test as designed will not do a very good job.

There is an obvious balancing act that must be played in designing experiments. We want both α and β to be small, and we want the power for detecting crucial differences to be high. This is accomplished in practice by choosing an appropriate sample size. Exercise 46 illustrates this idea in the context of testing a hypothesis on the average value of a distribution.

Remember that the hypothesis-testing procedure entails deciding on the level of significance (α) before the data are gathered and the test statistic is evaluated. That is, it involves presetting α. There are several reasons for wanting to do this. It gives a clear-cut way of making a decision. Once α is set, the critical region for the test is fixed also. If the observed value of the test statistic falls into this region,

Figure 7.7

| | **Actual situation** | |
Decision	H_0 true	H_1 true
Reject H_0	Type I error (probability α)	Correct decision (probability power)
Fail to reject H_0	Correct decision	Type II error (probability β)

we reject H_0; otherwise we do not. There is no room for debate after the data are gathered. Hence there can be no charge that the statisticians are manipulating the results to suit themselves. In addition, if the consequences of making a Type I error are very serious, then by presetting α we are able to specify *before the fact* exactly how large a risk we are willing to tolerate. The language underlying hypothesis testing is summarized in Fig. 7.7.

7.4 SIGNIFICANCE TESTING

In the last section we considered a method for deciding whether or not to reject a null hypothesis, called *hypothesis testing.* In this section we consider another method for doing so. This method, called *significance testing*, is coming into widespread use. This is due to its logical appeal and to the increasing use of computer packages in analyzing statistical data.

To understand why significance testing is so appealing, let us point out a bothersome aspect of hypothesis testing that might have occurred to you already. It is easy to spot the problem with a simple example. Suppose that we want to test

$$H_0: p \le .1$$
$$H_1: p > .1$$

based on a sample of size 20. The test statistic is X, the number of "successes" that are observed in the 20 trials. Since the null value is $p_0 = .1$, when $p = p_0$ the test statistic follows a binomial distribution with $E[X] = np_0 = 20(.1) = 2$. Values of X somewhat larger than 2 tend to lend credence to the alternative hypothesis. Suppose that we want α to be "very small," so we define the critical region to be $C = \{9, 10, 11, \ldots, 20\}$. For this test,

$$\alpha = P[\text{Type I error}]$$
$$= P[\text{reject } H_0 | p = p_0]$$
$$= P[X \text{ is in the critical region} | p = .1]$$
$$= P[X \ge 9 | p = .1]$$
$$= 1 - P[X < 9 | p = .1]$$
$$= 1 - P[X \le 8 | p = .1]$$
$$= 1 - .9999$$
$$= .0001$$

This is indeed a "very small value"! Now suppose that we conduct our test and observe 8 "successes." Via our rather rigid rules for hypothesis testing, we are unable to reject H_0, since 8 does not lie in the critical region. However, a little thought should make you a bit uneasy with this decision! Note that 8 is very close to 9, our rather arbitrarily selected lower boundary for the critical region. Let us see what the chances are of obtaining a value of 8 or more when $p = .1$:

$$P[X \geq 8|p = .1] = 1 - P[X < 8|p = .1]$$
$$= 1 - P[X \leq 7|p = .1]$$
$$= 1 - .9996$$
$$= .0004$$

This probability is certainly also "very small." It is hard to imagine a situation in which we would be willing to tolerate 1 chance in 10,000 of making a Type I error but would declare vehemently that 4 chances in 10,000 of making such an error is much too large to risk! There is so little difference between these probabilities that it seems a bit silly to insist that we adhere rigidly to our original cutoff point of 9.

The problem just demonstrated can be avoided by performing what is called a *significance test* rather than a hypothesis test. This method of deciding whether or not to reject H_0 entails setting up H_0 and H_1 exactly as before. However, we do not then preset α and specify a rigid critical region. Rather, we evaluate the test statistic and then determine the probability of observing a value of the test statistic at least as extreme as the value noted under the assumption that $\theta = \theta_0$. This probability is referred to by a variety of names, including the *critical level*, the *descriptive level of significance*, and the *probability*, or *P value* of the test. We use the term "*P* value" in this text. Note that the *P* value is the smallest level at which we could have preset α and still have been able to reject H_0. We reject H_0 if we consider this *P* value to be small.

Example 7.4.1. Automotive engineers are using more and more aluminum in the construction of automobiles in hopes of reducing the cost and improving gas mileage. For a particular model the number of miles per gallon obtained on the highway currently has a mean of 26 mpg with a standard deviation of 5 mpg. It is hoped that a new design, which utilizes more aluminum, will increase the mean mileage rating. Assume that σ is not affected by this change. These data are obtained during road testing:

33.8	24.3	18.8	23.7	25.3	29.6
24.9	31.5	34.4	28.0	20.5	36.7
30.3	33.5	27.4	27.6	22.5	30.7
28.6	27.1	28.8	16.5	32.7	25.2
33.1	37.5	25.1	34.5	29.5	26.8
30.0	28.4	25.6	19.8	28.9	27.7

Since our research hypothesis is taken as the alternative hypothesis, we are testing

$H_0: \mu \leq 26$

$H_1: \mu > 26$ (the new design increases gas mileage on the highway)

Since the sample mean is an unbiased estimator for the population mean, a logical test statistic is $\bar{X}$. Let us agree to reject H_0 in favor of H_1 if the observed value of the sample mean is "somewhat larger" than 26. By "somewhat larger" we mean too large to have reasonably occurred by chance if the true mean highway mileage is still 26 mpg.

The sample mean for these data is

$$\bar{x} = 28.04 \text{ mpg}$$

This value is larger than the null value for μ of 26 mpg. To see if there is enough difference to cause us to reject H_0, we find the *P* value for the test. That is, we compute the probability of observing a sample mean of 28.04 or larger if $\mu = 26$ and $\sigma = 5$.

This is done by noting if $\mu = 26$ and $\sigma = 5$, then the test statistic $\bar{X}$ is, by the Central Limit Theorem (Theorem 7.4.2), at least approximately normally distributed with mean $\mu = 26$ and standard deviation $\sigma/\sqrt{n} = 5/6$. Therefore

$$
\begin{aligned}
P[\bar{X} \geq 28.04 | \mu = 26, \sigma = 5] &= P\left[\frac{\bar{X} - 26}{(5/6)} \geq \frac{28.04 - 26}{(5/6)}\right] \\
&\doteq P[Z \geq 2.45] \\
&= 1 - P[Z \leq 2.45] \\
&= 1 - .9929 \qquad \text{(Table V, App. A)} \\
&= .0071
\end{aligned}
$$

There are two explanations for this very small probability. The null hypothesis is true, and we have observed a very rare sample that *by chance* has a large sample mean; the null hypothesis is not true, and the new process has, in fact, resulted in a higher mean mileage rating. We prefer the latter explanation! That is, we shall reject H_0 and report that the P value of our test is .0071.

There is a very easy way to deal with the difference between hypothesis tests and significance tests. For every test, simply calculate the P value. If an α level has been preset to ensure that a traditional or industry maximum acceptable level of risk is met, then compare the P value to the preset alpha value. *If $P \leq \alpha$, then we can reject the null hypothesis at the stated level of significance.* If one uses this technique, there is no need to find critical points and preset critical regions as was done in Example 7.3.3. This method is especially viable today when P values are available as a routine part of statistical packages and statistical calculator output.

Significance testing is a widely used concept. For right- or left-tailed tests the method of calculating the P value is clear. For a right-tailed test ($H_1: \theta > \theta_0$), the P value is the area to the right of the observed value of the test statistic; for a left-tailed test ($H_1: \theta < \theta_0$), it is the area to the left. However, one question still to be resolved is, "How do we compute a P value for a two-tailed test?" ($H_1: \theta \neq \theta_0$). If the distribution of the test statistic is symmetric, as it is for a Z or T statistic, then it is logical to double the apparent one-tailed P value. If the distribution is not symmetric, as with a chi-squared statistic, then presumably the two-tailed P value is nearly double the one-tailed value. This is only one of several proposed solutions to the problem, but it is the convention that we shall use.

7.5 HYPOTHESIS AND SIGNIFICANCE TESTS ON THE MEAN

One of the most commonly encountered problems is that of testing a hypothesis concerning the value of the mean. We have seen how this can be done if it is assumed that σ^2 is known. Since this assumption is usually not valid, we turn our attention to a method that can be used to test hypotheses concerning μ when σ^2 is unknown and must be estimated from the data at hand. Consider these examples.

Example 7.5.1. The maximum acceptable level for exposure to microwave radiation in the United States is an average of 10 microwatts per square centimeter. It is feared that a large television transmitter may be polluting the air nearby by pushing the level of microwave radiation above the safe limit. Since our research hypothesis is taken as the alternative, we are testing

$$H_0: \mu \leq 10$$
$$H_1: \mu > 10 \qquad \text{(unsafe)}$$

Example 7.5.2. Design engineers are working on a low-effort steering system that can be used in vans modified to fit the needs of disabled drivers. The old-type steering system required a force of 54 ounces to turn the van's 15-inch-diameter steering wheel. It is hoped that the new design will reduce the average force required to turn the wheel. In this case we are testing

$H_0: \mu \geq 54$

$H_1: \mu < 54$ (new system requires less force to operate than the old)

Example 7.5.3. A computer system currently has 10 terminals and uses a single printer. The average turnaround time for the system is 15 minutes. Ten new terminals and a second printer are added to the system. We want to determine whether or not the mean turnaround time is affected. To decide, we want to test

$H_0: \mu = 15$

$H_1: \mu \neq 15$ (the new equipment has an impact on turnaround time)

As you can see, a hypothesis on μ can take one of three general forms. With μ_0 denoting the null value of the mean, these are as follows:

Three Forms for Tests of Hypotheses on the Mean of a Distribution

I $H_0: \mu \leq \mu_0$	II $H_0: \mu \geq \mu_0$	III $H_0: \mu = \mu_0$
$H_1: \mu > \mu_0$	$H_1: \mu < \mu_0$	$H_1: \mu \neq \mu_0$
Right-tailed test	Left-tailed test	Two-tailed test

Form I is called a *right-tailed test* because when a hypothesis of this form is tested, the natural region leading to the rejection of H_0 is the upper- (or right-) tailed region of the distribution of the test statistic. This point is explained in Example 7.5.4. Similarly form II is a left-tailed test because the natural region of rejection of H_0 is the lower- (or left-) tailed region of the appropriate distribution. In a two-tailed test the critical region consists of both the lower- and upper-tail regions of the distribution of the test statistic. This is easy to remember because in a one-sided test, forms I and II, the inequality in the *alternative* hypothesis points toward the critical region.

There is one general statement to keep in mind when you test a hypothesis on any parameter:

> To test a hypothesis on a parameter θ, you must find a statistic whose probability distribution is known at least approximately under the assumption that $\theta = \theta_0$.

This statistic will serve as a test statistic. In the case at hand such a statistic is easy to find. From the discussion of Sec. 7.2 we know that if X is normal, the statistic $(\bar{X} - \mu_0)/(S/\sqrt{n})$ follows a T_{n-1} distribution. Tests based on this statistic are commonly called *T tests.*

Tests of hypotheses on μ are actually conducted by testing $H_0: \mu = \mu_0$ against one of the alternatives $\mu > \mu_0$, $\mu < \mu_0$ or $\mu \neq \mu_0$. It is safe to do this for reasons analogous to those discussed in Sec. 7.3. In particular, values of the test statistic that lead us to reject μ_0 and to conclude that $\mu > \mu_0$ will also lead us to reject any value less than μ_0; values of the test statistic that lead us to reject μ_0 and to conclude that $\mu < \mu_0$ will also lead us to reject any value greater than μ_0. For this reason, many statisticians prefer to express the three forms as

I	$H_0: \mu = \mu_0$	II	$H_0: \mu = \mu_0$	III	$H_0: \mu = \mu_0$
	$H_1: \mu > \mu_0$		$H_1: \mu < \mu_0$		$H_1: \mu \neq \mu_0$
	Right-tailed test		Left-tailed test		Two-tailed test

This emphasizes the fact that when performing a hypothesis test on μ, α is computed assuming that $\mu = \mu_0$; when performing a significance test on μ, the P value is computed under the assumption that $\mu = \mu_0$. We shall follow this notational convention in the remainder of this text.

Example 7.5.4. To determine whether a large television transmitter is polluting the nearby air (see Example 7.5.1), we intend to test

$$H_0: \mu = 10$$
$$H_1: \mu > 10$$

Notice that since the inequality associated with the *alternative* hypothesis points to the right, the test is right-tailed.

A sample of 25 readings is to be obtained at randomly selected times over a 1-week period. Our test statistic,

$$(\bar{X} - 10)/(S/\sqrt{25})$$

follows a T_{24} distribution if H_0 is true. Since $\bar{X}$ is an unbiased estimator for the mean, we expect the observed value of $\bar{X}$ to be close to 10 if H_0 is true. This forces the test statistic, $(\bar{X} - 10)$, to be small, causing the observed value of the test statistic to numerator of the be small also. However, if H_1 is true, we expect $\bar{X}$ to be larger than 10, forcing $\bar{X} - 10$ to be large and *positive*. This in turn results in a *large positive* value for the test statistic. Hence logically we should reject H_0 in favor of H_1 whenever the observed value of the test statistic is positive and too large to have reasonably occurred by chance.

Thus the natural critical region for the test is the right-tail, or upper, region of the T_{24} distribution. To decide how large a value is needed in order to reject H_0, let us preset α. If we make a Type I error, we shall shut down the transmitter unnecessarily; if we make a Type II error, we shall fail to detect a potential health hazard. We want α to be small but not so small as to force β to be extremely large. Let us choose α to be .1. The critical point for the test, read from Table VI of App. A and shown in Fig. 7.8, is 1.318. We shall reject H_0 in favor of H_1 if the observed value of the test statistic is 1.318 or larger.

When the experiment is conducted, it is found that $\bar{x} = 10.3$ and $s = 2$. The observed value of the test statistic is

$$(\bar{x} - 10)/(s/\sqrt{25}) = (10.3 - 10)/(2/5) = .75$$

Since this value falls below the critical point of 1.318, we are unable to reject H_0. These data do not support the contention that the transmitter is forcing the average microwave level above the safe limit.

We may present the arguments in a more systematic way. You will find that every hypothesis test is basically composed of four steps. They are

1. State the hypotheses. The research objective is the alternative hypothesis. The complement of alternative hypothesis is the null hypothesis.
2. Evaluate the value of the test statistic based on the data in the sample. The test statistic is selected according to the population parameter stated in the hypotheses.

Figure 7.8
Critical region for an $\alpha = .1$
level right-tailed test ($n = 25$).

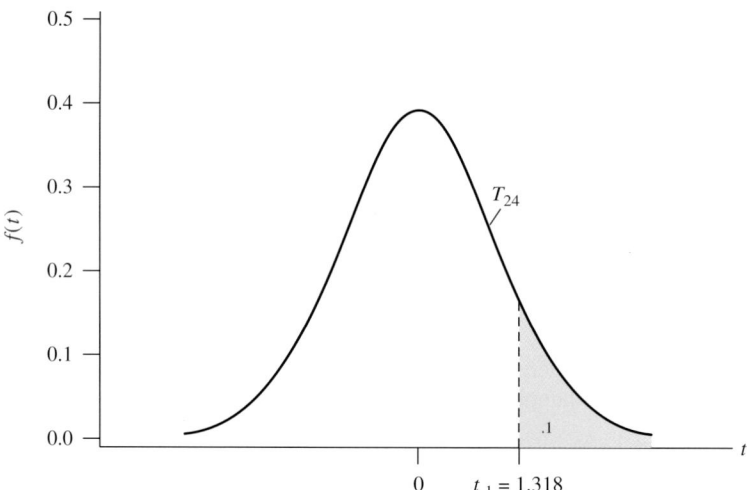

Figure 7.8
Critical region for an $\alpha = .1$
level right-tailed test ($n = 25$).

3. Find the critical value(s) according to the distribution of test statistic and the significance level α; or, find the P value of the test statistic.
4. Draw the conclusion by comparing the test statistic and critical value(s); or, draw the conclusion by comparing the P value and significance level α.

We use Example 7.5.4 to illustrate these four steps.

Example 7.5.4. (in four-step format)

[**Hypotheses**] Following Example 7.5.1, we test

$$H_0: \mu = 10$$
$$H_1: \mu \neq 10$$

[**Test statistic**] From the sample, we find that $\bar{x} = 10.3$ and $s = 2$. Applying the formula for the test statistic of T test, the value of the test statistic is

$$(\bar{x} - \mu_0)/(s/\sqrt{n}) = (10.3 - 10)/(2/\sqrt{25}) = .75$$

[**Critical value**] The degrees of freedom of the test statistic is $n - 1 = 25 - 1 = 24$. Thus, when the significance level is .1, according to T_{24} distribution, the critical value is

$$t_\alpha = t_{.1} = 1.318$$

[**Conclusion**] Since the value of test statistic is less then the critical value ($.75 < 1.318$), we are unable to reject H_0.

It should be pointed out that, in practice, it is not really necessary to find a critical point even if α has been preset. Rather, we can simply always evaluate the test statistic and find the P value. If the P value is at most equal to the preset α, then H_0 can be rejected at that α level. For instance, in the previous example α was preset at .10. The observed value of the test statistic is .75. This value and the associated P value is pictured in Fig. 7.9(*a*). From the T table with 24 degrees of freedom we see that this value lies between .685 and 1.318. [See Fig. 7.9(*b*).] The area to the right of .685 is .25; the P value is clearly smaller than .25. The area to the right of

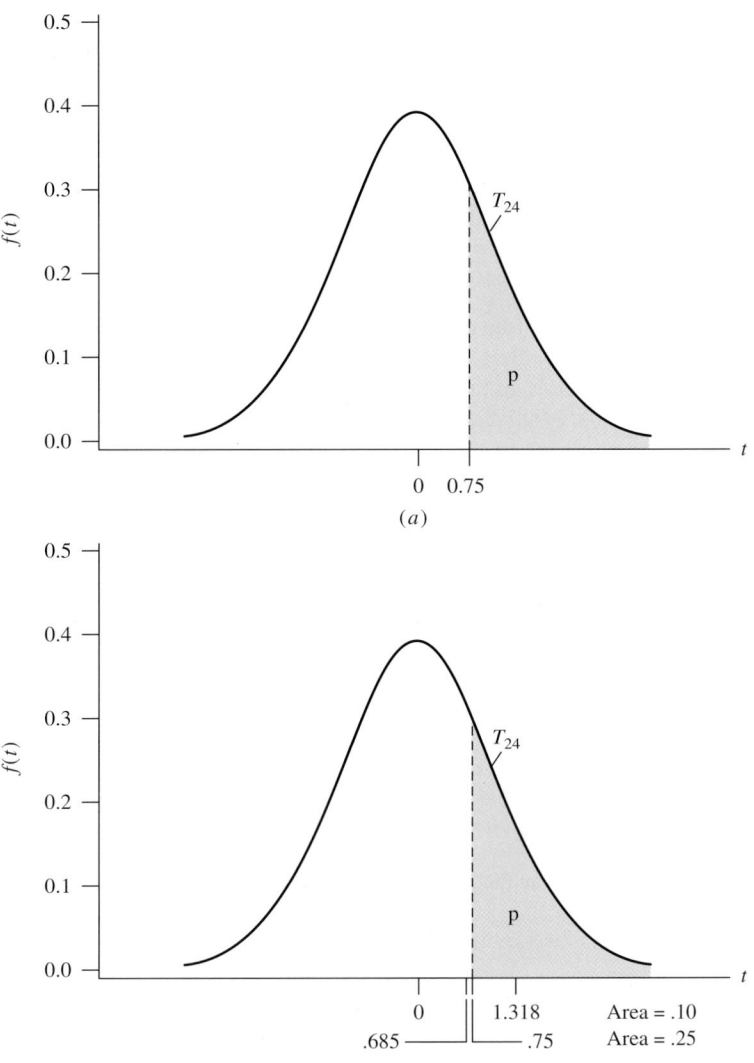

1.318 is .10; the P value is larger than this. By combining these results, we can conclude that $.10 < P < .25$. Since this P value exceeds the preset α of .05, H_0 cannot be rejected at this level. This technique is especially useful now, as the most serious data analysis is done using one of the commercially available computer packages. These packages typically report a P value automatically. To conduct a test in which α is preset, compare the report P value to α. If $P \le \alpha$, then H_0 can be rejected at the α level of significance; otherwise it cannot be rejected.

Some packages allow the user to indicate whether the test is right-, left-, or two-tailed; others automatically conduct a two-tailed test. In the latter case, the true P value is half that reported by the computer. Check the package documentation to be sure that you understand exactly what probability is being given.

The next example illustrates the use of significance testing in testing a two-tailed hypothesis.

Example 7.5.5. In studying the effect of adding 10 new terminals and one printer to an existing computer system (see Example 7.5.3), a sample of size 30 yields $\bar{x} = 14.0$ and $s = 3$. Suppose that the significance level is .1. We want to know whether the new equipment has an impact on turnaround time.

[Hypotheses] Since we do not have any preconceived notion as to whether the new equipment increases or decreases the mean turnaround time, we are conducting a two-tailed test. We are testing

$$H_0: \mu = 15$$
$$H_1: \mu \neq 15 \qquad \text{(the new equipment has an impact on turnaround time)}$$

That is, we shall reject H_0 in favor of H_1 if the observed value of the test statistic is too large in either the positive or negative sense to have occurred by chance.

[Test statistic] Applying the formula for the test statistic of T test, the observed value of the test statistic is

$$(\bar{x} - 15)/(s/\sqrt{30}) = (14 - 15)/(3/\sqrt{30}) \doteq -1.83$$

[Critical values] The degrees of freedom of the test statistic is $n - 1 = 30 - 1 = 29$. Thus, when the significance level is .1, according to T_{29} distribution, the critical values are

$$\pm t_{\alpha/2} = \pm t_{.05} = \pm 1.6999$$

[Conclusion] Since the value of test statistic is less than the left critical value ($-1.83 < -1.699$), we reject H_0 and conclude that the new equipment has an impact on turnaround time.

We can also estimate the P value of the test. From Table VI of App. A we see that

$$P[T_{29} \leq -1.699] = .05 \qquad \text{and} \qquad P[T_{29} \leq -2.045] = .025$$

Since -1.83 lies between -1.699 and -2.045, the probability of observing a value as large in the negative sense as that observed lies between .025 and .05. However, we were running a two-tailed test. This means that the P value of the test is the probability of observing a value as extreme as that observed in either the *positive or the negative* sense. That is, the P value is assumed to be double that computed above. We can report that for this test $.05 < P < .1$.

It should be emphasized that the statistic $(\bar{X} - \mu_0)/(S/\sqrt{n})$ follows the T_{n-1} distribution if X is *normal*. If X is not normal, then care must be taken. It has been found that for samples of moderate to large size ($n \geq 25$), violating this assumption does not seriously affect the distribution of the test statistic in that the probability of committing a Type I and a Type II error is not appreciably changed [6]. This property is called *robustness*. However, if the sample size is small, then T tests should not be run on nonnormal data. Many statistical software packages include some sort of test of normality. In this case, we are testing

H_0: data are drawn from a distribution that is normally distributed

H_1: data are drawn from a distribution that is not normally distributed

The results of such a test along with sample size considerations can be used to decide whether to proceed with a T test or turn to one of the nonparametric tests described in Sec. 7.7.

7.6 HYPOTHESIS TESTS ON THE VARIANCE

We now turn our attention to testing hypotheses on the value of σ^2 or σ. These tests take the same general form as tests on the mean. These are summarized below with σ_0^2 denoting the null value of the population variance.

Three Forms for Tests of Hypotheses on the Variance of a Distribution

I $H_0: \sigma^2 = \sigma_0^2$	II $H_0: \sigma^2 = \sigma_0^2$	III $H_0: \sigma^2 = \sigma_0^2$
$H_1: \sigma^2 > \sigma_0^2$	$H_1: \sigma^2 < \sigma_0^2$	$H_1: \sigma^2 \neq \sigma_0^2$
Right-tailed test	Left-tailed test	Two-tailed test

The test statistic used to test each of these is $(n-1)S^2/\sigma_0^2$. When sampling from a normal distribution, this statistic is known to follow a chi-squared distribution with $n-1$ degrees of freedom provided $\sigma^2 = \sigma_0^2$. As expected, the critical regions for right- and left-tailed tests are the upper- and lower-tail regions of the X_{n-1}^2 distribution, respectively; the critical region for the two-tailed test consists of both the upper- and lower-tail regions of the distribution.

Example 7.6.1. One random variable studied while designing the front-wheel-drive half-shaft of a new model automobile is the displacement (in millimeters) of the constant velocity (CV) joints. With the joint angle fixed at 12°, 20 simulations were conducted, resulting in the following data:

6.2	1.9	4.4	4.9	3.5
4.6	4.2	1.1	1.3	4.8
4.1	3.7	2.5	3.7	4.2
1.4	2.6	1.5	3.9	3.2

Engineers designing the front-wheel-drive half-shaft claim that the standard deviation in the displacement of the CV shaft is less than 1.5 millimeters. The estimated standard deviation based on the given 20 observations is 1.41 millimeters. Do these data support the contention of the engineers? Assume that the significance level $\alpha = .05$.

[**Hypotheses**] To answer this question, we test

$$H_0: \sigma = 1.5$$
$$H_1: \sigma < 1.5$$

This is equivalent to testing

$$H_0: \sigma^2 = (1.5)^2$$
$$H_1: \sigma^2 < (1.5)^2$$

[**Test statistic**] For these data $s = 1.41$. The observed value of the test statistic is

$$\frac{(n-1)s^2}{\sigma_0^2} = \frac{19(1.41)^2}{(1.5)^2} = 16.79$$

[**Critical value**] Since the test is left-tailed, we reject H_0 if this value is too small to have occurred by chance when H_0 is true. The degree of freedom is $n-1 = 20-1 = 19$. From the X_{19}^2 distribution, the critical value is

$$\chi_{.05}^2 = 30.14$$

[**Conclusion**] Since the critical value is greater than the value of test statistic (30.14 > 16.79), we are unable to reject H_0. These data did not support the claim that $\sigma < 1.5$ millimeters.

The P value is estimated as follows. From the chi-squared table we see that

$$P[X_{19}^2 \leq 14.6] = .25 \qquad \text{and} \qquad P[X_{19}^2 \leq 18.3] = .50$$

Since the observed value of the test statistic, 16.79, lies between 14.6 and 18.3, the P value of the test lies between .25 and .50.

Recall that when sample sizes are moderate to large ($n \geq 25$), the T statistic can be used to make inference on μ even though the normality assumption may be violated. It is when sample sizes are small that this becomes a serious problem. Unfortunately, the same cannot be said concerning the use of the X_{n-1}^2 statistic for making inferences on σ^2 and σ. For this reason, when constructing confidence intervals on σ^2 or testing hypotheses on the value of this parameter, a check for normality must be made. If the data are nonnormal then these methods should not be used.

7.7 ALTERNATIVE NONPARAMETRIC METHODS

We have seen how to use the Z and T statistics to test hypotheses concerning the mean of a normal distribution. The procedures presented assume that either we are sampling from a normal distribution or sample sizes are large enough so that deviations from the normality assumption do not seriously affect our results. In reality, experimenters often obtain data for which it is clearly unreasonable to assume an underlying normal distribution and for which sample sizes are small. When this occurs, usually the experimenter is advised to use a "nonparametric" test for location rather than the usual Z or T test. In this section we examine the meaning of the term "nonparametric" test. We also present some nonparametric alternatives for the usual Z and T tests for location.

The terms "nonparametric" and "distribution free" are often used interchangeably. When we use the term "nonparametric test," we shall mean a test with the property that no assumption is being made concerning the specific distribution from which the sample is drawn. Although we usually assume that the distribution is continuous, we do not have to specify the family to which the random variable under study belongs. In particular, we shall no longer have to assume that the random variable being studied is normally distributed. Hence nonparametric methods are applicable to a larger class of distributions than their normal theory analogs.

When comparing two statistical procedures designed to test essentially the same thing, we look at two characteristics: the probability of committing a Type I error and the power of the test. We want α to be small, but at the same time we want a high probability of rejecting a false null hypothesis. Typically, for a fixed α level the normal theory procedures are more powerful than their nonparametric counterparts *when the assumptions underlying the normal theory test are met*. However, studies have shown that when these assumptions are not met, the use of normal theory procedures leads to tests that are approximate in the sense that the apparent α level is suspect. For example, if we run a chi-squared test for variance on data that is far from normal at an apparent α level of .05, the actual probability of rejecting a true null hypothesis may be far from .05. In some cases the approximations are excellent, but in others they are so bad as to be completely unacceptable. In any case, using a normal theory procedure in situations in which the normal theory assumptions are not valid is dangerous. In such cases we turn to nonparametric procedures. These methods are usually superior for analyzing data when the normal theory assumptions are not met; they compare very favorably to the normal theory tests even when the normal theory assumptions are met. The safe course of action is to follow the advice: when in doubt use a nonparametric test!

In this section we shall discuss the sign test and the Wilcoxon signed-rank test, both of which can be used to test for location in the form of population medians.

Sign Test for Median

Recall that for a continuous distribution the median for a random variable X is defined to be the value M such that

$$P(X < M) = P(X > M) = 1/2$$

That is, the median is the 50th percentile of the distribution. For a symmetric distribution such as the normal, the population mean and median are identical. We shall see that the sign test is simply a form of the binomial test, which was discussed in Sec. 7.3. Let X denote a continuous random variable with median M and let $X_1, X_2, \ldots, X_n$ denote a random sample of size n from this unspecified distribution. If M_0 denotes the hypothesized value of the population median, then the usual forms of the hypothesis to be tested can be stated as follows:

Three Forms for Tests of Hypotheses on the Median of a Distribution		
$H_0: M = M_0$	$H_0: M = M_0$	$H_0: M = M_0$
$H_1: M > M_0$	$H_1: M < M_0$	$H_1: M \neq M_0$
Right-tailed test	Left-tailed test	Two-tailed test

Under the assumption of a continuous distribution, each of the differences $X_i - M_0$ has probability 1/2 of being positive, probability 1/2 of being negative, and probability 0 of being zero.

Let Q_+ denote the number of positive differences obtained. If H_0 is true, Q_+ is binomially distributed with parameters n and 1/2 and the expected value of Q_+ is $n/2$. That is, if H_0 is true, half the differences should be positive and the rest are negative. Note that in running a left-tailed test we want to detect a situation in which the true median M lies below the hypothesized median M_0. If this is true, we expect more than half the differences to be negative. This creates fewer positive differences than expected. Thus a logical procedure is to reject $H_0: M = M_0$ in favor of $H_1: M < M_0$ if the observed value of Q_+ is too small to have occurred by chance. In conducting a right-tailed test, the situation is reversed. In this case we reject $H_0: M = M_0$ in favor of $H_1: M > M_0$ if the observed value of Q_-, the number of negative differences obtained, is too small to have occurred by chance. A two-tailed test is conducted by rejecting $H_0: M = M_0$ in favor of $H_1: M \neq M_0$ if the smaller of Q_+ and Q_- is too small to have occurred by chance. The next example illustrates the use of the sign test.

Example 7.7.1.   A standard method for completing a task on an assembly line yields a median completion time of 55 seconds. A new procedure is developed that should reduce the median time required. In order to test the new procedure, 15 subjects are asked to complete the task, and these observations are obtained on the random variable X, the time required:

35	65	48	40	70	50	58	36
47	41	49	39	34	33	31	

Suppose that the significance level $\alpha = .05$. Do we have evidence that the new procedure reduces the median time required to complete the task?

[**Hypotheses**] We want to test

$$H_0: M = 55$$
$$H_1: M < 55$$

The stem-and-leaf diagram for these data is shown in Fig. 7.10. Note that the diagram does suggest that X is not normally distributed. Since the sample size is rather small, we shall test for location using the nonparametric sign test.

[**Test statistic**] The test is left-tailed. Hence the test statistic is Q_+, the number of positive differences obtained when 55 is subtracted from each observation. From the stem-and-leaf diagram it is easy to see that only three observations exceed 55. Thus the observed value of the test statistic

$$Q_+ = 3$$

[**The P value**] Since the test statistic Q_+ follows binomial distribution, we do not find the critical value. Rather, we find the P value of the test, which is the probability of seeing a value this small or smaller under the assumption that Q_+ is binomially distributed with $n = 15$ and $p = 1/2$. From Table I of App. A,

$$P = P[Q_+ \leq 3 | n = 15, p = 1/2] = .0176$$

[**Conclusion**] Since this P value is small ($.0176 < .05 = \alpha$), we reject H_0. We do have strong statistical evidence that the new procedure reduces the median time required to complete the task.

Since we assume that the underlying distribution is continuous, theoretically zero differences should not occur when conducting a sign test. However, as you might guess, sometimes zeros do occur in practice. These occur for various reasons, but the primary problem is the lack of instruments capable of precise measurement of continuous phenomena such as time, length, speed, and volume. Treatment of zero differences has been considered extensively. Various recommendations as to how to treat those differences have resulted. These are our recommendations:

Handling Zeros in a Sign Test

1. Assign to the zero differences the algebraic sign least conducive to the rejection of the null hypothesis. Thus for a left-tailed test we would consider zero differences to be positive; for a right-tailed test they would be considered to be negative. In a two-tailed test we assign to zero differences the algebraic sign of

Figure 7.10
Stem-and-leaf diagram for the time required to complete a task on an assembly line: diagram suggests a nonnormal population.

3	569431
4	80719
5	08
6	5
7	0

the less frequently occurring difference. For example, if one observed 3 negative signs, 15 positive signs, and 6 zeros in running a two-tailed test, then the 6 zeros would all be treated as though they were negative. This procedure makes sense because a zero difference supports the null hypothesis that $M = M_0$. The suggested technique gives the null hypothesis the benefit of the doubt by making it harder to reject H_0.

2. If the number of zeros is small relative to the sample size n, discard these differences and reduce the sample size accordingly.

Occasionally a situation arises in which the differences $X_i - M_0$ are such that we can observe the algebraic sign of each difference but not its magnitude. In this case, the sign test is about the only choice available for testing location. Exercise 53 is an example of this type of problem. Usually, the actual numerical value of the differences can be obtained. Unfortunately, the sign test does not make use of this additional information. It treats a negative difference of $-.1$ in exactly the same way as it does a negative difference of -1000. For data in which the actual differences can be found, a second nonparametric test is available for testing for location. This test, the Wilcoxon signed-rank test, makes use of both the sign and magnitude of the observed differences $X_i - M_0$.

Wilcoxon Signed-Rank Test

In this test we assume that $X_1, X_2, \ldots, X_n$ is a random sample of size n from a continuous distribution that is symmetric about an unknown median M. Consider the set of differences $X_i - M_0$, $i = 1, 2, 3, \ldots, n$, where M_0 is the hypothesized median of the distribution from which the sample is drawn. The null hypothesis to be tested is $H_0: M = M_0$ versus the usual alternatives $H_1: M > M_0$, $H_1: M < M_0$, or $H_1: M \neq M_0$. If H_0 is true, the differences $X_i - M_0$ are drawn from a distribution that is symmetric about zero. It is assumed that the differences are such that the magnitude as well as the algebraic sign of each can be obtained. To conduct the test, we form the set of n absolute differences $|X_i - M_0|$. These are then ranked from 1 to n in order of absolute magnitude, with the smallest absolute difference receiving a rank of 1. These ranks, which we denote by $R_1, R_2, \ldots, R_n$, are then assigned the algebraic sign of the difference score that generated the rank. If H_0 is true, then each rank is just as likely to be assigned a positive sign as a negative one. Consider the statistics

Wilcoxon Test Statistic

$$W_+ = \sum_{\substack{\text{all} \\ \text{positive} \\ \text{ranks}}} R_i \quad \text{and} \quad |W_-| = \sum_{\substack{\text{all} \\ \text{negative} \\ \text{ranks}}} |R_i|$$

If H_0 is true, then we should expect W_+ and $|W_-|$ to be approximately equal. If $M > M_0$, then W_+ would tend to be too large and $|W_-|$ too small. Similarly, if $M < M_0$, we would expect the reverse to be true. Hence, we define our test statistic to be $W = \min(W_+, |W_-|)$. The exact distribution of W has been tabled for various values of the sample size n and significance level α. One such table is Table VIII of App. A. Using this table, we reject H_0 if the observed value of W is less than or equal to the stated critical value.

In practice, ties in the difference scores $X_i - M_0$ can occur. If ties occur, the values for each tied group should be given the midrank of the group. For example, suppose that we observe difference scores of 3, −3, and 3, which should occupy ranks 8, 9, and 10. We would assign each of the three values a rank of 9 and then assign the next largest difference score a rank of 11. Example 7.7.2 illustrates the idea.

Example 7.7.2. The melting point for a new lightweight material designed for use in automobile interiors is being investigated. It is known that due to impurities in the material, the melting point is a random variable uniformly distributed over a small temperature interval. It is thought that the median melting point is less than 120° C. Do these data support this contention at .05 significance level?

115.1	117.8	116.5	121.0
120.3	119.0	119.8	118.5

[Hypotheses] We are testing

$$H_0: M = 120$$

$$H_1: M < 120$$

[Test statistic] We first subtract 120 from each observation and then find the absolute value of each difference.

x_i	115.1	120.3	117.8	119.0	116.5	119.8	121.0	118.5
$x_i - 120$	−4.9	.3	−2.2	−1.0	−3.5	−.2	1.0	−1.5
$|x_i - 120|$	4.9	.3	2.2	1.0	3.5	.2	1.0	1.5

We next rank these absolute differences from 1 to 8. Note that the value 1.0 occurs twice in what would normally be positions 3 and 4. We assign a rank of 3.5 to each of these values. The algebraic sign attached to each rank is the same as that of the difference that generated the rank.

$|x_i - 120|$	4.9	.3	2.2	1.0	3.5	.2	1.0	1.5
Rank	8	2	6	3.5	7	1	3.5	5
Signed rank	−8	2	−6	−3.5	−7	−1	3.5	−5

For these data

$$W_+ = \sum_{\substack{\text{all} \\ \text{positive} \\ \text{ranks}}} R_i = 2 + 3.5 = 5.5$$

$$|W_-| = \sum_{\substack{\text{all} \\ \text{negative} \\ \text{ranks}}} |R_i| = 8 + 6 + 3.5 + 7 + 1 + 5 = 30.5$$

Since the test is a left-tailed test, the test statistic is W_+. We reject H_0 if the observed value of this statistic is too small to have occurred by chance.

[Critical value] From Table VIII of App. A with $n = 8$ we see that that the critical value is 6.

[Conclusion] Since the value of test statistic is less than the critical value (5.5 < 6), we reject H_0 and conclude that the median melting point of this material is below 120° C.

We can estimate the P value of the test. We can reject H_0 at the $\alpha = .05$ level (critical point = 6), but we are unable to reject H_0 at $\alpha = .025$ (critical point = 4). Thus the P value of the test lies between .025 and .05.

If the sample size n exceeds values given in Table VIII of App. A, a large sample normal approximation may be used.

The following theorem states the approximate distribution of the Wilcoxon signed rank statistic:

> **Theorem 7.7.1 (Approximate Distribution of W).** Let W denote the Wilcoxon signed rank statistic. For large sample sizes, W is approximately normally distributed with mean
>
> $$E[W] = \frac{n(n+1)}{4}$$
>
> and variance
>
> $$\text{Var } W = \frac{n(n+1)(2n+1)}{24}$$

To use this theorem, we simply standardize W by subtracting its mean and dividing by its standard deviation. P values can then be found via the standard normal or Z table. This approach can be used when sample sizes exceed those listed in Table VIII of App. A. Exercise 57 illustrates this approximation procedure.

The Wilcoxon signed-rank test is almost as sensitive to departures from the null hypothesis as the normal theory T test even when the underlying distribution is normal. For other symmetric distributions the signed-rank test is usually more powerful than the T test. Hence this test should be considered a strong competitor to the T test for practical problems. This is particularly true for small samples where violations of the normal theory tests assumptions are of greatest concern.

Note that although a Wilcoxon signed-rank test does not assume normality, it does assume symmetry. Procedures have been developed to test the validity of this assumption. One such test is given in [20].

CHAPTER SUMMARY

In this chapter we considered confidence interval estimation of the variance and standard deviation of a normal distribution. We also considered interval estimation of a mean when the population variance is unknown. This procedure entails the use of the Student-t or T distribution. We discussed this new continuous distribution in detail and saw that its properties are similar to those of the Z or standard normal distribution. In particular, we saw that for large sample sizes t points are well approximated by z points.

We next turned our attention to methods used in testing a statistical hypothesis. We found that we are always dealing with two hypotheses, the null hypothesis H_0 and its alternative H_1. The point of view of the researcher is stated as the alternative hypothesis. Thus we hope that our data will allow us to reject H_0, thereby accepting H_1. We design our tests in such a way so that we always know the probability of rejecting a true null hypothesis. We found that we are always subject to error when testing a hypothesis. If we reject a true null hypothesis, we commit a Type I error; if we fail to reject a false null hypothesis, a Type II error is committed. Two methods were described for deciding whether or not to reject H_0. The first method is referred to as hypothesis testing. In conducting a hypothesis test, we preset α. This is done by setting up a rejection or critical region prior to data collection. We reject H_0 if the observed value of the test statistic falls into this critical region. The second method for deciding whether to reject H_0 is called significance testing. Here no critical region is set prior to data gathering. Rather, we evaluate the test statistic and find the probability or P value of the test. The P value is the probability of observing a value of the test statistic as unusual or more unusual than that observed if the null value of the parameter θ is correct. Thus the P value is the smallest value at which we could have preset α and still have been able to reject H_0. We reject H_0 if the P value is deemed to be small. There are advantages and disadvantages to each method. You should be familiar with both as they are both used extensively.

We considered in some detail what are commonly called T *tests*. These are tests specifically designed to test a hypothesis on the mean of a normal distribution. We saw that these tests require that sampling be from a normal distribution and that this restriction is especially important for small samples. In Sec. 7.7 we presented some nonparametric alternatives to the T test if the normality assumption appears to be invalid. Nonparametric tests are tests that make no assumption as to the family of distribution from which sampling is done.

Finally, we considered a method for testing a hypothesis on the variance or standard deviation of a normal distribution.

We introduced and discussed many new important terms and concepts that you should know. Some of these are:

Student-t distribution	Null hypothesis
Alternative hypothesis	Research hypothesis
Null value	Test statistic
Type I error	Type II error
α	β
Power	Size of test
Critical or rejection region	Level of significance
Significance test	Hypothesis test
Probability or P value	Critical level
Descriptive level of significance	Right-tailed test
Left-tailed test	Two-tailed test
Nonparametric test	Median

EXERCISES

Section 7.1

1. When programming from a terminal, one random variable of concern is the response time in seconds. These data are obtained for one particular installation:

1.48	1.26	1.52	1.56	1.48	1.46
1.30	1.28	1.43	1.43	1.55	1.57
1.51	1.53	1.68	1.37	1.47	1.61
1.49	1.43	1.64	1.51	1.60	1.65
1.60	1.64	1.51	1.51	1.53	1.74

(a) Construct a stem-and-leaf diagram. Does the assumption of normality appear reasonable?

(b) Find the unbiased point estimate for σ^2.

(c) Find a 95% confidence interval on σ^2.

(d) Find a 95% confidence interval on σ.

(e) Would you be surprised to hear the director of this installation claim that the standard deviation in response time is more than .2 second? Explain.

2. Highway engineers have found that the ability to see and read a sign at night depends in part on its "surround luminance." That is, it depends on the light intensity near the sign. These data are obtained on the surround luminance

(in candela per square meter) of 30 randomly selected highway signs in a large metropolitan area (based on "Use of Retroreflectors in the Improvement of Nighttime Highway Visibility," H. Waltman, *Color*, 1990, pp. 247–251):

10.9	1.7	9.5	2.9	9.1	3.2
9.1	7.4	13.3	13.1	6.6	13.7
1.5	6.3	7.4	9.9	13.6	17.3
3.6	4.9	13.1	7.8	10.3	10.3
9.6	5.7	2.6	15.1	2.9	16.2

(*a*) Find the sample variance for these data.

(*b*) Assume that the data are drawn from a normal distribution. Find a 90% confidence interval on the variance in the surround luminance in this area.

(*c*) Find a 90% confidence interval on the standard deviation in surround luminance.

(*d*) The normal probability rule (Sec. 4.5) implies that a normal random variable will lie within two standard deviations of its mean with probability .95. Use $\bar{X}$ and S to estimate the mean and standard deviation of the surround luminance in this area. Would it be unusual for the surround luminance for a randomly selected sign to exceed 18 cd/m² ? Explain.

3. X-ray microanalysis has become an invaluable method of analysis. With the electron microprobe, both quantitative and qualitative measures can be taken and analyzed statistically. One method for analyzing crystals is called the two-voltage technique. These measurements are obtained on the percentage of potassium present in a commercial product which theoretically contains 26.6% potassium by weight:

21.9	23.4	22.1	22.1	24.7	24.6
24.0	24.1	24.2	26.5	23.8	25.3
24.8	24.8	24.5	27.8	24.9	
27.2	25.1	25.5	23.7	26.5	
22.0	26.7	25.2	23.1	25.4	

(*a*) Check the reasonableness of the normality assumption by constructing a stem-and-leaf diagram for these data.

(*b*) Find the sample variance for these data.

(*c*) Find a 99% confidence interval for σ^2.

(*d*) Find a 99% confidence interval for σ. Note that this confidence interval is fairly long. Suggest a way to improve the interval estimate for σ based on these data. Try your suggestion to see if the new estimate is more informative than that given by the 99% confidence interval.

4. (*One-sided confidence interval on σ^2.*) Since variance is a measure of consistency, it is usually hoped that σ^2 will be small. For this reason, it is sometimes useful to construct what is called a *one-sided* confidence interval for σ^2. That is, we want to find an interval of the form $[0, L]$, where L is a statistic with the property that $P[\sigma^2 \le L] \doteq 1 - \alpha$. The formula for such an interval is

$$L = (n - 1)S^2/\chi^2_{1-\alpha}.$$

The point $\chi^2_{1-\alpha}$ is the lower-tailed chi-squared point with α area to its left and $1 - \alpha$ to the right. For example, to construct a 95% one-sided confidence interval on σ^2 the chi-squared point used would be that with $n - 1$ degrees of freedom and .05 area to the left. Use these data on X, the actual length of 63-mm nails, to find a 95% one-sided confidence interval on the variance in length:

63.0	63.1	63.0	63.0	62.9	63.0	63.0
63.1	62.8	63.1	63.1	63.0	62.9	63.2

The manufacturer wants to check to be sure that the population variance of the nails being produced does not exceed .03. Does this sample indicate that this is the case? Explain.

5. Robotic technology is an area of rapid growth. It was reported that 315,000 industrial robots would be in use in American industry by the year 1995. One important feature of a robot is its accuracy. In a study of a particular robot used to apply adhesive to a specified location, these data are obtained on the error (in inches) in the placement of the adhesive:

.001	.002	.003	.002	.002
.007	.003	.004	.003	.006
.006	.003	.005	.004	.004
.001	.008	.001	.004	.003
.001	.003	.003	.005	.006

(a) Construct a stem-and-leaf diagram. Does the assumption that the placement error is normally distributed appear reasonable?
(b) Find the sample variance for these data.
(c) Use Exercise 4 to find 90% one-sided confidence intervals on σ^2 and σ.
(d) This robot is acceptable if its standard deviation does not exceed .005 inch. Does this criteria appear to be met? Explain.

6. In Theorem 6.1.3 we showed that the sample variance is an unbiased estimator for σ^2 regardless of the distribution of the random variable X. If X is normal, this property is obtained more easily by making use of Theorem 7.1.1 and the properties of the chi-squared distribution given in Sec. 4.3. Use these results to show that for a normal random variable X, $E[S^2] = \sigma^2$ and Var $S^2 = 2\sigma^4/(n-1)$.

7. Recent research indicates that heating and cooling commercial buildings with groundwater-source heat pumps is economically sound. The crucial random variable being studied is the water temperature. A sample of 15 wells in the state of California yields a sample standard deviation of 7.5° F. Find a 95% confidence interval on the standard deviation in temperature of wells in California.

8. In pouring glass for use in automobile windshields uniformity of thickness is desirable to prevent distortion. Find a 95% one-sided confidence interval on the standard deviation in thickness if a sample of 10 windshields yields a sample standard deviation of 0.01 inch.

Section 7.2

9. Use the T table to find each of these points:
(a) $t_{.05}\ (\gamma = 8)$;
(b) $t_{.95}\ (\gamma = 8)$;
(c) $t_{.975}\ (\gamma = 12)$;
(d) $t_{.025}\ (\gamma = 12)$;
(e) $t_{.05}\ (\gamma = 121)$;
(f) $t_{.05}\ (\gamma = 150)$;
(g) Point t such that $P[-t \le T_{25} \le t] = .90$;
(h) Point t such that $P[-t \le T_{25} \le t] = .95$;
(i) Point t such that $P[T_{15} \ge t] = .05$;
(j) Point t such that $P[T_{20} \ge t] = .10$;
(k) Point t such that $P[T_{16} \le -t] = .05$;
(l) Point t such that $P[T_{30} \le -t] = .10$.

10. The "supergopher" is a device invented to drill through arctic pack ice. It is a cone-shaped apparatus 5 feet high, 4 feet wide, and wound with a copper coil. Water heated to 180° F is pumped through the coil. This allows the gopher to melt a vertical round shaft through the ice. Let X denote the distance or depth that the gopher can drill per hour. These data are obtained on 10 test holes (depth is in feet):

2.0	1.7	2.6	1.5	1.4
2.1	3.0	2.5	1.8	1.4

   (a) Use these data to find $\bar{x}$, s^2, and s.
   (b) Find a 90% confidence interval on the average distance that can be drilled in an hour. (Based on information from "The Lost Squadron," by Steven Petrow, *LIFE*, December, 1992.)

11. Metal conduits or hollow pipes are used in electrical wiring. In testing 1-inch pipes, these data are obtained on the outside diameter (in inches) of the pipe:

1.281	1.293	1.287	1.286
1.288	1.293	1.291	1.295
1.292	1.291	1.290	1.296
1.289	1.289	1.286	1.291
1.291	1.288	1.289	1.286

   (a) Find $\bar{x}$, s^2, and s for this sample.
   (b) Assume that sampling is from a normal distribution. Find a 95% confidence interval on the mean outside diameter of pipes of this type.
   (c) The makers of this type of pipe claim that the mean outside diameter is 1.29 inches. Does the confidence interval lead you to suspect this reported figure? Explain.

12. Lightweight hand-held, laser rangefinders are now used by civil engineers in hydrographic surveys. In testing one brand of rangefinder these data are obtained on the error (in meters) made in locating an object at a distance of 500 meters:

−.10	−.02	.10	−.03	.09
.01	−.05	.05	−.06	.01
.03	.06	.02	−.07	.03

   (a) Find point estimates for the mean and standard deviation in the error made by the laser.
   (b) Assume that these measurement errors are normally distributed. Find a 90% confidence interval on the mean measurement error.
   (c) A competitor claims that this particular model, on the average, overestimates the distance by at least .05 meter. Is there reason to doubt the claim based on the observed data? Explain.
   (d) Based on the normal probability rule (Sec. 4.5), would you consider it unusual for a single measurement error to be in excess of .15 meter? Explain.

13. One of the classic problems of operations research is the vehicle routing problem (VRP). This problem entails studying a system consisting of a given number of customers with known locations and demand for a commodity who are being supplied from a single depot by a number of vehicles with known capacity. The object of the study is to route the vehicles in such a way that the total distance traveled is minimized. The characteristics of a new algorithm are

being investigated. These data are obtained on the cpu time required to solve the problem:

2.0	1.4	3.5	2.3	3.2	3.6
.1	3.5	2.2	2.1	2.4	1.5
2.2	2.3	2.7	1.9	1.7	1.8
3.1	1.5	1.5	2.6	2.8	2.5
2.5	3.9	.8	1.8	3.3	3.7

(a) Estimate the mean and standard deviation in the time required to solve a problem via this algorithm.

(b) Find a 99% confidence interval on the mean time required to solve a problem.

(c) Another algorithm, written in a different language, requires an average of 6.6 seconds of cpu time. The solutions obtained are equivalent. Does the new algorithm appear to be more efficient than the other with respect to computing time? Explain.

14. To estimate the average number of pounds of copper recovered per ton of ore mined, a sample of 150 tons of ore is monitored. A sample mean of 11 pounds with a sample standard deviation of 3 pounds is obtained. Construct a 95% confidence interval on the mean number of pounds of copper recovered per ton of ore mined.

15. A certain amount of natural gas is produced with each barrel of crude oil. This gas escapes from the oil near the top of the well pipe. In an attempt to estimate the amount of natural gas available from wells in Kuwait these data are obtained on X, the number of cubic feet of gas obtained per barrel of crude oil (based on information found in "The Oil/Gas Separator: A New Cap for Quenching Oil Well Fires," *Energy and Technology*, December 1991, p. 1):

290	610	790	670	770
420	600	350	800	920
410	810	620	560	550
610	510	390	480	630
470	380	550	570	730
680	530	650	1000	720

(a) Construct a stem-and-leaf diagram for these data. Does the normality assumption that underlies the T procedures appear to be met? Explain.

(b) Construct a boxplot for these data. Are any data points flagged as outliers?

(c) Find a 99% confidence interval for the average volume of natural gas produced per barrel of crude oil by wells in Kuwait.

(d) If we wanted an interval based on these data that was shorter than the one found in part (c), what could be done to accomplish this?

16. Surface finishing for corrosion protection is usually the last manufacturing process that takes place before the sale or assembly of metal parts used in such things as automobiles and electrical appliances. This process is often done in shops that specialize in this procedure. A technique for applying bright zinc plating to steel is being tested. The variable under study is the thickness of the resulting coating in microns. These data are obtained on 25 test strips (based on figures found in "The Cinderella of Manufacturing," D. J. C. Hemsley, *Professional Engineering*, vol. 5, no. 7, July/August 1992, pp. 18–20):

6.4	8.3	7.9	7.5	6.9
8.5	7.0	7.4	7.2	6.8
7.1	8.1	7.5	7.7	8.5
7.8	7.3	8.4	8.0	7.8
7.5	7.8	7.6	8.4	9.9

(a) Construct a double stem-and-leaf diagram for these data. Comment on the possible distribution of this random variable.

(b) Construct a boxplot for these data, and identify the data point that is flagged as an outlier.

(c) Suppose that upon investigation it is found that the unusual data point discovered via your boxplot was for a strip that was inadvertently left in the coating solution longer than the procedure specified. What should be done with this data point?

(d) Construct a 95% confidence interval on the average thickness of the coating obtained via the new process.

(e) Would you be surprised to hear a claim that this average is 7.7 microns? Explain based on your confidence interval.

17. *(One-sided confidence interval on μ.)* A "one-sided" confidence interval can be used to approximate the maximum or minimum value of a population mean. An interval of the form $(-\infty, L]$ such that $P[\mu \leq L] \doteq 1 - \alpha$ allows us to place bounds on the maximum value of the population mean. The formula for such an interval is given by

$$L = \bar{X} + t_\alpha S/\sqrt{n}$$

An interval of the form $[L, \infty]$ allows us to place bounds on the minimum feasible value of the population mean. The formula for an interval of this type is

$$L = \bar{X} - t_\alpha S/\sqrt{n}$$

Use the following data on X, the time that a commercial airliner stays at the gate during a through flight, to find a 95% one-sided confidence interval that puts a bound on the minimum time in minutes expected for μ:

25   29   32   37   40   27   30   35   38   41   42   45   45   47   49   50   55   53   60

18. These data are obtained on the total nitrogen concentration (in ppm) of water drawn from a lake being considered for use as a source of drinking water for a locality:

.042	.023	.049	.036	.045	.025
.048	.035	.048	.043	.044	.055
.045	.052	.049	.028	.025	.039
.023	.045	.038	.035	.026	.059

Find a 95% one-sided confidence interval on the largest feasible value for μ. To be acceptable as a source of drinking water, the mean nitrogen content must lie below .07 ppm. Does this lake appear to meet this criterion? Explain.

19. *(Sample size required to estimate μ.)* Three factors determine the length of a confidence interval on μ. These are the confidence desired, the variability in the data, and the sample size. In an undesigned experiment it is possible that the resulting confidence interval is so long that it is almost useless. If σ is known or can be estimated from a small preliminary or "pilot" study, then it is possible to design an experiment in such a way that the resulting confidence interval will be short enough to be useful. This is done by selecting the sample size carefully.

(a) Let d denote the distance between $\bar{X}$, the center of the confidence interval, and $\bar{X} + z_{\alpha/2}\sigma/\sqrt{n}$, the upper confidence bound. Thus $d = z_{\alpha/2}\sigma/\sqrt{n}$. Note that the confidence interval itself is of length $2d$. Solve this equation for n to show that the sample size required to estimate μ to within d units with $100(1 - \alpha)\%$ confidence is

$$n \doteq \frac{(z_{\alpha/2})^2 \sigma^2}{d^2} \qquad \sigma \text{ known}$$

$$n \doteq \frac{(z_{\alpha/2})^2 \hat{\sigma}^2}{d^2} \qquad \sigma \text{ unknown}$$

(b) Reading digital displays in bright light poses a problem. Engineers want to design a filter to maximize both the luminance (brightness) and the chrominance (color) contrast. To do so, they intend to estimate the average number of footcandles in the cockpit of commercial airliners where the filter will be used. A preliminary pilot study is run, and an estimated standard deviation of 500 footcandles is obtained. How large a sample is needed to estimate μ to within 50 footcandles with 95% confidence?

(c) To determine whether or not the copper ore in a particular area is pure enough for open pit mining to be feasible, mining engineers must estimate the average grade of the ore. Past experience with this type of ore indicates that the grade ranges from 1% to 4% copper. The normal probability rule and Exercise 25, Chap. 6, imply that a rough estimate of σ is 1/4 of the range, or .75. How many test holes must be drilled to estimate μ to within .1% with 90% confidence?

20. A study is being designed to estimate the mean time required to assemble a panel of microprocessor chips for use in color television sets. An estimate of this mean is needed in order to set reasonable quotas for assembly line workers. A small pilot study is conducted, and these data are obtained on the assembly time in minutes:

1.0	1.5	2.2	3.0	2.7
2.0	2.4	2.6	2.3	1.7

(a) Based on these data, estimate σ.
(b) How large a sample is required to estimate μ to within .2 minute with 99% confidence?

Section 7.3

21. In 1969 in the United States, on average, 8% of household waste was metal. Because of the increase in recycling efforts, it is hoped that this figure has been reduced. An experiment is run to verify this contention.
(a) Set up the appropriate null and alternative hypotheses for the experiment.
(b) Explain in a practical sense what has occurred if a Type I error has been committed.
(c) Explain in a practical sense what has occurred if a Type II error has been committed.
(d) Explain in a practical sense what it means to say that H_0 has been rejected at the $\alpha = .05$ level of significance.

22. The mean level of background radiation in the United States is .3 rem per year. It is feared that as a result of the increased use of radioactive materials, this figure has increased.
(a) Set up the appropriate null and alternative hypotheses to document this claim.
(b) Explain in a practical sense the consequences of making a Type I and a Type II error.

23. As mentioned in Chap. 1, an important aspect of the engineering sciences is model building. Once a theoretical model is devised to explain a physical phenomenon, it must be tested to see that it yields results that are realistic. This testing is often done via computer simulation. In testing a model, we are testing

$$H_0: \text{model is credible}$$
$$H_1: \text{model is not credible}$$

(a) Explain in a practical sense what has occurred if a Type I error is committed. The probability of committing this error is referred to as the "model builder's risk." Do you see why this language is appropriate?

(b) Explain in a practical sense what has occurred if a Type II error is committed. The probability of committing an error of this type is called the "model user's risk." Does this seem appropriate?

24. A DNA test is conducted to see if the evidence can clear a suspect. From the suspect's perspective we are testing

$$H_0: \text{DNA is that of the suspect}$$
$$H_1: \text{DNA is not that of the suspect}$$

Suppose that a Type I error is made. In this setting, what has occurred? Suppose that the DNA test has high power. What does this mean?

25. Suppose we want to test

$$H_0: p \leq .4$$
$$H_1: p > .4$$

based on a sample of size 15.

(a) Find the critical region for an $\alpha \doteq .05$ level test.

(b) If when the data are gathered, $x = 11$, will H_0 be rejected? What type of error is possible at this point?

26. Suppose we want to test

$$H_0: p \geq .7$$
$$H_1: p < .7$$

based on a sample of size 10.

(a) Find the critical region for an $\alpha \doteq .05$ level test.

(b) If when the data are gathered, $x = 5$, will H_0 be rejected? What type of error might you be making?

27. It is a common practice to subject long-life items to larger than usual stress so that failure data can be obtained in a short amount of test time. Such tests are called accelerated life tests. Equipment used in computing makes use of metal oxide semiconductors (MOS). It is thought that "oxide short circuits" account for a majority of the early failures found in MOS integrated circuits. To verify this contention, a high-voltage screen test is applied to a number of circuits and 15 early failures are observed. Let X denote the number of failures due to oxide short circuits.

(a) Set up the appropriate null and alternative hypotheses.

(b) If H_0 is true and $p = .5$, what is the expected number of failures due to oxide short circuits in the 15 trials?

(c) Let us agree to reject H_0 in favor of H_1 if X is 11 or more. In this way we are presetting α at what level?

(d) Find β if $p = .6$; if $p = .7$; if $p = .8$; if $p = .9$.

(e) Find the power of the test if $p = .6$; if $p = .7$; if $p = .8$; if $p = .9$.

(f) If, when the data are gathered, we observe 12 early failures that are due to oxide short circuits, will H_0 be rejected? What type error might be committed?

(g) If, when the data are gathered, we observe 10 early failures due to oxide short circuits, will H_0 be rejected? What type error might be committed?

28. Quality and reliability are becoming important aspects of computer hardware and software. Past experience shows that the probability of failure during the first 1000 hours of operation for 16-kbit dynamic RAM produced by a United States firm is .2. It is hoped that new technology and stricter quality controls have reduced this failure rate. To verify this contention, 20 systems will be monitored for 1000 hours and the number of failures will be recorded.

(a) Set up the appropriate null and alternative hypotheses.

(b) Explain in a practical sense the consequences of making a Type I and a Type II error.

(c) If H_0 is true and $p = .2$, what is the expected number of failures during the first 1000 hours in the 20 trials?

(d) Let us agree to reject H_0 in favor of H_1 if the observed number of failures, X, is at most 1. In this way we are presetting α at what level?

(e) Suppose that it is essential that the test be able to distinguish between a failure rate of .2 and a failure rate of .1. Find the probability that the test as designed will be unable to do so. That is, find β if $p = .1$. Find the power of the test if $p = .1$.

(f) The results of part (e) indicate that the test as designed cannot distinguish well between $p = .1$ and $p = .2$. Keeping the sample size fixed at $n = 20$, can you suggest a way to modify the test that will lower β and to increase the power for detecting a failure rate of .1? Will α still be small enough to be acceptable? If not, can you suggest a way to redesign the experiment that will make both α and β low enough to be acceptable?

29. In Example 7.3.3 we test

$$H_0: p \leq .5$$

$$H_1: p > .5 \qquad \text{(majority of automobiles in operation have misaimed headlights)}$$

at the $\alpha = .0577$ level by agreeing to reject H_0 if at least 14 of the 20 cars sampled have misaimed headlights. We claim that values of X that are too large to occur by chance when $p = .5$ are also too large to occur by chance when $p < .5$. That is, if these values are rare when $p = .5$, they are even more rare when $p < .5$. To help see that this is true, find $P[X \geq 14]$ when $p = .4$; .3; .2; .1. Are each of these probabilities less than .0577 as expected?

30. A sample of size 9 from a normal distribution with $\sigma^2 = 25$ is used to test

$$H_0: \mu = 20$$

$$H_1: \mu = 28$$

The test statistic used is the sample mean, $\overline{X}$. Let us agree to reject H_0 in favor of H_1 if the observed value of $\overline{X}$ is greater than 25.

(a) If H_0 is true, what is the distribution of $\overline{X}$?

(b) In the diagram of Fig. 7.11, shade the region whose area is α.

(c) Find α. Remember that α is computed under the assumption that H_0 is true.

(d) If H_1 is true, what is the distribution of $\overline{X}$?

Figure 7.11

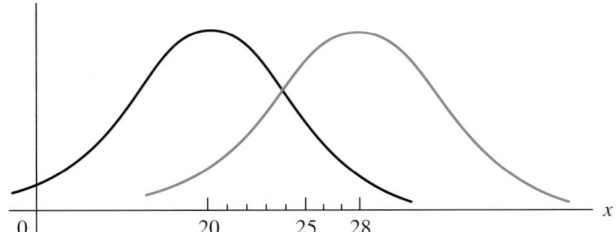

(e) In the diagram of Fig. 7.11, shade the region whose area is β. Remember that β is computed under the assumption that H_1 is true.

(f) Find β.

(g) Find the power of the test.

(h) If the sample size is increased, the standard deviation of $\overline{X}$ will decrease. What is the geometric effect of this on the two curves of Fig. 7.11?

(i) If the sample size is increased but the critical point is not changed, what will be the effect on α and β?

Section 7.4

31. Whenever a motorist encounters braking problems, especially an unpredictable pulling to one side, the villain is always held to be the brake pad. Trace elements, especially titanium, can combine with other elements to form minute particles of titanium carbonitride which alter the degree of friction between the pad and disc and lead to unequal wear. The percentage of titanium in a brake pad should not exceed 5%. A study is conducted to detect a situation in which the mean percentage of titanium in the brake pads being produced by a particular manufacturer exceeds 5%.

    (a) Set up the appropriate null and alternative hypotheses.

    (b) Discuss the practical consequences of making a Type I and a Type II error.

    (c) A sample of 100 brake pads yields a mean percentage of $\bar{x} = .051$. Assume that $\sigma = .008$. Find the P value of the test. Do you think that H_0 should be rejected? Explain. To what type of error are you now subject?

32. The current particulate standard for diesel car emission is .6 g/mi. It is hoped that a new engine design has reduced the emissions to a level below this standard.

    (a) Set up the appropriate null and alternative hypotheses for confirming that the new engine has a mean emission level below the current standard.

    (b) Discuss the practical consequences of making a Type I and a Type II error.

    (c) A sample of 64 engines tested yields a mean emission level of $\bar{x} = .5$ g/mi. Assume that $\sigma = .4$. Find the P value of the test. Do you think that H_0 should be rejected? Explain. To what type of error are you now subject?

33. It is thought that more than 15% of the furnaces used to produce steel in the United States are still open-hearth furnaces. To verify this contention, a random sample of 40 furnaces is selected and examined.

    (a) Set up the appropriate null and alternative hypotheses required to support the stated contention.

    (b) When the data are gathered, it is found that 9 of the 40 furnaces inspected are open-hearth furnaces. Use the normal approximation to the binomial distribution (Sec. 4.5) to find the P value of the test. Do you think that H_0 should be rejected? Explain. To what type of error are you now subject?

34. It is known that defective items will be produced even on automated assembly lines. A particular process typically produces 5% defectives. If the proportion of defectives exceeds 5%, then the line must be shut down and adjusted.

(a) Set up the appropriate null and alternative hypotheses needed to detect a situation in which the proportion of defectives produced exceeds .05.

(b) Discuss the practical consequences of committing a Type I and a Type II error.

(c) A random sample of 100 items is selected and tested. Of these, 7 are found to be defective. Use the normal approximation to the binomial distribution to find the P value of the test. Do you think that H_0 should be rejected?

Section 7.5

35. Find the critical point(s) for conducting a hypothesis test on the mean with σ^2 unknown for a

(a) left-tailed test with $n = 25$; $\alpha = .05$

(b) left-tailed test with $n = 150$; $\alpha = .10$

(c) right-tailed test with $n = 20$; $\alpha = .025$

(d) right-tailed test with $n = 16$; $\alpha = .01$

(e) two-tailed test with $n = 20$; $\alpha = .10$

(f) two-tailed test with $n = 30$; $\alpha = .05$

36. A new 8-bit microcomputer chip has been developed that can be reprogrammed without removal from the microcomputer. It is claimed that a byte of memory can be programmed in less than 14 seconds.

(a) Set up the appropriate null and alternative hypotheses needed to verify this claim.

(b) What is the critical point for an $\alpha = .05$ level test based on a sample of size 15?

(c) These data are obtained on X, the time required to reprogram a byte of memory:

11.6	14.7	12.9	13.3	13.2
13.1	14.2	15.1	12.5	15.3
13.3	13.4	13.0	13.8	12.3

Construct a stem-and-leaf diagram for these data. Does the normality assumption look reasonable?

(d) Test the null hypothesis. Can H_0 be rejected at the $\alpha = .05$ level? Interpret your result in a practical sense. To what type error are you now subject?

37. Ozone is a component of smog that can injure sensitive plants even at low levels. In 1979 a federal ozone standard of .12 ppm was set. It is thought that the ozone level in air currents over New England exceeds this level. To verify this contention, air samples are obtained from 30 monitoring stations set up across the region.

(a) Set up the appropriate null and alternative hypotheses for verifying the contention.

(b) What is the critical point for an $\alpha = .01$ level test based on a sample of size 30?

(c) When the data are analyzed, a sample mean of .135 and a sample standard deviation of .03 are obtained. Use these data to test H_0. Can H_0 be rejected at the $\alpha = .01$ level? What does this mean in a practical sense?

(d) What assumption are you making concerning the distribution of the random variable X, the ozone level in the air?

38. A model of Saudi Arabia's oil export strategy has been devised based on interviews with informed economists. The model is to be used to estimate the mean number of barrels of oil produced per day by this country. The usefulness of the model is to be partially checked by comparing the predicted mean for the year 1980 to its known value for that year, namely, 9.5 million barrels per day.

    (*a*) Find the critical points for testing

$$H_0: \mu = 9.5$$
$$H_1: \mu \neq 9.5$$

        at the $\alpha = .05$ level based on a sample of 50 simulations.

    (*b*) For the data collected, $\bar{x} = 9.8$ and $s = 1.2$. Test H_0. Can H_0 be rejected at the $\alpha = .05$ level? Based on these data, is there evidence that the model is not adequate? To what type of error are you now subject?

39. A low-noise transistor for use in computing products is being developed. It is claimed that the mean noise level will be below the 2.5-dB level of products currently in use.

    (*a*) Set up the appropriate null and alternative hypotheses for verifying the claim.

    (*b*) A sample of 16 transistors yields $\bar{x} = 1.8$ with $s = .8$. Find the P value for the test. Do you think that H_0 should be rejected? What assumption are you making concerning the distribution of the random variable X, the noise level of a transistor?

    (*c*) Explain, in the context of this problem, what conclusion can be drawn concerning the noise level of these transistors. If you make a Type I error, what will have occurred? What is the probability that you are making such an error?

40. The Elbe River is important in the ecology of central Europe, as it drains much of this region. Due to increased industrialization, it is feared that the mineral content in the soil is being depleted. This will be reflected in an increase in the level of certain minerals in the water of the Elbe. A study of the river conducted in 1982 indicated that the mean silicon level was 4.6 mg/l.

    (*a*) Set up the appropriate null and alternative hypotheses needed to gain evidence to support the contention that the mean silicon concentration in the river has increased.

    (*b*) A sample of size 28 yields $\bar{x} = 5.2$ with $s = 1.6$. Find the P value for the test. Do you think that H_0 should be rejected?

    (*c*) What practical conclusion can be drawn from these data?

41. Coal-handling maintenance is a very young technology. The emission standard for coal-burning plants is 4.8 pounds SO_2/per million Btu's/per 24-hour average. In an attempt to get emissions below this level, engineers are experimenting with burning a blend of high- and low-sulfur coal.

    (*a*) Set up the null and alternative hypotheses needed to support the contention that the new mixture falls below the emission standard set by the government.

    (*b*) Find the P value for the test if a sample of 200 readings yields a sample mean of 4.7 with a sample standard deviation of .5. Do you think that H_0 should be rejected? What does this mean in a practical sense?

42. Lasers are now used to detect structural movement in bridges and large buildings. These lasers must be extremely accurate. In laboratory testing of one such laser, measurements of the error made by the device are taken. The data obtained are used to test

$$H_0: \mu = 0$$
$$H_1: \mu \neq 0$$

A sample of 25 measurements yields $\bar{x} = .03$ millimeter over 100 meters and $s = .1$. Find the P value for this two-tailed test. Do you think that H_0 should be rejected? Interpret your result in a practical sense.

43. Clams, mussels, and other organisms that adhere to the water intake tunnels of electrical power plants are called *macrofoulants*. These organisms can, if left unchecked, inhibit the flow of water through the tunnel. Various techniques have been tried to control this problem, among them increasing the flow rate and coating the tunnel with Teflon, wax, or grease. In a year's time at a particular plant an unprotected tunnel accumulates a coating of macrofoulants that averages 5 inches in thickness over the length of the tunnel. A new silicone oil paint is being tested. It is hoped that this paint will reduce the amount of macrofoulants that adhere to the tunnel walls. The tunnel is cleaned, painted with the new paint, and put back into operation under normal working conditions. At the end of a year's time the thickness in inches of the macrofoulant coating is measured at 16 randomly selected locations within the tunnel. These data result (based on information from "Consider Non-fouling Coatings for Relief from Macrofouling," A. Christopher Gross, *Power*, October 1992, pp. 29–34):

4.2	4.5	4.1	4.6
4.4	4.0	4.7	4.3
5.0	6.2	3.6	4.5
5.1	3.5	3.0	2.8

(a) State the research hypothesis.
(b) Do these data support the contention that the new paint reduces the average thickness of the macrofoulants within this tunnel? Explain, based on the P value of the test.
(c) If α had been preset at .05, would H_0 have been rejected?
(d) Data in this problem are fictitious. Actually, the paint discussed in the journal article was much more effective than these data indicate. If these data had been real, do you think from a practical engineering point of view that the paint would be considered a major breakthrough in controlling the accumulation of macrofoulants? Explain.

44. Refineries, steel mills, food processing plants, and other industries separate oil and water using polyelectrolytes. These work better when pH is closely controlled. For example, chrome plating waste typically has a pH of 2.5. This wastewater must be neutralized before it is released into the environment. These data are obtained on the pH of wastewater samples that have been treated (based on a discussion found in "How to Choose a pH Measurement System," David M. Gray and Jeff Marshall, *Pollution Engineering*, November 1992, pp. 45–47):

6.2	6.5	7.6	7.7	7.0
7.0	7.2	6.8	7.5	8.1
7.1	7.0	7.1	7.8	8.5

Based on these data, is there evidence that the treatment process does not yield an average pH of 7 as desired? Explain based on the P value of the two-tailed test. If α had been preset at the .10 level, would H_0 have been rejected?

45. Due to the threat of terrorism there is a move to use "bag matching" on domestic flights in the United States. This means that a flight would not be permitted to leave whenever a passenger checks a bag but does not board the plane. It is thought that the average delay caused by such a check would be less than 7 minutes. These data are obtained on a sample of 100 flights in which bag checking was employed:

8.8	8.3	7.2	8.3	6.6	7.1	8.0	8.6	7.9	7.4
7.4	5.9	7.6	6.6	5.7	6.3	7.4	7.6	7.6	6.8
8.9	8.0	7.9	4.8	7.8	6.2	5.5	6.3	6.8	6.5
7.9	7.3	7.7	6.8	6.9	6.2	7.8	7.9	8.1	7.2
6.5	6.1	6.1	5.2	4.9	6.6	5.7	5.8	8.9	7.0
7.1	6.2	7.5	5.2	7.4	6.6	7.7	6.4	7.3	5.5
7.7	7.8	8.1	6.2	7.8	7.0	9.8	5.4	7.4	5.6
7.2	6.4	5.5	6.9	6.2	6.7	7.1	7.4	5.6	7.4
6.5	6.2	6.6	6.5	7.3	7.8	6.6	5.8	6.3	6.3
8.0	6.2	7.6	6.2	6.4	9.0	6.7	7.5	6.5	5.6

State the research hypothesis, and calculate the P value of the test. If a delay of an average of less than 7 minutes is acceptable to the public and would not cause undue disruption of schedules, would you advise that this procedure be implemented based on the results of this study?

46. *(Approximating sample sizes.)* In testing the hypothesis H_0: $\mu = \mu_0$, the experimenter can set α at any desired level. However, the value of β depends not only on the choice of α, but also on the difference between μ_0 and the alternative value μ_1. The farther apart these values lie, the more likely it is that we shall be able to distinguish them from one another. In designing an experiment, we want to pick a sample size that gives us a high probability of rejecting H_0 when there is a real practical difference between μ_0 and μ_1. That is, we want β to be small. Choosing the appropriate size for a T test is not easy. The problem is due to the fact that when H_0 is not true, our test statistic no longer follows a T distribution. Rather, it has what is called a noncentral T distribution. Fortunately, tables have been constructed using this distribution that allow us to determine the proper sample size for testing H_0: $\mu = \mu_0$ for various values of α, β, and Δ, where $\Delta = |\mu_0 - \mu_1|/\sigma$ and σ is the standard deviation of X. Table VII of App. A is one such table. Its use is illustrated here.

Example. Let us test H_0: $\mu = 10$ versus H_1: $\mu > 10$ at the $\alpha = .05$ level. Assume that we want to be 90% sure of detecting a situation in which μ has gotten as large as 12. Assume also that a pilot study has been run and that $\hat{\sigma} = 4$. Here

$$\Delta \doteq |\mu_0 - \mu_1|/\hat{\sigma} = |10 - 12|/4 = .5$$
$$\alpha = .05 \quad \text{and} \quad \beta = .1$$

From Table VII of App. A we see that for a one-sided test with these characteristics we need a sample of size $n = 36$.

(a) A pilot study indicates that the standard deviation of a particular random variable X is 1.25. How large a sample is required to test

$$H_0: \mu = 20$$
$$H_1: \mu > 20$$

at the $\alpha = .05$ level and $\beta = .05$ level if it is important to be able to distinguish between $\mu = 20$ and $\mu = 21$?

(b) In Exercise 37 we tested

$$H_0: \mu = .12$$
$$H_1: \mu > .12$$

at the $\alpha = .01$ level based on a sample of size 30. From this study we see that $\hat{\sigma} = .03$. Suppose that a mean ozone level of .14 is so serious that we must have a probability of .95 of detecting the situation. Approximately how large a sample is required?

(c) In Exercise 39 we tested

$$H_0: \mu = 2.5$$
$$H_1: \mu < 2.5$$

A sample of size 16 yielded $s = .8$. Assume that the new transistors are not financially worth marketing unless they reduce to mean noise level to at most 2 dB. Approximately how large a sample is needed to distinguish between a mean of 2.5 and a mean of 2.0 if $\alpha = .025$ and $\beta = .05$?

Section 7.6

47. A new process for producing small precision parts is being studied. The process consists of mixing fine metal powder with a plastic binder, injecting the mixture into a mold, and then removing the binder with a solvent. These data are obtained on parts that should have a 1-inch diameter and whose standard deviation should not exceed .0025 inch:

1.0030	.9997	.9990	1.0054	.9991
1.0041	.9988	1.0026	1.0032	.9943
1.0021	1.0028	1.0002	.9984	.9999

For these data $\bar{x} = 1.00084$ and $s = .00282$.
(a) Test

$$H_0: \mu = 1$$
$$H_1: \mu \neq 1$$

at the $\alpha = .05$ level.
(b) Test

$$H_0: \sigma = .0025$$
$$H_1: \sigma > .0025$$

at the $\alpha = .05$ level.

48. Indoor natatoriums or swimming pools are noted for their poor acoustical properties. The goal is to design a pool in such a way that the average time that it takes a low-frequency sound to die is at most 1.3 seconds with a standard deviation of at most .6 second. Computer simulations of a preliminary design are conducted to see whether these standards are exceeded. These data are obtained on the time required for a low-frequency sound to die:

1.8	3.7	5.0	5.3	6.1	.5
2.8	5.6	5.9	2.7	3.8	5.9
4.6	.3	2.5	1.3	4.4	4.6
5.3	4.3	3.9	2.1	2.3	7.1
6.6	7.9	3.6	2.7	3.3	3.3

For these data $\bar{x} = 3.97$ and $s = 1.89$.
(a) Test

$$H_0: \mu = 1.3$$
$$H_1: \mu > 1.3$$

at the $\alpha = .01$ level.
(b) Test

$$H_0: \sigma = .6$$
$$H_1: \sigma > .6$$

at the $\alpha = .01$ level. Does it appear that the design specifications are being met?

49. Incompatibility is always a problem when working with computers. A new digital sampling frequency converter is being tested. It takes the sampling frequency from 30 to 52 kilohertz word lengths of 14 to 18 bits and arbitrary formats and converts it to the output sampling frequency. The conversion error is thought to have a standard deviation of less than 150 picoseconds. These data are obtained on the sampling error made in 20 tests of the device:

133.2	−11.5	−126.1	17.9	139.4
−81.7	314.8	147.1	−70.4	104.3
56.9	44.4	1.9	−4.7	96.1
−57.3	−43.8	−95.5	−1.2	9.9

For these data $\bar{x} = 28.69$ and $s = 104.93$.
(a) Test

$$H_0: \mu = 0$$
$$H_1: \mu \neq 0$$

at the $\alpha = .1$ level.
(b) Test

$$H_0: \sigma = 150$$
$$H_1: \sigma < 150$$

at the $\alpha = .1$ level. Does the converter appear to be as accurate as claimed?

50. Use the data of Exercise 17 to test the null hypothesis that the standard deviation in gate time is less than 10 minutes.

Section 7.7

51. In each case, use the sign test to decide whether $H_0: M = M_0$ will be rejected in favor of the stated alternative at the $\alpha = .05$ level based on the data given. Do not discard zeros.
(a) $H_1: M > M_0$; $n = 15, Q_+ = 13$, no zeros
(b) $H_1: M > M_0$; $n = 20, Q_+ = 15$, no zeros
(c) $H_1: M > M_0$; $n = 20, Q_+ = 15$, three zeros
(d) $H_1: M < M_0$; $n = 10, Q_+ = 1$, no zeros
(e) $H_1: M < M_0$; $n = 10, Q_+ = 1$, one zero
(f) $H_1: M \neq M_0$; $n = 15, Q_+ = 2$, no zeros
(g) $H_1: M \neq M_0$; $n = 15, Q_+ = 2$, one zero
In each case above, what is the P value of the test?

52. Engineers are designing the safety devices for use in a new amusement-park ride. They think that the median height of patrons of rides of this sort exceeds 68 inches. Based on the sign test, do these data support this contention? Support your answer by finding the P value of the conservative sign test.

Height in inches				
65	73	72	71	68
74	74	66	68	69
70	66	72	67	73
69	70	73	70	74

53. Even with careful workmanship, digital scales may need some adjustment before being put into use. Unless there are systematic errors being made, the apparent zero of the scales before adjustment should fluctuate about true zero.

That is, some scales should weigh a little heavy, whereas others should give readings that are a little light. Ten such scales are randomly selected and tested. These data are obtained on the accuracy of the zero reading:

| heavy | light | heavy | heavy | heavy |
| light | light | light | heavy | heavy |

Based on these data, can we reject H_0: $M = 0$ in favor of H_1: $M > 0$ at the $\alpha = .05$ level?

54. In Example 7.7.2 we were able to reject

$$H_0: M = 120$$
$$H_1: M < 120$$

at the $\alpha = .05$ level. If we had used the sign test, which ignores the magnitude of the difference scores, could we have rejected H_0 at the $\alpha = .05$ level? Explain by finding $P[Q_+ \leq 2 | n = 8 \text{ and } p = 1/2]$.

55. An experiment for treating tar sand wastewater was conducted to determine whether a new treatment process removed more total organic carbon than a standard treatment process that is known to remove a median of 40 mg/l in a fixed detention time. Under the same experimental conditions the new process was replicated 10 times, yielding total organic carbon amounts removed of 38.8, 53.6, 39.0, 51.6, 40.1, 46.9, 40.9, 44.9, 41.0, and 43.2.
    (a) What is $E[W]$?
    (b) Using the signed-rank test, is there evidence that the new process removes significantly more total organic carbon than the standard process at the .05 level?

56. In an attempt to determine how many consultants are needed to answer questions of users at a computer center, these data are collected on X, the time in minutes required to answer a telephone inquiry:

1.5	1.0	5.0	1.9	3.0
1.3	2.1	1.7	6.5	4.2
6.3	5.6	5.1	2.5	6.9

    (a) What is $E[W]$?
    (b) Based on the signed-rank test, can we conclude that the median time required is less than 5 minutes? Explain, based on the P value of your test. (A zero score should be given the lowest rank and should be assigned the algebraic sign least conducive to rejecting the null hypothesis.)

57. A study of the expansion joints used in bridge beds is conducted. It is thought that these joints are expanding more than they were designed to expand, thus creating cracks in the pavement near the joint. The median design expansion at 95° F is 2 inches. Laboratory tests of 100 such joints are conducted at this temperature.
    (a) What is $E[W]$?
    (b) What is Var[W]?
    (c) Set up the appropriate null and alternative hypotheses.
    (d) If $|W_-| = 1600$, can H_0 be rejected? Explain, based on the P value of the test.

REVIEW EXERCISES

58. A consumer group wants to estimate the mean cost of the base system for a personal computer with certain specifications. It is thought that these computers range in price from $2390 to $4000.

(a) How large a sample should be taken to estimate μ to within $100 with 90% confidence?

(b) A random sample of size 50 yields these data (data in thousands of dollars):

2.43	2.86	2.74	2.75	2.69	2.64	2.91
2.89	3.18	3.00	3.21	3.07	3.72	3.24
3.17	3.57	3.37	3.56	3.30	2.32	3.09
2.99	3.20	3.25	3.70	3.45	2.82	2.88
2.71	3.25	2.86	2.93	3.45	3.11	3.86
2.96	3.00	2.88	3.19	3.56	3.21	3.33
3.39	3.14	2.90	3.49	3.02	3.56	2.87
2.32						

Construct a stem-and-leaf chart for these data. Use the digits 2 and 3 as stems 5 times each. Graph numbers beginning 2.0 and 2.1 on the first stem, those beginning 2.2 and 2.3 on the second stem, and so forth. Does the stem-and-leaf chart lead you to suspect that these data are not drawn from a distribution that is at least approximately normal?

(c) Find unbiased estimates for μ and σ^2 based on these data. Estimate σ. Is the estimate for σ unbiased?

(d) Find 90% confidence intervals on σ^2 and σ.

(e) Find a 90% confidence interval on μ.

59. Researchers are experimenting with a new compound used to bond Teflon to steel. The compounds currently in use require an average drying time of 3 minutes. It is thought that the new compound dries in a shorter length of time.

(a) Set up the null and alternative hypotheses needed to support the claim that the new compound dries faster than those currently in use.

(b) Discuss the practical consequences of making a Type I error; a Type II error.

(c) A pilot study shows that $\hat{\sigma} = .5$. Suppose that the new product is worth marketing if the average drying time can be shown to be 2.5 minutes or less. How large a sample is required to detect this situation with probability .95 with α set at .05?

(d) When the experiment is conducted, these data are obtained:

1.4	2.1	2.8	.9
2.4	1.7	3.7	2.7
2.6	1.9	2.8	2.8
2.2	2.2	3.4	1.9

Test the null hypothesis of part (a) at the $\alpha = .05$ level. Would you suggest marketing this new product?

60. It is thought that a majority of the procedures used in a statistical computer package run in less than .1 second. To verify this contention, a random sample of 20 programs that entail exactly one procedure is to be examined.

(a) Set up the appropriate null and alternative hypotheses needed to verify the claim.

(b) Let X denote the number of programs in which the procedure used runs in less than .1 second. Find the critical region for an $\alpha \doteq .025$ level test.

(c) When the test is conducted, 14 programs are found in which the procedure used runs in less than .1 second. Will H_0 be rejected? To what type error are you now subject?

(d) Find β if $p = .6$; if $p = .7$; if $p = .8$; if $p = .9$.

(e) Find the power of the test if $p = .6$; if $p = .7$; if $p = .8$; if $p = .9$.

61. Nickel powders are used in coatings used to shield electronic equipment from electromagnetic interference. It is thought that the mean size of the

individual nickel particles in one such coating is less than 3 micrometers. Do these data support this contention? Explain, based on the P value of the appropriate test.

3.26	3.07	2.46	1.76
1.89	2.95	3.35	3.82
2.42	1.39	1.56	2.42
2.03	3.06	1.79	2.96

62. We want to test

$$H_0: \mu = 5$$
$$H_1: \mu > 5$$

based on a random sample of size 25. The sample standard deviation is 2, and the observed value of the sample mean is 5.5. What is the P value for the test?

63. The accuracy of a tank's artillery is obviously affected by target size, distance of the tank from the target, and other random factors such as wind and terrain. A series of tests is conducted on a standard target of size 2.3 by 3.4 meters. This is the average size of targets that NATO tanks are likely to encounter. Each target is 3000 meters from the tank. The angular measure from the center of the target to the point of impact on the target is given in mils. A mil is an angle of size 1/6400th of a 360° circle. (See Fig. 7.12.) The system aiming error that can be tolerated and still hit the target is investigated. These data are obtained. Note that zero denotes a direct center hit (no error); positive errors result in a high hit; negative errors in a low hit (based on information found in "Tank Gun Accuracy," Major Bruce Held and Master Sergeant Edward Sunoski, *Armor*, January 1993, p. 6):

.8	−.3	.2	.2	0	0
−.1	.3	0	0	.2	.2
.1	0	0	0	.1	.1
−.1	−.1	−.1	.1	−.1	0
−.2	.1	.1	−.2	−.1	0
−.2	−.2	−.1	0	0	−.1
−.3					

Figure 7.12
(*a*) A direct hit on the center of the target; (*b*) any shell fired within the window shown should hit the target; (*c*) angular measure from center of target to actual impact point = aiming error that still allows for a hit.

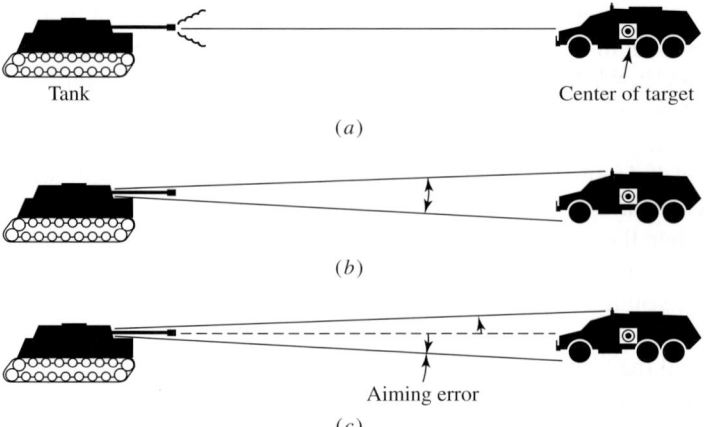

Tank

Center of target

(*a*)

(*b*)

Aiming error

(*c*)

(*a*) Sketch a stem-and-leaf diagram for these data. Are there any suspicious values in the data set?

(*b*) Sketch a boxplot for the data, and see if the one rather large data point qualifies as an outlier.

(*c*) It is thought that the "hit" that occurred when the outlier was obtained was, in fact, not a hit at all but, rather, an error in coding the data. For this reason, the outlier will be dropped from the data set in future analyses. Use the data given to find a 95% confidence interval on the average system error that occurs in firings that result in a hit.

64. A pattern recognition device involves 3000 bits. Each bit can be either on (the bit value is 1) or off (the bit value is 0). These bits are either fixed or changeable. If they are fixed, they cannot be reversed by the programmer. If more than 95% of the bits are fixed, then the system will crash. Bits will be tested sequentially until a changeable bit is found. We want to detect a situation in which the system will crash. The test statistic is X, the number of bits sampled in order to obtain the first changeable bit. Note that if the system will crash, then the probability of finding a fixed bit exceeds .95 and the probability of finding a changeable bit is less than .05. The random variable X is approximately geometrically distributed. We are testing

$$H_0: p \geq .05 \qquad \text{(system will not crash)}$$
$$H_1: p < .05 \qquad \text{(system will crash)}$$

(*a*) Explain why X is only approximately geometrically distributed. That is, what geometric property is not strictly met?

(*b*) If H_0 is true, what is $E[X]$?

(*c*) If H_1 is true, would you expect X to exceed $E[X]$ or to be smaller than $E[X]$?

(*d*) Find the critical point for the test if you want α to be between .05 and .10.

(*e*) On a particular run 30 consecutive fixed bits are found. Is this evidence yet that the system will crash using the α level of part (*d*)? In this case, what type error is possible? In the context of this problem, what are the consequences of making this error?

(*f*) On a particular run 60 consecutive fixed bits are found. What conclusion can you draw from this? What type error is possible? In the context of this problem, what are the consequences of making this error? (Based on a study conducted in 1993–1994 by Eyal Schwartz, Department of Computer Science, Radford University.)

65. To study the cost effectiveness of energy-saving programs, data are gathered on the cost of such programs. These data are obtained on the cost of various programs per kilowatt hour of electricity saved:

Residential programs (cost in cents)

3	8	7	4	7	7
6	4	8	8	5	8
6	9	9	11	8	8
10	8	3	8	9	9
7	9	6	11	13	5
10	12	7	12	8	8
8	10	11	10	14	6
181					

Commercial and industrial programs (cost in cents)

1	3	3	2	3
3	2	4	5	3
4	3	3	6	
3	4	5	3	

(a) Construct stem-and-leaf diagrams for each data set. Comment on the likelihood that the normality assumption underlying T statistics is satisfied in each case.

(b) Construct a boxplot for each data set. Identify the extreme outlier in the residential data set. Suppose that upon investigation it is found that this data point was obtained in a very atypical program in an affluent region of the country. Since the program is so unusual, it is decided not to include it in trying to estimate the cost of programs that could be put in place in most areas of the country. Use the remaining data to construct a 95% confidence interval on the average cost of residential energy-saving programs currently in use. If the outlier had been included, what effect would it have in the confidence interval obtained?

(c) Find a 95% confidence interval on the variance of the cost of residential energy-saving programs. Do not use the extreme outlier in your calculation. If the outlier were used, what would be the effect on the length of the confidence interval obtained?

(d) Find 95% confidence interval, on the mean, variance, and standard deviation of the cost of energy-saving programs in the industrial and commercial sectors.

(e) If the true average cost of electricity to residential customers is 8 cents per kilowatt hour, is there reason to question the cost effectiveness of residential energy-saving programs? Explain.

(f) If the true average cost of electricity to industrial and commercial customers is 5 cents per kilowatt hour, is there reason to question the cost effectiveness of commercial energy-saving programs? Explain. (Based on information found in "The Real Cost of Saving Electricity," *Technology Review*, February/March 1993, p. 12.)

66. Consider the information given in Exercise 6.35. Construct 95% confidence intervals on the average life span for each type of lamp. Based on these intervals, is there clear evidence that the mean life spans differ in value? Explain.

67. The Nuclear Regulatory Commission is responsible for monitoring companies using radioactive materials. Data obtained in a study of past accidents are given on the website. Variables in the data set are:

Number = accident number

Type = type of company with p = privately run, g = government not military, and m = military

Accident = type of accident with wb = whole body exposure and e = exposure to extremities only

Expose = exposure dose in rems

(a) Sort the data by accident, and obtain stem-and-leaf plots for the exposure dose for each type of accident.

(b) Find 90% confidence intervals for the mean exposure dose for each type of accident. Is there clear evidence that these means differ in value? Explain.

(c) Sort the data by accident and type to obtain 6 subgroups. Obtain descriptive statistics and boxplots for each subgroup. Discuss any similarities or differences that you observe from these descriptive tools.

Inferences on Proportions

In this chapter we discuss inferences on one proportion and the comparison of two proportions. As we have already seen, the binomial distribution can be used to test hypotheses on a proportion p when sample sizes are small. Here we see how to use the standard normal distribution to construct confidence intervals on p and test hypotheses concerning its value for large samples. We also begin our study of two sample problems by learning how to compare proportions based on samples drawn from two distinct populations.

8.1 ESTIMATING PROPORTIONS

The typical situation calling for the estimation of a proportion is as follows: There is a population of interest, a particular trait is being studied, and each member of the population can be classed as either having or failing to have the trait. We want to make inferences on p, the proportion of the population with the trait.

> **Example 8.1.1.** Quality and reliability are important aspects of software. The smallest of bugs in computer software once foiled a space shuttle launch; in Japan a signal malfunction in an electronic telephone exchanger shut down phone lines for hours. To estimate the reliability of 16-kilobit (kbit) dynamic RAMs being produced by a particular company, a sample of size 100 is to be drawn and tested. We are interested in estimating p, the proportion of circuits that operate correctly during the first 1000 hours of operation. Here the population consists of all 16-kbit dynamic RAMs produced by the company; the trait being studied is the ability of the circuit to function correctly during the first 1000 hours of use. Each circuit either will have the trait, that is, it will operate correctly, or else it will not.

To develop a logical point estimator for p, note that associated with a random sample of size n drawn from the population is a collection of n independent random variables $X_1, X_2, X_3, \ldots, X_n$ where

$$X_i = \begin{cases} 1 & \text{if the } i\text{th member of the sample has the trait} \\ 0 & \text{if the } i\text{th member of the sample does not have the trait} \end{cases}$$

For example, if we sample 100 circuits, we are dealing with a sample that would look something like this:

$$
\begin{array}{lll}
x_1 = 1 & x_4 = 1 & x_{98} = 1 \\
x_2 = 0 & x_5 = 0 \quad \cdots & x_{99} = 0 \\
x_3 = 0 & \vdots & x_{100} = 0
\end{array}
$$

In this case the first circuit sampled operates correctly during the first 1000 hours of use, so $x_1 = 1$; the second circuit does not operate correctly for this period of time, and so $x_2 = 0$, and so forth. Note that, in general, $X = \sum_{i=1}^{n} X_i$ gives the number of objects in the sample with the trait and that the statistic X/n gives the proportion of the sample with the trait. This statistic, called the *sample proportion,* is a logical point estimator for p:

> **Point estimator for p**
>
> $$\hat{p} = \frac{X}{n} = \frac{\text{number in sample with trait}}{\text{sample size}}$$

Example 8.1.2. Suppose that when the 100 tests mentioned in Example 8.1.1 are conducted, it is found that 91 of the 100 circuits tested perform properly during the first 1000 hours of operation. Thus 91 of the random variables $X_1, X_2, X_3, \ldots, X_{100}$ have value 1 and 9 assume the value 0. Based on these data, $\sum x_i = x = 91$ and

$$\hat{p} = x/n = 91/100 = .91$$

Confidence Interval on p

To develop a confidence interval on p, the distribution of $\hat{p}$ must be determined. This is accomplished by noticing that $\hat{p} = \sum X_i/n$ is actually nothing more than a very special sample mean. That is $\hat{p} = \bar{X}$ is the average of the point binomial or zero-one random variables X_i. By the Central Limit Theorem $\hat{p}$ is approximately normally distributed with the same mean as the X_i's and with variance equal to (Var X_i)/n. The mean and variance of X_i is determined easily. Since $X_i = 1$ only if an object with the trait is sampled and the true proportion of objects in the sample with the trait is p, $P[X_i = 1] = p$. Consequently, $P[X_i = 0] = 1 - p$. The density for X_i is as follows:

x_i	1	0
$f(x_i)$	p	$1 - p$

From this density it is easy to see that

$$E[X_i] = 1(p) + 0(1 - p) = p$$
$$E[X_i^2] = 1^2(p) + 0^2(1 - p) = p$$

and

$$\text{Var } X_i = E[X_i^2] - (E[X_i])^2 = p - p^2 = p(1 - p)$$

By the Central Limit Theorem we can conclude that $\hat{p}$ is approximately normally distributed with mean p and variance $p(1 - p)/n$. Notice that we have just shown that $\hat{p}$ has the properties desirable in a point estimator. It is unbiased for p and has a small variance for large sample sizes.

To obtain a random variable that involves p whose distribution is known to serve as a starting point for the confidence interval derivation, standardize $\hat{p}$. The resulting random variable,

$$(\hat{p} - p)/\sqrt{p(1 - p)/n}$$

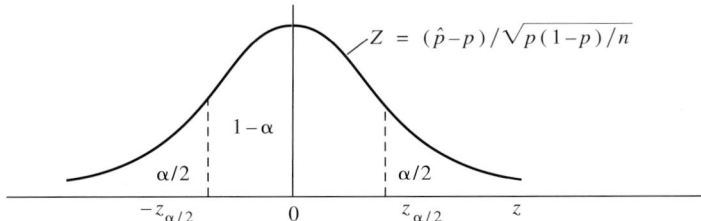

Figure 8.1
Partition of the Z curve needed
to construct a $100(1 - \alpha)\%$
confidence interval on p.

follows an approximate Z distribution for large sample sizes. The partition of the standard normal curve shown in Fig. 8.1 is needed to derive the bounds for a $100(1 - \alpha)\%$ confidence interval on p. From this diagram it can be seen that

$$P\left[-z_{\alpha/2} \le (\hat{p} - p)/\sqrt{p(1-p)/n} \le z_{\alpha/2}\right] \doteq 1 - \alpha$$

Isolating p in the middle of this inequality, we see that

$$P\left[\hat{p} - z_{\alpha/2}\sqrt{p(1-p)/n} \le p \le \hat{p} + z_{\alpha/2}\sqrt{p(1-p)/n}\right] \doteq 1 - \alpha$$

It appears that the confidence bounds for p are

$$\hat{p} \pm z_{\alpha/2}\sqrt{p(1-p)/n}$$

However, there is a problem here that has not been encountered before. The bounds for a confidence interval must be *statistics*. That is, they must be random variables whose expression contains no unknown parameters so that their numerical value can be obtained from a sample. Unfortunately, as written, the above bounds are not statistics, since the unknown parameter p appears in the expressions given. This means that we are attempting to use p to estimate p, a seeming impossible situation! The problem can be overcome easily. The obvious method is to replace p by its unbiased estimator $\hat{p}$, to yield these bounds:

$$\hat{p} \pm z_{\alpha/2}\sqrt{\hat{p}(1-\hat{p})/n}$$

A legitimate question to ask is, "Since we are replacing the true standard deviation of $\hat{p}$ by an estimator for this standard deviation, should we switch from Z to T_{n-1} as was done when estimating μ?" The answer to this question lies in considering the sample size. The derivation of the confidence bounds is based on the Central Limit Theorem, which assumes that a large sample is available. Furthermore, in experimental settings in which this formula is to be applied the sample size is expected to be large enough so that there is very little difference between a z and a t point. Thus we shall write the confidence bounds as

> **Confidence interval on p**
> $$\hat{p} \pm z_{\alpha/2}\sqrt{\hat{p}(1-\hat{p})/n}$$

and use z points when the formula is applied. Confidence bounds for samples of size 1 through 30 have been developed based on the binomial distribution, and should be used for samples this small. Tables for these are found in [10]. The use of this method is illustrated in the next example.

Example 8.1.3. The point estimate for the proportion of 16-kbit dynamic RAMs that function correctly for at least 1000 hours based on a sample of size 100 is .91. From the standard normal table the point required to construct a 95% confidence interval on p is $z_{.025} = 1.96$. The bounds for the confidence interval are

$$\hat{p} \pm z_{\alpha/2}\sqrt{\hat{p}(1 - \hat{p})/n}$$

or

$$.91 \pm 1.96\sqrt{.91(.09)/100}$$
$$.91 \pm .056$$

We can be approximately 95% confident that the true proportion of circuits that function correctly during the first 1000 hours of operation lies between .854 and .966. Converting to percentages, we can be approximately 95% confident that the true percentage of satisfactory circuits produced by this company lies between 85.4% and 96.6%. The word "approximately" is employed because we are approximating the distribution of $\hat{p}$ via the Central Limit Theorem and are also approximating p by $\hat{p}$ in finding the confidence bounds.

Sample Size for Estimating p

As when estimating a mean, it is possible that an experiment yields a confidence interval on p that is so long that it is virtually useless. This brings up one other important question: "How large a sample should be selected so that $\hat{p}$ lies within a specified distance d of p with a stated degree of confidence?" There are two ways to answer this question. The first is applicable when an estimate of p based on some prior experiment is available. Consider the diagram of Fig. 8.2.

Since we are $100(1 - \alpha)\%$ sure that p lies in the interval shown, we are $100(1 - \alpha)\%$ sure that $\hat{p}$ and p differ by at most d, where d is given by

$$d = z_{\alpha/2}\sqrt{\hat{p}(1 - \hat{p})/n}$$

This equation is solved for n to obtain the following formula for finding the sample size needed to estimate p with a stated degree of accuracy and confidence when a prior estimate of p is available:

Sample size for estimating p, prior estimate available

$$n \doteq \frac{z_{\alpha/2}^2\hat{p}(1 - \hat{p})}{d^2}$$

Example 8.1.4. How large a sample is required to estimate the proportion of 16-kbit dynamic RAMs that function properly during the first 1000 hours of use to within .01 (1 percentage point) with 95% confidence? We do have a prior estimate of p available, namely $\hat{p} = .91$. By the above formula

$$n \doteq \frac{z_{\alpha/2}^2\hat{p}(1 - \hat{p})}{d^2}$$

Figure 8.2
$100(1 - \alpha)\%$ confidence interval on p.

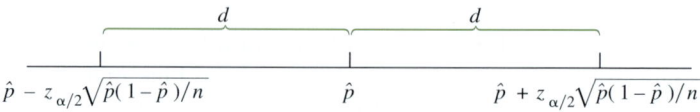

$$\hat{p} - z_{\alpha/2}\sqrt{\hat{p}(1 - \hat{p})/n} \qquad \hat{p} \qquad \hat{p} + z_{\alpha/2}\sqrt{\hat{p}(1 - \hat{p})/n}$$

Since we want 95% confidence, the point $z_{\alpha/2} = z_{.025} = 1.96$. The maximum desired difference between $\hat{p}$ and p is $d = .01$. Substituting, we obtain

$$n \doteq \frac{(1.96)^2(.91)(.09)}{(.01)^2} \doteq 3147$$

To get the desired accuracy, we need substantially more data than we now have available!

The second method for determining sample size for estimating proportions is based on a result from elementary calculus. It can be shown (Exercise 9) that $\hat{p}(1 - \hat{p})$ will never exceed 1/4. Therefore this term can be replaced by 1/4 in the previous sample size formula to obtain the following formula for use when no prior estimate of p is available:

> **Sample size for estimating p, no prior estimate available**
>
> $$n \doteq \frac{z_{\alpha/2}^2}{4d^2}$$

This expression will be very useful to you since in most applications no prior estimate of p is available.

Example 8.1.5.   A new method of precoating fittings used in oil, brake, and other fluid systems in heavy-duty trucks is being studied. How large a sample is needed to estimate the proportion of fittings that leak to within .02 with 90% confidence? Since no prior estimate of p is available,

$$n \doteq \frac{z_{\alpha/2}^2}{4d^2}$$

Here $z_{\alpha/2} = z_{.05} = 1.645$ and $d = .02$. Substituting, we have

$$n \doteq \frac{(1.645)^2}{4(.02)^2} \doteq 1692$$

It should be pointed out that sampling from a large finite population is usually done *without* replacement. Strictly speaking, the proportion of objects in the population with the given trait does vary from trial to trial. However, the change is so slight that its effect on our calculations is negligible. For this reason, the methods of this section can be used to study large populations even though the mathematical assumptions underlying the methods are not met completely.

8.2   TESTING HYPOTHESES ON A PROPORTION

When we have a preconceived idea of the value of a proportion or a percentage and we want statistical evidence to support our contention, we are in a hypotheses-testing situation. The hypotheses tested can assume any one of the usual three forms, depending on the purpose of the study. Let p_0 denote the null value of p. These forms are

I $H_0: p = p_0$	II $H_0: p = p_0$	III $H_0: p = p_0$
$H_1: p > p_0$	$H_1: p < p_0$	$H_1: p \neq p_0$
Right-tailed test	Left-tailed test	Two-tailed test

In Sec. 7.3 we saw how to test these hypotheses for *small* samples. The test statistic used is X, the number of objects in the sample with the trait of interest. When the null hypothesis is true, this statistic has a binomial distribution with parameters n and p_0. When sample sizes are large, appropriate binomial tables usually are not available. In this case we must find another logical test statistic.

Consider the random variable used to generate the confidence bounds for p. That is, consider the statistic

> **Test Statistic for Testing $H_0: p = p_0$**
>
> $$(\hat{p} - p_0)/\sqrt{p_0(1 - p_0)/n}$$

The statistic is a logical choice since it compares the unbiased point estimator for p, $\hat{p}$, to the null value p_0. Furthermore, if H_0 is true, then by the Central Limit Theorem this statistic has a standard normal distribution. Tests are conducted as you would expect. Namely, for a right-tailed test H_0 is rejected in favor of H_1 if the observed value of the test statistic is a large *positive* number; large *negative* numbers lead to rejection in a left-tailed test. In a two-tailed test H_0 is rejected for values of the test statistic that are too large in either the positive or the negative sense. These ideas are illustrated in the next example.

Example 8.2.1. The majority of faults on transmission lines are the result of external influences and are usually transitory. It is thought that more than 70% of all faults are caused by lightning. Data gathered over a year-long period show that 151 of 200 faults observed are due to lightning. Do we have sufficient evidence to support this contention at .05 significance level?

[Hypotheses] We test

$$H_0: p = .7$$
$$H_1: p > .7$$

[Test statistic] The observed value of the test statistic is

$$(\hat{p} - p_0)/\sqrt{p_0(1 - p_0)/n} = (151/200 - .7)/\sqrt{.7(.3)/200}$$
$$\approx 1.697$$

[Critical value] Since we are conducting a right-tailed test, we reject H_0 if this value is unusually large. Since the test statistic is normally distributed, the critical value is

$$z_{.05} = 1.645$$

from Table V of App. B.

[Conclusion] Since the value of test statistic is greater than the critical value (1.697 > 1.645), we shall reject H_0 and conclude that $p > .7$.

We can also estimate the P value and use it to decide the conclusion. From the standard normal table we see that $P[Z > 1.69] = .0455$ and $P[Z > 1.70] = .0446$. Since our observed value, 1.697, lies between 1.69 and 1.70, the P value lies between .0446 and .0455. There are two explanations for this small P value. The null hypothesis is true, and we have just observed a rare event, one that occurs only about 4 times in every 100 trials; or the null hypothesis is not true, and the true percentage of faults due to lightning exceeds 70%. As we have set significance level to be .05, the latter explanation seems more plausible, so we shall reject H_0 and conclude that $P > .7$.

This method for testing a hypothesis on p does assume that the sample size is "large." Following the guidelines given in Sec. 4.5, this is interpreted to mean that n and p_0 are such that $p_0 \leq .5$ and $np_0 > 5$ or $p_0 > .5$ and $n(1 - p_0) > 5$. These criteria are met in Example 8.2.1 since $p_0 = .7 > .5$ and $n(1 - p_0) = 200(.3) = 60 > 5$.

8.3   COMPARING TWO PROPORTIONS: ESTIMATION

The problem of comparing two proportions arises frequently in the engineering sciences. The general situation can be described as follows: There are two populations of interest, the same trait is studied in each population, each member of each population can be classed as either having the trait or failing to have it, and in each population the proportion having the trait is unknown. Random samples are drawn from each population. These samples are *independent* of one another in the sense that the objects drawn from one population do not determine in any way which objects are selected from the second population. Inferences are to be made on p_1, p_2, and $p_1 - p_2$, where p_1 and p_2 are the proportions in the first and second populations with the trait, respectively.

Example 8.3.1.   A study is conducted to compare computer usage in Canadian business to that of businesses in the United States. Interest centers on the proportion of businesses in each country with an on-site mainframe computer. Here the two populations being studied are "businesses" in Canada and "businesses" in the United States. Remember that before sampling is done we must clearly specify what constitutes a "business." That is, we must clearly define the target populations. The trait under study is that of having an on-site computer; each business sampled either does or does not own such equipment. We draw a sample at random from each population. We use the sample data to compare the proportion of Canadian businesses with an on-site mainframe computer to that of businesses in the United States. (See Fig. 8.3.)

Population I (Canadian businesses)

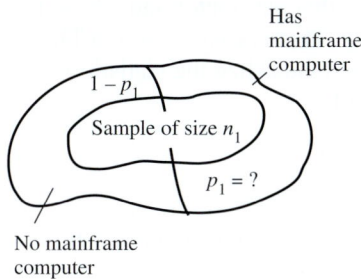

Population II (businesses in United States)

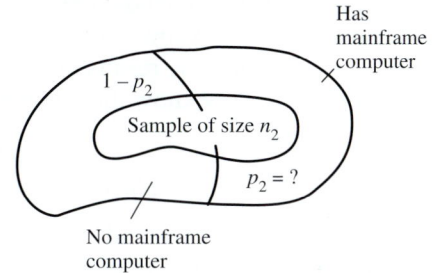

Figure 8.3
Independent samples drawn to estimate $p_1 - p_2$.

The problem of point estimation of the difference between two proportions is solved in the obvious way. We simply estimate p_1 and p_2 individually and take as our estimate for $p_1 - p_2$ the difference between the two. That is, our point estimator is

> **Point estimator for $p_1 - p_2$**
> $$\widehat{p_1 - p_2} = \hat{p}_1 - \hat{p}_2 = X_1/n_1 - X_2/n_2$$

where n_1 and n_2 are the sizes of the samples drawn from the two populations and X_1 and X_2 are the number of objects, respectively, in the samples with the trait.

Example 8.3.2. Independent random samples of size 375 are selected from the population of Canadian businesses and from the population of businesses in the United States. It is found that 221 of the Canadian firms and 232 of the firms in the United States have mainframe computers. For these data

$$\hat{p}_1 = x_1/n_1 = 221/375 = .589$$
$$\hat{p}_2 = x_2/n_2 = 232/375 = .619$$
$$\widehat{p_1 - p_2} = \hat{p}_1 - \hat{p}_2 = .589 - .619 = -.03$$

Confidence Interval on $p_1 - p_2$

To extend the point estimator $\hat{p}_1 - \hat{p}_2$ to an interval estimator, we must pause to consider the probability distribution of this statistic. Its approximate distribution is given in Theorem 8.3.1.

> **Theorem 8.3.1.** For large samples, the estimator $\hat{p}_1 - \hat{p}_2$ is approximately normal with mean $p_1 - p_2$ and variance $p_1(1 - p_1)/n_1 + p_2(1 - p_2)/n_2$.

Proof. We have shown in Sec. 8.1 that both $\hat{p}_1$ and $\hat{p}_2$ are approximately normal with means p_1 and p_2 and variances $p_1(1 - p_1)/n_1$ and $p_2(1 - p_2)/n_2$, respectively. Since the sum or difference of two normal random variables is normal, we can conclude that the statistic $\hat{p}_1 - \hat{p}_2$ is at least approximately normally distributed. Furthermore, by the rules for expectation and the rules for variance

$$E[\hat{p}_1 - \hat{p}_2] = E[\hat{p}_1] - E[\hat{p}_2] = p_1 - p_2$$

and

$$\text{Var } [\hat{p}_1 - \hat{p}_2] = \text{Var } \hat{p}_1 + \text{Var } \hat{p}_2 = p_1(1 - p_1)/n_1 + p_2(1 - p_2)/n_2$$

Note that Theorem 8.3.1 shows that the statistic $\hat{p}_1 - \hat{p}_2$ is an *unbiased* estimator for $p_1 - p_2$. To construct a $100(1 - \alpha)\%$ confidence interval on $p_1 - p_2$, we need a random variable whose expression involves this parameter and whose probability distribution is known at least approximately. This is easy to do via Theorem 8.3.1. We simply use the results of this theorem to standardize the statistic $\hat{p}_1 - \hat{p}_2$. In particular, we now know that the random variable

$$\frac{(\hat{p}_1 - \hat{p}_2) - (p_1 - p_2)}{\sqrt{p_1(1 - p_1)/n_1 + p_2(1 - p_2)/n_2}}$$

is at least approximately standard normal. Rather than repeat an algebraic argument given previously, let us consider three intervals that have been derived already and note their similarities.

Parameter being estimated	Began derivation with	Distribution	Bounds
$\mu(\sigma^2$ known)	$\dfrac{\bar{X} - \mu}{\sigma/\sqrt{n}}$	Z	$\bar{X} \pm z_{\alpha/2}\sigma/\sqrt{n}$
$\mu(\sigma^2$ unknown)	$\dfrac{\bar{X} - \mu}{S/\sqrt{n}}$	T	$\bar{X} \pm t_{\alpha/2}S/\sqrt{n}$
p	$\dfrac{\hat{p} - p}{\sqrt{p(1-p)/n}}$	$\sim Z$	$\hat{p} \pm z_{\alpha/2}\sqrt{\hat{p}(1-\hat{p})/n}$

The algebraic structure of each of the beginning variables is the same and is of the form

$$\frac{\text{Estimator} - \text{parameter}}{D}$$

where D is either the standard deviation of the estimator or an estimator for this standard deviation. This is also the algebraic form assumed by the variable

$$\frac{(\hat{p}_1 - \hat{p}_2) - (p_1 - p_2)}{\sqrt{p_1(1-p_1)/n_1 + p_2(1-p_2)/n_2}} \sim Z$$

The confidence bounds in the previous cases took the form

$$\text{Estimator} \pm \text{probability point} \cdot D$$

Applying the notion to the case at hand, we find that the proposed confidence bounds for a confidence interval on $p_1 - p_2$ will be

$$(\hat{p}_1 - \hat{p}_2) \pm z_{\alpha/2}\sqrt{p_1(1-p_1)/n_1 + p_2(1-p_2)/n_2}$$

Once again there is a slight problem. The proposed bounds are not *statistics*. They include the unknown population proportions p_1 and p_2. As in the one sample case, this problem can be overcome by replacing the population proportions with their estimators $\hat{p}_1$ and $\hat{p}_2$. This leads to the following formula for finding confidence intervals on the difference between two population proportions:

> **Confidence interval on $p_1 - p_2$**
>
> $$(\hat{p}_1 - \hat{p}_2) \pm z_{\alpha/2}\sqrt{\hat{p}_1(1-\hat{p}_1)/n_1 + \hat{p}_2(1-\hat{p}_2)/n_2}$$

Example 8.3.3. The point estimate for the difference in the proportion of businesses in Canada and the proportion of businesses in the United States with on-site mainframe computers is $\hat{p}_1 - \hat{p}_2 = .589 - .619 = -.03$. A 95% confidence interval for this difference is

$$(\hat{p}_1 - \hat{p}_2) \pm z_{\alpha/2}\sqrt{\hat{p}_1(1-\hat{p}_1)/n_1 + \hat{p}_2(1-\hat{p}_2)/n_2}$$

or

$$-.03 \pm 1.96\sqrt{(.589)(.411)/375 + (.619)(.381)/375}$$

$$-.03 \pm .07$$

That is, we are 95% confident that the true difference in proportions lies in the interval $[-.10, .04]$. Note that since this interval contains the number 0, it is possible that there is really no difference in the two population proportions p_1 and p_2.

The question of determining the sample size needed to estimate the difference between two proportions with a stated degree of accuracy and confidence is more complex than in the one sample case. However, if samples of equal size are chosen from each population, then the problem can be solved just as in the one sample case. The procedure is outlined in Exercise 21.

8.4 COMPARING TWO PROPORTIONS: HYPOTHESIS TESTING

Sometimes problems arise in which it is theorized prior to the experiment that one proportion or percentage differs from another by a specified amount. The purpose of the experiment is to gain statistical support for the contention. These hypotheses take any one of these three forms, where $(p_1 - p_2)_0$ represents the null value of the difference in proportions:

$$
\begin{array}{ll}
\text{I } H_0: p_1 - p_2 = (p_1 - p_2)_0 & \text{II } H_0: p_1 - p_2 = (p_1 - p_2)_0 \\
\quad H_1: p_1 - p_2 > (p_1 - p_2)_0 & \quad H_1: p_1 - p_2 < (p_1 - p_2)_0 \\
\quad \text{Right-tailed test} & \quad \text{Left-tailed test}
\end{array}
$$

$$
\begin{array}{l}
\text{III } H_0: p_1 - p_2 = (p_1 - p_2)_0 \\
\quad H_1: p_1 - p_2 \neq (p_1 - p_2)_0 \\
\quad \text{Two-tailed test}
\end{array}
$$

To test such hypotheses, a test statistic must be found. To derive such a statistic, consider the approximately standard normal random variable

$$
\frac{(\hat{p}_1 - \hat{p}_2) - (p_1 - p_2)_0}{\sqrt{p_1(1 - p_1)/n_1 + p_2(1 - p_2)/n_2}}
$$

that was used to construct confidence intervals on $p_1 - p_2$ in the previous section. This random variable is not a statistic, since it contains the unknown population proportions p_1 and p_2. We again overcome this problem in the logical way. In particular, we replace p_1 and p_2 by their unbiased estimators $\hat{p}_1$ and $\hat{p}_2$ to obtain the approximately standard normal test statistic

$$
\frac{(\hat{p}_1 - \hat{p}_2) - (p_1 - p_2)_0}{\sqrt{\hat{p}_1(1 - \hat{p}_1)/n_1 + \hat{p}_2(1 - \hat{p}_2)/n_2}}
$$

This is a logical choice for a test statistic, since it compares the estimated difference in proportion $\hat{p}_1 - \hat{p}_2$ with the hypothesized difference $(p_1 - p_2)_0$. If the hypothesized value is correct, then the estimated difference and the hypothesized difference should be close in value. This forces the numerator above to be close to zero and thus yields a small value for the test statistic. Large positive or large negative values of the test statistic indicate that the null hypothesis is not true and should be rejected in favor of an appropriate alternative.

Example 8.4.1. A corporation operates two foundries that are similar in size and that are engaged in the same production operations. An experimental safety program has been implemented at one location. Before expanding the program, the management wants to compare the proportion of workers injured during the trial period at the experimental site to that of its other plant. It is thought that the program is cost effective if these proportions differ by more than .05. When the trial period ends, it is found that 24 of the 263 workers at the control plant were injured, whereas only

5 of the 250 workers at the experimental site received injuries. Based on the data from the trial period, do we have the evidence to justify expanding the safety program? Since making a Type I error is costly, we preset $\alpha = .1$.

[Hypotheses] We are testing

$$H_0: p_1 - p_2 = .05$$
$$H_1: p_1 - p_2 > .05$$

where p_1 and p_2 denote the proportions of injured workers at the control and experimental plants, respectively.

[Test statistic] Based on the data from the trial period,

$$\hat{p}_1 = 24/263 = .091 \qquad \hat{p}_2 = 5/250 = .020 \qquad \hat{p}_1 - \hat{p}_2 = .071$$

The value of the test statistic is

$$\frac{(\hat{p}_1 - \hat{p}_2) - (p_1 - p_2)_0}{\sqrt{\hat{p}_1(1 - \hat{p}_1)/n_1 + \hat{p}_2(1 - \hat{p}_2)/n_2}} = \frac{.071 - .05}{\sqrt{(.091)(.909)/263 + (.02)(.98)/250}}$$
$$= 1.059$$

[Critical value] The critical value for this right-tailed test is

$$z_{.01} = 2.33$$

[Conclusion] Since the test statistic (1.059) does not exceed the critical point of 2.33, we are unable to reject the null hypothesis at the $\alpha = .01$ level. We do not have the evidence that is felt necessary to justify expanding the safety program.

Pooled Proportions

Although the hypothesized difference $(p_1 - p_2)_0$ can be any value at all, the most commonly proposed value is zero. In this case the hypotheses considered previously compare p_1 and p_2 and take these forms:

I $H_0: p_1 = p_2$	II $H_0: p_1 = p_2$	III $H_0: p_1 = p_2$
$H_1: p_1 > p_2$	$H_1: p_1 < p_2$	$H_1: p_1 \neq p_2$
Right-tailed test	Left-tailed test	Two-tailed test

Hypotheses of this sort can be tested via the previously developed test statistic with $(p_1 - p_2)_0$ set equal to zero. However, an alternative procedure is available. This alternative procedure, which is preferred by many statisticians, makes use of the fact that if H_0 is true, $\hat{p}_1$ and $\hat{p}_2$ are both estimators for the same proportion, which we denote by p. To see how to use this information, note that the variance of $\hat{p}_1 - \hat{p}_2$ is given by

$$p_1(1 - p_1)/n_1 + p_2(1 - p_2)/n_2$$

If H_0 is true, we can write this variance as

$$p(1 - p)/n_1 + p(1 - p)/n_2 = p(1 - p)(1/n_1 + 1/n_2)$$

We see that the random variable

$$\frac{\hat{p}_1 - \hat{p}_2}{\sqrt{p(1 - p)(1/n_1 + 1/n_2)}}$$

has a distribution that is approximately standard normal. We are now faced with the problem of estimating the unknown common population proportion p. Since $\hat{p}_1$ and $\hat{p}_2$ are both unbiased estimators for p, it makes sense to combine them in some way. We can simply average these estimators, but in so doing we ignore whatever differences might exist between the two sample sizes involved. To take these differences into account, we use a weighted average. Namely, we multiply each estimator by its corresponding sample size to obtain this "pooled" estimator for p:

> **Pooled estimator for p when $p_1 = p_2$**
> $$\hat{p} = \frac{n_1\hat{p}_1 + n_2\hat{p}_2}{n_1 + n_2}$$

The test statistic that results when p is replaced by $\hat{p}$ is

> **Test Statistic for Comparing Two Proportions**
> $$\frac{\hat{p}_1 - \hat{p}_2}{\sqrt{\hat{p}(1-\hat{p})(1/n_1 + 1/n_2)}}$$

The use of this statistic is demonstrated in our next example.

Example 8.4.2. Many consumers think that automobiles built on Mondays are more likely to have serious defects than those built on any other day of the week. To support this theory, a random sample of 100 cars built on Monday is selected and inspected. Of these, eight are found to have serious defects. A random sample of 200 cars produced on other days reveals 12 with serious defects. Do these data support the stated contention? Suppose that the significance level is .05.

[**Hypotheses**] To decide, we test

$$H_0: p_1 = p_2$$
$$H_1: p_1 > p_2$$

where p_1 and p_2 denote the proportions of cars with serious defects produced on Mondays and other days, respectively.

[**Test statistic**] Estimates for p_1 and p_2 are

$$\hat{p}_1 = x_1/n_1 = 8/100 = .08 \qquad \text{and} \qquad \hat{p}_2 = x_2/n_2 = 12/200 = .06$$

The pooled estimate for the common population proportion is

$$\hat{p} = \frac{n_1\hat{p}_1 + n_2\hat{p}_2}{n_1 + n_2} = \frac{100(.08) + 200(.06)}{100 + 200}$$
$$= 20/300$$
$$= .066$$

The observed value of the test statistic is

$$\frac{\hat{p}_1 - \hat{p}_2}{\sqrt{\hat{p}(1-\hat{p})(1/n_1 + 1/n_2)}} = \frac{.08 - .06}{\sqrt{.066(.934)(1/100 + 1/200)}}$$
$$= .658$$

[Critical value] From the standard normal table (see Table V in Appendix B), the critical value is

$$z_{.05} = 1.645$$

[Conclusion] Since the value of the test statistic is much less than the critical value ($.658 < 1.645$), we do not have sufficient statistical evidence to support the claim that cars built on Mondays are more likely to have serious defects than those built on other days.

From the standard normal table (see Table V of App. B), we see that the probability of observing a value this large or larger is approximately .2546. That is, the P value is approximately .2546. Since this probability is large, we shall not reject H_0.

Either one of the test statistics presented can be used to test $H_0: p_1 - p_2 = 0$ or $H_0: p_1 = p_2$, although the pooled statistic is preferable, since it is thought to be more powerful. To test $H_0: p_1 - p_2 = (p_1 - p_2)_0$, where $(p_1 - p_2)_0 \neq 0$, pooling is not appropriate because $\hat{p}_1$ and $\hat{p}_2$ are estimating different proportions. In this case the first statistic presented is the proper test statistic.

Note that we are comparing proportions based on *independent* random samples drawn from two populations.

CHAPTER SUMMARY

In this chapter, we considered methods that can be used to make inferences on a single proportion when sample sizes are large. We also saw how to determine the sample size required to estimate p to any desired degree of accuracy when we do and do not have prior estimates for p available.

We began our study of two sample problems by considering both point and interval estimation of the difference between two population proportions. The methods presented assume that samples are drawn independently. We also saw that $H_0: p_1 - p_2 = (p_1 - p_2)_0$ can be tested using as a test statistic the same random variable used to generate our confidence interval on $p_1 - p_2$, namely,

$$\frac{(\hat{p}_1 - \hat{p}_2) - (p_1 - p_2)_0}{\sqrt{\hat{p}_1(1 - \hat{p}_1)/n_1 + \hat{p}_2(1 - \hat{p}_2)/n_2}}$$

However, if $(p_1 - p_2)_0 = 0$, then a pooled procedure is preferable. This procedure makes use of the fact that if H_0 is true, $p_1 = p_2$. Since $\hat{p}_1$, and $\hat{p}_2$ are estimating the same thing, we pool them to form this estimator for the common population proportion p:

$$\hat{p} = \frac{n_1\hat{p}_1 + n_2\hat{p}_2}{n_1 + n_2} = \frac{X_1 + X_2}{n_1 + n_2}$$

Using this estimator, the test statistic used to test $H_0: p_1 - p_2 = 0$ is

$$\frac{\hat{p}_1 - \hat{p}_2}{\sqrt{\hat{p}(1 - \hat{p})(1/n_1 + 1/n_2)}}$$

We introduced the following term: Pooled estimator for p.

EXERCISES

Section 8.1

1. In order to be effective, reflective highway signs must be picked up by the automobile's headlights. To do so at long distances requires that the beams be on "high." A study conducted by highway engineers reveals that 45 of 50 randomly selected cars in a high-traffic-volume area have the headlights on low beam.

    (a) Find a point estimate for p, the proportion of automobiles in this type area that use low beams.

    (b) Find a 90% confidence interval on p.

    (c) How large a sample is required to estimate p to within .02 with 90% confidence?

2. A study of the electromechanical protection devices used in electrical power systems showed that of 193 devices that failed when tested, 75 were due to mechanical parts failures.

    (a) Find a point estimate for p, the proportion of failures that are due to mechanical failures.

    (b) Find a 95% confidence interval on p.

    (c) How large a sample is required to estimate p to within .03 with 95% confidence?

3. In 1980 the Bureau of Labor Statistics conducted a study of 1000 minor eye injuries received by workers in the workplace. The study revealed that 600 of the workers involved were not wearing eye protection at the time of the injury. It also revealed that 900 of the injuries received could have been prevented through the proper use of protective eyewear. Assume that current conditions in the workplace have not changed substantially from those encountered in 1980 relative to the use of eye protection.

    (a) Find a 90% confidence interval on the proportion of workers who receive minor eye injuries this year that will not be wearing eye protection at the time of the injury.

    (b) Find a 95% confidence interval on the proportion of minor eye injuries occurring this year that could be prevented through the proper use of protective eyewear.

4. A survey of companies using industrial robots showed that of 200 robots in use, 48 were used for loading and unloading.

    (a) Find a 95% confidence interval on p, the proportion of industrial robots currently being used for loading and unloading.

    (b) Would you be surprised to hear someone claim that a majority of the robots in use are used for loading and unloading? Explain.

5. One problem associated with the use of the supersonic transport (SST) is the sonic boom. In the late 1960s and early 1970s preliminary tests were run over Oklahoma City, St. Louis, and other areas. After the tests were run a survey was to be conducted to estimate the percentage of people who felt that they could not live with the sonic booms. How large a sample should have been chosen to estimate this percentage to within 3 percentage points with 95% confidence?

6. The Environmental Protection Agency recently identified 30,000 waste dumping sites in the United States that were considered to be at least potentially dangerous. How large a sample is needed to estimate the percentage of these sites that do pose a serious threat to health to within 2 percentage points with 90% confidence?

7. It is said that "doctors bury their mistakes, architects cover them with ivy, and engineers write long reports that never see the light of day." One area in which engineering mistakes are critical is dam-building. How large a sample is necessary to estimate the percentage of nonfederal earthen dams in the United States that are in need of immediate repair to within 1 percentage point with 90% confidence?

8. A market research study is to be conducted among users of a particular type of computer system. How many users should be sampled to estimate the percentage of users who plan to add terminals to within 4 percentage points with 90% confidence?

9. Consider the function $g(\hat{p}) = \hat{p}(1 - \hat{p})$.
   (a) Find $g'(\hat{p})$.
   (b) Find the critical point for g.
   (c) Find $g''(\hat{p})$, and use this to argue that g assumes its maximum value at the critical point.
   (d) What is the maximum value assumed by the function g?

Section 8.2

10. A poll of investment analysts taken earlier suggests that a majority of these individuals think that the dominant issue affecting the future of the solar energy industry is falling energy prices. A new survey is being taken to see if this is still the case. Let p denote the proportion of investment analysts holding this opinion.
    (a) Set up the appropriate null and alternative hypotheses.
    (b) When the survey is conducted, 59 of the 100 analysts sampled agreed that the major issue is falling energy prices. Is this sufficient to allow us to reject H_0? Explain, based on the P value of the test.
    (c) Interpret your results in the context of this problem.

11. A new computer network is being designed. The makers claim that it is compatible with more than 99% of the equipment already in use.
    (a) Set up the null and alternative hypotheses needed to get evidence to support this claim.
    (b) A sample of 300 programs is run, and 298 of these run with no changes necessary. That is, they are compatible with the new network. Can H_0 be rejected? Explain, based on the P value of the test.
    (c) What practical conclusion can be drawn on the basis of your test?

12. It is thought that the no defect rate for 64-K-RAM devices produced in Japan is less than 8%.
    (a) Set up the null and alternative hypotheses needed to support this claim.
    (b) A sample of 64 of these devices is tested, and 4 are found to have no defects. Can H_0 be rejected? Explain, based on the P value of the test.
    (c) In the context of this problem, what conclusion can be drawn from your data?

13. It is thought that over 60% of the business offices in the United States have a mainframe computer as part of their equipment.
    (a) Set up the appropriate null and alternative hypotheses for supporting this claim.
    (b) Find the critical point for an $\alpha = .05$ level test.
    (c) When data are gathered, it is found that 233 of the 375 offices studied have mainframe computers. Can H_0 be rejected at the $\alpha = .05$ level? To what type of error are you now subject?
    (d) Explain, in the context of this problem, the practical consequences of making the type of error to which you are subject.

14. Opponents of the construction of a dam on the New River claim that less than half the residents living along the river are in favor of its construction. A survey is conducted to gain support for this point of view.
    (a) Set up the appropriate null and alternative hypotheses.
    (b) Find the critical point for an $\alpha = .1$ level test.
    (c) Of 500 people surveyed, 230 favor the construction. Is this sufficient evidence to justify the claim of the opponents of the dam?
    (d) To what type of error are you now subject? Discuss the practical consequences of making such an error.

15. A battery-operated digital pressure monitor is being developed for use in calibrating pneumatic pressure gauges in the field. It is thought that 95% of the readings it gives lie within .01 lb/in^2 of the true reading. In a series of 100 tests, the gauge is subjected to a pressure of 10,000 lb/in^2. A test is considered to be a success if the reading lies within $10,000 \pm .01$ lb/in^2. We want to test

$$H_0: p = .95$$
$$H_1: p \neq .95$$

at the $\alpha = .05$ level.
    (a) What are the critical points for the test?
    (b) When the data are gathered, it is found that 98 of the 100 readings were successful. Can H_0 be rejected at the $\alpha = .05$ level? To what type error are you now subject?

16. Power line noise, voltage variations, and power outages all can affect computer performance. When noise enters a television set, the result is static and snow; when noise enters a computer, errors can occur and circuits can be damaged. It is thought that more than 80% of all line disturbances at a particular computer site are noise.
    (a) Set up the appropriate null and alternative hypotheses needed to verify this contention.
    (b) Find the critical point for an $\alpha = .01$ level test.
    (c) Of 150 line disturbances that occur during the study time, 133 are due to noise. Can H_0 be rejected at the $\alpha = .01$ level? Interpret your results in the context of this problem.

Section 8.3

17. A random sample of 500 workers engaged in research and development (R & D) last year is selected. Of these, 178 earn over $72,000 per year. Of the 450 workers in R & D studied during the current year, 220 earn in excess of $72,000 per year.
    (a) Let p_1 and p_2 denote the proportion of workers engaged in research and development who earned over $72,000 per year last year and this year, respectively. Find point estimates for p_1, p_2, and $p_1 - p_2$.
    (b) Find a 95% confidence interval for $p_1 - p_2$.
    (c) Would you be surprised to hear someone claim that the proportion of R & D workers earning over $72,000 was the same this year as it was last year? Explain, on the basis of the confidence interval of part (b).

18. Superplasticized concrete is formed by adding chemicals to conventional concrete to make it more fluid so that it can be placed more easily. Suppose that a sample of 50 new construction projects in the Dallas-Fort Worth area yields 15

that are using this type of concrete. A sample of 60 new projects in the Boston area also yields 15 using superplasticized concrete.

(a) Let p_1 and p_2 denote the proportion of new construction projects in Dallas-Fort Worth and Boston, respectively, that are using superplasticized concrete. Find point estimates for p_1, p_2, and $p_1 - p_2$.

(b) Find a 95% confidence interval for $p_1 - p_2$.

(c) Would you be surprised to hear someone claim that the proportion of Dallas-Fort Worth projects using this type of concrete is clearly larger than that in the Boston area? Explain, based on the confidence interval of part (b).

19. A study of the computer market is conducted. Random samples are drawn from among the users of the two leading mainframes. The purpose of the study is to estimate the proportion of users in each population that either do use or would like to use the small office system built by the mainframe supplier. These data result:

Type I	Type II
$n_1 = 200$	$n_2 = 190$
$x_1 = 62$	$x_2 = 76$

(a) Find point estimates for p_1, p_2, and $p_1 - p_2$.

(b) Find a 90% confidence interval for $p_1 - p_2$.

(c) Would you be surprised to hear someone claim that $p_1 = p_2$? Explain, based on the confidence interval of part (b).

20. The computer is expected to play an increasingly important role in crime control in the years to come. In 1983 the FBI had a noncomputerized Ident system containing the records of thousands of persons across the country. A random sample of 500 records shows that only 70% of these records include information on the disposition of the case. This is unfortunate, since approximately 1/3 of all cases are eventually dismissed. If the dismissal is not a part of the record, then an innocent person could be stigmatized.

(a) Assume that a new computerized criminal history system is developed and implemented. A random sample of size 500 is selected from the cases recorded in the new system. It is found that 410 of these include information on the disposition of the case. Estimate the proportion of cases in the new system that include information on the disposition of the case.

(b) Estimate the difference in proportions between the old Ident system and the new computerized system. (Subtract in the order New − Ident.)

(c) Find a 95% confidence interval on the difference in proportions.

(d) Is it safe to say that the new system is superior to Ident in the sense that it contains more "disposition of case" information? Explain, based on the confidence interval of part (c).

21. (Sample size for estimating $p_1 - p_2$.) The difference between two population proportions, $p_1 - p_2$, is to be estimated based on independent random samples drawn from the respective populations. Each of the samples is each to be of size n. Show that in order to estimate $p_1 - p_2$ to within d with $100(1 - \alpha)\%$ confidence, n is given by

> ### Sample size for estimating $p_1 - p_2$
>
> Sample size for estimating $p_1 - p_2$, prior estimates for p_1 and p_2 available
>
> $$n \doteq z_{\alpha/2}^2 \frac{[\hat{p}_1(1 - \hat{p}_1) + \hat{p}_2(1 - \hat{p}_2)]}{d^2}$$
>
> Sample size for estimating $p_1 - p_2$, no prior estimates for p_1 and p_2 available
>
> $$n \doteq \frac{z_{\alpha/2}^2}{2d^2}$$

22. What common sample size must we take from the populations of R & D workers last year and this year to estimate $p_1 - p_2$ to within .02 with 90% confidence? Use the data of Exercise 17 to obtain estimates for p_1 and p_2.

23. What common sample size should be selected from the Ident files and the new computer files to estimate $p_1 - p_2$ to within .03 with 95% confidence? Use the data of Exercise 20 to obtain estimates for p_1 and p_2.

24. A study is to be conducted to estimate the difference in the proportions of defective items produced during two different shifts of assembly line workers. What common sample size should be used to estimate this difference to within .04 with 90% confidence?

25. Automotive engineers want to compare the performance of their new six-cylinder front-wheel-drive automobiles to their four-cylinder model. Let p_1 and p_2 denote the proportion of automobiles experiencing engine problems during the first 5000 miles of use for the two models, respectively. What common sample size should be used to estimate $p_1 - p_2$ to within .05 with 90% confidence?

Section 8.4

26. The use of optical fibers in telecommunications, the military, and industry is increasing rapidly. These fibers must be strong, durable, able to operate over a wide temperature range, and insensitive to radiation. Most fiber failures are due to a brittle fracture that grows into a complete crack. Two different fiber-drawing heat sources are being studied. These are carbon furnaces and CO_2 laser heating. A company currently uses a carbon furnace but will switch to laser heating if it can be shown that the latter method reduces the proportion of failures by more than .02.

    (a) Let p_1 and p_2 denote the proportions of failures occurring using the carbon furnace and CO_2 laser heating, respectively. Set up the appropriate null and alternative hypotheses needed to support a move to the laser technique.

    (b) Find the critical point for an $\alpha = .05$ level test.

    (c) Of 100 test fibers produced using the carbon furnace, 5 failed, whereas only 1 of the 100 fibers produced using the laser technique resulted in failure. Estimate p_1, p_2, and $p_1 - p_2$. Can H_0 be rejected at the $\alpha = .05$ level? Would you recommend that the company switch production methods?

    (d) To what type of error are you now subject? Discuss the practical consequences of making this error.

27. The cost of correcting a defect in a bipolar digital integrated circuit depends on when the defect is discovered. If it is discovered before it is integrated into a computer system, the cost may be only pennies. However, if it is not found

until after the device is in the field it could cost thousands of dollars to repair. The electrical defect rate of two types of circuits produced by a particular company is being studied. It is suspected that the defect rate of their ALS circuits (advanced lower-power Schottky) is smaller than that of their LPS circuits (lower-power Schottky).

(a) Let p_1 and p_2 denote the proportions of ALS circuits and LPC circuits produced, respectively, that have electrical defects. Set up the null and alternative hypotheses needed to confirm their suspicions.

(b) What is the critical point for an $\alpha = .1$ level test?

(c) Two thousand circuits of each type are randomly selected and tested. It is found that three of the ALS and five of the LPS circuits have electrical defects. Estimate p_1, p_2, and $p_1 - p_2$. Based on these data, can H_0 be rejected at the $\alpha = .1$ level? .

28. Today's diesel engines require smoother surface finishes and better consistency than in the past. Two types of abrasives are being tested for use on the microfinishers that are used to polish crankshafts. The first uses a paper and cloth abrasive; the second, a coated abrasive film. Both come on rolls that can tear, causing downtime and delay in the polishing process. It is thought that the proportion of rolls that tear is higher for the paper-cloth abrasive than for the abrasive film. However, since the abrasive film is the more expensive of the two, the difference in these proportions must exceed .10 in order for the abrasive film to be economical.

(a) Set up the null and alternative hypotheses needed to support the contention that the abrasive film is economical. Let p_1 denote the proportion of rolls of the paper-cloth abrasive that tear during testing.

(b) What is the critical point for an $\alpha = .025$ level test?

(c) Fifteen of 50 rolls of the paper-cloth abrasive tear during testing, whereas only two of the 40 rolls of the abrasive film do so. Estimate p_1, p_2, and $p_1 - p_2$. Can H_0 be rejected at the $\alpha = .025$ level?

(d) To what type of error are you now subject? Discuss the practical consequences of committing such an error.

29. Two types of metal detectors are in use in airports around the world. One is called a continuous wave detector, and the other is called a pulse field wave detector. Both devices are equally efficient at detecting large metal objects such as guns or knives. However, it is thought that the continuous wave detector tends to be less efficient in that it can be triggered more easily by objects such as coins, lipstick holders, and other small harmless metal objects.

(a) Let p_1 and p_2 denote the proportions of passengers that pass through the continuous wave and the pulse wave detectors, respectively, that trigger the device. Set up the null and alternative hypotheses needed to support the contention that the continuous wave detector will be triggered by a higher proportion of passengers than will the pulse wave device.

(b) Random samples of 175 passengers are observed passing through each of these types of devices. Of those passing through the continuous wave device, 113 triggered a warning. However, only 4 of those passing through the pulse field detector activated an alarm. Do you think that H_0 should be rejected? What is the P value of the test? What practical conclusion can be drawn from these data?

30. Shot peening is used to compress the surface area of metal parts to make them more resistant to fractures. It is done by bombarding the surface with small particles hurled at high velocity. Each time a particle hits, it puts a small dent

in the surface and compresses the area directly beneath the surface. The bombardment continues until eventually the entire surface is compressed. Tests are conducted on a particular part to see if shot peening reduces the proportion of parts that fracture when put into use. These data result:

Not shot peened	Shot peened
$n_1 = 35$	$n_2 = 40$
number fractured $= 7$	number fractured $= 3$

Set up the appropriate null and alternative hypotheses. Based on these data, do you think that shot peening reduces the probability that a part will fracture when put into use? Explain, based on the P value of the test.

31. Show that $\hat{p} = (X_1 - X_2)/(n_1 + n_2)$. That is, show that $\hat{p}$ can be found by combining the two samples into one and by finding the usual sample proportion for the new sample. Verify this numerically using the data of Exercise 30.

32. Let X_1 and X_2 denote the number of objects with the trait of interest in independently drawn random samples of sizes n_1 and n_2, respectively. Assume that these random variables are binomially distributed with parameters p_1 and p_2.
    (a) Find the expected value of the pooled estimator $\hat{p}$.
    (b) Show that if $H_0: p_1 = p_2$ is true, then $\hat{p}$ is an unbiased estimator for the common population proportion p.

REVIEW EXERCISES

33. A survey of mining companies is to be conducted to estimate p, the proportion of companies that anticipate hiring either graduating seniors or experienced engineers during the coming year.
    (a) How large a sample is required to estimate p to within .04 with 94% confidence?
    (b) A sample of size 500 yields 105 companies that plan to hire such engineers. Find a point estimate for p. Find a 94% confidence interval for p.

34. It is thought that the majority of the mining engineers that graduated in 1970 from U.S. schools are now employed in the coal mining industry.
    (a) Set up the null and alternative hypotheses needed to gain statistical evidence to support this contention.
    (b) A random sample of 50 of these individuals is selected, and their current place of employment is determined. Twenty-six are working in the coal mining industry. Do you think that H_0 should be rejected? Explain, based on the P value of the test.

35. A procedure used to produce identical twins in cattle entails the microsurgical division of the embryo into two groups of cells followed by immediate embryo transfer. This procedure is thought to be more than 50% effective.
    (a) Set up the null and alternative hypotheses needed to support this claim.
    (b) Find the critical point for an $\alpha = .05$ level test based on a sample of size 100.
    (c) When the experiment is conducted, 55 of the transplants result in the birth of twins. Can H_0 be rejected at the $\alpha = .05$ level? Interpret your results in the context of this problem.

36. A programmable lighting control system is being designed. The purpose of the system is to reduce electricity consumption costs in buildings. The system eventually will entail the use of a large number of transceivers. Two types are

being considered. In life testing these data are gathered on the number of trans-
ceiver failures for each type:

Type I	Type II
$n_1 = 100$	$n_2 = 100$
$x_1 = 2$	$x_2 = 4$

(a) Find point estimates for $p_1 - p_2$, the difference in the failure rates for the
two types of transceivers.
(b) Find a 95% confidence interval for $p_1 - p_2$.
(c) Based on the interval of part (b), can we claim that $p_1 < p_2$? Explain.
(d) Is the interval found in part (b) short enough to give us a good idea of the
actual value of $p_1 - p_2$? What common sample size is needed to estimate
$p_1 - p_2$ to within .01 with 95% confidence?

37. One measure of quality and customer satisfaction is repeat business. A supplier
of paper used for computer printouts sampled 75 customer accounts last year
and found that 40 of these had placed more than one order during the year. A
similar survey conducted at the end of the current year revealed that 35 of 50
customers ordered again. Do these data support the contention that there has
been an increase in the proportion of repeat business over the 2-year period?
Explain, based on the P value of your test.

38. A company is experimenting with a new method for etching circuits that
should decrease the proportion of circuits that must be etched a second time.
To be cost effective the difference in proportions between the old and new
methods must exceed .1.
(a) Letting p_1 denote the proportion of circuits that must be redone using the
old method, set up the null and alternative hypotheses required to show
that the new method is cost effective.
(b) Find the critical point for $\alpha = .05$ level test of the hypothesis of part (a).
(c) These data are obtained on the number of circuits that must be reworked
using each method:

Old	New
$n_1 = 25$	$n_2 = 50$
$x_1 = 4$	$x_2 = 2$

Can H_0 be rejected at the $\alpha = .05$ level? To what type error are you now subject? What are the practical consequences of making such an error?

39. One source of water pollution is gasoline leakage from underground storage
tanks. A random sample of 100 gasoline stations is selected, and the tanks are
inspected. Twenty are found to have at least one leaking tank.
(a) Find a 95% confidence interval on the proportion of stations across the
country with a leakage problem.
(b) Assume that there are approximately 375,000 stations in the United States.
Find a 95% confidence interval on the number of stations with a leakage
problem.
(c) How large a sample is required to estimate the proportion of stations with
a leakage problem to within .02 with 95% confidence?

40. "The Desert Storm rules of engagement dictated that when an aircrew could
not locate or positively identify their primary or secondary targets they were to

return to base with their weapons." This rule was intended to minimize damage to civilian populations. During Desert Storm and Desert Shield 72,000 combat sorties were flown by allied forces. In 18,000 cases planes returned to base with their weapons. (Based on information taken from "Operations Law and the Rules of Engagement in Operations Desert Shield and Desert Storm," Lt. Col. John G. Humphries, *Airpower Journal,* Fall 1992, pp. 25–41.)

(*a*) Based on these data, find a point estimate for *p,* the proportion of combat missions which will return to base with their weapons in similar future engagements in which these rules of engagement are in force.

(*b*) Find a 95% confidence interval on *p.*

(*c*) Suppose that, in a future engagement, 10,000 combat missions are flown. Find a 95% confidence interval on the number of missions in which planes will return to base with their weapons.

CHAPTER 9

Comparing Two Means and Two Variances

In this chapter, we continue the study of two sample problems by considering methods for comparing the means of two populations. This problem is considered under two different experimental conditions, namely, when the samples drawn are independent and when the data are paired. These terms are explained in depth in the sections to come.

9.1 POINT ESTIMATION: INDEPENDENT SAMPLES

The general situation that we consider now is described as follows:

> There are two populations of interest, each with unknown mean. One random sample is drawn from the first population and one from the second in such a way that the objects selected from the first population have no bearing on those selected from the second. Samples selected in this way are said to be *independent* of one another. We want to estimate $\mu_1 - \mu_2$, the difference in population means, via a point estimator.

Example 9.1.1 illustrates this idea in a practical context.

Example 9.1.1. A study is conducted to compare the time required to inspect the wiring connections and insulation in two types of circuit breakers. Population I consists of all circuit breakers of the vacuum-interruptor type, and population II consists of all air-magnetic circuit breakers. A random sample is selected from each of these populations, and each circuit breaker chosen is inspected and the time in minutes required for the inspection is recorded. The samples are independent in the sense that the choice of a circuit breaker from population I has no effect whatsoever on the choice of circuit breakers from population II. We want to estimate $\mu_1 - \mu_2$, the difference in the mean times required to perform the inspection for the two populations. The study is visualized in Fig. 9.1.

The logical way to estimate $\mu_1 - \mu_2$ is to estimate each mean separately via its corresponding sample mean and then estimate $\mu_1 - \mu_2$ to be the difference between these sample means. That is, a logical point estimator for the difference in population means is the difference in sample means.

> **Point Estimator for the Difference Between Two Means**
> $$\widehat{\mu_1 - \mu_2} = \hat{\mu}_1 - \hat{\mu}_2 = \bar{X}_1 - \bar{X}_2$$

Figure 9.1
Independent samples of circuit breakers drawn from two different populations.

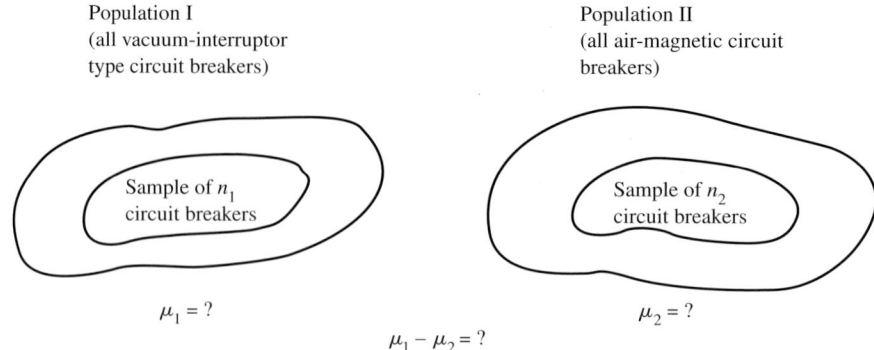

Population I
(all vacuum-interruptor type circuit breakers)

Population II
(all air-magnetic circuit breakers)

Sample of n_1 circuit breakers

Sample of n_2 circuit breakers

$\mu_1 = ?$

$\mu_2 = ?$

$\mu_1 - \mu_2 = ?$

Example 9.1.2. When the study of Example 9.1.1 is completed, these data result:

Vacuum-interruptor (I)				Air-magnetic (II)		
3.0	5.3	6.9	4.1	7.1	9.3	8.2
8.0	6.7	6.3		10.4	9.1	8.7
7.1	4.2	7.2		12.1	10.7	10.6
5.1	5.5	5.8		10.5	11.3	11.5

Based on these data,

$$\hat{\mu}_1 = \bar{x}_1 = 75.2/13 = 5.78 \text{ min}$$
$$\hat{\mu}_2 = \bar{x}_2 = 119.5/12 = 9.96 \text{ min}$$

The estimated difference in mean inspection times is

$$\widehat{\mu_1 - \mu_2} = \hat{\mu}_1 - \hat{\mu}_2 = \bar{x}_1 - \bar{x}_2 = 5.78 - 9.96 = -4.18$$

Based on these data, it appears that, on the average, the vacuum-interruptor circuit breaker can be inspected in about 4.18 minutes less time than the air-magnetic type breaker.

When finding the confidence intervals for $\mu_1 - \mu_2$ or when testing a hypothesis concerning the value of this difference, it is necessary to know the distribution of the random variable $\bar{X}_1 - \bar{X}_2$. The next theorem pinpoints its distribution under the assumption that both samples are drawn from normal distributions. The theorem also shows that the estimator $\bar{X}_1 - \bar{X}_2$ is an unbiased estimator for $\mu_1 - \mu_2$. We shall use this theorem to motivate many of the statistical procedures presented later.

> **Theorem 9.1.1 (Distribution of $\bar{X}_1 - \bar{X}_2$).** Let $\bar{X}_1$ and $\bar{X}_2$ be the sample means based on independent random samples of sizes n_1 and n_2 drawn from normal distributions with means μ_1 and μ_2 and variances σ_1^2 and σ_2^2, respectively. Then $\bar{X}_1 - \bar{X}_2$ is normal with mean $\mu_1 - \mu_2$ and variance $\sigma_1^2/n_1 + \sigma_2^2/n_2$.

As in the one-sample case, because of the Central Limit Theorem, it is safe to assume that for large sample sizes $\bar{X}_1 - \bar{X}_2$ is at least approximately normal even if the samples are drawn from populations that are not themselves normal.

9.2   COMPARING VARIANCES: THE *F* DISTRIBUTION

There are two opinions as to the best way to compare the means of two normal populations. This is due to the fact that there are two distinct possibilities. These are

1. σ_1^2 and σ_2^2 are unknown and equal.
2. σ_1^2 and σ_2^2 are unknown and unequal.

One philosophy is that the experimental data or past experience should be used as a guide to determine the prevailing situation. Then one of two possible test statistics is chosen to compare means, with the choice dependent on the perceived relationship between the population variances. A second philosophy disregards the relationship between the variances and uses the same test statistic to compare means in both cases. Since you will see both approaches used in research literature, we shall discuss them both. You can decide for yourself which you prefer.

The first philosophy mentioned requires that we develop a test for comparing the variances of two normal populations. Theoretically, tests on the relationship between two variances can take any of the usual three forms. In practice, only two are needed. These are:

$$\text{I } H_0: \sigma_1^2 = \sigma_2^2 \qquad \text{II } H_0: \sigma_1^2 = \sigma_2^2$$
$$H_1: \sigma_1^2 > \sigma_2^2 \qquad \quad H_1: \sigma_1^2 \neq \sigma_2^2$$
$$\text{Right-tailed test} \qquad \text{Two-tailed test}$$

where, in the right-tailed case, σ_1^2 denotes the population variance thought to be the *larger* of the two. To test either of these hypotheses, a test statistic must be developed. The statistic should be logical, but more importantly, it must be such that its probability distribution is known under the assumption that the null hypothesis is true. That is, its distribution must be known when it is assumed that the population variances are equal.

It is easy to find a logical statistic for comparing variances. Recall that the sample variances S_1^2 and S_2^2 are unbiased estimators for the population variances σ_1^2 and σ_2^2, respectively. Thus to compare σ_1^2 with σ_2^2, we simply compare S_1^2 with S_2^2. This is done not by looking at the difference of the two, but, rather, by looking at their ratio, S_1^2/S_2^2. If the null hypothesis is true and the population variances are really equal, then we expect S_1^2 and S_2^2 to be close in value, forcing S_1^2/S_2^2 to be close to 1.

If S_1^2/S_2^2 is much larger than 1, then we conclude that the population variances are different. When we use the phrase "much larger than 1" we are speaking in terms of probabilities. That is, an observed value of the statistic is much larger than 1 if it is too large to have reasonably occurred by chance if, in fact, the population variances are equal. To determine the probability of observing various values of the statistic S_1^2/S_2^2, we must know its probability distribution. We shall show that this statistic follows a distribution previously unencountered. In particular, if the population variances are equal, it follows what is called an *F* distribution. This distribution is defined in terms of a distribution previously studied, namely, the chi-squared distribution. In particular, any *F* random variable can be written as the ratio of two independent chi-squared random variables, each divided by their respective degrees of freedom. The formal definition of the *F* distribution is given in Definition 9.2.1.

Definition 9.2.1 (*F* distribution). Let $\mathbf{X}^2_{\gamma_1}$ and $\mathbf{X}^2_{\gamma_2}$ be independent chi-squared random variables with γ_1 and γ_2 degrees of freedom, respectively. The random variable

$$\frac{\mathbf{X}^2_{\gamma_1}/\gamma_1}{\mathbf{X}^2_{\gamma_2}/\gamma_2}$$

follows what is called an *F* distribution with γ_1 and γ_2 degrees of freedom.

The important properties of the family of *F* random variables are summarized as follows:

Properties of *F* Distributions

1. There are infinitely many *F* random variables, each identified by two parameters, γ_1 and γ_2, called *degrees of freedom*. These parameters are always positive integers: γ_1 is associated with the chi-squared random variable of the numerator of the *F* random variable, and γ_2 is associated with the chi-squared random variable of the denominator. The notation F_{γ_1, γ_2} denotes an *F* random variable with γ_1 and γ_2 degrees of freedom.

2. Each *F* random variable is continuous.

3. The graph of the density of each *F* random variable is an asymmetric curve of the general shape shown in Fig. 9.2.

4. *F* random variables cannot assume negative values.

A partial summary of the cumulative distribution for *F* random variables with selected degrees of freedom is given in Table IX of App. A. In the table γ_1, the degrees of freedom for the numerator, appears as column headings; γ_2, the degrees of freedom for the denominator, appears as row headings. *F* points for degrees of freedom that exceed 120 may be approximated well via row or column 120. Once again, we use the notational convention of denoting the point of the F_{γ_1, γ_2} curve with area *r* to its right by f_r. Example 9.2.1 illustrates the use of Table IX.

Example 9.2.1. Consider $F_{10, 15}$, the *F* random variable with 10 and 15 degrees of freedom.

(a) Find $P[F_{10, 15} \leq 2.544]$. This probability can be read directly from Table IX. Simply scan the numbers in column 10 and row 15 until you locate 2.544. It can be seen that $P[F_{10, 15} \leq 2.544] = F(2.544) = .95$.

(b) Find $P[F_{10, 15} > 2.059]$. Since the *F* distribution is continuous, this probability is $1 - F(2.059)$. From Table IX, $F(2.059) = .90$. Hence $P[F_{10, 15} > 2.059] = .10$.

(c) Via our notational convention we can say that $f_{.05} = 2.544$ and $f_{.10} = 2.059$.

Figure 9.2
A typical *F* density.

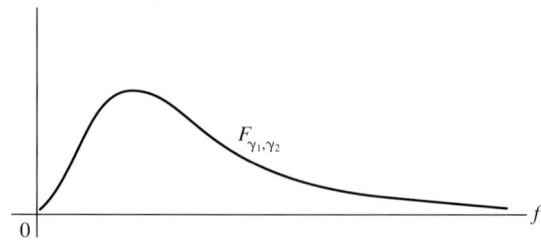

We are now in a position to verify that our proposed statistic for testing $H_0: \sigma_1^2 = \sigma_2^2$ does indeed follow an F distribution when H_0 is true.

> **Theorem 9.2.1 (Distribution of S_1^2/S_2^2).** Let S_1^2 and S_2^2 be sample variances based on independent random samples of sizes n_1 and n_2 drawn from normal populations with means μ_1 and μ_2 and variances σ_1^2 and σ_2^2, respectively. If $\sigma_1^2 = \sigma_2^2$, then the statistic S_1^2/S_2^2 follows an F distribution with $n_1 - 1$ and $n_2 - 1$ degrees of freedom.

Proof. We have already shown that the random variable $(n - 1)S^2/\sigma^2$ follows a chi-squared distribution with $n - 1$ degrees of freedom. (Theorem 7.1.1.) Applying this result here, we can conclude that the random variables $(n_1 - 1)S_1^2/\sigma_1^2$ and $(n_2 - 1)S_2^2/\sigma_2^2$ are chi-squared random variables with $n_1 - 1$ and $n_2 - 1$ degrees of freedom, respectively. Furthermore, since sampling is independent, these chi-squared random variables are independent. By Definition 9.2.1 the random variable

$$\frac{\dfrac{(n_1 - 1)S_1^2/\sigma_1^2}{(n_1 - 1)}}{\dfrac{(n_2 - 1)S_2^2/\sigma_2^2}{(n_2 - 1)}} = \frac{\sigma_2^2 S_1^2}{\sigma_1^2 S_2^2}$$

follows an F distribution with $n_1 - 1$ and $n_2 - 1$ degrees of freedom. If $\sigma_1^2 = \sigma_2^2$, then the above ratio reduces to S_1^2/S_2^2 as desired.

Note that the degrees of freedom associated with the statistic S_1^2/S_2^2 are $n_1 - 1$ and $n_2 - 1$. That is, the number of degrees of freedom for the numerator is 1 less than the size of the sample drawn from population I; that of the denominator is 1 less than the size of the sample drawn from population II.

There are several things to realize concerning this F test.

Assumptions Underlying the F Test for Equal Variances

1. Normality is assumed, and the test is sensitive to violations of this assumption. If it appears from the stem-and-leaf diagram or a histogram that either population does not have at least an approximate bell shape, then the test should not be used.

2. The test for equality of variances performs best when sample sizes are equal. If they are very different and there is any doubt concerning the normality of the two sampled populations, then the test should not be used.

3. The test is not very powerful. That is, the null hypothesis that $\sigma_1^2 = \sigma_2^2$ will not be rejected fairly often when, in fact, the variances are different. To minimize this problem, it is suggested that the test be performed at a relatively high α level. (α levels as high as .20 are satisfactory.)

These restrictions on the use of the F test partially explain the preference of some statisticians for the second philosophy mentioned for comparing means.

Example 9.2.2. A study of two types of materials used in electrical conduits, tubes used to house electrical wires, is to be conducted. The purpose of the study is to compare the strength of one to the other. Strength is to be assessed by measuring the load in pounds required to crush a 6-inch piece of material to 40% of its original diameter. Two questions are posed. Each is to be answered statistically, based on information obtained from independently drawn samples of the two materials.

The primary question is, "Does material A on the average withstand a heavier load than material B?" That is, "Is $\mu_A > \mu_B$?" However, before this question can be answered, we want to consider the question, "Is $\sigma_A^2 = \sigma_B^2$?" From an experiment, these data are obtained:

Material A	Material B
$n_A = 25$	$n_B = 16$
$\bar{x}_A = 380$ lb	$\bar{x}_B = 370$ lb
$s_A^2 = 100$	$s_B^2 = 400$

Suppose that the significance level is .1.

[Hypotheses] We wish first to test

$$H_0: \sigma_A^2 = \sigma_B^2$$
$$H_1: \sigma_A^2 \neq \sigma_B^2$$

[Test statistic] To compare variances, we form the ratio S_1^2/S_2^2 where S_1^2 is the *larger* of the two sample variances. In this case S_1^2 is the sample variance for material B, 400, and S_2^2 is the sample variance for material A, 100. The observed value of the test statistic is

$$s_B^2/s_A^2 = 4$$

Since this value is somewhat larger than 1, there is some evidence that the variances of the two materials differ.

[Critical value] The number of degrees of freedom associated with the test statistic are $n_B - 1 = 16 - 1 = 15$ and $n_A - 1 = 25 - 1 = 24$. This test is two-tailed. We enter Table IX of App. A with 15 and 24 degrees of freedom. The right critical value is

$$f_{\alpha/2} = f_{.05} = 2.108$$

[Conclusion] Since the value of test statistic is much larger than critical value, we can reject the null hypothesis of equal variances and conclude that the two variances are different. To compare averages, we should use a test procedure that does *not* assume that population variances are the same.

9.3 COMPARING MEANS: VARIANCES EQUAL (POOLED TEST)

Suppose that the primary objective of a study is to compare means and, after considering the information at hand, we have no reason to believe that population variances are unequal. In this case we can use a procedure called the *pooled, independent,* or *uncorrelated T test* to compare μ_1 to μ_2. The comparison can be done via confidence interval estimation or by means of a hypothesis or a significance test. We begin by developing the bounds for a $100(1 - \alpha)\%$ confidence interval on the differences in population means.

Confidence Interval on $\mu_1 - \mu_2$: Pooled

It has been shown that $\bar{X}_1 - \bar{X}_2$ is an unbiased estimator for $\mu_1 - \mu_2$. To extend this point estimator to a confidence interval, once again, we must find a random variable whose expression involves the parameter of interest, in this case $\mu_1 - \mu_2$, whose distribution is known. Such a random variable is provided by Theorem 9.1.1. This theorem states that when normal populations are sampled, the random variable

$\bar{X}_1 - \bar{X}_2$ is normal with mean $\mu_1 - \mu_2$ and variance $\sigma_1^2/n_1 + \sigma_2^2/n_2$. By standardizing this random variable, it can be concluded that the random variable

$$\frac{(\bar{X}_1 - \bar{X}_2) - (\mu_1 - \mu_2)}{\sqrt{\sigma_1^2/n_1 + \sigma_2^2/n_2}}$$

is standard normal. If the population variances have been compared and no difference has been detected, then we assume that they are equal. Let σ^2 denote this common population variance. That is, let $\sigma_1^2 = \sigma_2^2 = \sigma^2$. Substituting into the above expression, we conclude that

$$\frac{(\bar{X}_1 - \bar{X}_2) - (\mu_1 - \mu_2)}{\sqrt{\sigma^2(1/n_1 + 1/n_2)}}$$

is standard normal. Since σ^2 is unknown, it must be estimated from the data. This is one by a *pooled* sample variance. Note that we already have two unbiased estimators for σ^2, namely, S_1^2 and S_2^2. The idea is to pool, or combine, these estimators to form a single unbiased estimator for σ^2 in such a way that sample sizes are taken into account. It is natural to want to attach greater importance, or "weight," to the sample variance associated with the larger sample. The pooled variance, as defined now, does exactly this.

> **Definition 9.3.1 (Pooled variance).** Let S_1^2 and S_2^2 be the sample variances based on independent samples of sizes n_1 and n_2, respectively. The *pooled variance,* denoted by S_p^2, is given by
>
> $$S_p^2 = \frac{(n_1 - 1)S_1^2 + (n_2 - 1)S_2^2}{n_1 + n_2 - 2}$$

Note that we weight S_1^2 and S_2^2 by multiplying by $n_1 - 1$ and $n_2 - 1$, respectively. The more natural way to weight is to multiply by the corresponding sample sizes n_1 and n_2. We choose to weight in this somewhat odd way so that the random variable $(n_1 + n_2 - 2)S_p^2/\sigma^2$ will follow a chi-squared distribution. This is necessary so that the test statistic that we use to test for equality of means will follow a T distribution.

Example 9.3.1. Consider a sample variance $s_1^2 = 24$ based on a sample of size 16 and a second sample variance $s_2^2 = 20$ based on a sample size of 121. The value of the ratio s_1^2/s_2^2 is $24/20 = 1.20$. Based on these sample variances, the population variances σ_1^2, and σ_2^2 cannot be declared to be different even with α set at .2 ($f_{,1}1.545$). The pooled estimate for the common population variance is

$$\hat{\sigma}^2 = s_p^2 = \frac{(n_1 - 1)s_1^2 + (n_2 - 1)s_2^2}{n_1 + n_2 - 2}$$

$$= \frac{15(24) + 120(20)}{16 + 121 - 2}$$

$$= \frac{2760}{135} = 20.44$$

Note that this estimate is quite different from 22, the value obtained by ignoring sample sizes and arithmetically averaging s_1^2 and s_2^2.

To obtain a random variable that can be used to construct a $100(1 - \alpha)\%$ confidence interval on $\mu_1 - \mu_2$, we replace the unknown population variance σ^2 in the Z random variable

$$\frac{(\bar{X}_1 - \bar{X}_2) - (\mu_1 - \mu_2)}{\sqrt{\sigma^2(1/n_1 + 1/n_2)}}$$

by the pooled estimator S_p^2, to obtain the random variable

$$\frac{(\bar{X}_1 - \bar{X}_2) - (\mu_1 - \mu_2)}{\sqrt{S_p^2(1/n_1 + 1/n_2)}}$$

As in the one-sample case, replacing the population variance by its estimator does affect the distribution. The former random variable is a Z random variable; the latter has a T distribution with $n_1 + n_2 - 2$ degrees of freedom. The algebraic structure of this random variable is the same as that encountered previously, namely,

$$\frac{\text{Estimator} - \text{parameter}}{D}$$

Therefore the confidence interval on $\mu_1 - \mu_2$ takes the same general form as most of the intervals encountered previously. These bounds are given in Theorem 9.3.1.

Theorem 9.3.1 (Confidence interval on $\mu_1 - \mu_2$: Pooled variance). Let $\bar{X}_1$ and $\bar{X}_2$ be sample means based on independent random samples drawn from normal distributions with means μ_1 and μ_2, respectively, and common variance σ^2. Let S_p^2 denote the pooled sample variance. The bounds for a $100(1 - \alpha)\%$ confidence interval on $\mu_1 - \mu_2$ are

$$(\bar{X}_1 - \bar{X}_2) \pm t_{\alpha/2}\sqrt{S_p^2(1/n_1 + 1/n_2)}$$

where the point $t_{\alpha/2}$ is found relative to the $T_{n_1 + n_2 - 2}$ distribution.

Example 9.3.2. A study is conducted to estimate the difference in the mean occupational exposure to radioactivity in utility workers in the years 1973 and 1979. These data based on independent samples of workers for the 2 years are obtained:

1973	1979
$n_1 = 16$	$n_2 = 16$
$\bar{x}_1 = .94$ rem	$\bar{x}_2 = .62$ rem
$s_1^2 = .040$	$s_2^2 = .028$

We first check for equality of variances at the $\alpha = .2$ level.

[Hypotheses] We test

$$H_0: \sigma_1^2 = \sigma_2^2$$
$$H_1: \sigma_1^2 \neq \sigma_2^2$$

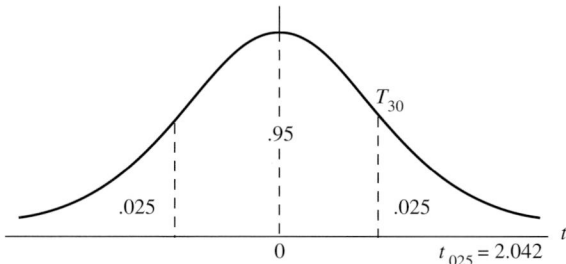

[**Test statistic**] Since the larger sample variance is the numerator of the test statistic, we see that the observed value of the test statistic is

$$s_1^2/s_2^2 = .040/.028 \doteq 1.43$$

[**Critical value**] We enter the F table, Table IX, with $n_1 - 1 = 15$ and $n_2 - 1 = 15$ degrees of freedom. It can be seen that

$$f_{\alpha/2} = f_{.10} = 1.972$$

[**Conclusion**] Since the value of test statistic is less than the critical value ($1.43 < 1.972$), we are unable to reject the null hypothesis of equal variances, even at an α level of .20.

We do not have strong evidence that the population variances differ. We, therefore, pool s_1^2 and s_2^2 and estimate σ^2 by

$$\hat{\sigma}^2 = s_p^2 = \frac{15(.040) + 15(.028)}{16 + 16 - 2} = .034$$

To compare means, let us find a 95% confidence interval on $\mu_1 - \mu_2$. The partition of the $T_{16 + 16 - 2} = T_{30}$ curve needed is shown in Fig. 9.3. The bounds for the confidence interval are

$$(\bar{x}_1 - \bar{x}_2) \pm t_{\alpha/2}\sqrt{s_p^2(1/n_1 + 1/n_2)} = (.94 - .62) \pm 2.042\sqrt{.034(1/16 + 1/16)}$$
$$= .32 \pm .13$$

We can be 95% confident that the difference in mean occupational exposure to radioactivity for the 2 years in between .19 and .45 rem. This interval does not contain the number 0 and is positive-valued throughout, an indication that the mean exposure in 1973 was, in fact, higher than in 1979.

Pooled T Test

As in previous instances, the random variable used to derive confidence bounds for a parameter also serves as a test statistic for testing various hypotheses concerning the parameter. In this case the following random variable serves as a test statistic for testing any of the usual hypotheses, where $(\mu_1 - \mu_2)_0$ denotes the hypothesized difference in population means:

> **Pooled T Test Statistic**
> $$\frac{(\bar{X}_1 - \bar{X}_2) - (\mu_1 - \mu_2)_0}{\sqrt{S_p^2(1/n_1 + 1/n_2)}} = T_{n_1 + n_2 - 2}$$

The hypothesized difference can be any value whatsoever. However, the most commonly encountered hypothesized value is zero. In this case the purpose is to determine whether the population means differ and, if so, which is the larger. Such hypotheses take these forms:

I	H_0: $\mu_1 = \mu_2$	II	H_0: $\mu_1 = \mu_2$	III	H_0: $\mu_1 = \mu_2$
	H_1: $\mu_1 > \mu_2$		H_1: $\mu_1 < \mu_2$		H_1: $\mu_1 \neq \mu_2$
	Right-tailed test		Left-tailed test		Two-tailed test

We can distinguish between H_0 and H_1 by presetting α and performing a hypothesis test or by performing a significance test and then reporting its P value, leaving to the researcher the final decision of whether or not to reject H_0.

Example 9.3.3. The tensile strength of a material is the ability that the material possesses to resist deformation when a force or a load is applied to it. A study of the tensile strength of ductile iron annealed or strengthened at two different temperatures is conducted. It is thought that the lower temperature will yield the higher mean tensile strength. These data result:

1450°F	1650°F
$n_1 = 10$	$n_2 = 16$
$\bar{x}_1 = 18{,}900$ psi	$\bar{x}_2 = 17{,}500$ psi
$s_1^2 = 1600$	$s_2^2 = 2500$

We first ensure that pooling is appropriate. The significance level of the test for variances is .2.

[Hypotheses] We test

$$H_0: \sigma_1^2 = \sigma_2^2$$
$$H_1: \sigma_1^2 \neq \sigma_2^2$$

[Test statistic] By using the larger sample variance as the numerator of the test statistic we obtain

$$s_1^2/s_2^2 = 2500/1600 \approx 1.5625$$

as the observed value of the F test statistic.

[Critical value] This test is two-tailed. The number of degrees of freedom associated with the statistic are 15 and 9. From Table IX of App. A we see that

$$f_{\alpha/2} = f_{.10} = 2.340$$

which is the critical value.

[Conclusion] Since the value of the test statistic is less than the critical value ($1.5625 < 2.340$), we are unable to reject the null hypothesis and therefore we shall pool s_1^2 and s_2^2 to estimate the common population variance.

Next, we carry out the test for population means. In this part, we assume that the significance level is .05.

[Hypotheses] We test

$$H_0: \mu_1 = \mu_2$$
$$H_1: \mu_1 > \mu_2$$

[**Test statistic**] We have justified pooling the sample variances in the above. In this case

$$s_p^2 = \frac{(n_1 - 1)s_1^2 + (n_2 - 1)s_2^2}{n_1 + n_2 - 2} = \frac{9(1600) + 15(2500)}{10 + 16 - 2} = 2162.5$$

The observed value of the test statistic is

$$\frac{(\bar{x}_1 - \bar{x}_2) - (\mu_1 - \mu_2)_0}{\sqrt{s_p^2(1/n_1 + 1/n_2)}} = \frac{(18,900 - 17,500) - 0}{\sqrt{2162.5(1/10 + 1/16)}} = 74.68$$

[**Critical value**] Based on the $T_{10 + 16 - 2} = T_{24}$ distribution, the critical value is

$$t = t_{.05} = 1.711$$

[**Conclusion**] Since the value of test statistic is much larger than the critical value ($74.68 > 1.711$), H_0 is rejected. We have very strong evidence that the mean tensile strength of iron annealed at $1450°$ F is higher than the mean strength of that annealed at $1650°F$.

We can also estimate the P value of the test. Based on the the $T_{10 + 16 - 2} = T_{24}$ distribution, the P value, which is the probability of observing a value of 74.68 or larger if $\mu_1 = \mu_2$, is less than $.0005(t_{.0005} = 3.745)$.

9.4   COMPARING MEANS: VARIANCES UNEQUAL

If a difference is detected when the population variances are compared, then pooling is inappropriate. It is still possible to compare means using an approximate T statistic. Again, the desired statistic is found by modifying the Z random variable in a logical way.

$$\frac{(\bar{X}_1 - \bar{X}_2) - (\mu_1 - \mu_2)}{\sqrt{\sigma_1^2/n_1 + \sigma_2^2/n_2}}$$

Since now there is evidence that $\sigma_1^2 \neq \sigma_2^2$, each population variance is estimated separately; these estimates are *not* combined. Instead, the population variances in the Z random variable above are replaced by their respective estimators, S_1^2 and S_2^2, to obtain this test statistic:

Unequal Variance Test Statistic

$$\frac{(\bar{X}_1 - \bar{X}_2) - (\mu_1 - \mu_2)_0}{\sqrt{S_1^2/n_1 + S_2^2/n_2}}$$

As in the past, making this change results in a change in distribution from Z to an approximate T. This time, however, the number of degrees of freedom must be estimated from the data. Several methods have been suggested for doing this. Here we demonstrate the *Smith-Satterthwaite* procedure. According to this procedure, γ, the number of degrees of freedom, is given by

Smith-Satterthwaite Degrees of Freedom

$$\gamma \doteq \frac{[S_1^2/n_1 + S_2^2/n_2]^2}{\dfrac{[S_1^2/n_1]^2}{n_1 - 1} + \dfrac{[S_2^2/n_2]^2}{n_2 - 1}}$$

The value for γ will not necessarily be an integer. If it is not, we round it *down* to the nearest integer. We round down rather than up in order to take a conservative approach. As the number of degrees of freedom associated with T random variables increases, the corresponding bell-shaped curves become more compact. Practically speaking, this means that, for example, the point $t_{.05}$ associated with the T_{10} curve (1.812) is a little larger than the point $t_{.05}$ associated with the T_{11} curve (1.796). If we can reject a null hypothesis based on the T_{10} distribution, it will also be rejected based on the T_{11} distribution. The converse does not necessarily hold.

The Smith-Satterthwaite procedure is illustrated in a significance testing context in the next example.

Example 9.4.1.   In Example 9.2.2 we began a study of the load-bearing properties of two materials used in electrical conduits. The primary question posed was, "Is material A, on the average, better able to withstand a heavy load than material B?" That is, "Is $\mu_A > \mu_B$?" These data were gathered:

Material A	Material B
$n_A = 25$	$n_B = 16$
$\bar{x}_A = 380$ lb	$\bar{x}_B = 370$ lb
$s_A^2 = 100$	$s_B^2 = 400$

Suppose that the significance level is .05.

[Hypotheses] We test

$$H_0: \mu_A = \mu_B$$
$$H_1: \mu_A > \mu_B$$

[Test statistic] We tested for equality of variances and found evidence that $\sigma_A^2 \neq \sigma_B^2$. Therefore, we do *not* pool s_A^2 and s_B^2. Rather, we use the Smith-Satterthwaite procedure. The observed value of the test statistic is

$$\frac{(\bar{x}_A - \bar{x}_B) - (\mu_A - \mu_B)_0}{\sqrt{s_A^2/n_A + s_B^2/n_B}} = \frac{(380 - 370) - 0}{\sqrt{100/25 + 400/16}} = 1.857$$

[Critical value] The degrees of freedom required are

$$\gamma \doteq \frac{[s_A^2/n_A + s_B^2/n_B]^2}{\dfrac{[s_A^2/n_A]^2}{n_A - 1} + \dfrac{[s_B^2/n_B]^2}{n_B - 1}}$$

$$= \frac{[100/25 + 400/16]^2}{\dfrac{[100/25]^2}{25 - 1} + \dfrac{[400/16]^2}{16 - 1}}$$

$$= 19.86$$

This value is rounded *down* to 19. Based on the T_{19} distribution, the critical value is

$$t_{.05} = 1.729$$

[Conclusion] Since the value of test statistic is greater than the critical value (1.857 > 1.729), we can reject H_0 and conclude that material A is capable of withstanding heavier loads on the average than is material B.

The P value of the test can be estimated as the follows. Based on the T_{19} distribution, $t_{.05} = 1.729$ and $t_{.025} = 2.093$. Since the observed value of our test statistic lies between these two values, the P value of our test lies between .025 and .05.

The Smith-Satterthwaite procedure can be used to construct confidence bounds on $\mu_1 - \mu_2$ when the population variances are unequal. The use of these bounds is outlined in Exercise 27.

We mentioned earlier that there are two opinions concerning the best course of action when comparing two means. The first, which entails the use of a preliminary F test run at a high α level to decide whether or not to pool, has been demonstrated in the last two sections. It embraces a philosophy that might be called a "sometimes pool" point of view. The second philosophy makes use of a very nice property of the Smith-Satterthwaite procedure. Namely, recent simulation studies have shown that not only does it perform well when variances are unequal, but it yields results that are virtually equivalent to those obtained with the pooled T test when variances are equal. For this reason, there seems to be no real need to pool; simply use the Smith-Satterthwaite procedure in all cases. Neither philosophy is clearly "best." However, the latter may be the safer road to take, as it avoids the pitfalls inherent in the use of the F test for variances. You are free to choose the procedure that appeals to you.

9.5 COMPARING MEANS: PAIRED DATA

In many instances problems arise in which two random samples are available but they are *not* independent; rather, each observation in one sample is naturally or by design paired with an observation in the other. To see what we mean, consider Example 9.5.1.

Example 9.5.1. One important aspect of computing is the cpu time required by a particular algorithm to solve a problem. A new algorithm is developed to solve zero-one multiple objective problems in linear programming. It is thought that the new algorithm will solve problems faster than the algorithm currently used. To obtain statistical evidence to support this research hypothesis, a number of problems will be selected at random. Each problem will be solved twice; once using the current algorithm and once using the newly developed one. Thus each test problem generates two observations, and we have two data sets. One data set represents a random sample of cpu times using the old algorithm; the other represents a sample of cpu times for the new one. These data sets are *not* independent; they are based on the same problems solved by two different methods and so are paired by design. The idea is illustrated in Fig. 9.4.

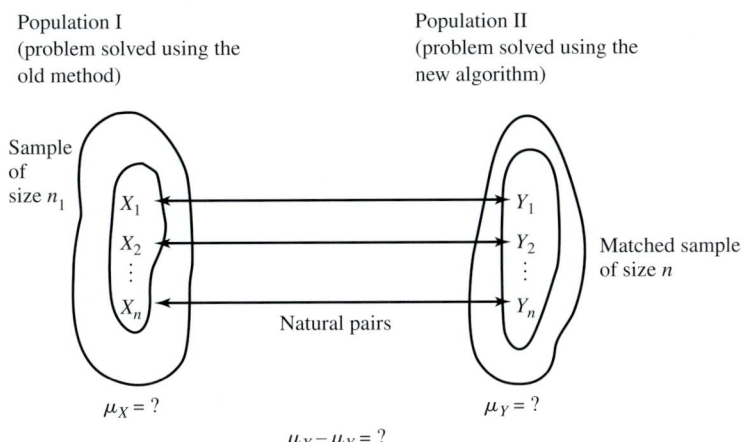

Population I
(problem solved using the old method)

Population II
(problem solved using the new algorithm)

Sample of size n_1

X_1 Y_1

X_2 Y_2

X_n Y_n

Matched sample of size n

Natural pairs

$\mu_X = ?$ $\mu_Y = ?$

$\mu_X - \mu_Y = ?$

Figure 9.4
Matched or paired samples of cpu times drawn from two different populations.

When pairing as we just illustrated occurs, the methods of Secs. 9.3 and 9.4 are no longer applicable. Rather, a procedure for answering the question, "What is $\mu_X - \mu_Y$?" must be developed that takes into account the fact that the observations are paired. This is done easily. Note that when data are paired, we can define a new random variable D by $D = X - Y$. The n differences $D_i = X_i - Y_i$, $i = 1, 2, 3, \ldots, n$ constitute a set of observations on D; that is, they constitute a random sample of size n drawn from the population of differences. Since, by the rules for expectation,

$$\mu_X - \mu_Y = E[X] - E[Y] = E[X - Y] = E[D] = \mu_D$$

the original question, "What is $\mu_X - \mu_Y$?" is equivalent to, "What is μ_D?" We are reduced from the original two-sample problem to the *one*-sample problem of making an inference on the mean of the population of differences. This problem is not new, and it can be handled using the methods of Chap. 7. In particular, the formula for the $100(1 - \alpha)\%$ confidence bounds on $\mu_X - \mu_Y = \mu_D$ is

Confidence bounds on $\mu_X - \mu_Y$ for paired data

$$\bar{D} \pm t_{\alpha/2} S_d / \sqrt{n}$$

In this formula $\bar{D}$ and S_d are the sample mean and sample standard deviation of the sample of difference scores, respectively, and $t_{\alpha/2}$ is the appropriate point relative to the T_{n-1} distribution.

Paired T Test

The null hypothesis $\mu_X = \mu_Y$ is equivalent to the hypotheses $\mu_D = 0$. The test statistic for testing this hypothesis based on the sample of difference scores is

Paired T Test Statistic

$$\frac{\bar{D} - 0}{S_d / \sqrt{n}}$$

which follows a T distribution with $n - 1$ degrees of freedom if H_0 is true. The use of this statistic is now illustrated.

Example 9.5.2. When the experiment described in Example 9.5.1 is conducted, these data result:

cpu time, s

Program	Old (x)	New (y)	Difference $d = x - y$
1	8.05	.71	7.34
2	24.74	.74	24.00
3	28.33	.74	27.59
4	8.45	.77	7.68
5	9.19	.80	8.39
6	25.20	.83	24.37
7	14.05	.82	13.23
8	20.33	.77	19.56
9	4.82	.71	4.11
10	8.54	.72	7.82

If the significance level is .05, do we have sufficient evidence to say that the new algorithm is faster than the old?

[Hypotheses] We want to test

$$H_0: \mu_X = \mu_Y$$
$$H_1: \mu_X > \mu_Y$$

This is equivalent to test

$$H_0: \mu_D = 0$$
$$H_1: \mu_D > 0$$

[Test statistic] For these data $\bar{d} = 14.409$ and $s_d = 8.653$, the observed value of the test statistic is

$$\frac{\bar{d} - 0}{s_d/\sqrt{n}} = \frac{14.409 - 0}{8.653/\sqrt{10}} = 5.266$$

[Critical value] The degrees of freedom of the test statistic is $n - 1 = 9$. From the Table VI of App. A, the critical value is

$$t = t_{.05} = 1.833$$

[Conclusion] Since the critical value is less than the value of test statistic ($1.833 < 5.266$), we reject H_0 and conclude that, on the average, the new algorithm is faster than the old.

Based on the $T_{n-1} = T_9$ distribution, the P value for this test is less than .0005 ($t_{.0005} = 4.781$).

In using the "paired T" procedures, it is assumed that the random variable $D = X - Y$ is at least approximately normally distributed. Note that in the case of a paired comparison we do not need to check for equality of variances. This is due to the fact that we are actually studying a single population, the population of differences. We are concerned only with the variance of this one population.

9.6 ALTERNATIVE NONPARAMETRIC METHODS

In Sec. 9.3 the T test for testing equality of population means for two independent samples was discussed. Under the assumptions of normally distributed random variables with equal but unknown population variances, this is the most powerful test for testing means. However, as one might expect, these rather restrictive assumptions are not always reasonable to assume in applications. For such situations an alternative nonparametric test is available that is almost as good as the T test even when all the necessary assumptions are met and may be considerably superior to the T test when the assumptions are clearly not met. If the sample sizes are reasonably large, the T test is quite robust. That is, the test is not very sensitive to departures from normality. However, for small samples, and particularly when the variances are unequal, the T test can lead to invalid conclusions. Under these circumstances a nonparametric test should be strongly considered as an alternative approach for testing equality of location for two populations. The most widely used such test is the *Wilcoxon rank-sum* test.

Wilcoxon Rank-Sum Test

Let X and Y be continuous random variables. Let $X_1, X_2, \ldots, X_m$ and $Y_1, Y_2, \ldots, Y_n$ be independent random samples of size m and n from the underlying distribution of X and Y, respectively. For convenience we assume that the X sample represents the smaller sample, and hence $m \leq n$. The null hypothesis to be tested is that the X and Y populations are identical. However, the test that we use is especially sensitive to differences in location. For this reason, the null hypothesis is usually stated in terms of equal population medians. Thus the three forms that hypotheses may take are:

$H_0\colon M_X = M_Y$	$H_0\colon M_X = M_Y$	$H_0\colon M_X = M_Y$
$H_1\colon M_X > M_Y$	$H_1\colon M_X < M_Y$	$H_1\colon M_X \neq M_Y$
Right-tailed test	Left-tailed test	Two-tailed test

To perform the test, the $m + n$ observations are pooled to form a single sample with the group identity of each observation retained. These observations are then ordered smallest to largest and ranked from 1 to $N = m + n$. If ties occur, each tied value receives the average group rank as in previous Wilcoxon procedures. The test statistic, denoted by W_m, is the sum of the ranks associated with the observations that originally constituted the smaller sample (X values). The logic behind this choice of test statistic is this. If the X population is located below the Y population, then the smaller ranks will tend to be associated with the X values. This produces a small value of W_m. If the reverse is true, then W_m will tend to be large. Thus, logically, we should reject $H_0\colon M_X = M_Y$ in favor of $H_1\colon M_X < M_Y$ for small values of W_m; we reject $H_0\colon M_X = M_Y$ in favor of $H_1\colon M_X > M_Y$ for large values of W_m. Upper and lower critical points for selected values of m, n, and α are found in Table X of App. A. Example 9.6.1 demonstrates the use of this table.

Example 9.6.1. An experiment is conducted on two brands of kerosene heaters. The manufacturer of brand A claims that his model will heat an 8-foot room from 60 to 70°F in less time than a competitor's brand B. A random sample of 12 heaters is selected from brand B; an independent sample of 15 heaters is selected from brand A. The observations are timed in seconds to raise the room temperature the 10° specified.

Brand B		Brand A	
69.3	52.6	28.6	30.6
56.0	34.4	25.1	31.8
22.1	60.2	26.4	41.6
47.6	43.8	34.9	21.1
53.2		29.8	36.0
48.1		28.4	37.9
23.2		38.5	13.9
13.8		30.2	

Test the manufacturer's claim at .05 significance level.

[**Hypotheses**] The following hypothesis is to be tested:

$$H_0\colon M_B = M_A$$
$$H_1\colon M_B > M_A$$

[Test statistic] Ordering the pooled observations from the smallest to the largest, retaining group identity, we obtain the following corresponding ranks:

Observation	13.8	13.9	21.1	22.1	23.2	25.1	26.4	28.4	28.6
Brand	B	A	A	B	B	A	A	A	A
Rank	1	2	3	4	5	6	7	8	9

Observation	29.8	30.2	30.6	31.8	34.4	34.9	36.0	37.9	38.5
Brand	A	A	A	A	B	A	A	A	A
Rank	10	11	12	13	14	15	16	17	18

Observation	41.6	43.8	47.6	48.1	52.6	53.2	56.0	60.2	69.3
Brand	A	B	B	B	B	B	B	B	B
Rank	19	20	21	22	23	24	25	26	27

Brand B is the smaller sample ($m = 12$), and hence the test statistic W_m is

$$W_m = 1 + 4 + 5 + 14 + 20 + 21 + 22 + 23 + 24 + 25 + 26 + 27 = 212$$

[Critical value] From Table X, for $m = 12$, $n = m + 3 = 15$ and $\alpha = .05$, the critical value for a right-tailed test is 202.

[Conclusion] Since $W_m = 212 > 202$, we reject H_0 and conclude that brand A heaters do, in fact, raise the temperature in less time than brand B.

When the sample sizes m or n exceed the values in Table X, a large sample normal approximation can be used to test H_0. The test statistic is

$$\frac{W_m - E(W_m)}{\sqrt{\text{Var } W_m}}$$

This statistic is approximately distributed as a standard normal random variable, where

$$E(W_m) = [m(m + n + 1)/2]$$

and $$\text{Var } W_m = mn(m + n + 1)/12$$

Several things should be pointed out concerning the Wilcoxon statistic. First, although the null hypothesis is stated in terms of medians, if the distributions of X and Y are symmetric, we are also testing equality of means. Thus for normal populations the Wilcoxon statistic is analogous to the normal theory T test for independent samples. Second, the Wilcoxon statistic can be used with data that cannot be measured but that can, nevertheless, be ranked. Examples of data of this sort are given in Exercises 39 and 40.

Wilcoxon Signed-Rank Test for Paired Observations

In Sec. 9.5 the T test for paired data was discussed. The signed-rank test for paired observations is the nonparametric analog when the normal assumptions are not met. This test is almost as good as the paired T test even when the underlying distribution is normal, and it is usually preferred to the paired T test for other distributions.

We discussed the Wilcoxon signed-rank test for a single sample in Sec. 7.7. The corresponding test for paired data is a simple modification of the method given in Sec. 7.7. Here we let X and Y be continuous random variables that are assumed to have symmetric distributions. We want to test the hypothesis that the medians of these two distributions are equal. Thus our hypotheses takes the form

H_0: $M_X = M_Y$	H_0: $M_X = M_Y$	H_0: $M_X = M_Y$
H_1: $M_X > M_Y$	H_1: $M_X < M_Y$	H_1: $M_X \neq M_Y$
Right-tailed test	Left-tailed test	Two-tailed test

Consider a random sample $(X_1, Y_1), (X_2, Y_2), \ldots, (X_n, Y_n)$ of paired observations on X and Y. We first form the differences $X_1 - Y_1, X_2 - Y_2, \ldots, X_n - Y_n$. If the null hypothesis is true, the population of difference scores is symmetric about 0. Thus to test H_0: $M_X = M_Y$, we test H_0: $M_{X-Y} = 0$. The test is performed exactly as before. We first order the absolute values of the differences from the smallest to the largest and rank them from 1 to n. Tied scores are assigned the average group rank. Each rank is assigned the sign of the difference that generated the rank. Once again, the test statistics used are

$$W_+ = \sum_{\substack{\text{all} \\ \text{positive} \\ \text{ranks}}} R_i \qquad \text{and} \qquad W_- \sum_{\substack{\text{all} \\ \text{negative} \\ \text{ranks}}} R_i$$

Right-tailed tests are conducted via $|W_-|$, and left-tailed tests utilize W_+ as the test statistic. In each case we reject H_0 for values that are too small to have occurred by chance based on the critical points found in Table VIII of App. A.

The next example should refresh your memory of the Wilcoxon signed-rank procedure.

Example 9.6.2. An experiment is conducted to compare the amount of memory required to analyze a data set using the two leading statistical packages. These data are obtained:

Program	Package X	Package Y	Difference $X - Y$
1	512K	500K	12
2	650K	600K	50
3	890K	890K	0
4	410K	400K	10
5	1050K	1025K	25
6	1500K	1400K	100
7	600K	625K	−25
8	750K	710K	40

At the $\alpha = .1$ level, let us test whether there are differences in the medians of the amount of memory required.

[Hypotheses] Let us test

$$H_0: M_X = M_Y$$
$$H_1: M_X \neq M_Y$$

[Test statistic] We order the absolute values of the differences from the smallest to the largest and rank them from 1 to 8. We then assign to each rank the algebraic sign of the difference that generated the rank. The zero difference is assigned the algebraic sign that is least conductive to the rejection of H_0. In this case the zero difference is considered to be negative. We thus obtain these signed ranks:

$|X - Y|$	0	10	12	25	25	40	50	100
Rank	1	2	3	4.5	4.5	6	7	8
Signed Rank	−1	2	3	−4.5	4.5	6	7	8

For these data

$$W_+ = 2 + 3 + 4.5 + 6 + 7 + 8 = 30.5$$
$$|W_-| = |-1| + |-4.5| = 5.5$$

For a two-tailed test the test statistic is W, the smaller of W_1 and $|W_-|$.

[Critical value] From Table VIII of App. A we see that the critical point for a two-tailed test at the $\alpha = .1$ level is 6.

[Conclusion] Since $5.5 < 6$, we can reject H_0 and claim that there are differences in the medians of these two populations.

One other comment should be made. If the differences are such that we only know whether a difference is positive or negative, then the null hypothesis can be tested via the sign test. This idea was discussed in the one-sample context in Sec. 7.7. An example of data of this sort is given in Exercise 44.

9.7   A NOTE ON TECHNOLOGY

As you probably suspect, most of the techniques that have been demonstrated in this chapter can be implemented by available technology. However, in order to use the technology tools properly, the material in this chapter must be thoroughly understood or the tools can be misused and abused. In this section, we discuss briefly some of these statistical computing aids.

The TI83 calculator has been especially designed with statisticians and users of statistics in mind. It will perform many of the functions discussed in this chapter. In particular, it can perform a preliminary F test to compare variances using either raw data or summary statistics. The test chosen can be either right-tailed, two-tailed, or left-tailed. As demonstrated in this chapter, the latter test is not needed if the test is performed in such a way that the larger of the two sample variances is chosen as the numerator of the test statistic. The output generated will include the exact P value of the test. This P value can be compared to a preset α level if so desired. Based on the results of this test, either pooled or Smith-Satterthwaite T type confidence intervals or T tests can be found or conducted. These can be implemented using either raw data or summary statistics. In each case, you will be asked whether you want to pool variances.

There are many statistical packages on the market. Some, like SAS, include a preliminary F test and automatically include the results of the two-tailed F test for comparing variances as part of the output of the two-sample means comparison procedure. SAS, by default, runs both the pooled and Smith-Satterthwaite tests for comparing means. It is the responsibility of the user to choose the proper test based on the reported results of the F test for comparing variances. Example 9.7.1 gives the SAS output and its interpretation for the data of Exercise 26.

Example 9.7.1. In this example, the ability of a plasma coating to reduce wear in rotary valves used in the pulp and paper industry is investigated. Two samples of sizes 8 (coated valves) and 10 (uncoated valves) are selected. It is thought that the coating will reduce wear and hence the primary test is a left-tailed test to compare means. It is given by

$$H_0: \mu_c = \mu_u$$
$$H_1: \mu_c < \mu_u$$

where C denotes coated values and U denotes uncoated ones. The preliminary F test is two-tailed and is given by

$$H_0: \sigma_c^2 = \sigma_u^2$$
$$H_1: \sigma_c^2 \neq \sigma_u^2$$

The output of the SAS procedure and its interpretation is given below.

TESTING FOR EQUALITY OF MEANS AND VARIANCES
T TEST PROCEDURE

VARIABLE: WEAR

GROUP	N	MEAN	STD DEV	STD ERROR	MINIMUM	MAXIMUM
C	8	0.08300000	0.00924276	0.00326781	0.07200000	0.09900000
U	10	0.10420000	0.03342587	0.01057019	0.05200000	0.15600000

VARIANCES	T		DF	PROB> !T!	
UNEQUAL	−1.9162	③	10.7	④ 0.0825	⑤
EQUAL	−1.7320		16.0	0.1025	

FOR HO: VARIANCES ARE EQUAL, $F' = 13.08$ WITH 9 AND 7 DF

① PROB $> F' = 0.0027$
②

The value of the F statistic used to compare variances is 13.08. This value is shown in ①. The P value for the two-tailed F test is .0027. This is the P value listed in ②. Since this value is small, we conclude that $\sigma_c^2 \neq \sigma_u^2$, and use the Smith-Satterthwaite procedure to compare means. The T statistic and its corresponding degrees of freedom are shown in ③ and ④, respectively. The P value for the two-tailed test is .0825. This value is shown in ⑤. The one-tailed P value is .0825/2 = .04125.

Since this P value is small, we reject the null hypothesis and conclude that the average wear for coated valves is less than that for uncoated ones. You should compare these results with those you obtained by hand earlier.

In Example 9.7.2, you will see the MINITAB output for the same data as that analyzed in Example 9.7.1.

Example 9.7.2. The MINITAB package does not have an option to run a preliminary F test to compare variances, and it does not do so by default. This test can be run by hand, or a more informal approach can be taken. Since it is safer to not pool when pooling is appropriate than to pool when it should not be done, the easiest path to take is simply to never pool. This is the MINITAB default option. If it is thought that pooling is appropriate, then MINITAB allows you to select a pooling option. You must indicate whether your means test is to be right-, left-, or two-tailed. The MINITAB output for the data of Exercise 26 is shown below. Note that it includes a 95% confidence interval on the difference in means shown at ① as well as the results of the left-tailed test to compare means given at ②. You should compare the values given here with those shown on the SAS output.

Two Sample T Test and Confidence Interval

Two sample T for c vs u

	N	Mean	StDev	SE Mean
c	8	0.08300	0.00924	0.0033
u	10	0.1042	0.0334	0.011

① 95% CI for mu c − mu u: (−0.0459, 0.003)
② T-Test mu c = mu u (vs <): T = −1.92   P = 0.042   DF = 10

CHAPTER SUMMARY

In this chapter we continued our study of two sample problems by learning how to compare the means, variances, and medians of two populations. We considered two different experimental settings, namely, those problems in which independent samples are drawn from the two populations and problems in which data are paired.

To compare variances based on independent samples, it was necessary to introduce a new continuous distribution called the F distribution. Although some studies are designed specifically to compare variances, more often variances are compared as a first step in comparing means. If there is statistical evidence based on the F test that the population variances are not equal, then we use the Smith-Satterthwaite T procedure to compare means. Otherwise we can use either a pooled T procedure or the Smith-Satterthwaite procedure. The choice is yours. Each of these procedures assumes that sampling is from normal distributions. A nonparametric alternative to these tests was presented. This alternative procedure, called the Wilcoxon rank-sum test, does not require normality, and no knowledge of population variances is necessary.

When data are paired, we do not need to consider the individual population variances. In this case we work with a population of difference scores. It is assumed that this population is normally distributed and inferences are made on the difference in population means via a one-sample "paired" T test. The Wilcoxon signed-rank test for paired data was introduced as a nonparametric test for location when it is evident that the population of difference scores is not normally distributed.

We introduced and defined important terms that you should know. These are:

F distribution   Smith-Satterthwaite test
Pooled estimator for σ^2 Paired T test
Pooled T test

REAL WORLD APPLICATION OF COMPARING TWO MEANS

Obesity is a big problem in many developed countries since it is the cause of a lot of serious health problems. Many people are seeking for an effective weight loss method. Diet regimen is one of them.

The effectiveness of a diet regimen depends on many factors. Many kinds of weight loss modalities have demonstrated discordance in weight loss achieved between black and white populations. A study on this topic was carried out by the subjects who participated in the very low calorie diet (VLCD) weight loss program at the Obesity Center of the University of California between 1999 and 2001. A total of 304 (152 in each group) subjects were selected in the study. Their weight, BMI, blood pressure, serum lipid profile, and fasting glucose level were measured at baseline and after 12-week diet intervention.

The following table shows the mean weights for black and white patients at baseline.

	Black	Weight
Mean weight (kg)	105	104

In this study, Student's t-test was used to compare two means. It was found that the P value of the test is 0.907. So there was no significant difference in the weights between black and white subjects at baseline.

Following treatment, means of weight loss achieved in black and white subjects were measured and shown in the following table.

	Black	White
Mean weight loss (kg)	20.9	23.0

Paired t-test was used to determine the significance of the weight reduction of each group. The P values of the tests for black and white subjects were both found to be less than 0.0001. They were statistically significant and we conclude that the VLCD weight loss program effectively reduced the weights of both black and white participants.

EXERCISES

Section 9.1

1. A firm receives integrated circuits in lots of 100 from two different suppliers. These data are obtained on the number of defective items found per lot:

Supplier I					Supplier II				
3	8	5	7		0	2	3	1	3
2	3	8	1		1	1	5	4	0
5	0	6	2						

Estimate μ_1, μ_2, and $\mu_1 - \mu_2$.

2. Many gold and silver deposits that were once considered uneconomical are now being exploited, thanks to improvements in methods for recovering precious metals from ore. In a study to compare the potential of two different open-pit gold mines, ore samples are obtained from each mine. The mean number of ounces of gold recovered per ton of ore is .233 for the first mine and .127 for the second. Estimate the difference in the mean number of ounces of gold per ton of ore for these two mines.

3. A press used to remove water from copper-bearing materials is being tested using two different types of filter plates. These data are obtained on the percentage of moisture remaining in the material after treatment:

Regular chamber (I)			Diaphragm chamber (II)		
8.10	8.16	8.16	7.58	7.65	7.69
7.96	7.98	7.93	7.66	7.67	7.67
7.97	8.08	8.06	7.58	7.62	7.65
8.02	7.87	7.94	7.65	7.58	7.71
7.82	8.11	7.92	7.63	7.54	
8.15	7.91	8.00	7.46	7.40	

Estimate μ_1, μ_2, and $\mu_1 - \mu_2$.

4. Let $\bar{X}_1$ be the sample mean based on a sample of size 25 drawn from a normal distribution with mean 8 and variance 16. Let $\bar{X}_2$ be the sample mean based on a sample of size 36 drawn from a normal distribution with mean 5 and variance 9. What is the distribution of each of these random variables?
   (a) $\bar{X}_1$
   (b) $\bar{X}_2$
   (c) $(\bar{X}_1 - 8)/(4/5)$
   (d) $(\bar{X}_2 - 5)/(3/6)$

Figure 9.5

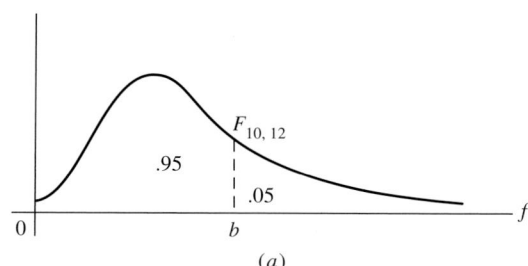

(a)

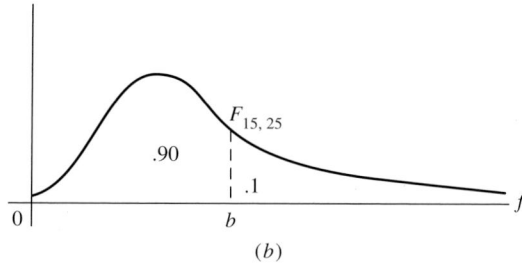

(b)

(e) $\bar{X}_1 - \bar{X}_2$

(f) $\dfrac{(\bar{X}_1 - \bar{X}_2) - (8 - 5)}{\sqrt{16/25 + 9/36}}$

Section 9.2

5. Use Table IX of App. A to find each of the following:

(a) $P[F_{10,9} \le 2.416]$ (b) $f_{.1}(10, 9 \text{ df})$

(c) $P[F_{20,4} \ge 5.803]$ (d) $f_{.05}(30, 30 \text{ df})$

(e) $P[F_{20,120} \ge 1.659]$ (f) $f_{.05}(20, 4 \text{ df})$

6. In each part of Fig. 9.5, find the point indicated.

7. In each case, test for equality of variances at the indicated level.

(a) $n_1 = 10$ $n_2 = 8$ $\alpha = .20$
     $s_1^2 = .25$ $s_2^2 = .05$

(b) $n_1 = 13$ $n_2 = 20$ $\alpha = .10$
     $s_1^2 = 4$ $s_2^2 = 2$

(c) $n_1 = 121$ $n_2 = 150$ $\alpha = .10$
     $s_1^2 = 6$ $s_2^2 = 4.4$

8. The cost of repairing a fiberoptic component may depend on the stage of production at which it fails. These data are obtained on the cost of repairing parts that fail when installed in the system and on the cost of repairing parts that fail after the system is installed in the field:

System failure	Field failure
$n_1 = 21$	$n_2 = 25$
$\bar{x}_1 = \$65$	$\bar{x}_2 = \$120$
$s_1^2 = 25$	$s_2^2 = 100$

(a) It is thought that the variance in cost of repairs made in the field is larger than the variance in cost of repairs made when the component is placed into the system. Set up the null and alternative hypotheses needed to gain statistical evidence to support this contention.

(b) Use the given data to test H_0 at the $\alpha = .10$ level.

9. A study of the sodium content in a 6-fluid ounce serving of a soft drink is conducted. These data are obtained on various types of ginger ales and cola drinks:

Ginger ale	Cola
$n_1 = 10$	$n_2 = 10$
$\bar{x}_1 = 9.6$	$\bar{x}_2 = 9.9$
$s_1^2 = 10.89$	$s_2^2 = 11.90$

(a) It is thought that the variability in the sodium content in ginger ales is smaller than that of colas. Set up the null and alternative hypotheses needed to support this contention.

(b) Use the given data to test H_0 at the $\alpha = .1$ level.

10. Prices for regular unleaded gasoline can vary widely from day to day and location to location. These data were obtained on June 1, 2001, from a sample of stations across the respective state (price is in dollars per gallon):

South Carolina					Michigan				
1.46	1.47	1.42	1.51	1.55	1.69	1.79	1.72	1.76	1.80
1.52	1.48	1.47	1.53	1.50	1.59	1.89	1.72	1.63	1.55
					1.91	1.71			

Use these data to test for equality of variances. What is the P value of your test, and what conclusion do you draw?

11. A study is conducted to compare the variability in the number of hours that a rechargable flashlight will operate after its battery has been fully charged. These data are obtained for two different brands of batteries:

Brand X	Brand Y
$n_1 = 25$	$n_2 = 21$
$s_1^2 = .021$	$s_2^2 = .018$

Use these data to test for equality of variances. What is the P value of your test? Can you conclude that the variances are unequal at the $\alpha = .2$ level? at the $\alpha = .1$ level?

Section 9.3

12. (a) Let $s_1^2 = 42, s_2^2 = 37, n_1 = 10, n_2 = 14$. Find s_p^2.

(b) Let $s_1^2 = 28, s_2^2 = 30, n_1 = 20, n_2 = 20$. Find s_p^2. Do not use your calculator!

(c) Let $s_1^2 = 20, s_2^2 = 40, n_1 = 10, n_2 = 50$. Find s_p^2. Why is s_p^2 closer in value to s_2^2 than to s_1^2?

13. A study of report writing by engineers is conducted. A scale that measures the intelligibility of engineers' English is devised. This scale, called an "index of confusion," is devised so that low scores indicate high readability. These data are obtained on articles randomly selected from engineering journals and from unpublished reports written in 1979:

Journals				Unpublished reports			
1.79	1.75	1.67	1.65	2.39	2.51	2.86	2.14
1.87	1.74	1.94		2.56	2.29	2.49	
1.62	2.06	1.33		2.36	2.58	2.33	
1.96	1.69	1.70		2.62	2.41	1.94	

(a) Test $H_0: \sigma_1^2 = \sigma_2^2$ at the $\alpha = .2$ level to be sure that pooling is appropriate.

(b) Find s_p^2.

(c) Find a 90% confidence interval on $\mu_1 - \mu_2$.

(d) Does there appear to be a difference between μ_1 and μ_2? Explain, based on the confidence interval of part (c).

14. To decide whether or not to purchase a new hand-held laser scanner for use in inventorying stock, tests are conducted on the scanner currently in use and on the new scanner. These data are obtained on the number of 7-inch bar codes that can be scanned per second:

New	Old
$n_1 = 61$	$n_2 = 61$
$\bar{x}_1 = 40$	$\bar{x}_2 = 29$
$s_1^2 = 24.9$	$s_2^2 = 22.7$

(a) Test $H_0: \sigma_1^2 = \sigma_2^2$ at the $\alpha = .2$ level to be sure that pooling is appropriate.

(b) Find s_p^2.

(c) Find a 90% confidence interval on $\mu_1 - \mu_2$.

(d) Does the new laser appear to read more bar codes per second on the average? Explain.

(e) Since the number of bar codes that can be scanned per second is discrete, we have not satisfied the normality requirement. What theorem justifies the use of the pooled T procedure in this case?

15. Environmental testing is an attempt to test a component under conditions that closely simulate the environment in which the component will be used. An electrical component is to be used in two different locations in Alaska. Before environmental testing can be conducted, it is necessary to determine the soil composition in these localities. These data are obtained on the percentage of SiO_2 by weight of the soil:

Anchorage	Kodiak
$n_1 = 10$	$n_2 = 16$
$\bar{x}_1 = 64.94$	$\bar{x}_2 = 57.06$
$s_1^2 = 9$	$s_2^2 = 7.29$

(a) Test $H_0: \sigma_1^2 = \sigma_2^2$ at the $\alpha = .2$ level.

(b) Find s_p^2.

(c) Find a 99% confidence interval on $\mu_1 - \mu_2$.

(d) Based on the interval of part (c), does there appear to be a difference between μ_1 and μ_2? Explain.

16. Show that $E[S_p^2] = \sigma^2$, thus proving that the pooled variance is an unbiased estimator for the common population variance.

17. During a total solar eclipse the temperature drops quickly as the moon passes between the earth and the sun. These data are obtained on the drop in temperature in degrees Fahrenheit at two types of locations in southern Africa during the June 2001 eclipse:

Mountainous terrain						River-level terrain					
15	12	16	16	13	15	13	17	19	16	15	19
11	19	15				18	20	21	22	24	

Is there evidence at the $\alpha = .20$ level of significance that there is a difference in the variances in temperature drop seen in these two terrains?

18. Use the data of Exercise 17 to form a 95% confidence interval on the difference in the average temperature drop between the two types of terrain. Based on this interval, is there evidence that a real difference exists? If so, which region appears to exhibit the greatest average change? Explain, based on the interval that you constructed.

19. The time in seconds required to connect to the Internet via a dial-in service is influenced by a variety of factors such as number of phone lines available in the local calling area, time of day, day of the week, number of users in the area, and so on. These data are obtained in a given area at two different times of the day but always on the same day of the week:

Morning (9:00 A.M. to 11:00 A.M.)							Night (10:00 P.M. to midnight)						
10	20	31	42	44	44	15	10	11	21	31	15	40	27
33	35	47	47	45	22	33	20	32	38	36	22	41	39
21	51	53	52	37	28	35	24	33	25	35	36	42	43
56	63	60	48	49	55	57	52	51					
62	61												

(*a*) Is pooling appropriate? Explain by comparing variances.

(*b*) Find a 99% confidence interval on the difference in the average time required to access the Internet during these two time periods. Which time period appears to give the fastest average access time?

(*c*) Why is the interval that you obtained so long? How could you use these same data to obtain a shorter interval?

Section 9.4

20. Calculate the number of degrees of freedom for a Smith-Satterthwaite procedure based on these data:

(*a*) $n_1 = 9$ $\qquad\qquad$ $n_2 = 16$
$\qquad$ $s_1^2 = 38.07$ $\qquad$ $s_2^2 = 16.89$

(*b*) $n_1 = 25$ $\qquad\qquad$ $n_2 = 25$
$\qquad$ $s_1^2 = .42$ $\qquad\qquad$ $s_2^2 = 1$

21. Strontium-90, a radioactive element produced by nuclear testing, is closely related to calcium. In dairy lands, strontium-90 can make its way into milk via the grasses eaten by dairy cows. It then becomes concentrated in the bones of those who drink the milk. In 1959 a study was conducted to compare the mean concentration of strontium-90 in the bones of children to that of adults. It was thought that the level in children was higher because the substance was present during their formative years.

(*a*) Set up the null and alternative hypotheses needed to verify this contention.

(*b*) Based on these data, is pooling appropriate?

Children	Adults
$n_1 = 121$	$n_2 = 61$
$\bar{x}_1 = 2.6$ picocuries per gram	$\bar{x}_2 = .4$ picocurie per gram
$s_1^2 = 1.44$	$s_2^2 = .0121$

(c) Test the null hypothesis of part (a). Can H_0 be rejected? Explain, based on the P value of your test. What practical conclusion can be drawn from these data?

22. Water and other nonaqueous volatiles are present in differing concentrations in coal from different seams. To measure the percentage by weight of these substances for a particular seam, readings are taken at two different temperatures. These data result:

Water

105° C			160° C		
15.11	15.30	15.44	15.14	15.33	15.40
15.23	15.32	15.48	15.28	15.34	15.77
15.27	15.37	15.36	15.26	15.38	15.52

Nonaqueous volatiles

105° C			160° C		
.343	.601	.676	1.533	1.780	1.625
.481	.543	.541	1.190	1.636	1.692
.475	.108	.106	2.015	1.464	1.991

(a) Use the water data to test

$$H_0: \mu_1 = \mu_2$$
$$H_1: \mu_1 \neq \mu_2$$

at the $\alpha = .05$ level. Does the temperature at which the readings are taken appear to affect the mean reading of the water concentration of the coal? Explain. Be ready to defend your choice of a test statistic.

(b) Use the nonaqueous volatiles data to test

$$H_0: \mu_1 = \mu_2$$
$$H_1: \mu_1 \neq \mu_2$$

at the $\alpha = .05$ level. Does the temperature at which the readings are taken appear to affect the mean reading of the concentration of nonaqueous volatiles in the coal? Explain.

23. It is thought that the gas mileage obtained by a particular model of automobile will be higher if unleaded premium gasoline is used in the vehicle rather than regular unleaded gasoline. To gather evidence to support this contention, 10 cars are randomly selected from the assembly line and tested using a specified brand of premium gasoline; 10 others are randomly selected and tested using the brand's regular gasoline. Tests are conducted under identical controlled conditions. These data result:

Premium		Regular	
35.4	31.7	29.7	34.8
34.5	35.4	29.6	34.6
31.6	35.3	32.1	34.8
32.4	36.6	35.4	32.6
34.8	36.0	34.0	32.2

(a) Set up the null and alternative hypotheses needed to compare the mean mileage for these two gasolines.

(b) Decide whether or not to reject H_0 via a significance test. What is the approximate P value of the test? Be ready to defend your choice of a test statistic.

(c) Interpret your results in the context of this problem.

24. A new coal liquefaction process is being studied. It is claimed that the new process results in a higher yield of distillate synthetic fuel than the current process. These observations are obtained on the number of kilograms of distillate synthetic fuel produced per kilogram of hydrogen consumed in the process:

New				Old			
16.4	12.8	15.4	17.0	11.1	10.5	10.9	10.1
17.7	12.2	18.7		12.8	13.2	12.6	
15.9	14.7	19.1		12.1	14.5	15.6	
11.3	14.1	16.5		14.2	15.3	14.2	

(a) Set up the null and alternative hypotheses needed to support the stated claim.

(b) Since putting the new process into production is very expensive, a Type I error is costly. To compensate for this, test the null hypothesis of part (a) at the $\alpha = .01$ level. Would you recommend that the new process be used? Explain.

25. A study is conducted to compare the tensile strength of two types of roof coatings. It is thought that, on the average, butyl coatings are stronger than acrylic coatings. These data are gathered:

Tensile strength, lb / in^2

Acrylic				Butyl			
246.3	247.7	287.5	248.3	340.7	263.4	272.6	271.4
255.0	246.3	284.6	243.7	270.1	341.6	332.6	303.9
245.8	214.0	268.7	276.7	371.6	307.0	362.2	324.7
250.7	242.7	302.6	254.9	306.6	319.1	358.1	360.1

(a) Set up the null and alternative hypotheses needed to verify the research hypothesis.

(b) Is pooling appropriate? Explain.

(c) Test the null hypothesis of part (a). Can H_0 be rejected? Explain, based on the P value of your test. What practical conclusion can be drawn in this case?

26. It is thought that the application of a plasma coating that contains submicron particles of tungsten carbide will reduce wear to rotary valves used in the pulp and paper industry. Tests are conducted to compare the wear in coated and uncoated valves. These data are gathered on the wear of the part in millimeters over the test period:

Coated		Uncoated		
.075	.099	.095	.074	.104
.078	.082	.096	.149	.052
.092	.088	.136	.081	
.078	.072	.156	.099	

(a) Set up the null and alternative hypotheses needed to support the contention that the plasma coating on the average reduces the wear in these valves.

(b) Based on these data, is pooling appropriate?

(c) Test the null hypothesis of part (a). Does it appear that the coating is effective in reducing wear? Explain, based on the P value of your test.

27. (*Confidence interval on* $\mu_1 - \mu_2$: *Variances unequal.*) The lower and upper bounds for a $100(1 - \alpha)\%$ confidence interval on $\mu_1 - \mu_2$ when $\sigma_1^2 \neq \sigma_2^2$ are given by

$$(\bar{X}_1 - \bar{X}_2) \pm t_{\alpha/2}\sqrt{S_1^2/n_1 + S_2^2/n_2}$$

Consider the data of Exercise 10. Use these data to find a 95% confidence interval on the difference in the average gasoline price per gallon in South Carolina and Michigan on the day that the data were collected. Does this interval provide good evidence that the average price was higher in Michigan on this day than it was in South Carolina? Explain, based on your confidence interval.

28. A manufacturer of power-steering components buys hydraulic seals from two sources. Samples are selected from among the seals obtained from these two suppliers, and each seal is tested to determine the amount of pressure that it can withstand. These data result:

Supplier I	Supplier II
$n_1 = 10$	$n_2 = 10$
$\bar{x}_1 = 1350$ lb/in^2	$\bar{x}_2 = 1338$ lb/in^2
$s_1^2 = 100$	$s_2^2 = 29$

(a) We want to find a 95% confidence interval on $\mu_1 - \mu_2$. Is pooling appropriate?
(b) Construct a 95% confidence interval on $\mu_1 - \mu_2$.
(c) Is there evidence based on the confidence interval that, on the average, the seals from supplier I can withstand higher pressures than those from supplier II? Explain.

29. Aseptic packaging of juices is a method of packaging that entails rapid heating followed by quick cooling to room temperature in an air-free container. Such packaging allows the juices to be stored unrefrigerated. Two machines used to fill aseptic packages are compared. These data are obtained in the number of containers that can be filled per minute:

Machine I	Machine II
$n_1 = 25$	$n_2 = 25$
$\bar{x}_1 = 115.5$	$\bar{x}_2 = 112.7$
$s_1^2 = 25.2$	$s_2^2 = 7.6$

(a) Is pooling appropriate?
(b) Find a 90% confidence interval on $\mu_1 - \mu_2$.
(c) Is there evidence based on the confidence interval that machine I is faster, on the average, than machine II? Explain.

30. These data are obtained on the power output in kilowatts of two new diesel motors for small cars:

Direct fuel injection				Indirect fuel injection			
38.5	38.2	39.2	38.5	38.9	38.3	38.4	39.0
38.9	38.0	39.1	39.1	37.7	37.2	38.4	
37.4	37.6	39.0	38.0	38.2	37.0	37.9	
39.0	37.7	38.1	37.4	38.2	37.5	39.7	

Construct a 95% confidence interval on $\mu_1 - \mu_2$ using the appropriate T procedure. Based on your confidence interval, does there appear to be a difference in the mean power of these two engines? Explain.

Section 9.5

31. Information about ocean weather can be extracted from radar returns with the aid of a special algorithm. A study is conducted to estimate the difference in wind speed as measured on the ground and via the Seasat satellite. To do so, wind speeds are measured using the two methods simultaneously at 12 specified times. These data result:

	Windspeed, m/s		
Time	Ground (x)	Satellite (y)	Differences $(d = x - y)$
1	4.46	4.08	
2	3.99	3.94	
3	3.73	5.00	
4	3.29	5.20	
5	4.82	3.92	
6	6.71	6.21	
7	4.61	5.95	
8	3.87	3.07	
9	3.17	4.76	
10	4.42	3.25	
11	3.76	4.89	
12	3.30	4.80	

(a) Find the difference scores for the above data subtracting in the order indicated.
(b) Find $\bar{d}$ and s_d.
(c) Find a 95% confidence interval on the mean difference in measurements taken by these methods. Based on this interval, is there reason to believe that, on the average, the satellite measurements differ from those taken on the ground? Explain.

32. A study is conducted to estimate the average difference in the cost of analyzing data using two different statistical packages. To do so, 15 data sets are used. Each is analyzed by each package, and the cost of the analysis is recorded. These observations result:

Program	Package I $	Package II $	Program	Package I $	Package II $
1	.26	.29	9	.19	.33
2	.24	.32	10	.25	.33
3	.26	.24	11	.29	.33
4	.22	.33	12	.25	.28
5	.25	.28	13	.23	.30
6	.23	.27	14	.20	.24
7	.18	.25	15	.25	.32
8	.25	.26			

(a) Find the set of difference scores subtracting in the order package I minus package II.
(b) Find $\bar{d}$ and s_d.
(c) Find a 90% confidence interval on the mean difference in the cost of running a data analysis using the two packages.

33. Post Three Mile Island regulations require provisions by which people within 10 miles of a nuclear power plant can be notified promptly in the event of a general nuclear emergency. In a study of one such system the sound level at 69 locations within 10 miles of the plant is first simulated and then field tested. Subtracting in the order measured siren level minus simulated siren level, it is found that $\bar{d} = .04$ decibels and $s_d = 2.43$. Find a 95% confidence interval on the mean difference between the actual siren level and the simulated level. Based on this interval, is there reason to suspect that a difference exists? Explain.

34. A new method for measuring the concentration of Pu^{239} based on the registration of α-particles and fission-fragment tracts is studied. Test solution media of various concentrations are obtained, and each is split into two portions. The concentration of the first portion is determined using the new method; the second portion is measured using the standard technique. It is thought that the new procedure tends to give a higher average reading than the standard techniques. Do the following data support this research hypothesis? Explain, based on the P value of your test.

Concentration of Pu^{239} (μ/ml)

Sample number	New	Old
1	3.78	3.35
2	3.58	3.60
3	3.77	3.41
4	3.82	3.69
5	3.67	3.48
6	3.66	3.50
7	3.48	3.33
8	3.63	3.64
9	3.88	3.65
10	3.53	3.64

35. Highway engineers studying the effects of wear on dual-lane highways suspect that more cracking occurs in the travel lane of the highway than in the passing lane. To verify this contention, 30 one-hundred-feet-long test strips are selected, paved, and studied over a period of time. It is found that the mean difference in the number of major cracks is 4.5 with a sample standard deviation of 8.1. Do these data support the research hypothesis? Explain, based on the P value of the test.

36. Two different compilers are compared for efficiency. The comparison is done by running 25 randomly selected programs using each compiler. These data on the compile time in seconds are obtained:

Program	Compiler I	Compiler II	Program	Compiler I	Compiler II
1	3.76	4.28	14	4.02	4.25
2	4.78	3.89	15	4.34	4.28
3	4.66	3.30	16	4.10	3.35
4	3.38	3.53	17	3.25	3.83
5	2.52	2.73	18	4.52	3.82
6	3.46	3.19	19	4.24	3.75
7	4.19	3.34	20	5.33	3.66
8	4.15	4.71	21	3.84	4.14
9	3.61	4.21	22	3.95	3.68
10	3.91	3.76	23	4.32	3.09
11	4.47	3.62	24	4.22	3.12
12	3.53	3.26	25	4.25	3.97
13	4.14	2.87			

(*a*) It is thought that the second compiler is the faster of the two. Set up the null and alternative hypotheses needed to support this contention.

(*b*) What is the critical point for an $\alpha = .05$ level test of this hypothesis? Can H_0 be rejected at the .05 level? Interpret your results in the context of this problem.

Section 9.6

37. A study is conducted to determine the effect of acid rain and other industrial pollutants on lake water. Random samples are drawn from 10 lakes in a heavily industrialized area and from eight lakes in a primitive forested area. These data are obtained on the pH of the water:

Industrial area (*I*)		Primitive area (*P*)	
6.9	7.0	7.0	6.8
6.2	6.5	6.9	7.1
6.3	6.6	6.7	7.0
5.9	5.5	7.1	7.2
6.0	7.3		

At the $\alpha = .025$ level, can we claim that the pH of the water in the industrialized area tends to be lower than that in the primitive area?

38. Polychlorinated biphenyls (PCB) are worldwide environmental contaminants of industrial origin that are related to DDT. They are being phased out in the United States, but they will remain in the environment for many years. An experiment is run to study the effects of PCB on the reproductive ability of screech owls. The purpose is to compare the shell thickness of eggs produced by birds exposed to PCB to that of birds not exposed to the contaminant. It is thought that shells of the former group will be thinner than those of the latter. Do these data support this research hypothesis? Explain.

Shell thickness, mm			
Exposed to PCB (*E*)		**Free of PCB (*F*)**	
.21	.226	.22	.27
.223	.215	.265	.18
.25	.24	.217	.187
.19	.136	.256	.23
.20		.20	

39. An automobile manufacturer is experimenting with a new type of paint designed to resist corrosion. Five automobile hoods are painted with the new paint; seven are painted using the old mixture. All hoods are subjected to identical accelerated life testing. At the end of the testing period an impartial judge is asked to rank the hoods from 1 to 12, with lower ranks indicating less corrosion. These data result (N = new. O = old):

Rank	1	2	3	4	5	6	7	8	9	10	11	12
Type	*N*	*O*	*N*	*N*	*N*	*O*	*N*	*O*	*O*	*O*	*O*	*O*

At the $\alpha = .05$ level, can we reject $H_0: M_N = M_O$ and conclude that the new paint resists corrosion better than the old?

40. A study is conducted to compare a new drill tip to be used in drilling oil wells to the drill tips currently in use. Four new tips (N) are field tested. The length of time each is usable is recorded. After comparison with file data on the old tips (O), these data are obtained (a lower rank indicates a longer lasting drill tip):

Rank	1	2	3	4	5	6	7	8	9
Type	N	O	O	N	N	N	O	O	O

At the $\alpha = .05$ level, can we claim that the new drill tips tend to last longer than the old ones?

41. Manufacturers of brand A mainframe computers claim that maintenance costs are lower for their equipment than for that of their nearest competitor. Before purchasing brand A, a company makes an independent investigation of this claim. Samples of repair records are obtained from users of the two types of equipment. These data result:

Brand A	Competitor
$m = 75$	$n = 100$
$W_m = 5937$	

 (a) What is $E[W_m]$?
 (b) What is Var W_m?
 (c) Do the data support the claim of the makers of brand A equipment? Explain, based on the P value of your test.

42. A study of visual and auditory reaction time is conducted for a group of college basketball players. Visual reaction time is measured by the time needed to respond to a light signal, and auditory reaction time is measured by the time needed to respond to the sound of an electric switch. Fifteen subjects were measured with time recorded to the nearest millisecond:

Subject	Visual	Auditory
1	161	157
2	203	207
3	235	198
4	176	161
5	201	234
6	188	197
7	228	180
8	211	165
9	191	202
10	178	193
11	159	173
12	227	137
13	193	182
14	192	159
15	212	156

Is there evidence that the visual reaction time tends to be slower than the auditory reaction time?

43. A firm has two possible sources for its computer hardware. It is thought that supplier X tends to charge more than supplier Y for comparable items. Do these data support this contention at the $\alpha = .05$ level?

Item	Price (X), \$	Price (Y), \$
1	6000	5900
2	575	580
3	15,000	15,000
4	150,000	145,000
5	76,000	75,000
6	5650	5600
7	10,000	9975
8	850	870
9	900	890
10	3000	2900

Would the sign test have yielded the same results? If not, explain the discrepancy.

44. An experiment was conducted to compare the appearance of two types of paint on houses after normal exposure for a period of 2 years. Twenty pairs of similarly constructed homes were selected, and brand A paint was applied to one house of each pair and brand B was applied to the other member of each pair. After 2 years a paint expert was asked to judge the appearances of the two brands for each pair of houses. $A > B$ and $B > A$ denotes the order of preferred appearance of A preferred to B and B preferred to A, respectively. The outcome of the experiment is as follows:

	Pair		Pair		Pair		Pair
1	$A > B$	6	$B > A$	11	$A > B$	16	$A > B$
2	$A > B$	7	$B > A$	12	$A > B$	17	$B > A$
3	$B > A$	8	$A > B$	13	$A > B$	18	$A > B$
4	$A > B$	9	$B > A$	14	$B > A$	19	$A > B$
5	$A > B$	10	$A > B$	15	$A > B$	20	$A > B$

Using an appropriate test, test the hypothesis that the two brands of paint are equally preferred in terms of appearance after 2 years.

REVIEW EXERCISES

45. Researchers are experimenting with the use of microprocessors to help reduce fuel and power consumption in furnaces used to process magnetite ore. A particular system is designed to maintain gas flow through the machine in such a way as to ensure that sufficient heat is available to raise the raw ore pellets to 1300° C. A study is conducted to compare the temperature setting needed to accomplish this using the computerized system to that setting needed using the conventional method. It is thought that the computerized system will result in a lower average required setting with a smaller variability in settings than the conventional system.
    (a) We are interested in testing two null hypotheses. State these null hypotheses and their alternatives.
    (b) Sample runs yield these data:

Computerized	Conventional
$n_1 = 25$	$n_2 = 25$
$\bar{x}_1 = 733°$ C	$\bar{x}_2 = 822°$ C
$s_1 = 10°$ C	$s_2 = 50°$ C

Assuming normality, test the null hypotheses of part (*a*). Be ready to defend your choice of test statistics. Do these data support the two contentions stated concerning the computerized system? Explain.

46. Dross is scum that forms on the surface of molten metal during processing. A new technique is being developed to reduce the formation of this substance. To be profitable, the reduction must amount to an average of more than 15 kilograms per ton over the current method.

    (*a*) Set up the null and alternative hypotheses needed to support the contention that the new process will be profitable.

    (*b*) Trial runs produce these data:

Old method	New method
$n_1 = 10$	$n_2 = 10$
$\bar{x}_1 = 20$ kg/t	$\bar{x}_2 = 1$ kg/t
$s_1 = 2.5$ kg/t	$s_2 = .5$ kg/t

    Assuming normality, test the null hypothesis of part (*a*). Be ready to defend your choice of test statistics. Does it appear that the new process will be profitable? Explain.

47. A composite of 6/6 nylon and steel is being studied for possible use in cam gears. Sixteen gears of different types are produced, and the noise level obtained using these gears is compared to that of an identical gear made of cast iron. These data are obtained:

Gear	Cast iron	Composite
	Noise level in dB	
1	75	74
2	90	88
3	80	81
4	60	60
5	110	107
6	95	92
7	93	90
8	88	84
9	70	66
10	65	64
11	91	86
12	100	97
13	85	83
14	50	44
15	62	60
16	67	64

    (*a*) Construct a stem-and-leaf diagram for the differences in the noise levels for the two gears. Subtract in the order cast iron minus composite. Does it appear that these differences are approximately normally distributed?

    (*b*) Find a 95% confidence interval on the mean difference in the reduction in the noise level. Does it appear that gears made from the composite have a lower average decibel level than those made from cast iron? Explain.

48. Chains have long been used in kilns in cement plants to help reduce heat consumption. A study is conducted to determine if chains will have the same effect when using cheaper raw materials with high sulfur and chlorine

content. The purpose of the study is to estimate the difference in specific heat consumption in kilns with and without the use of chains. Independent samples of sizes 14 and 16, respectively, are used in the study. These data result:

Without chains	With chains
$n_1 = 16$	$n_2 = 14$
$\bar{x}_1 = 6150$ kJ/kg	$\bar{x}_2 = 5250$
$s_1 = 80$ kJ/kg	$s_2 = 75$

Find a 95% confidence interval on $\mu_1 - \mu_2$. Be ready to defend your choice of the confidence bounds used. Does it appear that the chains are effective? Explain.

49. It is thought that the heat loss in glass pipes is smaller than that in steel pipes of the same size. To verify this contention, nine pairs of pipes of assorted diameters are obtained. Various liquids at identical starting temperatures are run through 50-meter segments of each type of pipe, and the heat loss is measured in each case. These data result:

	Heat loss (in °C)	
Pair	Steel	Glass
1	4.6	2.5
2	3.7	1.3
3	4.2	2.0
4	1.9	1.8
5	4.8	2.7
6	6.1	3.2
7	4.7	3.0
8	5.5	3.5
9	5.4	3.4

Assuming normality, can we conclude that the mean heat loss is higher in steel pipes than in those made of glass? Explain, based on the P value of your test.

50. A study is conducted to compare the total printing time in seconds of two brands of laser printers on various tasks. Data below are for the printing of charts. (Based on information found in MACWORLD, March 1993, p. 1980.)

Task	Brand 1 time	Brand 2 time
1	21.8	36.5
2	22.6	35.2
3	21.0	36.2
4	19.7	34.0
5	21.9	36.4
6	21.6	36.1
7	22.5	37.5
8	23.1	38.0
9	22.2	36.3
10	20.1	35.9
11	21.4	35.7
12	20.5	34.9
13	22.7	37.1
14	20.5	34.2
15	21.3	35.4

(a) Estimate the average difference in printing time for these two lasers.

(b) Find a 95% confidence interval on the average difference in printing times.

(c) Based on the confidence interval found in part (b), would you be surprised to hear a claim that these two printers are equally fast in printing charts? Explain.

51. Two drugs, amantadine (A) and rimantadine (R), are being studied for use in combatting the influenza virus. A single 100-milligram dose is administered orally to healthy adults. The variable studied is T_{max}, the time in minutes required to reach maximum plasma concentration. These data are obtained (based on information found in "Drug Therapy", Gordon Douglas, Jr., *New England Journal of Medicine,* vol. 322, February 1990, pp. 443–449):

$T_{max}(A)$			$T_{max}(R)$		
105	123	12.4	221	227	280
126	108	134	261	264	238
120	112	130	250	236	240
119	132	130	230	246	283
133	136	142	253	273	516
145	156	170	256	271	
200					

(a) Construct a boxplot for each data set, and identify outliers.

(b) Assume that the outlier 12.4 of set A is the result of a misplaced decimal point. Replace the outlier 12.4 with the true value 124. Test for equality of variances at the $\alpha = .20$ level.

(c) Construct a 95% confidence interval on the difference in the average time required to reach maximum plasma concentration for these two drugs.

(d) Based on the confidence interval of part (c), can it be concluded that there is a difference in means? Explain.

52. A study is conducted to help understand the effect of smoking on sleep patterns. The random variable considered is X, the time in minutes that it takes to fall asleep. Samples of smokers and nonsmokers yield these observations on X:

Nonsmokers						Smokers					
17.2	19.7	18.1	15.1	18.3	17.6	15.1	20.5	17.7	21.3	16.0	24.8
16.2	19.9	19.8	23.6	24.9	20.1	16.8	21.2	18.1	22.1	15.9	25.2
19.8	22.6	20.0	24.1	25.0	21.4	22.8	22.4	19.4	25.2	18.3	25.0
21.2	18.9	22.1	20.6	23.3	20.2	25.8	24.1	15.0	24.1	21.6	16.3
21.1	16.9	23.0	20.1	17.5	21.3	24.3	25.7	15.2	18.0	23.8	17.9
21.8	22.1	21.1	20.5	20.4	20.7	23.2	25.1	16.1	17.2	24.9	19.9
19.5	18.8	19.2	22.4	19.3	17.4	15.7	15.3	19.9	23.1	23.0	25.1

(a) Construct a stem-and-leaf diagram for each of these data sets. Use the integers from 15 to 25 inclusive as stems.

(b) Would you be surprised to hear someone claim that there is no difference in the distribution of X for the two groups? Explain.

(c) Perform any statistical tests that you believe are appropriate to detect differences that might exist. They can be either normal theory or nonparametric.

53. Recycling has become important as landfills become harder to obtain. A study of white paper disposal is conducted. These data are obtained on the amount of

white paper thrown out per year by bank employees and employees in other businesses. (Data are in hundreds of pounds.)

Bank employees		Other Businesses	
3.1	2.6	6.9	5.3
2.9	2.0	6.4	5.2
3.8	3.2	4.7	5.1
3.3	2.4	4.3	5.9
2.7	2.3	5.1	5.8
3.0	3.1	6.3	4.9
2.8	2.1	5.9	4.8
2.5	3.4	5.4	4.0
2.0	1.9	5.2	4.0
2.9	2.5	5.7	5.2
2.1	2.5	6.2	5.0
2.7	2.3	4.2	4.1
2.2	2.3	5.0	3.9
1.8	1.5	3.7	3.7
1.9	1.2	5.1	3.4
1.7	1.7		

(Based on information found in "White Paper Recycling," *MIT Technology, Review,* August/September 1992, p. 20.)

Do these data support the contention that, on the average, bank employees dispose of more white paper per year than do employees in other businesses? Explain by conducting appropriate statistical tests (either normal theory or nonparametric). Be ready to defend your choice of tests.

54. A builder has a choice of two fairly comparable building sites. Since a septic system is to be installed, it is essential that each site be tested for its ability to perk, or absorb, water. Test holes are dug at randomly selected locations and filled with water. The variable of interest is the time in seconds that it takes for the water to drain from the hole. Use the MINITAB output given below to answer each of the following questions.

(*a*) How many holes were dug at each site?
(*b*) How many degrees of freedom would be associated with a pooled T test?
(*c*) How many degrees of freedom did MINITAB use in conducting the means comparison?
(*d*) Do you think that it would have been an acceptable approach to pool in this case? Explain.
(*e*) Is the T test a right-, left-, or two-tailed test?
(*f*) What is the P value of the test?
(*g*) Can it be concluded that the average perk time differs at the $\alpha = .05$ level? Could this conclusion be reached at the $\alpha = .10$ level?

Two-Sample T Test and Confidence Interval

Two-sample T for site 1 vs. site 2

	N	Mean	StDev	SE Mean
site 1	17	13.00	1.58	0.38
site 2	13	14.15	1.57	0.44

95% CI for mu site 1 − mu site 2: (−2.35, 0.04)
T test mu site 1 = mu site 2 (vs. not =):
$T = -1.99$ $P = 0.058$ DF = 26

55. In Example 9.5.2 a paired T test was run to compare the mean cpu times for two computing algorithms. It was thought that the old algorithm (X) runs slower than the newer algorithm (Y). Thus, the research hypothesis is

$$H_1: \mu_X > \mu_Y \quad \text{or} \quad H_1: \mu_D > 0$$

where $D = X - Y$.

Use the following SAS output to answer each of the questions posed:

PAIRED T TEST

VARIABLE	MEAN	STANDARD DEVIATION	STD ERROR OF MEAN	T	PR > !T!
DIFF	14.40900000 ④	8.65276635 ③	2.73624497 ⑤	5.27 ①	0.0005 ②

(a) Identify what each of the numbered values ①–⑤ represents, and compare these values to those found by hand earlier.

(b) Based on these data, has H_1 been supported at the $\alpha = .05$ level? At the $\alpha = .10$ level?

CHAPTER 10

Simple Linear Regression and Correlation

In this chapter we shall assume that the variable X is *not* a random variable. Rather, it is a mathematical variable—an entity that can assume different values but whose value at the time under consideration is not determined by chance. To illustrate, suppose that we are developing a model to describe the temperature of the water off the continental shelf. Since the temperature depends in part on the depth of the water, two variables are involved. These are X, the water depth, and Y, the water temperature. We are not interested in making inferences on the depth of the water. Rather, we want to describe the behavior of the water temperature under the assumption that the depth of the water is known precisely in advance. Even if the depth of the water is fixed at some value x, the water temperature will still vary due to other random influences. For example, if several temperature measurements are taken at various places each at a depth of $x = 1000$ feet, these measurements will vary in value. For this reason, we must admit that for a given x we are really dealing with a "conditional" random variable, which we denote by $Y|x$ (Y given that $X = x$). This conditional random variable has a mean denoted by $\mu_{Y|x}$. It is obvious that the average temperature of ocean water depends in part on the depth of the water; we do not expect the average temperature at $x = 1000$ feet to be the same as that at $x = 5000$ feet. That is, it is reasonable to assume that $\mu_{Y|x}$ is a function of x. We call the graph of this function the *curve of regression of Y on X*. Since we assume that the value of X is known in advance and that the value assumed by Y depends in part on the particular value of X under consideration, Y is called the *dependent* or *response* variable. The variable X whose value is used to help predict the behavior of $Y|x$ is called the *independent* or *predictor* variable or the *regressor*.

Our immediate problem is to estimate the form of $\mu_{Y|x}$ based on data obtained at some selected values $x_1, x_2, x_3, \ldots, x_n$ of the predictor variable X. The actual values used to develop the model are not overly important. If a functional relationship exists, it should become apparent regardless of which X values are used to discover it. However, to be of practical use, these values should represent a fairly wide range of possible values of the independent variable X. Sometimes the values used can be preselected. For example, in studying the relationship between water temperature and water depth, we might know that our model is to be used to predict water temperature for depths from 1000 to 5000 feet. We can choose to measure water temperatures at any depths that we wish within this range. For example, we might take measurements at 1000-foot increments. In this way we preset our X values at $x_1 = 1000$, $x_2 = 2000$, $x_3 = 3000$, $x_4 = 4000$, and $x_5 = 5000$ feet. When the X values used to develop the regression equation are preselected, the study is said to be *controlled.* Sometimes the X values used to develop the equation are chosen via some random mechanism. For example, in studying the effect of air quality on the pH of rainwater, we shall be forced to select a sample of days, record the air quality reading for the day, and measure the pH of the rainwater. In this case the values of X used to develop the regression equation are not preselected by the researcher. They do represent a set of typical X values. Studies of this sort are called *observational studies.* Regardless of how the X values for study are selected, our random sample is properly viewed as taking the form

$$\{(x_1, Y|x_1), (x_2, Y|x_2), (x_3, Y|x_3), \dots, (x_n, Y|x_n)\}$$

Note that the first member of each ordered pair denotes a value of the independent variable X; it is a *real number.* The second member of each pair is a random variable.

In this chapter we learn to estimate the curve of regression of Y on X when the regression is considered to be linear. In this case the equation $\mu_{Y|x}$ is given by

> ### Linear Curve of Regression of Y on X
>
> $$\mu_{Y|x} = \beta_0 + \beta_1 x$$

where β_0 and β_1 denote real numbers.

Much of the theory behind the techniques presented depends on linear algebra. For this reason, we cannot prove some of the results based on material from this text. Where it is possible to verify results we shall do so.

10.1   MODEL AND PARAMETER ESTIMATION

Description of Model

Recall from elementary algebra that the equation for a straight line is $y = b + mx$, where b denotes the y intercept and m denotes the slope of the line. In the simple linear regression model

$$\mu_{Y|x} = \beta_0 + \beta_1 x$$

β_0 denotes the intercept and β_1 the slope of the regression line. To estimate the regression line, we must find a logical way to estimate the theoretical parameters β_0 and β_1. To understand how this is done, we first rewrite our model in an alternative form.

In conducting a regression study, we shall be observing the variable X at n points $x_1, x_2, x_3, \dots, x_n$. These points are assumed to be measured without error. When they are preselected by the experimenter, we say that the study is a *controlled* study; when they are observed at random, then the study is called an *observational* study. Both situations are handled in the same way mathematically. In either case we shall be concerned with the n random variables $Y|x_1, Y|x_2, Y|x_3, \dots, Y|x_n$. Recall that a random variable varies about its mean value. Let E_i denote the random difference between $Y|x_i$ and its mean, $\mu_{Y|x_i}$. That is, let

$$E_i = Y|x_i - \mu_{Y|x_i}$$

Solving this equation for $Y|x_i$, we conclude that

$$Y|x_i = \mu_{Y|x_i} + E_i$$

In this expression it is assumed that the random difference E_i has mean 0. Since we are assuming that the regression is linear, we can conclude that $\mu_{Y|x_i} = \beta_0 + \beta_1 x_i$. Substituting, we see that

$$Y|x_i = \beta_0 + \beta_1 x_i + E_i$$

It is customary to drop the conditional notation and to denote $Y|x_i$ by Y_i. Thus an alternative way to express the simple linear regression model is

---

Simple Linear Regression Model

$$Y_i = \beta_0 + \beta_1 x_i + E_i \qquad (10.1)$$

---

where E_i is assumed to be a random variable with mean 0.

Our data consist of a collection of n pairs (x_i, y_i), where x_i is an observed value of the variable X and y_i is the corresponding observation for the random variable Y. The observed value of a random variable usually differs from its mean value by some random amount. This idea is expressed mathematically by writing

$$y_i = \beta_0 + \beta_1 x_i + \varepsilon_i \qquad (10.2)$$

In this equation ε_i denotes a realization of the random variable E_i when Y_i takes on the value y_i.

In a regression study it is useful to plot the data points in the xy plane. Such a plot is called a *scattergram*. We do not expect these points to lie exactly in a straight line. However, if linear regression is applicable, then they should exhibit a linear trend. These theoretical ideas are illustrated in Fig. 10.1 in the context of our water temperature study. Note that since we do not know the true values for β_0 and β_1, we shall not know the true value for ε_i, the vertical distance from the point (x_i, y_i) to the true regression line.

Once β_0 and β_1 have been approximated from the available data, we can replace these theoretical parameters by their estimated values in the regression model.

Figure 10.1
(*a*) A scattergram of hypothetical data on depth of water (*x*) versus its temperature (*y*)—the data exhibits a linear trend, indicating that linear regression is reasonable; (*b*) theoretical and unknown line of regression passes through the data points; (*c*) ε_i is the distance from y_i to its mean value, $\mu_{Y|x_i}$.

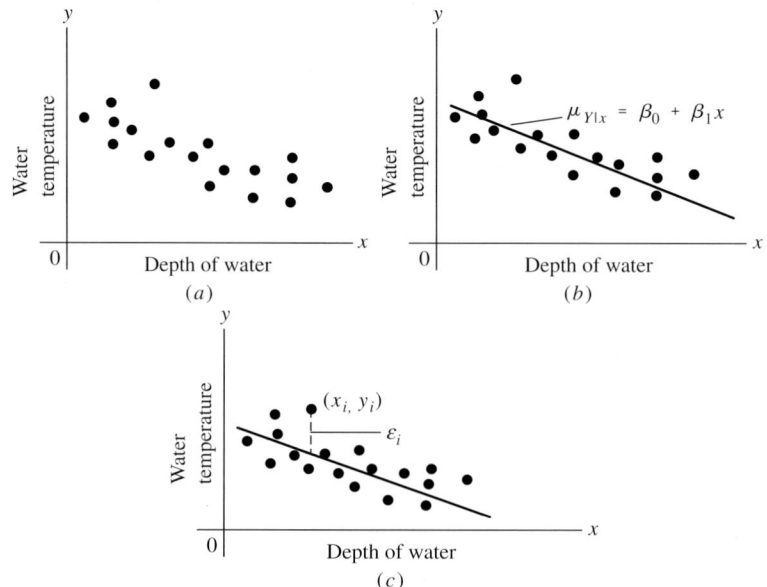

Letting b_0 and b_1 denote the estimates for β_0 and β_1, respectively, the estimated line of regression takes the form

$$\hat{\mu}_{Y|x} = b_0 + b_1 x$$

Just as the data points do not all lie on the theoretical line of regression, they also do not all lie on this estimated regression line. If we let e_i denote the vertical distance from a point (x_i, y_i) to the estimated regression line, then each data point satisfies the equation

$$y_i = b_0 + b_1 x_i + e_i$$

The term e_i is called the *residual*. Figure 10.2 illustrates this idea and points out the difference between ε_i and e_i graphically.

Least-Squares Estimation

The parameters β_0 and β_1 are estimated by the method of *least squares*. The reasoning behind this method is quite simple. From the many straight lines that can be drawn through a scattergram we wish to pick the one that "best fits" the data. The fit is "best" in the sense that the values of b_0 and b_1 chosen are those that minimize the sum of the squares of the residuals. In this way we are essentially picking the line that comes as close as it can to all data points simultaneously. For example, if we consider the sample of five data points shown in Fig. 10.3, then the least-squares procedure selects that line which causes $e_1^2 + e_2^2 + e_3^2 + e_4^2 + e_5^2$ to be as small as possible.

The residuals are squared before summing for a very practical reason. Notice that the residual for a data point that lies above the estimated regression line is positive; for a point that lies below the line the residual is negative. If the residuals themselves

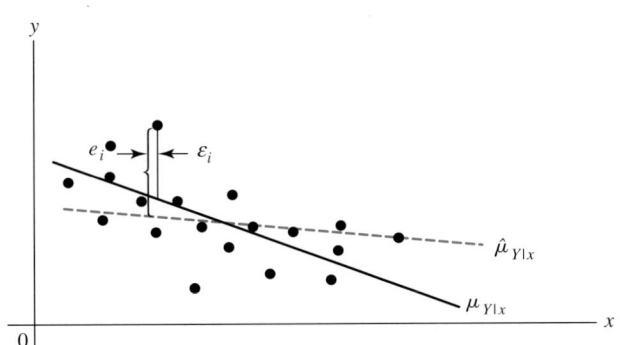

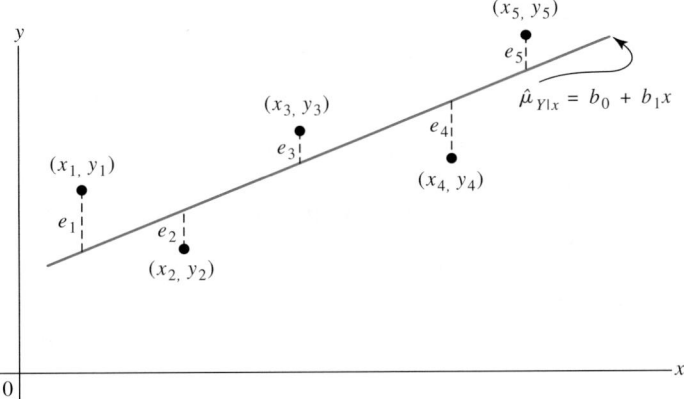

Figure 10.2
ε_i is the vertical distance from the point (x_i, y_i) to the true regression line $\mu_{Y|x} = \beta_0 + \beta_1 x$; e_i is the vertical distance from the point (x_i, y_i) to the estimated regression line $\hat{\mu}_{Y|x} = b_0 + b_1 x$.

Figure 10.3
The least-squares procedure minimizes the sum of the squares of the residuals e_i.

are summed, the negative and positive values will counteract one another and the sum will always be 0. You are asked to verify this fact in Exercise 5.

The general derivation of the least-squares estimates for β_0 and β_1 depends on the minimization technique studied in elementary calculus. In particular, we shall express the sum of squares of the residuals as a function of the two variables b_0 and b_1, differentiate this function with respect to these variables, set these derivatives equal to 0, and solve the resulting equations for b_0 and b_1. Before presenting the derivation, let us note that the residual e_i is sometimes called the residual *error*. For this reason the sum of squares of the residuals often is called the *error sum of squares* and is denoted by SSE (sum of squares error). Since the word "error" tends to suggest that a mistake has been made, this language is somewhat misleading. However, it is recognized widely, and so we shall adhere to its use.

The sum of squares of the errors about the estimated regression line is given by

$$\text{SSE} = \sum_{i=1}^{n} e_i^2 = \sum_{i=1}^{n} (y_i - b_0 - b_1 x_i)^2$$

Differentiating SSE with respect to b_0 and b_1, we obtain

$$\frac{\partial \text{SSE}}{\partial b_0} = -2 \sum_{i=1}^{n} (y_i - b_0 - b_1 x_i)$$

$$\frac{\partial \text{SSE}}{\partial b_1} = -2 \sum_{i=1}^{n} (y_i - b_0 - b_1 x_i) x_i$$

We now set these partial derivatives equal to 0 and use the rules of summation to obtain the equations

$$n b_0 + b_1 \sum_{i=1}^{n} x_i = \sum_{i=1}^{n} y_i$$

$$b_0 \sum_{i=1}^{n} x_i + b_1 \sum_{i=1}^{n} x_i^2 = \sum_{i=1}^{n} x_i y_i$$

These equations are called the *normal* equations. They can be solved easily to obtain these estimates for β_0 and β_1:

Least-squares estimates for β_0 and β_1

$$b_1 = \frac{n \sum_{i=1}^{n} x_i y_i - \left(\sum_{i=1}^{n} x_i \right) \left(\sum_{i=1}^{n} y_i \right)}{n \sum_{i=1}^{n} x_i^2 - \left(\sum_{i=1}^{n} x_i \right)^2}$$

$$b_0 = \bar{y} - b_1 \bar{x}$$

Before illustrating these ideas, let us point out a very practical aspect of regression that we have not yet mentioned. Namely, even though the regression equation actually estimates the mean value of Y for a given value x, it is used extensively to estimate the value of Y itself. Common sense tells us that a logical choice for the predicted value of Y for a given value x is its estimated average value $\hat{\mu}_{Y|x}$. For example, if asked to predict the ocean water temperature at a depth of 1000 feet, a logical choice is the average temperature at this depth. To emphasize this use of the estimated regression line, we rewrite it in the form

$$\hat{y} = \hat{\mu}_{Y|x} = b_0 + b_1 x$$

Example 10.1.1. Since humidity influences evaporation, the solvent balance of water-reducible paints during sprayout is affected by humidity. A controlled study is conducted to examine the relationship between humidity (X) and the extent of solvent evaporation (Y). Knowledge of this relationship will be useful in that it will allow the painter to adjust his or her spraygun setting to account for humidity. These data are obtained:

Observation	(x) Relative humidity, (%)	(y) Solvent evaporation, (%) wt
1	35.3	11.0
2	29.7	11.1
3	30.8	12.5
4	58.8	8.4
5	61.4	9.3
6	71.3	8.7
7	74.4	6.4
8	76.7	8.5
9	70.7	7.8
10	57.5	9.1
11	46.4	8.2
12	28.9	12.2
13	28.1	11.9
14	39.1	9.6
15	46.8	10.9
16	48.5	9.6
17	59.3	10.1
18	70.0	8.1
19	70.0	6.8
20	74.4	8.9
21	72.1	7.7
22	58.1	8.5
23	44.6	8.9
24	33.4	10.4
25	28.6	11.1

Summary statistics for these data are

$$n = 25 \qquad \sum x = 1314.90 \qquad \sum y = 235.70$$
$$\sum x^2 = 76{,}308.53 \qquad \sum y^2 = 2286.07 \qquad \sum xy = 11{,}824.44$$

To estimate the simple linear regression line, we estimate the slope β_1 and intercept β_0. These estimates are

$$\hat{\beta}_1 = b_1 = \frac{n\sum xy - \left[\left(\sum x\right)\left(\sum y\right)\right]}{n\sum x^2 - \left(\sum x\right)^2}$$

$$= \frac{25(11{,}824.44) - [(1314.90)(235.70)]}{25(76{,}308.53) - (1314.90)^2}$$

$$= -.08$$

$$\hat{\beta}_0 = b_0 = \bar{y} - b_1\bar{x}$$

$$= 9.43 - (-.08)(52.60)$$

$$= 13.64$$

Hence the estimated regression equation is

$$\hat{\mu}_{Y|x} = \hat{y} = 13.64 - .08x$$

The graph of this equation is shown in Fig. 10.4. To predict the extent of solvent evaporation when the relative humidity is 50%, we substitute the value 50 for x in the equation

$$\hat{y} = 13.64 - .08x$$

to obtain $\hat{y} = 13.64 - .08(50) = 9.64$. That is, when the relative humidity is 50%, we predict that 9.64% of the solvent, by weight, will be lost due to evaporation.

In modern statistical analysis the computer is routinely used. There is pedagogical merit in going through the methods of calculations as we do in this text. However, in practice, we recommend the use of modern statistical software packages. Some of the major packages are SAS (Statistical Analysis System), MINITAB, BMDPC (Biomedical Computer Programs) and SPSS (Statistical Package for the Social Sciences). We will present here some typical outputs using SAS. It would be helpful to compare the calculations with the SAS output. Note that the estimated regression line slope and intercept are given at ① and ② respectively. We will refer to ③ and ④ in Example 10.3.1.

ESTIMATED LINE
OF REGRESSION

GENERAL LINEAR MODELS PROCEDURE

DEPENDENT VARIABLE: Y

SOURCE	DF	SUM OF SQUARES	MEAN SQUARE	F VALUE
MODEL	1	45.82966081	45.82966081	58.36
ERROR	23	18.06073919	0.78524953	PR > F
CORRECTED TOTAL	24	63.89040000		0.0001

R-SQUARE	C.V.	ROOT MSE	Y MEAN
0.717317	9.3991	0.88614306	9.42800000

SOURCE	DF	TYPE I SS	F VALUE	PR > F
X	1	45.82966081	58.36	0.0001

SOURCE	DF	TYPE III SS	F VALUE	PR > F
X	1	45.82966081	58.36	0.0001

PARAMETER	ESTIMATE	T FOR HO: PARAMETER = 0	PR > !T!	STD ERROR OF ESTIMATE
INTERCEPT	13.63886687 ②	23.56	0.0001	0.57898306
X	−0.08006059 ①	−7.64 ③	0.0001 ④	0.01047971

Recall from elementary calculus that the slope of a line gives the change in y for a unit change in x. If the slope is positive, then as x increases so does y; as x decreases, so does y. If the slope is negative, things operate in reverse. An increase in x signals a decrease in y, whereas a decrease in x yields an increase in y. In the previous example the slope is $-.08$. If the relative humidity increases by 1 percentage point, then the mean solvent evaporation decreases by .08. If the relative humidity decreases by 3 percentage points, then the mean solvent evaporation should increase by $3(.08) = .24$.

We end this section with a word of caution. A given data set gives evidence of linearity only over those values of X spanned by the data set. For values of X beyond those covered there is no evidence of linearity. Thus it is dangerous to use an estimated regression line to predict values of Y corresponding to values of X that lie far beyond the range of the X values included in the data set.

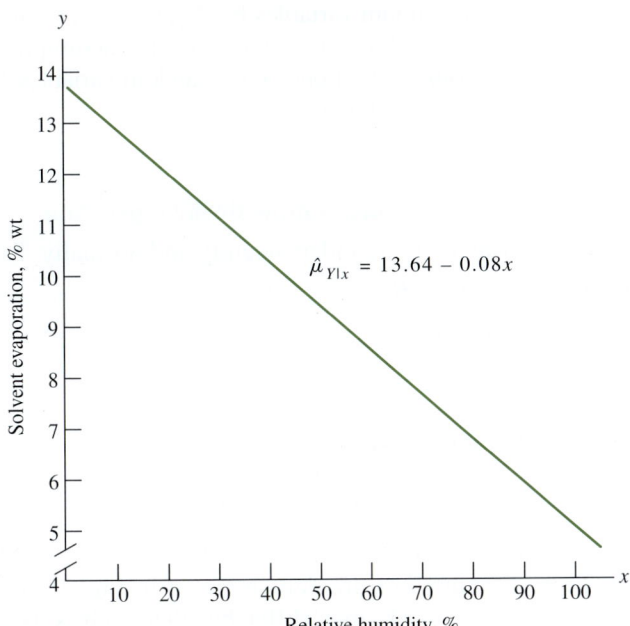

Figure 10.4
A graph of the estimated line
of regression of Y, the extent
of evaporation on X, the
relative humidity.

10.2   PROPERTIES OF LEAST-SQUARES ESTIMATORS

For a given set of observations on (X, Y) the method of least squares yields estimates b_0 and b_1 for β_0 and β_1, the intercept and slope of the true regression line, respectively. Since the values obtained for b_0 and b_1 vary from data set to data set, it is evident that they are actually observed values of random variables, which we denote by β_1 and β_2. These random variables are estimators for β_0 and β_1 and are given by

> **Least-squares estimators for β_0 and β_1**
>
> $$B_1 = \hat{\beta}_1 = \frac{n\sum\limits_{n=1}^{n} x_i Y_i - \left(\sum\limits_{i=1}^{n} x_i\right)\left(\sum\limits_{i=1}^{n} Y_i\right)}{n\sum\limits_{i=1}^{n} x_i^2 - \left(\sum\limits_{i=1}^{n} x_i\right)^2}$$
>
> $$B_0 = \hat{\beta}_0 = \bar{Y} - B_1 \bar{x}$$

In this section we derive the mathematical properties of these estimators. Knowledge of these properties will allow us to find confidence intervals on $\beta_0, \beta_1, \mu_{Y|x}$, and $Y|x$ as well as to test hypotheses on the values of β_0 and β_1.

   Recall that one way to express the simple linear regression model is

$$Y_i = \beta_0 + \beta_1 x_i + E_i$$

where E_i is assumed to be a random variable with mean 0. To determine the properties of B_0 and B_1, we must make certain other assumptions concerning E_i. In particular, we assume that $E_1, E_2, E_3, \ldots, E_n$ is a random sample from a distribution that is normal with mean 0 and variance σ^2. We express this by writing

$$E_i \sim N(0, \sigma^2)$$

Note that this implies that the random variables $E_1, E_2, E_3, \ldots, E_n$ are independent. Since our model expresses Y_i as a linear function of E_i, the assumptions concerning $E_1, E_2, E_3, \ldots, E_n$ impose some restrictions on the random variables $Y_1, Y_2, Y_3, \ldots, Y_n$. Namely, we are assuming the following:

> ### Model assumptions: Simple linear regression
> 1. The random variables Y_i are independently and normally distributed.
> 2. The mean of Y_i is $\beta_0 + B_1 x_i$.
> 3. The variance of Y_i is σ^2.

We express these assumptions by writing

$$Y_i \sim N(\beta_0 + \beta_1 x_i, \, \sigma^2)$$

Notice that σ^2 is a measure of the variability of the responses about the true regression line. These assumptions are demonstrated in Fig. 10.5. Note that the mean values of $Y_1, Y_2, Y_3, \ldots, Y_n$ may differ but that each is assumed to have the same variance. Thus the associated normal curves may differ in location, but all of them have the same shape.

Before using the assumptions just made to determine the distribution of B_0 and B_1, we pause to state some results that will make our work simpler. These results can be verified easily by applying the rules governing the behavior of the summation symbol.

> ### Some properties of summation
> 1. $\displaystyle\sum_{i=1}^{n} (x_i - \bar{x}) = 0$
>
> 2. $\displaystyle\sum_{i=1}^{n} (x_i - \bar{x})(Y_i - \bar{Y}) = \sum_{i=1}^{n} (x_i - \bar{x}) Y_i$
>
> 3. $\displaystyle\sum_{i=1}^{n} (x_i - \bar{x})(Y_i - \bar{Y}) = \left(n \sum_{i=1}^{n} x_i Y_i - \sum_{i=1}^{n} x_i \sum_{i=1}^{n} Y_i \right) \Big/ n$
>
> 4. $\displaystyle\sum_{i=1}^{n} (x_i - \bar{x})^2 = \sum_{i=1}^{n} (x_i - \bar{x}) x_i$
>
> 5. $\displaystyle\sum_{i=1}^{n} (x_i - \bar{x})^2 = \left[n \sum_{i=1}^{n} x_i^2 - \left(\sum_{i=1}^{n} x_i \right)^2 \right] \Big/ n$

Distribution of B_1

To develop confidence intervals or test hypotheses or the slope of a regression line, we need to know the distribution of B_1, the estimator for this slope. We shall show that the model assumptions on Y_i ensure the B_1 is normally distributed with $E[B_1] = \beta_1$ and $\text{Var } B_1 = \sigma^2 / \sum_{i=1}^{n} (x_i - \bar{x})^2$. Notice that this implies that the least-squares estimator for β_1 is an unbiased estimator for this parameter.

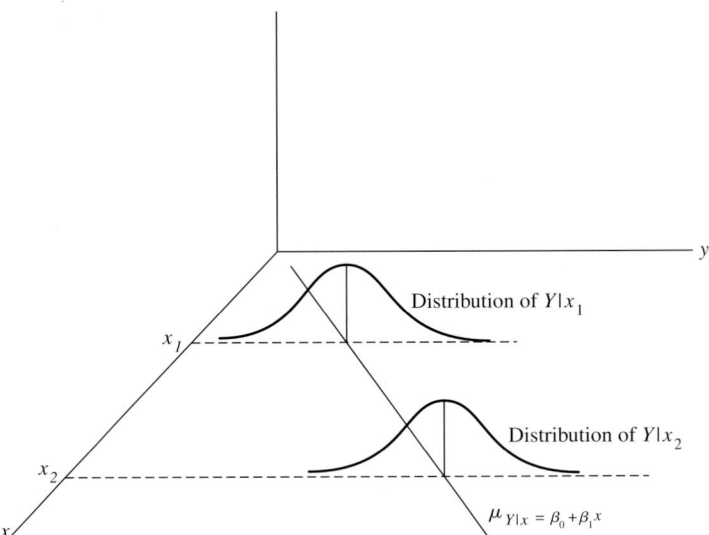

Figure 10.5
For each i, Y_i is normally
distributed with mean
$\mu_{Y|x_i} = \beta_0 + \beta_1 x_i$ and
variance σ^2.

To derive the distribution of B_1, we first use properties 2, 3, and 5 above to rewrite the estimators as shown:

$$B_1 = \frac{n \sum_{i=1}^{n} x_i Y_i - \left(\sum_{i=1}^{n} x_i \right)\left(\sum_{i=1}^{n} Y_i \right)}{n \sum_{i=1}^{n} x_i^2 - \left(\sum_{i=1}^{n} x_i \right)^2}$$

$$= \frac{\sum_{i=1}^{n} (x_i - \bar{x})(Y_i - \bar{Y})}{\sum_{i=1}^{n} (x_i - \bar{x})^2}$$

$$= \frac{\sum_{i=1}^{n} (x_i - \bar{x}) Y_i}{\sum_{i=1}^{n} (x_i - \bar{x})^2}$$

Letting

$$c_j = \frac{(x_j - \bar{x})}{\sum_{i=1}^{n} (x_i - \bar{x})^2} \qquad j = 1, 2, 3, \ldots, n$$

we have expressed B_1 in the form

$$B_1 = c_1 Y_1 + c_2 Y_2 + \cdots + c_n Y_n$$

That is, we have expressed B_1 as a linear function of the independent normal random variables $Y_1, Y_2, \ldots, Y_n$. Since any linear function of independent normal random variables is normally distributed, we can conclude that B_1 is normal. Using the rules for expectation, we see that

$$E[B_1] = E[c_1Y_1 + c_2Y_2 + \ldots + c_nY_n]$$

$$= E\left[\frac{(x_1 - \bar{x})Y_1 + (x_2 - \bar{x})Y_2 + \cdots + (x_n - \bar{x})Y_n}{\sum\limits_{i=1}^{n}(x_i - \bar{x})^2}\right]$$

$$= \frac{\sum\limits_{i=1}^{n}(x_i - \bar{x})E[Y_i]}{\sum\limits_{i=1}^{n}(x_i - \bar{x})^2}$$

For each i, $E[Y_i] = \beta_0 + \beta_1 x_i$. Substituting, we see that

$$E[B_1] = \frac{\sum\limits_{i=1}^{n}(x_i - \bar{x})(\beta_0 + \beta_1 x_i)}{\sum\limits_{i=1}^{n}(x_i - \bar{x})^2}$$

$$= \frac{\sum\limits_{i=1}^{n}(x_i - \bar{x})\beta_0 + \beta_1\sum\limits_{i=1}^{n}(x_i - \bar{x})x_i}{\sum\limits_{i=1}^{n}(x_i - \bar{x})^2}$$

By summation of properties 1 and 4,

$$E[B_1] = \beta_1$$

This result shows that B_1 is an unbiased estimator for β_1. We apply the rules of variance to find Var B_1 as follows:

$$\text{Var } B_1 = \text{Var}\left[\frac{\sum\limits_{i=1}^{n}(x_i - \bar{x})Y_i}{\sum\limits_{i=1}^{n}(x_i - \bar{x})^2}\right]$$

$$= \left[\frac{1}{\sum\limits_{i=1}^{n}(x_i - \bar{x})^2}\right]^2 \text{Var }\sum\limits_{i=1}^{n}(x_i - \bar{x})Y_i$$

$$= \left[\frac{1}{\sum\limits_{i=1}^{n}(x_i - \bar{x})^2}\right]^2 \sum\limits_{i=1}^{n}\text{Var}(x_i - \bar{x})Y_i$$

$$= \left[\frac{1}{\sum\limits_{i=1}^{n}(x_i - \bar{x})^2}\right]^2 \sum\limits_{i=1}^{n}(x_i - \bar{x})^2\text{Var }Y_i$$

Since Var Y_i is assumed to be σ^2 for each i, we can substitute to obtain

$$\text{Var } B_1 = \left[\frac{1}{\sum\limits_{i=1}^{n} (x_i - \bar{x})^2} \right]^2 \sum_{i=1}^{n} (x_i - \bar{x})^2 \sigma^2$$

$$= \frac{\sigma^2}{\sum\limits_{i=1}^{n} (x_i - \bar{x})^2}$$

These results are summarized by writing

> **Distribution of B_1**
>
> $$B_1 \sim N\left(\beta_1, \sigma^2 \Big/ \sum_{i=1}^{n} (x_i - \bar{x})^2 \right)$$

Distribution of B_0

Confidence intervals on the intercept of the regression line and hypothesis tests on this parameter are based on knowledge of the distribution of B_0, the estimator for this intercept. We shall show that this estimator is normally distributed with $E[B_0] = \beta_0$ and $\text{Var } B_0 = \sigma^2 \sum_{i=1}^{n} x_i^2 / n \sum_{i=1}^{n} (x_i - \bar{x})^2$. Once again, the least-squares estimator for β_0 is an unbiased estimator for this parameter.

To derive the distribution of the estimator B_0, we note first that it can be shown that $\bar{Y}$ and B_1 are independent. (See Exercise 14.) Since

$$B_0 = \bar{Y} - B_1 \bar{x}$$

B_0 is a linear function of independent normal random variables and therefore is itself normally distributed. Using the rules of expectation, we see that

$$\begin{aligned}
E[B_0] &= E[\bar{Y} - B_1 \bar{x}] \\
&= E[(Y_1 + Y_2 + \cdots + Y_n)/n - B_1 \bar{x}] \\
&= (E[Y_1] + E[Y_2] + \cdots + E[Y_n])/n - \bar{x} E[B_1] \\
&= [(\beta_0 + \beta_1 x_1) + (\beta_0 + \beta_1 x_2) + \cdots + (\beta_0 + \beta_1 x_n)]/n - \bar{x} E[B_1] \\
&= \left(n\beta_0 + \beta_1 \sum_{i=1}^{n} x_i \right)\Big/ n - \bar{x} E[B_1] \\
&= \beta_0 + \bar{x}\beta_1 - \bar{x}\beta_1 \\
&= \beta_0
\end{aligned}$$

This result shows that B_0 is an unbiased estimator for β_0. The variance of B_0 is given by

$$\begin{aligned}
\text{Var } B_0 &= \text{Var}(\bar{Y} - B_1 \bar{x}) \\
&= \text{Var } \bar{Y} + \bar{x}^2 \text{ Var } B_1
\end{aligned}$$

Note that

$$\begin{aligned}
\text{Var } (\bar{Y}) &= \text{Var}(Y_1 + Y_2 + \cdots + Y_n)/n \\
&= \frac{\text{Var } Y_1 + \text{Var } Y_2 + \cdots + \text{Var } Y_n}{n^2} \\
&= \frac{n\sigma^2}{n^2} = \frac{\sigma^2}{n}
\end{aligned}$$

By substituting, we see that

$$\text{Var } B_0 = \frac{\sigma^2}{n} + \frac{\bar{x}^2 \sigma^2}{\displaystyle\sum_{i=1}^{n}(x_i - \bar{x})^2}$$

$$= \frac{\sigma^2 \displaystyle\sum_{i=1}^{n}(x_i - \bar{x})^2 + n\bar{x}^2 \sigma^2}{n \displaystyle\sum_{i=1}^{n}(x_i - \bar{x})^2}$$

$$= \frac{\sigma^2 \left[\dfrac{n \displaystyle\sum_{i=1}^{n} x_i^2 - \left(\displaystyle\sum_{i=1}^{n} x_i\right)^2}{n} + \dfrac{\left(\displaystyle\sum_{i=1}^{n} x_i\right)^2}{n} \right]}{n \displaystyle\sum_{i=1}^{n}(x_i - \bar{x})^2}$$

$$= \frac{\displaystyle\sum_{i=1}^{n} x_i^2}{n \displaystyle\sum_{i=1}^{n}(x_i - \bar{x})^2} \, \sigma^2$$

To summarize, we have shown that

Distribution of B_0

$$B_0 \sim N\left[\beta_0, \ \frac{\displaystyle\sum_{i=1}^{n} x_i^2}{n \displaystyle\sum_{i=1}^{n}(x_i - \bar{x})^2} \, \sigma^2 \right]$$

Estimator of σ^2

To test hypotheses and construct confidence intervals on various parameters, we must estimate the unknown variance σ^2. Recall that σ^2 denotes the variability of each of the random variables Y_i about the *true* regression line. To estimate this variability, we use information concerning the variability of the data points about the *fitted* regression line.

Since the residual measures the unexplained or random deviation of a data point from the estimated line of regression, the residuals are used to estimate σ^2. That is, our estimate makes use of SSE, the sum of the squares of the residuals. In particular, we shall estimate σ^2 by

Estimator for σ^2

$$S^2 = \hat{\sigma}^2 = \text{SSE}/(n - 2)$$

We divide SSE by $n - 2$ so that the estimate will be unbiased for σ^2. (See Exercise 13.)

Summary of Theoretical Results

Before closing this section, let us introduce some notation that will make the results obtained here easier to remember. Namely, we shall denote $\sum_{i=1}^{n}(x_i - \bar{x})^2$ by S_{xx}. The symbol S_{yy} will denote $\sum_{i=1}^{n}(y_i - \bar{y})^2$ or $\sum_{i=1}^{n}(Y_i - \bar{Y})^2$. Whether we are dealing with the random variables Y_i or their observed values y_i should be clear from the context in which the symbol is used. Similarly, S_{xy} will denote either $\sum_{i=1}^{n}(x_i - \bar{x})(y_i - \bar{y})$ or $\sum_{i=1}^{n}(x_i - \bar{x})(Y_i - \bar{Y})$ and SSE will denote $\sum_{i=1}^{n}(y_i - b_0 - b_1 x_i)^2$ or $\sum_{i=1}^{n}(Y_i - B_0 - B_1 x_i)^2$. This notation can be used to rewrite the error sum of squares as follows:

$$
\begin{aligned}
\text{SSE} &= \sum_{i=1}^{n} (Y_i - B_0 - B_1 x_i)^2 \\
&= \sum_{i=1}^{n} (Y_i - \bar{Y} + B_1 \bar{x} - B_1 x_i)^2 \\
&= \sum_{i=1}^{n} [(Y_i - \bar{Y}) - B_1(x_i - \bar{x})]^2 \\
&= \sum_{i=1}^{n} (Y_i - \bar{Y})^2 - 2B_1 \sum_{i=1}^{n} (x_i - \bar{x})(Y_i - \bar{Y}) + B_1^2 \sum_{i=1}^{n} (x_i - \bar{x})^2 \\
&= S_{yy} - 2B_1 S_{xy} + B_1^2 S_{xx}
\end{aligned}
$$

Note that

$$
B_1 = \frac{\sum_{i=1}^{n}(x_i - \bar{x})(Y_i - \bar{Y})}{\sum_{i=1}^{n}(x_i - \bar{x})^2} = \frac{S_{xy}}{S_{xx}}
$$

By substituting, we see that

$$
\begin{aligned}
\text{SSE} &= S_{yy} - 2B_1 S_{xy} + B_1 \frac{S_{xy}}{S_{xx}} S_{xx} \\
&= S_{yy} - B_1 S_{xy}
\end{aligned}
$$

Let us summarize the theoretical results that we have obtained in this section. We have shown that

1. $\sum_{i=1}^{n}(x_i - \bar{x})^2 = S_{xx}$
2. $\sum_{i=1}^{n}(Y_i - \bar{Y})^2 = S_{yy}$
3. $\sum_{i=1}^{n}(x_i - \bar{x})(Y_i - \bar{Y}) = S_{xy}$
4. $B_1 = S_{xy}/S_{xx}$ is an unbiased estimator for β_1. This estimator is normally distributed with variance $\sigma_{B_1}^2 = \sigma^2/S_{xx}$.
5. $B_0 = \bar{Y} - B_1 \bar{x}$ is an unbiased estimator for β_0. This estimator is normally distributed with variance $\sigma_{B_0}^2 = (\sum_{i=1}^{n} x_i^2 \sigma^2)/n S_{xx}$.
6. $S^2 = \text{SSE}/(n-2)$ is an unbiased estimator for σ^2.

10.3 CONFIDENCE INTERVAL ESTIMATION AND HYPOTHESIS TESTING

In the previous sections we considered point estimation procedures for the parameters associated with the simple linear regression model. We showed that the estimators given are unbiased. With this information alone, we can estimate a regression line from a sample of paired observations (x_i, y_i) and predict the value of Y or estimate the mean value of Y for a given value x. As in the past, we do not end our study with point estimation. We continue by developing pertinent confidence intervals and by learning how to test hypotheses on the model parameters. In this section we consider these topics:

1. Hypothesis testing and confidence interval estimation on the slope of the regression line
2. Hypothesis testing and confidence interval estimation on the intercept of the regression line
3. Confidence interval estimation on the mean value of Y for a given value x
4. Prediction interval estimation on the value of Y itself for a given value x

We consider these ideas in the order listed.

Inferences about Slope

One of the first questions that a scientist wants to answer is, "Is the regression 'significant'?" The term "significant regression" as used here means that there is sufficient statistical evidence to conclude that the slope of the true regression line is not zero. Note that if $\beta_1 = 0$, then our regression model is

$$Y_i = \beta_0 + E_i$$

This implies that the variation in Y is due solely to random fluctuations about the line $Y = \beta_0$. If $\beta_1 \neq 0$, then at least some of the variation in Y is explained by the fact that Y is being observed at different x values. In the latter case our regression model is helpful in estimating $\mu_{Y|x}$ and predicting $Y|x$.

To develop a test statistic for testing $H_0: \beta_1 = 0$, we reconsider B_1, the point estimator for β_1. Recall that

$$B_1 \sim N(\beta_1, \sigma^2/S_{xx})$$

By standardizing, we can conclude that the random variable

$$(B_1 - \beta_1)/(\sigma/\sqrt{S_{xx}})$$

is standard normal. It can be shown that the random variable $(n - 2)S^2/\sigma^2 = \text{SSE}/\sigma^2$ has a chi-squared distribution with $n - 2$ degrees of freedom and that B_1 and S^2 are independent [19]. By applying Definition 7.2.1, the definition of a T random variable, we can conclude that the random variable

$$\frac{(B_1 - \beta_1)/(\sigma/\sqrt{S_{xx}})}{\sqrt{(n-2)S^2/\sigma^2(n-2)}} = \frac{B_1 - \beta_1}{S/\sqrt{S_{xx}}}$$

has a T distribution with $n - 2$ degrees of freedom. If $\beta_1 = 0$, then this random variable can be used to test for significant regression:

Test Statistic $H_0: \beta_1 = 0$

$$T_{n-2} = \frac{B_1}{S/\sqrt{S_{xx}}}$$

This statistic serves as the test statistic for testing any of the usual three hypotheses:

$H_0: \beta_1 = 0$	$H_0: \beta_1 = 0$	$H_0: \beta_1 = 0$
$H_1: \beta_1 > 0$	$H_1: \beta_1 < 0$	$H_1: \beta_1 \neq 0$
Right-tailed test	Left-tailed test	Two-tailed test

The null hypothesis is rejected for large positive values of the test statistic in conducting a right-tailed test; large negative values lead to rejection of H_0 in a left-tailed test. In a two-tailed test H_0 is rejected for large values in either the positive or negative direction. The three cases for the regression line slope of $\beta_1 > 0, \beta_1 < 0$, and $\beta_1 = 0$ are illustrated in Fig. 10.6.

One other point needs to be made. We have considered the null value to be 0 because this is the value most often encountered in practice. We can test $H_0: \beta_1 = \beta_1^0$, where β_1^0 denotes any hypothesized value for the slope of the regression line. The test statistic for this generalized null hypothesis is

> **Test Statistic for Inferences on the Slope**
>
> $$T_{n-2} = \frac{(B_1 - \beta_1^0)}{S/\sqrt{S_{xx}}}$$

Example 10.3.1. In Example 10.1.1 we estimated the regression equation of Y, the extent of solvent evaporation while spray painting, on X, the relative humidity, to be

$$\hat{\mu}_{Y|x} = 13.64 - .08x$$

We now determine whether the regression is significant. That is, we test

$$H_0: \beta_1 = 0$$
$$H_1: \beta_1 \neq 0$$

Summary statistics for the data given previously are

$$n = 25 \qquad \sum x = 1314.90 \qquad \sum y = 235.70$$
$$\sum x^2 = 76,308.53 \qquad \sum y^2 = 2286.07 \qquad \sum xy = 11,824.44$$

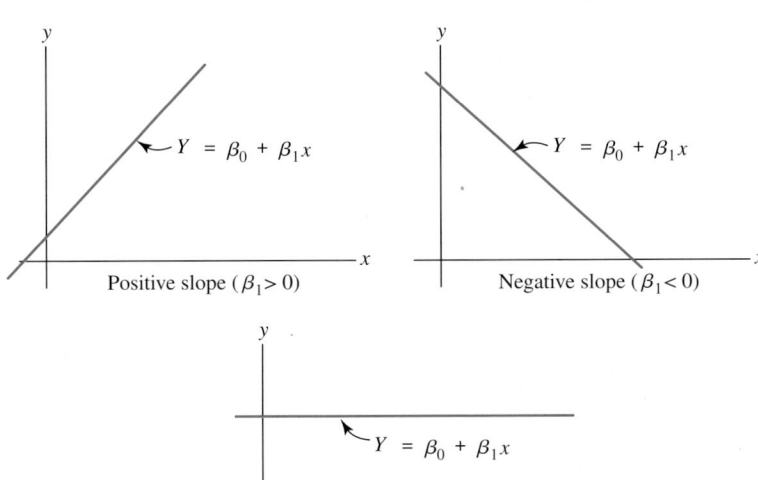

Figure 10.6
Relative positive, negative, and zero slopes for a linear regression line.

For these data

$$S_{xx} = \left[n \sum x^2 - \left(\sum x \right)^2 \right] \Big/ n$$

$$= [25(76{,}308.53) - (1314.90)^2]/25$$

$$= 7150.05$$

$$S_{yy} = \left[n \sum y^2 - \left(\sum y \right)^2 \right] \Big/ n$$

$$= [25(2286.07) - (235.70)^2]/25$$

$$= 63.89$$

$$S_{xy} = \left[n \sum xy - \sum x \sum y \right] \Big/ n$$

$$= [25(11{,}824.44) - (1314.90)(235.70)]/25$$

$$= -572.44$$

Using these data, we obtain

$$\text{SSE} = S_{yy} - b_1 S_{xy}$$

$$= 63.89 - (-.08)(-572.44)$$

$$= 18.09$$

Hence

$$s^2 = \text{SSE}/(n-2)$$

$$= 18.09/23$$

$$= .79$$

The observed value of the $T_{n-2} = T_{23}$ test statistic is

$$t = \frac{b_1}{s/\sqrt{S_{xx}}}$$

$$= \frac{-.08}{\sqrt{.79}/\sqrt{7150.05}}$$

$$= -7.62$$

From Table VI of App. A we see that $P[T_{23} \le -7.62] < .0005$. Since this is a two-tailed test, $P < 2(.0005) = .001$. We can reject H_0 and conclude that the slope of the true regression line is not zero. That is, knowledge of the x value does help in estimating $\mu_{Y|x}$ and predicting $Y|x$. Refer also to the SAS output following Example 10.1.1. The T statistic for testing $\beta_1 = 0$ is given at ③, and the corresponding P value is given at ④. To derive the bounds for a confidence interval on the slope, note that the random variable

$$T_{n-2} = \frac{B_1 - \beta_1}{S/\sqrt{S_{xx}}}$$

is of the form

$$\frac{\text{Estimator} - \text{parameter}}{D}$$

where D is the estimator for the standard deviation of B_1. This is the same algebraic structure encountered several times in the past. (See Secs. 8.3 and 9.3.) The resulting confidence interval for β_1 assumes the familiar form

$$\text{Estimator} \pm \text{probability point} \cdot D$$

In this case the confidence interval is

> ### Confidence interval on β_1, the slope of the regression line
>
> $$B_1 \pm t_{\alpha/2} S / \sqrt{S_{xx}}$$
>
> where $t_{\alpha/2}$ is the appropriate point based on the T_{n-2} distribution.

Inferences about Intercept

Hypothesis tests on β_0, the intercept of the true regression line, are conducted by noting that since

$$B_0 \sim N(\beta_0, \sigma^2 \sum x^2 / n S_{xx})$$

the random variable

$$\frac{B_0 - \beta_0}{\left(\sigma \sqrt{\sum x^2}\right) / \left(\sqrt{n \cdot S_{xx}}\right)}$$

is standard normal. It can be shown that B_0 and S^2 are independent. Thus the random variable

$$\frac{(B_0 - \beta_0) / \left(\sigma \sqrt{\sum x^2} / \sqrt{n S_{xx}}\right)}{\sqrt{(n-2) S^2 / \sigma^2 (n-2)}} = \frac{B_0 - \beta_0}{\left(\dfrac{S \sqrt{\sum x^2}}{\sqrt{n S_{xx}}}\right)}$$

follows a T distribution with $n - 2$ degrees of freedom.

The test statistic for testing $H_0\colon \beta_0 = 0$ is

> ### Test Statistic $H_0\colon \beta_0 = 0$
>
> $$T_{n-2} = \frac{B_0}{\left(\dfrac{S \sqrt{\sum x^2}}{\sqrt{n S_{xx}}}\right)}$$

Confidence intervals on the value of β_0 are found as follows:

> ### Confidence interval on β_0, the intercept of the regression line
>
> $$B_0 \pm t_{\alpha/2} \frac{S \sqrt{\sum x^2}}{\sqrt{n S_{xx}}}$$
>
> where $t_{\alpha/2}$ is the appropriate point based on the T_{n-2} distribution.

The next example illustrates the use of these confidence intervals.

Example 10.3.2. We continue the analysis of the data on the extent of solvent evaporation during spray painting and relative humidity by finding confidence intervals on β_0 and β_1. These summary statistics, found earlier, are needed:

$$s^2 = .79 \qquad \sum x^2 = 76{,}308.53 \qquad b_0 = 13.64$$
$$S_{xx} = 7150.05 \qquad b_1 = -.08 \qquad n = 25$$

A 99% confidence interval on the slope of the regression line is given by

$$b_1 \pm t_{.005} s / \sqrt{S_{xx}} \qquad \text{or} \qquad -.08 \pm 2.807\sqrt{.79}/\sqrt{7150.05}$$

The point $t_{.005}$ is based on the $T_{n-2} = T_{23}$ distribution. Completing the calculations, we see that we can be 99% confident that the slope of the true regression line lies in the interval $[-.109, -.051]$. Note that this interval does not contain 0. This is expected, since we rejected $H_0: \beta_1 = 0$ in our last example.

A 90% confidence interval on the intercept of the regression line is given by

$$b_0 \pm t_{.05} s \sqrt{\sum x^2} / \sqrt{n S_{xx}} \quad \text{or} \quad 13.64 \pm 1.714\sqrt{.79}\sqrt{76{,}308.53}/\sqrt{25(7150.05)}$$

We can be 90% confident that the true regression line crosses the y axis between the points $y = 12.64$ and $y = 14.64$.

Inferences about Estimated Mean

In addition to finding a point estimate for $\mu_{Y|x}$, the mean value of Y for a given x value, it is useful to be able to obtain a confidence interval on this parameter. To do so, we consider the distribution of the point estimator for $\mu_{Y|x}$ by rewriting this estimator in the form

$$\hat{\mu}_{Y|x} = B_0 + B_1 x$$
$$= \bar{Y} - B_1 \bar{x} + B_1 x$$
$$= \bar{Y} + B_1(x - \bar{x})$$

Since $\bar{Y}$ and B_1 are both normally distributed and independent, $\hat{\mu}_{Y|x}$ is normal. In Exercise 12 we found that this estimator is unbiased for $\mu_{Y|x}$. The only other information needed is its variance. Using the rules for variance, we see that

$$\text{Var}(\hat{\mu}_{Y|x}) = \text{Var}[\bar{Y} + B_1(x - \bar{x})]$$
$$= \text{Var}\,\bar{Y} + (x - \bar{x})^2 \text{Var}\,B_1$$
$$= \sigma^2/n + (x - \bar{x})^2 \sigma^2 / S_{xx}$$
$$= \left[1/n + \frac{(x - \bar{x})^2}{S_{xx}} \right] \sigma^2$$

To summarize, we can conclude that

Distribution of $\mu_{Y|x}$

$$\hat{\mu}_{Y|x} \sim N\left\{ \mu_{Y|x}, \left[1/n + \frac{(x - \bar{x})^2}{S_{xx}} \right] \sigma^2 \right\}$$

By the standardization process it can be shown that

$$\frac{\hat{\mu}_{Y|x} - \mu_{Y|x}}{\sigma \sqrt{\dfrac{1}{n} + \dfrac{(x - \bar{x})^2}{S_{xx}}}}$$

is standard normal. Dividing by $\sqrt{(n-2)S^2/\sigma^2(n-2)} = S/\sigma$, we find that the random variable

$$\frac{\hat{\mu}_{Y|x} - \mu_{Y|x}}{S\sqrt{\dfrac{1}{n} + \dfrac{(x-\bar{x})^2}{S_{xx}}}}$$

follows a T distribution with $n-2$ degrees of freedom. Since the random variable is of the same algebraic form as those encountered earlier, confidence intervals on $\mu_{Y|x}$ are found using this formula:

> **Confidence interval on $\mu_{Y|x}$, the mean value of Y when $X = x$**
>
> $$\hat{\mu}_{Y|x} \pm t_{\alpha/2}S\sqrt{\frac{1}{n} + \frac{(x-\bar{x})^2}{S_{xx}}}$$
>
> where $t_{\alpha/2}$ is the appropriate point based on the T_{n-2} distribution.

This formula can be used to construct what is called a *confidence band* about the estimated regression line. To do so, one simply constructs $100(1-\alpha)\%$ confidence intervals at several selected points and then joins the endpoints of these intervals with a smooth curve. The true regression line should lie within the band. Figure 10.7 illustrates this idea.

Inferences about a Single Predicted Value

One of the primary uses of the estimated regression line is to predict the value of Y itself for a specified value x. We know that the point estimator for $Y|x$ is the same as the point estimator for $\mu_{Y|x}$, namely,

$$\hat{Y}x = \hat{\mu}_{Yx} = B_0 + B_1 x$$

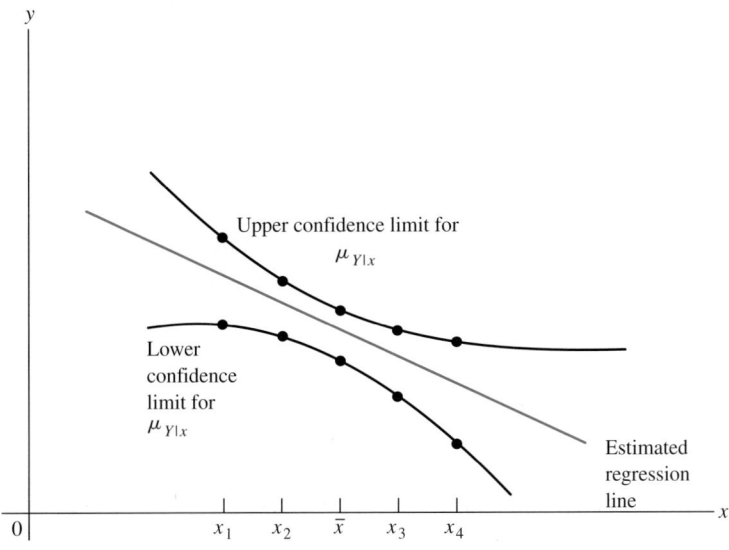

Figure 10.7
95% confidence band on $\mu_{Y|x}$.

Note that $Y|x$ is a random variable, not an unknown constant. When we ask for a "prediction interval" on $Y|x$, we are asking for two statistics L_1 and L_2 with the property that

$$P[L_1 \leq Y|x \leq L_2] \doteq 1 - \alpha$$

That is, we are asking for two statistics that will trap the observed value of $Y|x$ between them $(1 - \alpha)100\%$ of the time. To find these statistics, we use the guideline for constructing a confidence interval given in Chap. 6. This guideline requires that we find a random variable whose expression involves $Y|x$ and whose distribution we know. Recall that

$$\hat{\mu}_{Y|x} \sim N\left\{\mu_{Y|x}, \left[\frac{1}{n} + \frac{(x - \bar{x})^2}{S_{xx}}\right]\sigma^2\right\}$$

and that, via our model assumptions

$$Y|x \sim N(\mu_{Y|x}, \sigma^2)$$

It can be shown that the random variable $\hat{Y}|x - Y|x$ is normally distributed [19]. Using the rules for expectation, we obtain

$$E[\hat{Y}|x - Y|x] = E[\hat{Y}|x] - E[Y|x]$$
$$= \mu_{Y|x} - \mu_{Y|x} = 0$$

Similarly, the rules for variance are used to show that

$$\text{Var}[\hat{Y}|x - Y|x] = \text{Var } \hat{Y}|x + \text{Var } Y|x$$
$$= \left[\frac{1}{n} + \frac{(x - \bar{x})^2}{S_{xx}}\right]\sigma^2 + \sigma^2$$
$$= \left[1 + \frac{1}{n} + \frac{(x - \bar{x})^2}{S_{xx}}\right]\sigma^2$$

In conclusion, it can be seen that

$$(\hat{Y}x - Yx) \sim N\left\{0, \left[1 + \frac{1}{n} + \frac{(x - \bar{x})^2}{S_{xx}}\right]\sigma^2\right\}$$

In this case standardization and division by S/σ results in the T random variable

$$T_{n-2} = \frac{\hat{Y}|x - Y|x}{S\sqrt{1 + \frac{1}{n} + \frac{(x - \bar{x})^2}{S_{xx}}}}$$

The algebraic structure of this random variable parallels that seen earlier. For this reason, we can conclude that a $100(1 - \alpha)\%$ "prediction interval" on $Y|x$ is given by

---

Prediction interval on $Y|x$, the value of Y when $X = x$

$$\hat{Y}|x \pm t_{\alpha/2}S\sqrt{1 + \frac{1}{n} + \frac{(x - \bar{x})^2}{S_{xx}}}$$

where $t_{\alpha/2}$ is the appropriate point based on the T_{n-2} distribution.

By evaluating the prediction limits at several x values, we can construct a prediction band on $Y|x$. Note that the confidence limits for $\mu_{Y|x}$ and $Y|x$ are similar. The difference is that the former entails the term

$$\sqrt{\frac{1}{n} + \frac{(x - \bar{x})^2}{S_{xx}}}$$

whereas the corresponding term in the latter is a little larger, namely,

$$\sqrt{1 + \frac{1}{n} + \frac{(x - \bar{x})^2}{S_{xx}}}$$

This is to be expected, since we should be able to estimate an average response more precisely than we can predict an individual observation. Graphically, the confidence band on $\mu_{Y|x}$ will be contained in the corresponding prediction band for $Y|x$. This idea is illustrated in Fig. 10.8.

The next example should demonstrate clearly the difference between these two types of intervals.

Example 10.3.3. An investigation is conducted to study gasoline mileage in automobiles when used exclusively for urban driving. Ten properly tuned and serviced automobiles manufactured during the same year are used in the study. Each automobile is driven for 1000 miles, and the average number of miles per gallon

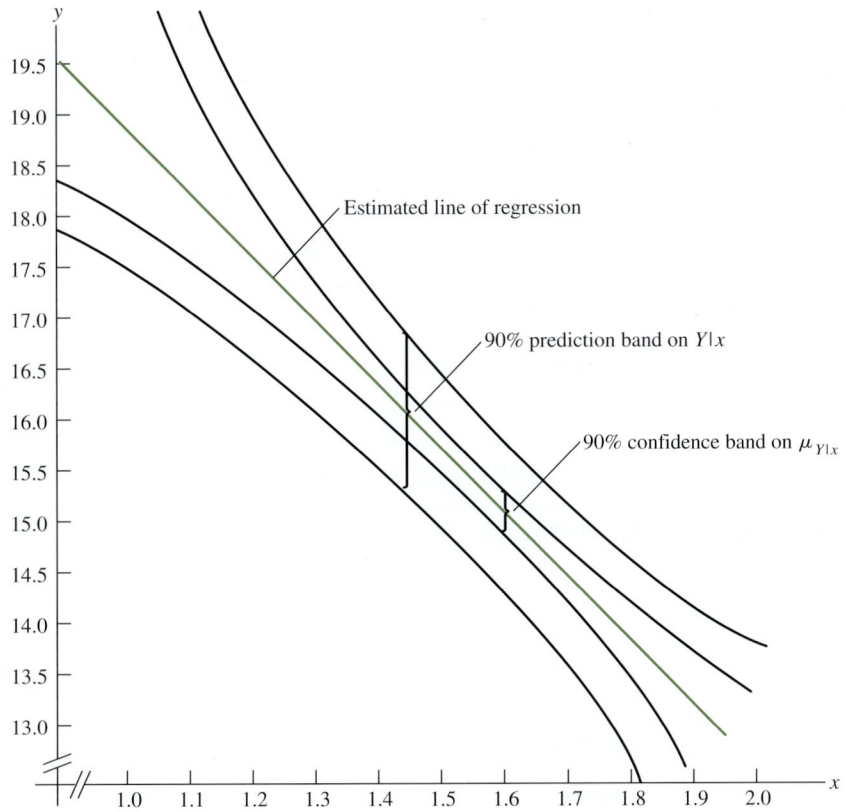

Figure 10.8
Relative positions of a 90% confidence band on $\mu_{Y|x}$ and a 90% prediction band on $Y|x$.

(mi/gal) obtained (Y) and the weight of the car in tons (X) are recorded. These data result:

Car number	1	2	3	4	5	6	7	8	9	10
Miles per gallon (y)	17.9	16.5	16.4	16.8	18.8	15.5	17.5	16.4	15.9	18.3
Weight in tons (x)	1.35	1.90	1.70	1.80	1.30	2.05	1.60	1.80	1.85	1.40

Summary statistics for these data are

$$n = 10 \qquad \sum x^2 = 28.6375 \qquad \sum y^2 = 2900.46 \qquad S_{xx} = .581 \qquad S_{xy} = -2.345$$
$$\sum x = 16.75 \qquad \sum y = 170.0 \qquad \sum xy = 282.405 \qquad S_{yy} = 10.46$$

$$\hat{\beta}_1 = b_1 = \frac{n\sum xy - \sum x \sum y}{n\sum x^2 - \left(\sum x\right)^2}$$
$$= \frac{10(282.405) - (16.75)(170.0)}{10(28.6375) - (16.75)^2}$$
$$= -4.03$$
$$\hat{\beta}_0 = b_0 = \bar{y} - b_1\bar{x}$$
$$= 17.0 - (-4.03)(1.675)$$
$$= 23.75$$

The estimated line of regression is

$$\hat{\mu}_{Y|x} = b_0 + b_1 x = 23.75 - 4.03x$$

The reader can verify that $H_0: \beta_1 = 0$ can be rejected with $P < .0001$. Thus the regression is significant; the model is useful in predicting gasoline mileage based on automobile weight. Suppose that we are interested in all cars weighing 1.7 tons. The estimated average mileage for these cars is

$$\hat{\mu}_{Y|x = 1.7} = 23.75 - 4.03(1.7) = 16.899 \text{ mi/gal}$$

This estimate is not very useful without some idea of its accuracy. To pinpoint the accuracy, we construct a 90% confidence interval on $\hat{\mu}_{Y|x = 1.7}$. To do so, we must compute SSE and s^2 for these data:

$$\text{SSE} = S_{yy} - b_1 S_{xy}$$
$$= 10.46 - (-4.03)(-2.345)$$
$$= 1.01$$
$$s^2 = \text{SSE}/(n - 2) = 1.01/8 = .126$$

A 90% confidence interval on $\mu_{Y|x}$ is

$$\hat{\mu}_{Y|x} \pm t_{\alpha/2} s \sqrt{\frac{1}{n} + \frac{(x - \bar{x})^2}{S_{xx}}}$$

or

$$16.899 \pm 1.86\sqrt{.126}\sqrt{\frac{1}{10} + \frac{(1.7 - 1.675)^2}{.581}}$$
$$16.899 \pm .21$$

We can be 90% confident that the average gas mileage for cars weighing 1.7 tons lies between 16.689 and 17.109 mi/gal.

To predict the gas mileage for a single car weighing 1.7 tons, we use the interval

$$\hat{y}|x \pm t_{\alpha/2}s\sqrt{1 + \frac{1}{n} + \frac{(x - \bar{x})^2}{S_{xx}}}$$

For these data this interval is

$$16.899 \pm 1.86\sqrt{.126}\sqrt{1 + \frac{1}{10} + \frac{(1.7 - 1.675)^2}{.581}}$$

$$16.899 \pm .69$$

We can be 90% confident that the gas mileage for any individual automobile weighing 1.7 tons lies between 16.209 and 17.589 mi/gal. As expected, the prediction interval used to predict the gas mileage for a single auto is wider than that used to predict the average mileage for a group of automobiles.

We should note here that the width of a confidence or prediction band is a function of x. To see why this is true, consider the formula for constructing a prediction interval of $Y|x$. The width of the interval is determined in part by the term

$$\sqrt{1 + \frac{1}{n} + \frac{(x - \bar{x})^2}{S_{xx}}}$$

It is evident that this term is smallest when $x = \bar{x}$. Hence we can predict the value of Y more precisely for values of x that are near the average value $\bar{x}$. This fact is evident graphically in the confidence bands shown in Fig. 10.8. In this figure the bands are narrowest at $\bar{x} = 1.675$.

10.4   RESIDUAL ANALYSIS

Recall that to construct confidence intervals on β_0, β_1, and $\mu_{Y|x}$ and prediction intervals on $Y|x$ or to test hypotheses concerning β_0 and β_1, we need to make some model assumptions. When the simple linear regression model is written as

> **Simple Linear Regression Model**
> $$Y_i = \beta_0 + \beta_1 x_i + E_i$$

then the model assumptions are expressed as assumptions concerning the behavior of the random variables $E_1, E_2, \ldots, E_n$. In particular, it is assumed that these random variables are independent, normally distributed random variables with mean 0 and common variance σ^2. Before a fitted regression line is used to make predictions in practice, an effort should be made to check the validity of these assumptions. In this section we present some graphical techniques that can be used to do so. These procedures utilize the residuals, whose behavior under ideal conditions should mirror that of the random variables E_i.

Residual Plots

Recall that e_i, the ith residual, is the vertical distance from the ith data point to the fitted regression line. Therefore

$$e_i = y_i - [b_0 + b_1 x_i]$$

For a given value x_i the predicted response is found by substitution into the regression equation. That is, $\hat{y}_i = b_0 + b_1 x_i$. It can be seen that the ith residual can be written as

$$e_i = y_i - \hat{y}_i$$

The ith residual is the difference between the ith observed response and its predicted value. To check the model assumptions, we construct residual plots. A residual plot is a scattergram of the points (x_i, e_i). In such a plot we are plotting the regressor value (horizontal axis) versus the residual value (vertical axis). A residual plot can be used to help answer two questions:

1. Do the model assumptions underlying simple linear regression appear to be met?

2. If the model assumptions do not appear to be met, then which assumptions fail?

Since residual plots are useful in pinpointing or diagnosing problems that might exist, they are sometimes referred to as "diagnostic tools."

Figure 10.9(a) shows a plot of a data set for which simple linear regression is appropriate. The data points exhibit an upward linear trend; they cluster tightly about the estimated line of regression; and the spread of the data points at each value of the regressor is about the same. The residual plot associated with an ideal data set of this sort is shown in Fig. 10.9(b). Notice that the residual plot depicts a set of points that scatter randomly about 0. The fact that these points vary about 0 is to be expected, since it has been shown that the average value of the residuals is always 0. However, notice also that the spread of the residuals is about the same throughout the plot. This is expected whenever the assumption of a common variance σ^2 is valid.

Figure 10.9
(a) The scattergram of a data set for which simple linear regression is appropriate. Points exhibit a linear trend with uniform spread about the estimated line of regression; (b) a residual plot for the ideal case. Residuals scatter randomly about 0 with a uniform spread.

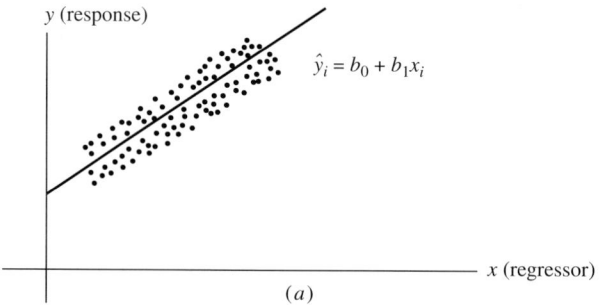

y (response)

$\hat{y}_i = b_0 + b_1 x_i$

x (regressor)

(a)

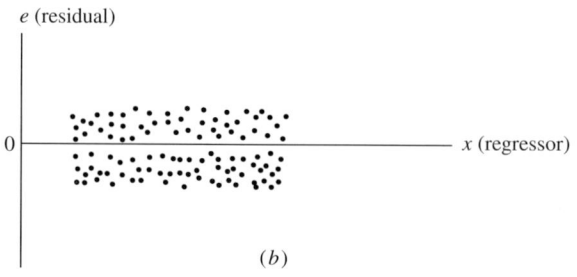

e (residual)

0

x (regressor)

(b)

In Fig. 10.10(*a*) a data set that signals trouble is shown. Notice that although there is an upward linear trend, the spread of the responses appears to increase as *x* increases. This is an indication that the common variance assumption might not be met. That is, the variance in response for small values of the regressor seems to be different from that for large values of *x*. How is this problem seen on a residual plot? Probably just as you suspect—the residual plot will show a random scatter about 0, with the spread of the points increasing as the value of *x* becomes larger. See Fig. 10.10(*b*).

Two other problems can be spotted with the help of a residual plot. They are model misspecification and gaps in the data. Model misspecification occurs when we try to fit a straight line to data that is not linear; gaps in the data occur due to poor experimental design or perhaps due to the loss of some data during experimentation. Both problems make the use of a linear prediction equation risky at best. Figure 10.11(*a*) illustrates a set of data that is clearly not linear together with a "regression line" that has nevertheless been forced through the data. Figure 10.11(*b*) shows how this error is seen on a residual plot. Notice that the residuals do vary about 0, but there appears to be a pattern in the residuals. The scatter is not random. This lack of randomness is what we are looking for, since it signals the possibility that simple linear regression does not adequately describe the relationship between the regressor and the response.

In Fig. 10.12(*a*) a data set that contains gaps with respect to the regressor values is shown. Even though it is possible to fit a line to these data, to do so is risky. We are assuming that the linear trend suggested by the responses for low and high values of the regressor continues in the midrange of *x*. We really have no evidence

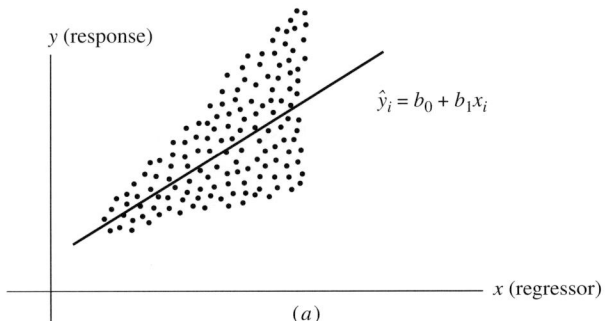

(*a*)

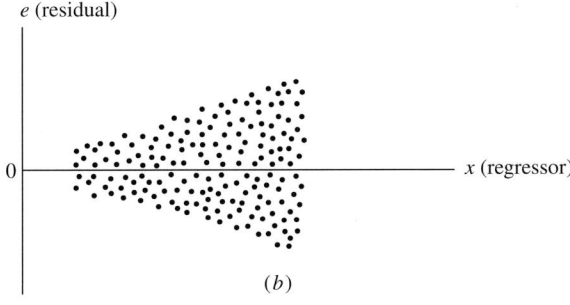

(*b*)

Figure 10.10
(*a*) A data set that signals that the common variance assumption is probably not valid; (*b*) a residual plot that throws doubt on the validity of the common variance assumption. Residuals scatter randomly about 0, but the spread of the points is not uniform.

Figure 10.11
(*a*) A data set for which the simple linear regression model is inappropriate; (*b*) a residual plot in which a linear prediction equation has been fitted to a set of data points that do not exhibit a linear trend. The residuals exhibit a pattern rather than a random scatter about 0.

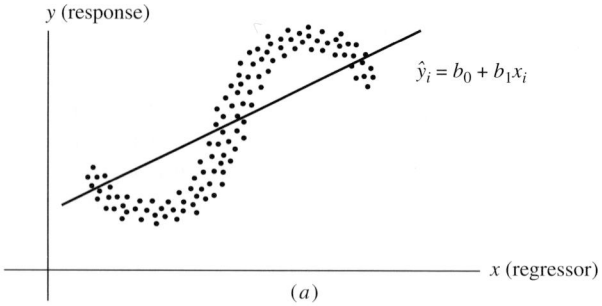

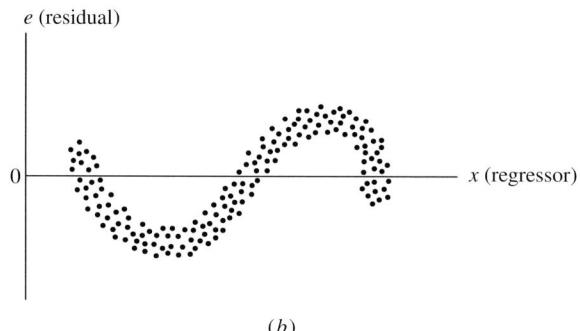

that this is the case, and it is not appropriate to assume this. Figure 10.12(*b*) shows the residual plot for a data set with a definite gap in the data.

Residual plots are helpful in spotting potential problems. However, they are not always as easy to interpret as are those given in Figs. 10.9 through 10.12. Since patterns are hard to spot with small data sets except in extreme cases, residual plots are most useful with fairly large collections of data. Furthermore, to get a clear picture of the validity of the variance assumption, we should design experiments in such a way that multiple observations are taken at each distinct value of the regressor.

Checking for Normality: Stem-and-Leaf Plots and Boxplots

One model assumption that has not been investigated yet is that of normality. This assumption can be checked visually as was done in previous chapters or analytically by using a standard software package such as SAS. Here we consider two visual checks that do not require formal testing. They work best for fairly large samples, since they both require that some value judgments be made concerning shape. Such judgments are hard to make with small samples because patterns do not appear in such data sets unless the violations in the assumptions are extreme.

The first visual technique is one that should come to mind immediately. Namely, construct a stem-and-leaf diagram of the residuals as explained in Chap. 5. If the normality assumption is valid, we expect the plot to exhibit the approximate bell shape indicative of a normal curve.

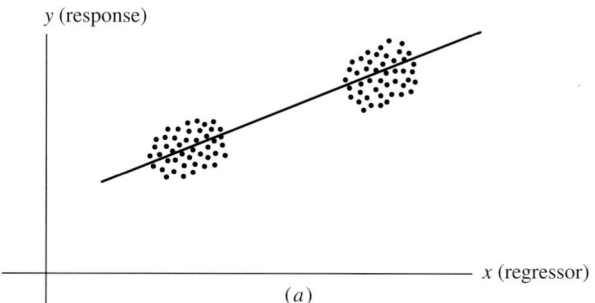

(a)

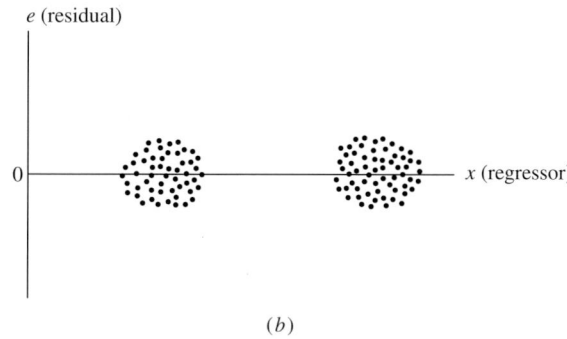

(b)

Figure 10.12
(a) A data set with a midrange gap. Simple linear regression is risky. (b) A residual plot showing gaps in the regressor values.

The boxplot can also be used as a diagnostic tool. It allows us to check for possible violations of the normality assumption and also to detect the presence of outliers. A boxplot for residuals obtained when the normality assumption is valid should be symmetric with the median line near or at 0. Outliers are not expected, since these values are extremely rare whenever the random variable involved is normally distributed. (See Exercise 26 of Chap. 5.) Thus asymmetry or the presence of outliers signals that the normality assumption might not hold.

Outliers are very troublesome in regression studies. They can greatly influence the regression line in that the line tends to be pulled toward the outlier. This can cause the fitted line not to pass through the center of the bulk of the data as is desired. If outliers are detected via a boxplot, then they must be investigated. If the data point is found to be an error or suspect in some way, then it should not be used in the analysis.

Example 10.4.1 illustrates the use of residual and stem-and-leaf plots in a regression study. In this example you will see that the plots associated with real data are not always as easy to interpret as you would like.

Example 10.4.1. Consider the problem described in Example 10.1.1 in which we found an equation by which the extent of solvent evaporation (Y) can be predicted based on knowledge of the humidity (X). The scattergram for the data given in Example 10.1.1 is shown in Fig. 10.13(a). The line shown in the picture is the graph of the estimated line of regression. Its equation is given by

$$\hat{\mu}_{Y|x} = 13.64 - .08x$$

There is a linear trend to the data, and the data points appear to lie reasonably close to the estimated line of regression. Figure 10.13(b) shows the residual plot. Notice

Figure 10.13
(*a*) Scattergram of data with an
estimated line of regression.
(*b*) A residual plot for the data
of Examples 10.1.1 and 10.4.1.
Residuals show no obvious
pattern that would signal
model misspecification.

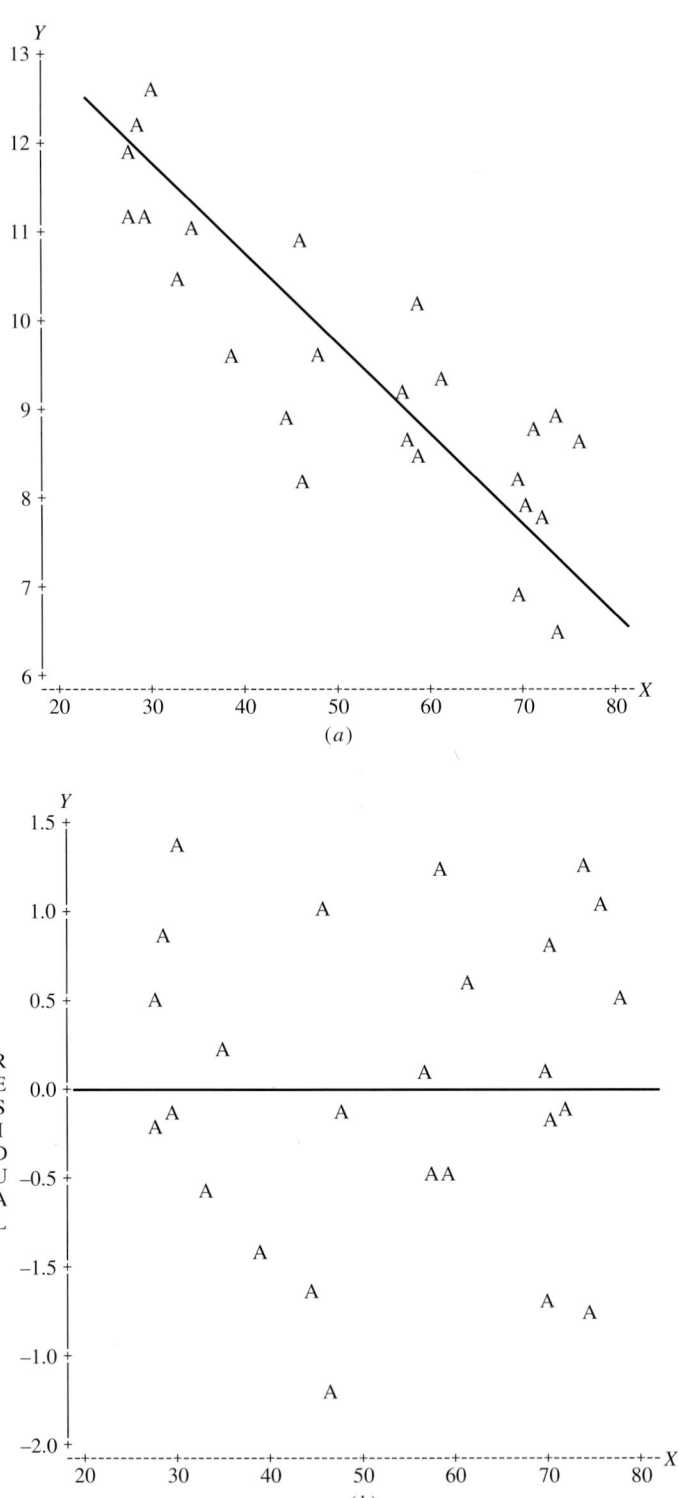

that this plot does not reveal a pattern that might indicate that the linear model is not
appropriate; the points do appear to scatter randomly about 0 as desired. There are
no obvious differences in spread as *x* increases. However, the experiment was not
designed with multiple observations at distinct values of the regressor, so this
assumption is not easy to verify. To check for normality, we construct a stem-and-leaf

Table 10.1

Observed response (Y), predicted response (Predict.), and residual (Resid.) for the 25 data points of Examples 10.1.1 and 10.4.1

Observation number	Y	Predict.	Resid.
1	11.0	10.8127	0.18727
2	11.1	11.2611	−0.16107
3	12.5	11.1730	1.32700
4	8.4	8.9313	−0.53130
5	9.3	8.7231	0.57685
6	8.7	7.9305	0.76945
7	6.4	7.6824	−1.28236
8	8.5	7.4982	1.00178
9	7.8	7.9786	−0.17858
10	9.1	9.0354	0.06462
11	8.2	9.9241	−1.72406
12	12.2	11.3251	0.87488
13	11.9	11.3892	0.51084
14	9.6	10.5085	−0.90850
15	10.9	9.8920	1.00797
16	9.6	9.7559	−0.15593
17	10.1	8.8913	1.20873
18	8.1	8.0346	0.06537
19	6.8	8.0346	−1.23463
20	8.9	7.6824	1.21764
21	7.7	7.8665	−0.16650
22	8.5	8.9873	−0.48735
23	8.9	10.0682	−1.16816
24	10.4	10.9648	−0.56484
25	11.1	11.3491	−0.24913

```
 1   3  0  0  2  2
 0   5  7  8  5
 0   1  0  0
-0   1  1  1  1  4  2
-0   5  9  5
-1   2  2  1
-1   7
```

Figure 10.14
A stem-and-leaf diagram for the residuals of Table 10.1. The diagram uses double stems, with a "leaf" being the first decimal place of the residual.

plot of the residuals. The residuals are given in Table 10.1, and the stem-and-leaf plot for these residuals is shown in Fig. 10.14. In the plot we use double stems, with the "leaf" being the first decimal place of the residual. For example, the value .18727 is graphed as $0 \mid 1$ on the "low" 0 stem. Does this plot give convincing evidence of normality? Does it clearly exhibit the bell-shape characteristic of a normal curve? The answer to these questions is probably "no." The shape is rather nondescript. To determine whether there is enough evidence to indicate that the normality assumption is probably not valid, a formal test is needed. Figure 10.15 gives the SAS printout for PROC UNIVARIATE. This printout includes the observed value of the statistic W used to test

H_0: data are from a normal distribution

H_1: data are from a nonnormal distribution

The value of the statistic, .958717, is shown at (1); its P value, .4028, is shown at (2). Since this P value is large, H_0 should not be rejected. Based on these residuals, there

Figure 10.15
An analytic test for normality.

SAS

UNIVARIATE PROCEDURE

Variable = RESID Residual

Moments

N	25	Sum Wgts	25
Mean	0	Sum	0
Std Dev	0.867485	Variance	0.752531
Skewness	−0.17648	Kurtosis	−0.84038
USS	18.06074	CSS	18.06074
CV	.	Std Mean	0.173497
T:Mean = 0	0	Prob > |T|	1.0000
Sgn Rank	−0.5	Prob > |S|	0.9896
Num ^ = 0	25		
① W:Normal	0.958717	Prob < W	0.4028 ②

Quantiles (Def = 5)

100% Max	1.326999	99%	1.326999
⑤ 75% Q3	0.769453	95%	1.217641
③ 50% Med	−0.15593	90%	1.208726
④ 25% Q1	−0.5313	10%	−1.23463
0% Min	−1.72406	5%	−1.28236
		1%	−1.72406

Range	3.051055
Q3 − Q1	1.300757
Mode	−1.72406

Extremes

Lowest	Obs	Highest	Obs
−1.72406(	11)	1.00178(	8)
−1.28236(	7)	1.007969(	15)
−1.23463(	19)	1.208726(	17)
−1.16816(	23)	1.217641(	20)
−0.9085 (	14)	1.326999(	3)

Stem	Leaf	#	Boxplot
1	00223	5	|
0	5689	4	+−−−−+
0	112	3	| |
−0	22222	5	⑥ *−−−−−*
−0	9655	4	+−−−−+
−1	322	3	|
−1	7	1	

```
----+----+----+----+
```

is no reason to suspect that the residuals do not follow a normal distribution. Notice that SAS gives a stem-and-leaf diagram that is different from that given in Fig. 10.14. Remember that there are no set rules in defining leaves. The SAS plot has defined a "leaf" by first rounding the residual to one decimal place and then plotting the residuals. Does this clarify the question of normality? Probably not. There is still no clearly defined bell visible.

In the previous example interest centered on checking for normality. In the next example the boxplot is used to check for symmetry and the presence of outliers. A definite lack of symmetry or the presence of outliers are both signals of possible violation of the normality assumption.

Example 10.4.2. To construct a quick boxplot for the residuals given in Table 10.1, we again retain only the first decimal place of the number, as was done in constructing the stem-and-leaf plot shown in Fig. 10.14. For these date $n = 25$, the median location is $(n + 1)/2 = 13$, and the median value is $-.1$. Quartile locations are at $(13 + 1)/2 = 7$. The quartile values are $q_1 = -.5$ and $q_3 = .7$. The interquartile range is $q_3 - q_1 = 1.2$. Inner fences are located at

$$f_1 = q_1 - 1.5iqr = -.5 - 1.5(1.2) = -2.3$$

and

$$f_3 = q_3 + 1.5iqr = .7 + 1.5(1.2) = 2.5$$

Since no residual values lie beyond the inner fences, the set of residuals does not contain any outliers. This is good news, for outliers are very rare (occurring with probability .007) when sampling from a normal distribution. The boxplot for the residuals is shown in Fig. 10.16. Notice that it does not exhibit the perfect symmetry expected from a normal distribution, but it is not skewed enough to signal a clear violation of the normality assumption. Values for the median, q_1, and q_3 based on the actual residual values are given by SAS in Fig. 10.15 at (3), (4), and (5), respectively. The SAS boxplot is shown at (6). The + in the box is at 0, the average value

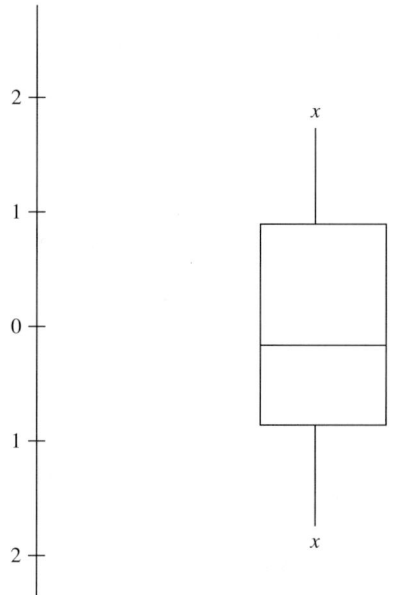

Figure 10.16
A boxplot for the residuals of Table 10.1 based on the stem-and-leaf plot of Fig. 10.14. No outliers are detected. Although the plot does not exhibit perfect symmetry, it is not skewed enough to reject the normality assumption.

of the residuals. In a perfectly symmetric plot, the ideal plot, this + would coincide with the median. Since our data set is not perfect, as is usually the case with real data, these two values differ slightly.

Regression is an art as well as a science. Real-life data sets are seldom perfect. You will be called upon to make some value judgments as to the appropriateness of linear regression. The tools presented in this section will help you make these judgments.

10.5 CORRELATION

Thus far in this chapter we have considered problems related to simple linear regression. Our primary problem has been to express the *mean* value of a random variable Y as a linear function of a *nonrandom* variable X. There are two important differences between the regression studies that we have been considering and the correlation studies that we shall consider now. First, in a correlation study both X and Y must be *random variables.* Second, we are not looking for a linear relationship between X and the mean of Y; rather we are trying to measure the strength of the linear relationship that exists between X and Y itself.

The theoretical parameter used to measure the linear relationship between X and Y is the Pearson coefficient of correlation ρ. This parameter is defined by

> **Pearson correlation coefficient**
>
> $$\rho = \frac{\text{Cov}(X, Y)}{\sqrt{(\text{Var } X)(\text{Var } Y)}}$$

The parameter ρ assumes values between -1 and 1 inclusive. Values of 1 or -1 indicate perfect positive or negative linear relationships, respectively. A value of 0 indicates no linear relationship. When this occurs, we say that X and Y are uncorrelated. Figure 10.17 illustrates the graphical interpretation of ρ.

Previously we found the theoretical value of ρ based on knowledge of the joint density function for X and Y. Unfortunately, these densities are seldom known in practice. For this reason, the job of the researcher is to estimate ρ based on a set $\{(x_i, y_i): i = 1, 2, 3, \ldots, n\}$ of observations on the random variable (X, Y). It is easy to see how this can be done. We must estimate Var X, Var Y, and Cov(X, Y). We shall use the maximum likelihood estimators for variance. That is,

$$\widehat{\text{Var } X} = \sum_{i=1}^{n} (X_i - \bar{X})^2/n = S_{xx}/n$$

$$\widehat{\text{Var } Y} = \sum_{i=1}^{n} (Y_i - \bar{Y})^2/n = S_{yy}/n$$

To estimate Cov(X, Y), note that

$$\text{Cov}(X, Y) = E[(X - \mu_X)(Y - \mu_Y)]$$

We estimate Cov(X, Y) by averaging products analogous to that on the right-hand side of the above equation. Therefore

$$\widehat{\text{Cov}(X, Y)} = \sum_{i=1}^{n} (X_i - \bar{X})(Y_i - \bar{Y})/n = S_{xy}/n$$

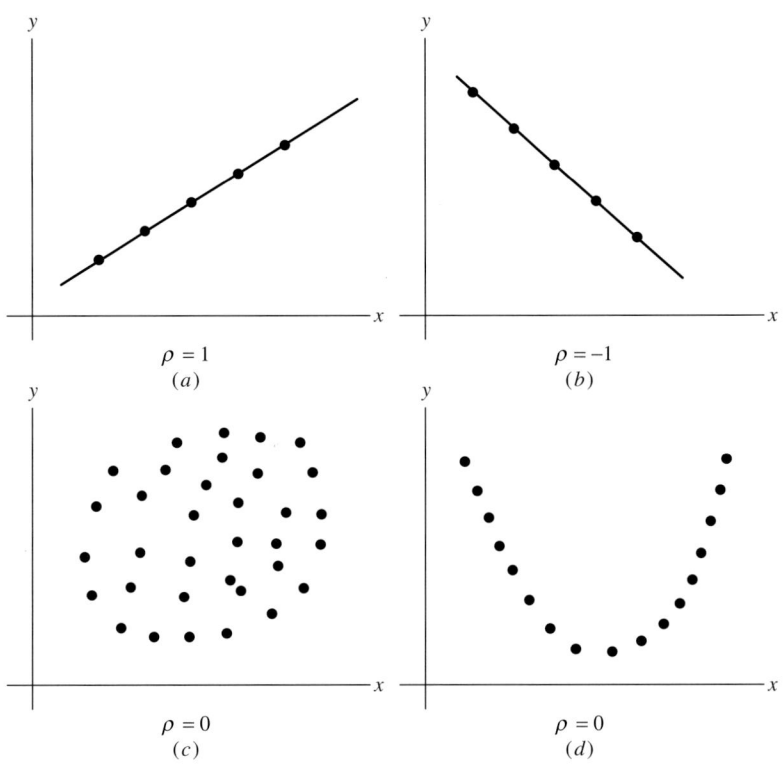

Figure 10.17
(a) $\rho = 1$, perfect positive relationship; (b) $\rho = -1$, perfect negative relationship; (c) $\rho = 0$, no relationship exists; (d) $\rho = 0$, a relationship exists but it is not linear.

When we combine these estimators, the estimator for ρ is given by

Estimator for ρ, the Pearson correlation coefficient

$$\hat{\rho} = R = \frac{S_{xy}}{\sqrt{S_{xx}S_{yy}}}$$

Many calculators will compute $\hat{\rho}$ for you automatically. If you have such a calculator, you should use it to compute $\hat{\rho}$. Otherwise, the following computational formula is useful:

Computational formula for r, the estimated Pearson correlation coefficient

$$\hat{\rho} = r = \frac{n\Sigma xy - \Sigma x \Sigma y}{\sqrt{[n\Sigma x^2 - (\Sigma x)^2][n\Sigma y^2 - (\Sigma y)^2]}}$$

Example 10.5.1 In studying the effect of sewage effluent on a lake, researchers take measurements of the nitrate concentration of the water. An older manual method has been used to monitor this variable. However, a new automated method has been devised. If a high positive correlation exists between the measurements taken by using the two methods, then the automated method will be put into routine use. These data are obtained on the nitrate concentration in micrograms of nitrate per liter of water:

x (manual)	y (automated)
25	30
40	80
120	150
75	80
150	200
300	350
270	240
400	320
450	470
575	583

The scattergram for these data is shown in Fig. 10.18. Since these points exhibit a fairly well-defined increasing trend, we expect r to be positive and close in value to 1. Summary statistics for these data are

$$n = 10 \qquad \sum x^2 = 900{,}775 \qquad \sum y^2 = 919{,}489$$
$$\sum x = 2405 \qquad \sum y = 2503 \qquad \sum xy = 902{,}475$$
$$S_{xx} = 322{,}372.5 \qquad S_{xy} = 300{,}503.5$$
$$S_{yy} = 292{,}988.1$$

The estimated correlation between X and Y is

$$\hat{\rho} = r = \frac{S_{xy}}{\sqrt{S_{xx}S_{yy}}}$$
$$= \frac{300{,}503.5}{\sqrt{(322{,}372.5)(292{,}988.1)}}$$
$$\doteq .978$$

As expected, there appears to be a strong positive linear relationship between X and Y.

Figure 10.18
A scattergram of manual readings versus automated readings.

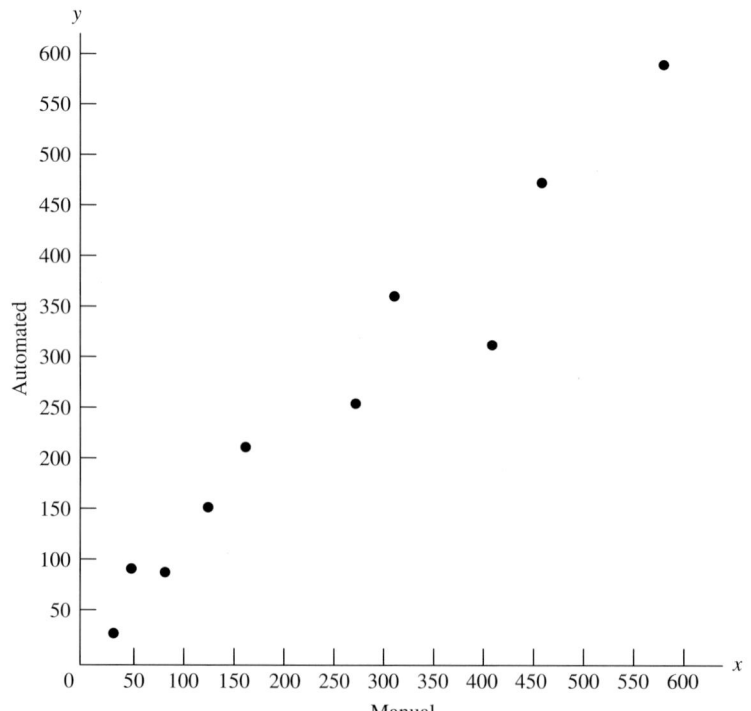

Interval Estimation and Hypothesis Tests on ρ

It is almost always possible to develop a logical point estimator for a parameter θ based on its definition alone. However, before confidence intervals can be constructed or hypothesis tests conducted, it is usually necessary to make some assumptions concerning the distribution of the random variable under study. This is true here. We have a logical point estimator for ρ. To draw statistical inferences concerning its value, we must assume a probability distribution for the two-dimensional random variable (X, Y). The distribution assumed is the *bivariate normal* distribution. The joint density for such a random variable is given by

Bivariate normal density

$$f(x, y) = k \exp\left\{-\frac{1}{2(1-\rho^2)}\left[\left(\frac{x-\mu_X}{\sigma_X}\right)^2 - 2\rho\left(\frac{x-\mu_X}{\sigma_X}\right)\left(\frac{y-\mu_Y}{\sigma_Y}\right) + \left(\frac{y-\mu_Y}{\sigma_Y}\right)^2\right]\right\}$$

where $k = \dfrac{1}{2\pi\sigma_X\sigma_Y\sqrt{1-\rho^2}}$

$-\infty < x < \infty$

$-\infty < y < \infty$

$-\infty < \mu_X < \infty$

$-\infty < \mu_Y < \infty$

$\sigma_X > 0$

$\sigma_Y > 0$

$-1 < \rho < 1$

This distribution has many interesting theoretical properties. Among them are the following:

1. The marginal distributions for both X and Y are normal. The parameters μ_X, μ_Y, σ_X, and σ_Y that appear in the expression for $f(x, y)$ are the means and standard deviations for X and Y, respectively.
2. The parameter ρ that appears in the expression for $f(x, y)$ is the correlation coefficient between X and Y.
3. If $\rho = 0$, then X and Y are independent.
4. The curves of regression of X on Y and Y on X are both linear. The latter is given by

$$\mu_{Y|x} = \mu_Y + \rho\frac{\sigma_Y}{\sigma_X}(x - \mu_X)$$

Although we shall not be overly concerned with these theoretical properties, they will make it easier to understand the relationship between correlation and regression. In assuming that (X, Y) has a bivariate normal distribution, we are assuming a linear regression model. That is, we are assuming that

$$\mu_{Y|x} = \beta_0 + \beta_1 x$$

where $\beta_1 = (\sigma_Y/\sigma_X)\rho$. Since σ_Y and σ_X are both positive, it is easy to see that the slope of the regression line and the correlation coefficient have the same algebraic sign. It is also easy to see that $\rho = 0$ if and only if $\beta_1 = 0$. Thus to test H_0: $\rho = 0$ against any one of the usual alternatives, we use the same test statistic as that used earlier to test H_0: $\beta_1 = 0$, namely, $B_1/(S/\sqrt{S_{xx}})$. Since we shall have a point estimate for ρ available when we test H_0: $\rho 0$, it is convenient to express our test statistic in the alternative form

> **Test Statistic H_0: $\rho = 0$**
>
> $$T_{n-2} = \frac{R\sqrt{n-2}}{\sqrt{1-R^2}}$$

(see Exercise 52).

We illustrate the use of this statistic in the next example.

Example 10.5.2.   In our previous example we estimated the correlation between X, the manual nitrate reading, and Y, the automated reading, by $r = .978$. Although intuition certainly leads us to suspect that we have strong evidence that $\rho \neq 0$, we must remember that the sample size is small with $n = 10$. For this reason, we should carry out the hypothesis testing. Suppose that the significance level is .05.

[Hypotheses] We test

$$H_0: \rho = 0$$
$$H_1: \rho \neq 0$$

[Test statistic] The observed value of the test statistic

$$T_{n-2} = \frac{R\sqrt{n-2}}{\sqrt{1-R^2}}$$

is

$$t = \frac{.978\sqrt{10-2}}{\sqrt{1-(.978)^2}} = 13.26$$

[Critical values] Based on the T_8 distribution, the critical values are ± 2.306.

[Conclusion] Since the value of the test statistic is greater than the right critical value ($2.306 < 13.26$), we reject H_0 and conclude that $\rho \neq 0$.

We can also estimate the P value of the test. Since $t_{.0005} = 5.041$ and the test is two-tailed, $P < .001$. We do have strong evidence that $\rho \neq 0$.

The exact distribution of R depends on the true value of ρ. Furthermore, for large values of ρ this distribution is decidedly nonnormal. Fortunately, there exists a simple change of variable that results in a random variable whose distribution is approximately normal. In particular, it can be shown that when (X, Y) has a bivariate normal distribution, then the random variable

$$\frac{1}{2}\ln\left(\frac{1+R}{1-R}\right)$$

is approximately normally distributed with

$$\mu = \frac{1}{2}\ln\left(\frac{1+\rho}{1-\rho}\right) \qquad \text{and} \qquad \sigma^2 = \frac{1}{n-3}$$

This result, due to R. A. Fisher, was first published in 1921. Standardizing, we can conclude that the random variable

$$Z = \frac{\frac{1}{2}\ln\left(\frac{1+R}{1-R}\right) - \frac{1}{2}\ln\left(\frac{1+\rho}{1-\rho}\right)}{\sqrt{\frac{1}{n-3}}}$$

is approximately standard normal. To develop confidence bounds on ρ we note that

$$P\left[-z_{\alpha/2} \le \frac{\frac{1}{2}\ln\left(\frac{1+R}{1-R}\right) - \frac{1}{2}\ln\left(\frac{1+\rho}{1-\rho}\right)}{\sqrt{\frac{1}{n-3}}} \le z_{\alpha/2}\right] \doteq 1 - \alpha$$

Although the algebraic argument is a bit messy, this inequality can be solved for ρ to obtain these bounds for a $100(1-\alpha)\%$ confidence interval on ρ:

Confidence interval on ρ, the Pearson correlation coefficient

$$\text{Lower bound} = \frac{(1+R) - (1-R)\exp(2z_{\alpha/2}/\sqrt{n-3})}{(1+R) + (1-R)\exp(2z_{\alpha/2}/\sqrt{n-3})}$$

$$\text{Upper bound} = \frac{(1+R) - (1-R)\exp(-2z_{\alpha/2}/\sqrt{n-3})}{(1+R) + (1-R)\exp(-2z_{\alpha/2}/\sqrt{n-3})}$$

To see how to evaluate these bounds, consider the next example.

Example 10.5.3. We know that a point estimate for ρ, the correlation between the manual nitrate reading and the automated reading, is .978. To find a 95% confidence interval on ρ, we first note that $z_{.025} = 1.96$ and $n = 10$. The lower bound for the confidence interval is

$$\frac{(1+r) - (1-r)\exp(2z_{\alpha/2}/\sqrt{n-3})}{(1+r) + (1-r)\exp(2z_{\alpha/2}/\sqrt{n-3})}$$

$$= \frac{(1+.978) - (1-.978)\exp(2(1.96)/\sqrt{7})}{(1+.978) + (1-.978)\exp(2(1.96)/\sqrt{7})}$$

$$= \frac{(1+.978) - .022(4.4)}{(1+.978) + .022(4.4)}$$

$$= \frac{1.881}{2.075} \doteq .907$$

Substituting, we find that the upper bound is .995. We are 95% confident that the true value of the correlation coefficient lies in the interval [.907, .995]. Since we rejected $H_0: \rho = 0$, it is not surprising that 0 is not in this interval.

Although the usual null hypothesis concerning ρ is $H_0: \rho = 0$, other null values can be tested via the Fisher transformation. Letting ρ_0 denote any null value for ρ, we see that the Z statistic

$$\boxed{\begin{array}{c} \textbf{Test Statistic } H_0\colon \rho = \rho_0 \\[6pt] Z = \dfrac{\dfrac{1}{2}\ln\left(\dfrac{1+R}{1-R}\right) - \dfrac{1}{2}\ln\left(\dfrac{1+\rho_0}{1-\rho_0}\right)}{\sqrt{\dfrac{1}{n-3}}} \end{array}}$$

serves as the test statistic for testing $H_0\colon \rho = \rho_0$.

Coefficient of Determination

Strictly speaking, one should not use the techniques of simple linear regression presented in this chapter and the correlation techniques given here on the same data set. The former assumes that X is not a random variable; the latter requires that it be a random variable. Even so, R can be useful in a regression study. As we shall show, it is an indicator of the adequacy of the simple linear regression model. To see why this is true, note that

$$\text{SSE} = S_{yy} - B_1 S_{xy}$$

Dividing each side of this equation by S_{yy} and replacing B_1 with S_{xy}/S_{xx}, we see that

$$\frac{\text{SSE}}{S_{yy}} = 1 - \frac{S_{xy}^2}{S_{xx}S_{yy}}$$

Since $R = S_{xy}/\sqrt{S_{xx}S_{yy}}$, we may conclude that

$$\frac{\text{SSE}}{S_{yy}} = 1 - R^2$$

or that $R^2 = 1 - \text{SSE}/S_{yy}$. This equation can be rewritten as

$$R^2 = \frac{S_{yy} - \text{SSE}}{S_{yy}}$$

Since S_{yy} measures the total variability in Y and SSE measures the random variability in Y about the estimated regression line, $S_{yy} - \text{SSE}$ measures the variability in Y explained by the linear regression model. The random variable R^2 represents the proportion of the variability in Y explained by the model. When this proportion is multiplied by 100%, we obtain a statistic called the *coefficient of determination*. If R lies close to 1 or -1, then R^2 will also be close to 1, yielding a coefficient of determination near 100%. When R is near 0, then the coefficient of determination is also near 0. Thus the relative size of $R^2 \times 100\%$ is a good descriptive measure of the adequacy of the model.

Although there are no hard and fast rules concerning the interpretation of R and R^2, the charts given in Fig. 10.19 are useful. Keep in mind the fact that the interpretation of these statistics is somewhat subject matter dependent. An R^2 value of 50% might be considered very large in a social science setting where human

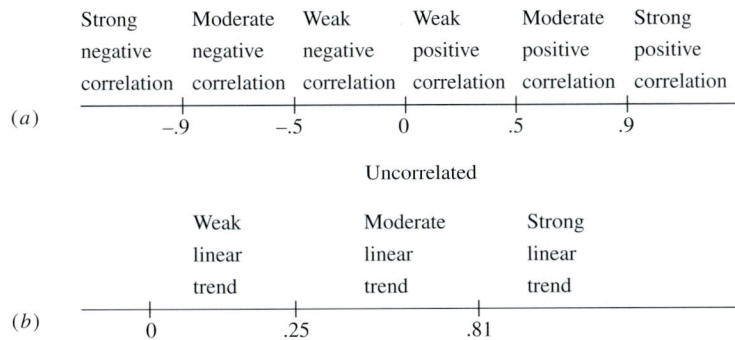

Figure 10.19
(*a*) A suggested interpretation
of R; (*b*) a suggested
interpretation of R^2.

subjects are involved; however, the same figure could be considered very small in a designed engineering experiment. The interpretation of R and R^2 must be left to the discretion of the subject matter expert.

CHAPTER SUMMARY

In this chapter we have considered most of the important aspects of simple linear regression and correlation. A verbal and mathematical description of the regression model was given along with the least-squares method for estimating the slope and intercept parameters of the model. We saw that under minimal assumptions these estimators were unbiased. When the random error E_i was assumed to be normally distributed, the distribution of $Y|x_i$, $\hat{\beta}_0$, and $\hat{\beta}_1$ was given. Utilizing these distributional properties, we considered methods for testing hypotheses and estimating confidence intervals about the slope β_1 and the intercept β_0. We carefully distinguished between predicting the mean response of the dependent variable Y at a fixed value of the independent variable x and predicting a single value of the dependent variable Y at x. Methods were given for constructing interval estimates for both cases, and we observed that prediction of the mean led to a shorter interval (more precise) than for a single value.

In addition to simple linear regression, we considered the Pearson correlation coefficient. Methods were given for estimating the true correlation, testing hypothesis about the correlation, and estimating confidence intervals for the correlation coefficient ρ.

We also introduced and defined terms that you should know. These are:

Linear regression	Least-squares estimation
Least-squares properties	Independent variable
Dependent variable	Slope of regression
Intercept of regression	Pearson correlation
Pure error	Bivariate normal distribution
Residual error	SSE
Significant correlation	Coefficient of determination
Observational study	Response variable
Designed study	Regressor
Scattergram	Predictor variable
Residual	

EXERCISES

Section 10.1

1. Consider the following observations on the independent variable X and the dependent variable Y:

x	y	x	y
80	5.0	100	6.7
75	4.7	115	4.2
75	4.5	110	5.5
70	3.2	105	6.4
65	2.0	105	6.8
100	6.5	100	7.2
95	6.0	130	2.0
90	5.5	125	2.9
85	5.2	120	3.8

(a) Plot the scattergram for these data.
(b) Does it appear reasonable that a linear regression could be used for these data?
(c) Sketch, by eye, a linear regression line on the scattergram in part (a).

2. For each of the three following data sets, plot a scattergram and subjectively state whether it appears that a linear regression will (i) fit the data well, (ii) give only a fair fit, or (iii) fit the data poorly:

(a)

x	5	15	25	35	45	50
y	10	18	20	25	32	45

(b)

x	5	10	20	30	40	50
y	15	22	32	35	30	15

(c)

x	10	15	20	30	40	50
y	40	35	30	22	14	7

3. The normal equations were given in this section. Solve the normal equations for b_0 and b_1, and show that your solution can be written in the form given as the least-squares estimates for β_0 and β_1.

4. Consider any arbitrary data set $(x_1, y_1), (x_2, y_2), \ldots, (x_n, y_n)$. Let $\bar{x}$ and $\bar{y}$ denote the respective sample means for the independent variable X and the dependent variable Y. For the estimated linear regression equation $\hat{\mu}_{Y|x} = b_0 + b_1 x$, show that the point $(\bar{x}, \bar{y})$ always lies on the estimated regression line.

5. Verify that $\Sigma e_i = 0$. *Hint:* Write e_i as $y_i - (b_0 + b_1 x_i)$, and remember that $b_0 = \bar{y} - b_1 \bar{x}$.

6. For each of the data sets of Exercise 2, estimate β_0 and β_1. Find the residuals in each case, and verify that, apart from round-off error, the residuals sum to 0.

7. The relationship between energy consumption and household income was studied, yielding the following data on household income X (in units of \$1000/year) and energy consumption Y (in units of 10^8 Btu/year).

Energy consumption (y)	Household income (x)
1.8	20.0
3.0	30.5
4.8	40.0
5.0	55.1
6.5	60.3
7.0	74.9
9.0	88.4
9.1	95.2

(a) Plot a scattergram of these data.

(b) Estimate the linear regression equation $\mu_{Y|x} = \beta_0 + \beta_1 x$.

(c) If $x = 50$ (household income of \$50,000), estimate the average energy consumed for households of this income. What would your estimate be for a single household?

(d) How much would you expect the change in consumption to be if any household income increases \$2000/year (2 units of \$1000)?

(e) How much would you expect consumption to change if any household income decreases \$2000/year?

8. Consider the data in Exercise 7.

(a) Write the normal equations for these data.

(b) Solve the normal equations for b_0 and b_1, and verify that your results are the same as those you obtained in part (b) of Exercise 7.

9. Connectors used in computers are subject to simultaneous multidimensional stresses such as high temperatures and mechanical stresses. A study is conducted to identify and quantify interface stresses. Experiments are conducted to investigate the relationship between pitch and connector length. These data are obtained:

Connector length, x (inches)	Pitch (millimeters)
.150	3.81
.100	2.54
.098	2.50
.079	2.00
.050	1.27
.040	1.02
.039	1.00
.032	0.80
.020	0.50
.016	0.40
.010	0.25
.005	0.13

(a) Plot a scattergram for these data.

(b) Estimate the regression line.

(c) Calculate the residuals, and show that, apart from round-off error, they sum to 0.

(d) Estimate the average pitch for all connectors of length 0.03 inch (in.).

(e) Estimate the pitch of a particular connector of length 0.03 in.

(f) By how much would you expect the pitch to change if the connector length increased by 0.1 in.? By .05 in.?

(*g*) Would it be reasonable to expect to be able to use the estimated regression line to predict the pitch well for connectors of length .175 in.? .39 in.? 1.0 in.? Explain.

10. A particular type of power brush is a wheel made of wire strands extending outward around a hub. It is used for many purposes such as finishing aluminum bicycle rims, producing a matte finish on plastic, and removing burrs from gear teeth. The shorter the wire length and the coarser the wire, the more severe is the buffing action. A study is conducted to develop a chart for suggested use of the wheel. Tests are conducted on a 2-in.-brush-diameter wheel. These data are obtained:

x (rpm $\times$ 1000)	y (surface ft/min covered in removing burrs)
1.0	525, 520, 527
1.5	785, 780, 790
1.75	915, 900, 922
2.5	1300, 1295, 1310
3.0	1575, 1565, 1582
4.0	2100, 2110, 2090
6.0	3125, 3120, 3133
10.0	5250, 5256, 5245

(*a*) Sketch a scattergram for these data.
(*b*) Estimate the regression line.
(*c*) Estimate the surface feet per minute that can be covered when a wheel of this sort is used at 3450 revolutions per minute (rpm).

Section 10.2

11. Verify the following summation properties:

(*a*) $\sum_{i=1}^{n} (x_i - \bar{x}) = 0.$

(*b*) $\sum_{i=1}^{n} (x_i - \bar{x})(Y_i - \bar{Y}) = \sum_{i=1}^{n} (x_i - \bar{x})Y_i.$

(*c*) $\sum_{i=1}^{n} (x_i - \bar{x})(Y_i - \bar{Y}) = \left(n\sum_{i=1}^{n} x_i Y_i - \sum_{i=1}^{n} x_i \sum_{i=1}^{n} Y_i \right) \Big/ n.$

(*d*) $\sum_{i=1}^{n} (x_i - \bar{x})^2 = \sum_{i=1}^{n} (x_i - \bar{x})x_i.$

(*e*) $\sum_{i=1}^{n} (x_i - \bar{x})^2 = \left[n\sum_{i=1}^{n} x_i^2 - \left(\sum_{i=1}^{n} x_i \right)^2 \right] \Big/ n.$

12. The estimator of the true mean of the dependent variable Y was given by $\hat{\mu}_{Y|x} = B_0 + B_1 x$. Show that $E(\hat{\mu}_{Y|x}) = \mu_{Y|x}$, and hence that $\hat{\mu}_{Y|x}$ is an unbiased estimator, for $\mu_{Y|x}$.

13. The proof that $S^2 = \text{SSE}/(n-2)$ is an unbiased estimator for σ^2 is tricky. The steps in the proof are outlined below:

(*a*) Show that $\text{SSE} = S_{yy} - S_{xx}B_1^2.$

(*b*) Show that $\text{SSE} = \sum_{i=1}^{n} Y_i^2 - n\bar{Y}^2 - S_{xx}B_1^2.$

(c) Show that $E[\text{SSE}] = \sum_{i=1}^{n} E[Y_i^2] - nE[\bar{Y}^2] - S_{xx}E[B_1^2]$.

(d) Show that $E[Y_i^2] = \text{Var } Y_i + (E[Y_i])^2$

$$= \sigma^2 + (\beta_0 + \beta_1 x_i)^2$$

$$E[\bar{Y}^2] = \text{Var } \bar{Y} + (E[\bar{Y}])^2$$

$$= \sigma^2/n + (\beta_0 + \beta_1 \bar{x})^2$$

$$E[B_1^2] = \text{Var } B_1 + (E[B_1])^2$$

$$= \sigma^2/S_{xx} + \beta_1^2.$$

(e) Substitute and simplify to show that $E[\text{SSE}] = (n - 2)\sigma^2$.

(f) Show that $E[S^2] = \sigma^2$.

14. Recall the estimators for the parameters μ_Y and β_1 are denoted by $\bar{Y}$ and B_1, respectively. Prove that $\text{Cov}(\bar{Y}, B_1) = 0$. *Hint:* It is assumed that Y_i and Y_j are uncorrelated. Write $\bar{Y}$ and B_1 as linear combinations of Y_i ($\bar{Y} = \sum_{i=1}^{n} a_i Y_i$ and $B_1 = \sum_{i=1}^{n} c_i Y_i$). Note that $a_i = 1/n$ and $c_i = (x_i - \bar{x})/\sum_{i=1}^{n}(x_i - \bar{x})^2$.

15. Suppose that the *true* regression equation is known to be $\mu_{Y|x} = 10 + 2.5x$. Under the assumption of normality we have seen that the estimator for β_1, B_1 is also normally distributed with mean β_1 and variance $\sigma^2/\sum_{i=1}^{n}(x_i - \bar{x})^2$. Suppose that it is also known that $\text{Var } B_1 = 1.2$. For a sample of 25 observations, (x_i, y_i), find the probability that the estimate of β_1 will be greater than 3.5.

Section 10.3

In production flow-shop problems, performance is often evaluated by minimum make-span, the total elapsed time from starting the first job on the first machine until the last job is completed on the last machine. For a particular flow-shop the make-span was evaluated with respect to the number of jobs to be done. Let the independent variable X denote the number of jobs and the dependent variable Y denote the make-span (in standardized units):

Number of jobs (x)	4	5	6	7	8	9
Make-span (y)	3.75	4.90	4.88	7.2	7.3	9.1
	10	11	12	13	14	15
	9.0	11.9	11.5	14.1	13.9	17.5

Refer to these data for Exercises 16 through 18.

16. (a) Estimate the linear regression equation $\mu_{Y|x} = \beta_0 + \beta_1 x$.
   (b) Plot the estimated regression equation.
17. Test for a significant linear regression at the $\alpha = .05$ level of significance.
18. (a) At $x = \bar{x}$, compute a 95% confidence interval for $\mu_{Y|x}$, and verbally explain the answer.
   (b) At $x = 12$, compute a 95% confidence interval for $\mu_{Y|x}$, and verbally explain the answer.
   (c) How do you explain the different widths of the intervals in parts (a) and (b)?

Refer to the data in Exercise 7 for Exercises 19 through 22.

19. Test H_0: $\beta_0 = 2$ at the .01 level of significance.
20. Calculate a 95% confidence interval for the true intercept β_0.
21. Test for significant linear regression; that is, test H_0: $\beta_1 = 0$ at the .05 level.
22. (a) If $x = /50$, estimate Y, a single predicted value of Y when $x = 50$.
   (b) Calculate a 95% prediction interval for $Y|x = 50$, and interpret your answer.

Let x denote the number of lines of executable SAS code, and let Y denote the execution time in seconds. Use the following summary information to do Exercises 23 through 28.

$$n = 10 \qquad \sum_{i=1}^{10} x_i = 16.75 \qquad \sum_{i=1}^{10} y_i = 170 \qquad \sum_{i=1}^{10} x_i^2 = 28.64$$

$$\sum_{i=1}^{10} y_i^2 = 2898 \qquad \sum_{i=1}^{10} x_i y_i = 285.625$$

23. Estimate and plot the line of regression.
24. (a) Estimate Var $Y_i = \sigma^2$.
    (b) Estimate the standard deviation of B_1.
    (c) Estimate the standard deviation of B_0.
25. Test the hypothesis $\beta_1 = 0$ at the .01 level, and verbally state the conclusion.
26. Test the hypothesis $\beta_0 = 25$ at the .05 level, and discuss the conclusion in the context of the problem.
27. If significant regression is found, estimate the average time required to run a SAS program with 15 lines of executable code.
28. If regression is not significant, what does this mean mathematically? Can you think of a practical reason from a computing standpoint that regression might not be significant in this case?
29. The following data represent carbon dioxide (CO_2) emissions from coal-fired boilers (in units of 1000 tons) over a period of years between 1965 and 1977. The independent variable (year) has been standardized to yield the following table:

Year (x)	0	5	8	9	10	11	12
CO_2 emission (y)	910	680	520	450	370	380	340

(a) Estimate the linear regression equation $\mu_{Y|x} = \beta_0 + \beta_1 x$.
(b) Is there a significant linear trend in CO_2 emission over this time span? That is, test $H_0: \beta_1 = 0$ at the .01 level of significance.
(c) Would it be wise to use the estimated regression line to estimate the average CO_2 emissions from coal-fired boilers for the year 2000? Explain.

The following data represent the known weights of calcium oxide (CaO) from nine different samples and the corresponding weights determined by a standard chemical procedure. The known weight is treated as the independent variable X.

CaO present (x)	3.0	7.0	11.5	15.0	19.0
CaO found (y)	2.7	6.8	11.1	14.6	18.8
	24.0	30.0	35.0	39.0	39.0
	23.5	29.7	34.5	38.4	38.5

Use these data to do Exercises 30 through 32.

30. Find the linear regression line used to estimate $\mu_{Y|x}$, the average weight of CaO found for a known weight x.

31. Compute an unbiased estimate of the variance of Y about the true linear regression line.

32. (*a*) If $x = 15$, estimate $\mu_{Y|x}$.

(*b*) Compute and interpret a 90% confidence interval for $\mu_{Y|x}$ when $x = 15$.

33. An experiment was completed to study the relationship between concentrations of estrone in saliva and in free plasma. The following data were obtained:

Subject	Estrone in saliva (x)	Estrone in free plasma (y)
1	7.4	30.0
2	7.5	25.0
3	8.5	31.5
4	9.0	27.5
5	9.0	39.5
6	11.0	38.0
7	13.0	43.2
8	14.0	49.0
9	14.5	55.0
10	16.0	48.5

(*a*) Plot a scattergram of the data.

(*b*) Estimate the line of regression of Y on X.

(*c*) If the estrone level is 12.1, predict the level of estrone in free plasma.

(*d*) Test for a significant linear regression at the .10 level.

Section 10.4

34. Reconsider Exercise 1. Even though the scattergram of the data suggests that simple linear regression is not appropriate, a straight line can be forced through the data via least squares.

(*a*) Estimate b_0 and b_1 to force a line through the data of Exercise 1.

(*b*) Use the estimated line of regression to find $\hat{y}_i$ for $i = 1$ to 18.

(*c*) Find the 18 residuals for these data, and verify that, apart from round-off error, these residuals sum to 0.

(*d*) Form a residual plot, and notice that it does not exhibit the ideal pattern expected when simple linear regression is appropriate.

(*e*) Can you suggest an equation that would probably describe the pattern seen in the raw data much better than does a straight line?

(*f*) Sketch and interpret the boxplot for the residuals.

35. Sketch residual plots for each of the data sets given in Exercise 2. (The residuals were found in Exercise 6.) Which, if any, of the plots suggest that the assumptions underlying simple linear regression are not met?

36. Consider the residuals found in Exercise 9.

(*a*) Sketch and interpret the residual plot.

(*b*) Sketch and interpret the boxplot of the residuals.

37. In the earliest stages of the development of electronic technology solders were used to assemble components. Due to the fact that they are applied hot, stress such as creeping, distortion, and metal fatigue can result. A study of the use of amalgams as alternatives to solder is conducted. An amalgam is an alloy between a liquid metal and a powder formed at room temperature. These data are obtained on the curing time in minutes (x) and the hardness rating (y) of a gallium/nickel/copper amalgam:

x (curing time)	y (hardness in durometers, D)
5	9, 9, 10
1500	68, 70, 72
1800	82, 80, 83
2000	87, 86, 86
3500	91, 90, 90
4200	91, 91, 92
5800	95, 96, 93

(Based on information from "Amalgams for Improved Electronics Interconnection," Colin A. MacKay, *IEEE MICRO*, April 1993, pp. 46–58.)

(*a*) Sketch a scattergram for these data.

(*b*) Even though simple linear regression is not appropriate, force a regression line through the data. Graph this line on the scattergram.

(*c*) Estimate each residual visually, and sketch a rough residual plot. Discuss the plot.

(*d*) If you have SAS or some other computer software available, find the exact values of the residuals and form a residual plot by computer.

38. Figure 10.20 shows residual plots for various data sets. In each case, identify any model assumptions that might be violated.

Figure 10.20
Residual plots.

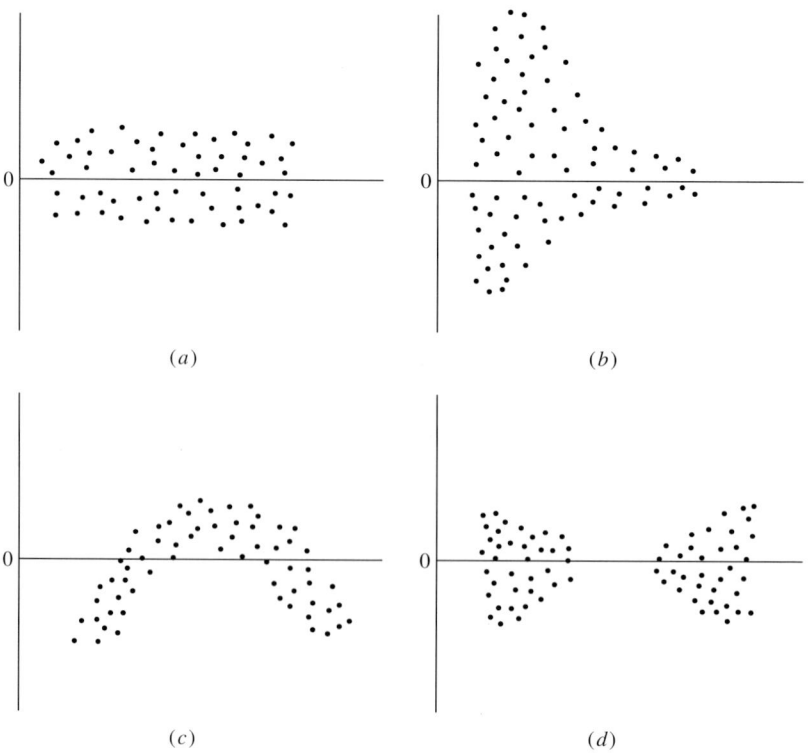

Section 10.5

Pesticides used in food production can be found in food consumed by humans. A study focusing on chickens exposed to malaoxon was conducted. The chickens were also exposed to a liver enzyme inducer to determine whether liver detoxification of the pesticide is affected. The following data were reported as a percentage of normal pesticide detoxification (y) and percentage of normal liver enzyme levels (x):

Enzyme level (x)	Detoxification level (y)
95	108
110	126
118	102
124	121
145	118
140	155
185	158
190	178
205	159
222	184

Refer to these data for Exercises 39 through 42.

39. (*a*) Plot a scattergram of the data.
   (*b*) Estimate ρ, the correlation between X and Y.
40. Test the null hypothesis that X and Y are uncorrelated at the 0.10 level. That is, test H_0: $\rho = 0$. Discuss your conclusion.
41. Find a 90% confidence interval on ρ.
42. Test H_0: $\rho = .8$ at the $\alpha = .05$ level of significance.
43. These data are obtained in the random variables x, the percentage copper of a sample, and its Rockwell hardness rating y:

x	y
.01	58.0
.03	66.0
.01	55.0
.02	63.2
.10	58.3
.08	57.9
.12	69.3
.15	70.1
.10	65.2
.11	62.3

   (*a*) Plot a scattergram of these data.
   (*b*) Find a point estimate for ρ.
   (*c*) Find a 95% confidence interval for ρ and discuss your conclusion.

44. Show that $B_1/(S/\sqrt{S_{xx}}) = R\sqrt{n-2}/\sqrt{1-R^2}$. *Hint:* Use the fact that $B_1 = S_{xy}/S_{xx}$ and $R = S_{xy}/\sqrt{S_{xx}S_{yy}}$ to show that $B_1/(S/\sqrt{S_{xx}}) = \sqrt{S_{yy}}R/S$. Then use the fact that SEE $= S_{yy} - B_1S_{xy}$ and $S^2 = \text{SSE}/(n-2)$.
45. Show that if the random variable (X, Y) has a bivariate normal distribution with $\rho = 0$, then X and Y are independent.
46. Does a correlation coefficient of zero always imply that X and Y are independent?

47. Show that if the random variable (X, Y) has a bivariate normal distribution, then the point (μ_X, μ_Y) lies on the true line of regression of Y on X.
48. Find and interpret the coefficient of determination for the data sets of Exercise 2.
49. Find and interpret the coefficient of determination for the data of Exercise 7.
50. Find and interpret the coefficient of determination for the data of Exercise 10.
51. Find and interpret the coefficient of determination for the data of Exercise 1.

REVIEW EXERCISES

The following data represent the fuel gas temperature [in degrees Fahrenheit (°F)] and unit heat rate [in BTU's per kilowatt hour (Btu/kWh)] for a combustion turbine to be used in coal gasification:

Gas temperature, °F (x)	Heat, Btu/kWh Units of 100 (y)
100	99.1
150	98.5
200	98.2
250	98.0
300	97.8
350	97.6
400	97.5
450	97.0
500	96.8

Use these data for Exercises 52 through 56.

52. Estimate the regression curve $\mu_{Y|x} = \beta_0 + \beta_1 x$.
53. Test $H_0: \beta_1 = 0$ versus $H_1: \beta_1 < 0$. Use $\alpha = .05$.
54. Estimate the coefficient of determination as a measure of goodness of fit of the linear regression curve.
55. Calculate a 90% confidence interval on β_0, and discuss your results in the context of the data.
56. Calculate a 95% confidence interval on β_1, and discuss your results in the context of the data.

An engineer wishes to investigate the recovery of heat normally lost to the environment in the form of exhaust gases from furnaces. Her experiment is designed by fixing flow speed past heat pipes [in meters per second (m/sec)] and then measuring the recovery ratio. The study yielded the following data:

Flow speed, m/sec (x)	Recovery ratio (y)
1	.740
1.5	.745
2	.718
2.5	.678
3	.652
3.5	.627
4	.607
4.5	.507
5	.545

Refer to these data for Exercises 57 through 59.

57. Estimate the curve of regression $\mu_{Y|x} = \beta_0 + \beta_1 x$.
58. Test for significant regression at the .05 level.
59. (a) If the flow speed is fixed at 3.25 (m/sec), predict $\mu_{Y|x\,=\,3.25}$ and $Y|x = 3.25$.
    (b) Calculate and interpret the 95% confidence interval on $\mu_{Y|x\,=\,3.25}$.
    (c) Calculate and interpret the 95% prediction interval on $Y|x = 3.25$.
    (d) How do you explain the differing widths of the intervals calculated in parts (b) and (c)?

In studying the effect of air quality on a lake, the experimenter takes observations on the pH of the water and the air quality as measured on an air quality index. The index goes from 0 to 100 with larger numbers representing high pollution. These data are obtained:

pH (x)	4.5	4.1	4.8	4.0	5.0	6.0	3.5	4.9	3.2	6.1
Air quality (y)	40	50	30	60	20	10	70	30	85	15

Refer to these data for Exercises 60 through 62.

60. (a) Plot the data on the xy plane.
    (b) Estimate the correlation coefficient ρ.
61. Test for a significant negative correlation at the .05 level of significance.
62. Calculate and interpret the 90% confidence interval on ρ.
63. Suppose that a set of 10 pairs of data (x, y) yield an estimated correlation of $r = .3$.
    (a) Give the approximate smallest P value for testing H_0: $\rho = 0$ versus H_1: $\rho \neq 0$.
    (b) Suppose again that $r = .3$, but for $n = 50$ observations. What is the approximate smallest P value for testing the same hypothesis as given in part (a)? How do you explain the difference between P values in parts (a) and (b)?
64. (a) What relationship does the size (in absolute value) of the correlation coefficient have to the slope of the linear regression line of Y on x?
    (b) What relationship does the size (in absolute value) of the correlation coefficient have to the closeness of the points to the linear regression line of Y on x?
65. When processing flow-shops involve semiautomatic or manual operators, processing times can be regarded as random variables. An investigator decided to study the correlation between make-spans (the time elapsed until the last job is completed on the last machine) for two different systems. The study yielded the following bivariate observations for a random selection of 10 sets of jobs:

Job	System 1 (x)	System 2 (y)
1	4.1	3.9
2	5.0	5.1
3	4.9	5.0
4	5.3	4.9
5	13.5	13.3
6	12.0	13.2
7	19.2	21.3
8	10.0	9.1
9	24.1	23.0
10	6.9	8.1

(a) Estimate the Pearson correlation coefficient.
(b) Compute a 95% confidence interval on the true correlation ρ.
(c) Test for a significant correlation at the .05 level. Do parts (b) and (c) tend to agree?
(d) Calculate the coefficient of determination. Explain its meaning.

66. The effect of acid type and pH on the weight loss in western red cedar was studied. Sulfurous acid was used at pH levels of 2.0, 2.5, 3.0, 3.5, and 4.0 with distilled water (pH 5.6) as a control. An accelerated weathering chamber is used for a total of 200 hours. Red cedar wafers of identical size are obtained and weighed. Each wafer is then soaked in the acid solution for one hour and then placed in the weathering chamber for 25 hours. This process is repeated until weathering time reaches 200 hours, at which time a final weight is obtained. The following data were obtained:

Obs. no.	pH	Start wt.	Final wt.	Wt. loss
1	2.0	696	661	35
2	2.0	696	664	32
3	2.0	694	664	30
4	2.5	699	668	31
5	2.5	696	668	28
6	2.5	692	666	26
7	3.0	697	673	24
8	3.0	698	675	23
9	3.0	698	674	24
10	3.5	699	677	22
11	3.5	698	678	20
12	3.5	699	674	25
13	4.0	698	677	21
14	4.0	698	678	20
15	4.0	698	673	25
16	5.6	698	677	21
17	5.6	698	678	20
18	5.6	696	672	24

(a) Plot the graph for pH versus weight loss. What does this suggest in terms of levels of sulfurous acid and effect on western cedar weight loss?
(b) Compute the Pearson correlation for pH versus weight loss.
(c) Test for a significant correlation at the $\alpha = .05$ level of significance.
(d) Compute a 95% confidence interval for the true value of the correlation ρ.

An electrical engineer is concerned with predicting power demand based on temperature of the current day. This would enable the company to buy and transfer power based on short-term weather predictions, and, hence, brownouts could be reduced or avoided. A demand scale was devised from zero to ten, with zero representing very low demand and ten representing maximum demand. A random sample of 40 days over the 365-day year was obtained, yielding the following data:

Obs. no.	Temperature	Demand	Obs. no.	Temperature	Demand
1	30	2.9	21	67	1.4
2	11	5.1	22	33	3.1
3	97	4.0	23	81	1.9
4	41	2.0	24	101	3.5
5	53	1.6	25	84	1.5
6	105	4.5	26	36	1.2
7	68	.3	27	98	3.1
8	1	8.0	28	79	1.0
9	48	.6	29	98	3.6
10	106	5.6	30	3	7.8
11	98	3.3	31	86	2.7
12	33	2.6	32	34	2.2
13	10	5.2	33	108	4.5
14	63	.9	34	14	4.9
15	50	1.3	35	89	2.9
16	2	7.1	36	53	1.3
17	45	1.4	37	55	.7
18	59	.6	38	15	4.7
19	7	6.3	39	87	2.2
20	96	3.5	40	87	1.9

Refer to these data for Exercises 67 through 69.

67. (*a*) Plot the data for the independent variable (temperature) versus the response variable (demand). Do you believe that a simple linear regression line will predict demand well for this case? Why?

(*b*) Make two separate plots using only temperature values equal to or less than 60 degrees for one graph and a separate graph for temperature values greater than 60 degrees. Do you think that using two separate regression lines to predict demand would work well?

(*c*) Can you think of another way to model these data for predicting demand? Discuss possible alternatives.

68. (*a*) Estimate the linear regression line for temperature values equal to or less than 60 degrees.

(*b*) Test for a significant linear regression at the $\alpha = .05$ level of significance.

(*c*) Using the regression equation in part (*a*), predict "demand" for temperature equal to 15 degrees.

(*d*) Compute a 95% confidence interval on the average "demand" when the temperature is 15 degrees.

69. Repeat Exercise 68 (*a*)–(*d*) using temperature values greater than 60 degrees. Also predict demand, and compute your confidence interval when the temperature is 90 degrees.

CHAPTER 11

Analysis of Variance

In Chap. 7 we discussed the problem of hypothesis testing on the mean of a single population. The problem was extended to testing the equality of two population means in Chap. 9. In the latter case, we were concerned primarily with comparing means based on independent samples drawn from normal populations. We used either the pooled T test or the Satterthwaite procedure. We also considered the paired T test, a method for comparing means based on paired data. In this chapter these problems are extended to that of comparing several population means via a statistical methodology called *analysis of variance* (*ANOVA*). This is a procedure in which the total variation in a measured response is partitioned into components that can be attributed to recognizable sources of variation. These individual components are useful in testing pertinent hypotheses.

In this chapter we touch on an area of statistics called *experimental design*. Experimental design is a broad and important area of applied statistics that deals with the practical and theoretical aspects of designing experimental studies. There are three major phases of such a study. These are *problem formulation,* the *design* of the experiment, and the *analysis* of the data collected.

In phase 1 the researcher carefully states the problem to be solved. This process should include gathering all information currently known about the problem, a consideration of the point of view of others, a determination of the scope of the study, and a clear statement of the purpose of the study.

Phase 2, the design of the study, includes choosing the response variable(s) and trying to anticipate which other variables might have an influence on the response. Techniques for controlling or at least measuring the influence of these variables are determined. Cost, time, and other physical constraints are considered. Ultimately, decisions are made concerning the number of observations to be taken, the order of experimentation, and the method of randomization to be used. A statistical model is then formulated. Ideally, this model is one that describes the experimental design, is simple enough to be understood and analyzed by available analysis of variance techniques, and allows the questions posed in the formulation phase to be answered statistically.

If the experiment has been well designed, then phase 3, the analysis of the data collected, is not difficult, since the proper analysis has been anticipated.

In this chapter we develop the ANOVA techniques for frequently encountered "single-factor" experimental designs. By "single-factor" we mean designs in which interest centers on a single primary factor that can influence the response. For example, in conducting an experiment to study the viscosity of a particular motor oil, we can think of the temperature of the oil as a factor; in a study of the speed with which a sorting algorithm is able to sort a random array the computer scientist could view the degree to which the array is out of sort at the outset as a factor.

11.1 ONE-WAY CLASSIFICATION FIXED-EFFECTS MODEL

Assume that we are interested in comparing the means of k populations. The experimental situation may be either of the following:

1. We have k populations, each identified by some common characteristic to be studied in the experiment. Independent random samples of sizes $n_1, n_2, \ldots, n_k$ are selected from each of the k populations, respectively. Differences observed in the measured response are attributed to basic differences among the k populations.

2. We have a collection of N homogeneous experimental units and wish to study the effects of k different treatments. These units are randomly divided into k subgroups of sizes $n_1, n_2, \ldots, n_k$, and each subgroup receives a different experimental treatment. The k subgroups are viewed as constituting independent random samples of size $n_1, n_2, \ldots, n_k$ drawn from k populations.

Although the above experimental situations are different, they are similar in that each results in independent random samples drawn from populations with means $\mu_1, \mu_2, \ldots, \mu_k$. Our interest is in testing the null hypothesis that the population means are equal. That is, we want to test

$$H_0: \mu_1 = \mu_2 = \cdots = \mu_k$$
$$H_1: \mu_i \neq \mu_j \quad \text{for some } i \text{ and } j$$

(at least two of the means are not equal)

As you can see, this is an extension of the two sample problems based on independent samples studied in Chap. 9.

The model that we develop is called a *one-way classification fixed-effects model*. The term "one-way classification" refers to the fact that only one factor or attribute is being studied in the experiment. The factor is studied at k different *levels*. In the second experimental situation described we usually use the word "treatments" rather than factor levels. The term "fixed-effects" refers to the fact that the treatments or levels of the factor involved are specifically selected by the experimenter because they are of particular interest. They are not randomly selected from a larger group of possible treatments or levels. Example 11.1.1 should make the meaning of these terms clear.

Example 11.1.1. A study is designed to investigate the sulfur content of the five major coal seams in a certain geographical region. Core samples are taken at randomly selected points within each seam, and the measured response is the percentage of sulfur per core sample. We want to detect any differences that might exist in the average sulfur content for these five seams. Each seam constitutes a population. We want to compare population means by testing

$$H_0: \mu_1 = \mu_2 = \mu_3 = \mu_4 = \mu_5$$
$$H_1: \mu_i \neq \mu_j \quad \text{for some } i \text{ and } j$$

based on independent samples drawn from these populations. The one factor under study is the coal seam involved. The factor is being studied at five levels. These levels are not selected at random, Rather, we have intentionally chosen to study the five major seams in the region. The design is a fixed-effects design. This study is an example of the first experimental situation described earlier. Figure 11.1 illustrates the idea.

Figure 11.1
Random samples of sizes $n_1, n_2,$ $n_3, n_4,$ and n_5 independently selected from the five major coal seams in a particular geographical region.

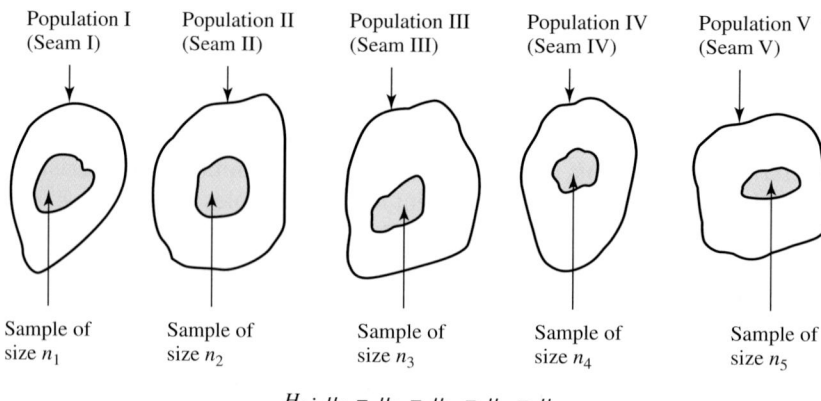

Population I (Seam I) Population II (Seam II) Population III (Seam III) Population IV (Seam IV) Population V (Seam V)

Sample of size n_1 Sample of size n_2 Sample of size n_3 Sample of size n_4 Sample of size n_5

$$H_0: \mu_1 = \mu_2 = \mu_3 = \mu_4 = \mu_5$$

Table 11.1

Data layout one-way classification

	Treatment or factor level					
	1	2	3	$\cdots$	k	
	Y_{11}	Y_{21}	Y_{31}		Y_{k1}	
	Y_{12}	Y_{22}	Y_{32}		Y_{k2}	
	Y_{13}	Y_{23}	Y_{33}		Y_{k3}	
	$\vdots$	$\vdots$	$\vdots$		$\vdots$	
	Y_{1n_1}	Y_{2n_2}	Y_{3n_3}		Y_{kn_k}	
Total	$T_1.$	$T_2.$	$T_3.$	$\cdots$	$T_k.$	$T..$
Sample mean	$\bar{Y}_1.$	$\bar{Y}_2.$	$\bar{Y}_3.$	$\cdots$	$\bar{Y}_k.$	$\bar{Y}..$

Notationally, we let Y_{ij} denote the jth response for the ith treatment or factor level $i = 1, 2, \ldots, k$ and $j = 1, 2, \ldots, n_i$. Here n_i represents the size of the sample drawn from the ith population. The total number of observations for the k samples combined is $N = n_1 + n_2 + \cdots + n_k$. The data collected in a single-factor experiment as well as some important sample statistics are displayed conveniently as shown in Table 11.1. A dot in the following notation indicates the subscript over which summation is being conducted. Note that

$$T_i. = \text{total of the } i\text{th treatment responses}$$
$$= \sum_{j=1}^{n_i} Y_{ij}$$

$$\bar{Y}_i. = \text{sample mean for the } i\text{th treatment}$$
$$= T_i./n_i$$
$$T.. = \text{total of all responses}$$
$$= \sum_{i=1}^{k} T_i. = \sum_{i=1}^{k} \sum_{j=1}^{n_i} Y_{ij}$$

$$\bar{Y}.. = \text{sample mean of all responses}$$
$$= T../N$$

Example 11.1.2 illustrates the use of this notation.

Example 11.1.2. These data and summary statistics are obtained on the sulfur content of the five major coal seams in a particular geographical region:

	Factor (coal seam)			
1	2	3	4	5
1.51	1.69	1.56	1.30	.73
1.92	.64	1.22	.75	.80
1.08	.90	1.32	1.26	.90
2.04	1.41	1.39	.69	1.24
2.14	1.01	1.33	.62	.82
1.76	.84	1.54	.90	.72
1.17	1.28	1.04	1.20	.57
	1.59	2.25	.32	1.18
		1.49		.54
				1.30

$T_1. = 11.62 \quad T_2. = 9.36 \quad T_3. = 13.14 \quad T_4. = 7.04 \quad T_5. = 8.8 \quad T.. = 49.96$

$\bar{Y}_1. = 1.66 \quad \bar{Y}_2. = 1.17 \quad \bar{Y}_3. = 1.46 \quad \bar{Y}_4. = .88 \quad \bar{Y}_5. = .88 \quad \bar{Y}.. = 1.189$

We know that the five sample means $\bar{Y}_1., \bar{Y}_2., \bar{Y}_3., \bar{Y}_4.,$ and $\bar{Y}_5.$ are unbiased estimators for the population means $\mu_1, \mu_2, \mu_3, \mu_4,$ and μ_5. By inspection we see that there are some differences among the sample means. The question to be answered is, "Are these differences extreme enough to conclude that there is a real difference in the average sulfur content among these five coal seams?" To answer this question, we need to develop an analytic method for testing $H_0: \mu_1 = \mu_2 = \mu_3 = \mu_4 = \mu_5$ based on these data.

The Model

To see how to test the null hypothesis of equal treatment means, we must devise a statistical model. To begin, note that each response can be expressed as

$$Y_{ij} = \mu_i + E_{ij}$$

where μ_i denotes the theoretical mean of the ith population and E_{ij} represents the random difference between the jth observation taken from the ith population and the mean of that population. That is, $E_{ij} = Y_{ij} - \mu_i$. An alternative way to write this model is obtained by letting $\alpha_i = \mu_i - \mu$, where

$$\mu = \sum_{i=1}^{k} n_i \mu_i / N$$

In a practical sense, μ represents an overall mean effect found by pooling the k individual population means. Note that if the sample sizes are equal, then μ is just the average of the k population means. Since α_i is the difference between the overall mean μ and the mean of the ith population, α_i measures the effect of the ith treatment. Note that

$$\sum_{i=1}^{k} n_i \alpha_i = \sum_{i=1}^{k} n_i (\mu_i - \mu) = \sum_{i=1}^{k} n_i \mu_i - N\mu = 0$$

By substitution the one-way classification model with fixed effects can be expressed in any of the three ways given below:

> **One-way classification, fixed-effects model**
>
> $$Y_{ij} = \mu_i + E_{ij}$$
> $$Y_{ij} = \mu + (\mu_i - \mu) + (Y_{ij} - \mu_i)$$
> $$Y_{ij} = \mu + \alpha_i + E_{ij}$$

These models are demonstrated in Fig. 11.2.

The latter models express mathematically the idea that each response can be partitioned into three recognizable components as follows:

Response of jth experimental unit to ith treatment	overall mean response	deviation from overall mean due to the fact that unit received ith treatment	random deviation from ith population mean due to random influences
=	+	+

That is,

$$Y_{ij} \qquad = \mu \qquad + (\mu_i - \mu \text{ or } \alpha_i) \qquad + (Y_{ij} - \mu_i \text{ or } E_{ij})$$

The null hypothesis of equal treatment means can be expressed in an alternative form by noting that if $\mu_1 = \mu_2 = \cdots = \mu_k$, then

$$\mu = \sum_{i=1}^{k} n_i \mu_i / N = N\mu_i / N = \mu_i \qquad \text{for each } 1, 2, \ldots, k$$

and $\alpha_1 = \mu_1 - \mu = 0$ for each i. This implies that testing

$$H_0: \mu_1 = \mu_2 = \cdots = \mu_k$$

is equivalent to testing

$$H_0: \alpha_1 = \alpha_2 = \cdots = \alpha_k = 0$$

Figure 11.2
(a) $Y_{ij} = \mu_i + E_{ij}$; the jth observation in the ith sample is viewed as being partitional into two recognizable components. These are μ_i, the average for the ith treatment, and E_{ij}, a deviation from this average caused by random influences acting upon the jth experimental unit;
(b) $Y_{ij} = \mu + (\mu_i - \mu) + (Y_{ij} - \mu_i)$ or $Y_{ij} = \mu + \alpha_i + E_{ij}$; the jth observation in the ith sample is partitioned into three recognizable components. These are μ, an overall average, $\mu_i - \mu$, an effect due to the fact that the ith treatment is involved, which could cause μ_i to differ from the overall average, and E_{ij}; the deviation due to random influences.

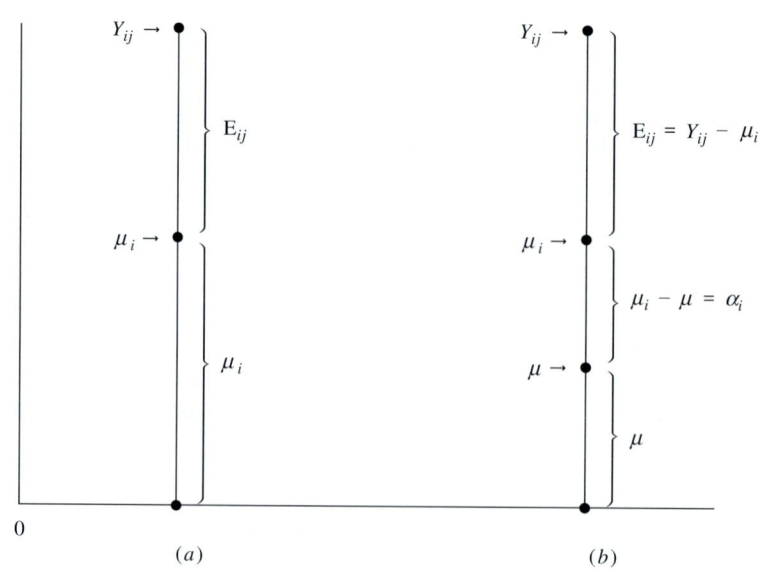

Testing H_0

To derive a test statistic, we must make some assumptions concerning the random differences E_{ij}. These assumptions are similar to those we made in the regression models considered earlier. In particular, we are assuming that the random differences E_{ij} are independent, normally distributed random variables, each with mean 0 and variance σ^2. In more easily understood terms we are assuming that:

1. The k samples represent independent samples drawn from k specific populations with unknown means $\mu_1, \mu_2, \ldots, \mu_k$.
2. Each of the k populations is normally distributed.
3. Each of the k populations has the same variance, σ^2.

When expressed in this form, it is easy to see that these assumptions parallel those made in Chap. 9 relative to the pooled T test for comparing two population means.

Analysis of variance has been defined as a procedure whereby the total variation in some measured response is subdivided into components that can be attributed to recognizable sources. Since $\mu, \mu_1, \mu_2, \ldots, \mu_k$ are theoretical population means, the model does this in only the theoretical sense. To partition an observation in a practical way, these theoretical means are replaced by their unbiased estimators $\bar{Y}.., \bar{Y}_1., \bar{Y}_2., \ldots, \bar{Y}_k.$, respectively. By replacing the theoretical means by their estimators in the model, we obtain the following identity:

$$Y_{ij} = \bar{Y}.. + (\bar{Y}_i. - \bar{Y}..) + (Y_{ij} - \bar{Y}_i.)$$

Note that $\bar{Y}..$ is an estimator for μ, the overall pooled mean effect; $\bar{Y}_i. - \bar{Y}..$ is an estimator for $\alpha_i = \mu_i - \mu$, the effect of the ith treatment; and $Y_{ij} - \bar{Y}_i.$ is an estimator for $E_{ij} = Y_{ij} - \mu_i$, the random error. The term $Y_{ij} - \bar{Y}_i.$ is usually called a *residual*. This identity is equivalent to

$$Y_{ij} - \bar{Y}.. = (\bar{Y}_i. - \bar{Y}..) + (Y_{ij} - \bar{Y}_i.)$$

If each side of the identity is squared and summed over all possible values of i and j, we get

$$\sum_{i=1}^{k} \sum_{j=1}^{n_i} (Y_{ij} - \bar{Y}..)^2 = \sum_{i=1}^{k} \sum_{j=1}^{n_i} [(\bar{Y}_i. - \bar{Y}..) + (Y_{ij} - \bar{Y}_i.)]^2$$

$$= \sum_{i=1}^{k} \sum_{j=1}^{n_i} (\bar{Y}_i. - \bar{Y}..)^2 + 2 \sum_{i=1}^{k} \sum_{j=1}^{n_i} (\bar{Y}_i. - \bar{Y}..)(Y_{ij} - \bar{Y}_i.)$$

$$+ \sum_{i=1}^{k} \sum_{j=1}^{n_i} (Y_{ij} - \bar{Y}_i.)^2$$

The middle term is 0 since

$$\sum_{j=1}^{n_i} (Y_{ij} - \bar{Y}_i.) = \sum_{j=1}^{n_i} Y_{ij} - n_i \bar{Y}_i. = 0$$

Noting that

$$\sum_{i=1}^{k} \sum_{j=1}^{n_i} (\bar{Y}_i. - \bar{Y}..)^2 = \sum_{i=1}^{k} n_i (\bar{Y}_i. - \bar{Y}..)^2$$

we get what is called the *sum of squares* identity for the one-way classification analysis of variance.

Sum of squares identity

$$\sum_{i=1}^{k} \sum_{j=1}^{n_i} (Y_{ij} - \bar{Y}..)^2 = \sum_{i=1}^{k} n_i (\bar{Y}_i. - \bar{Y}..)^2 + \sum_{i=1}^{k} \sum_{j=1}^{n_i} (Y_{ij} - \bar{Y}_i.)^2$$

Each of the components of this identity can be interpreted in a meaningful way. In particular,

$$\sum_{i=1}^{k} \sum_{j=1}^{n_i} (Y_{ij} - \bar{Y}..)^2 = \text{measure of the total variability in the data}$$

$$= \text{total sum of squares } (SS_{\text{Tot}})$$

$$\sum_{i=1}^{k} n_i (\bar{Y}_i. - \bar{Y}..)^2 = \text{measure of variability in data attributed to the fact that different factor levels or treatments are used}$$

$$= \text{treatment sum of squares } (SS_{\text{Tr}})$$

$$\sum_{i=1}^{k} \sum_{j=1}^{n_i} (Y_{ij} - \bar{Y}_i.)^2 = \text{measure of variability in data attributed to random fluctuation among subjects with the same factor level}$$

$$= \text{residual or error sum of squares } (SS_E)$$

Symbolically, the sum of squares identity can be written as

Conceptual sum of squares identity one way classification

$$SS_{\text{Tot}} = SS_{\text{Tr}} + SS_E$$

If there are differences in the population means, then we expect most of the variation in the responses to be due to the fact that different treatments are being used. That is, we expect SS_{Tr} to be large relative to SS_E. The analysis of variance procedure uses this idea to test the null hypothesis of equal treatment means by comparing the between treatment variation (SS_{Tr}) to the within treatment variation (SS_E) via an appropriate F ratio.

To construct an appropriate F ratio, we must consider the expected values of the statistics SS_{Tr} and SS_E. To do so, we use the model assumption that the random errors E_{ij} are independent normally distributed random variables, each with mean 0 and variance σ^2. To begin, note that for each i,

$$\bar{Y}_i. = \sum_{j=1}^{n_i} (\mu + \alpha_i + E_{ij})/n_i$$

$$= \frac{n_i \mu + n_i \alpha_i + \sum_{j=1}^{n_i} E_{ij}}{n_i}$$

$$= \mu + \alpha_i + \bar{E}_i.$$

Also, since $\sum_{i=1}^{k} n_i \alpha_i = 0$,

$$\bar{Y}.. = \sum_{i=1}^{k} \sum_{j=1}^{n_i} Y_{ij}/N$$

$$= \sum_{i=1}^{k} \sum_{j=1}^{n_i} (\mu + \alpha_i + E_{ij})/N$$

$$= \frac{N\mu + \sum_{i=1}^{k} n_i \alpha_i + \sum_{i=1}^{k} \sum_{j=1}^{n_i} E_{ij}}{N}$$

$$= \mu + \bar{E}..$$

Substituting, we can rewrite SS_{Tr} as shown:

$$SS_{\text{Tr}} = \sum_{i=1}^{k} n_i (\bar{Y}_i - \bar{Y}..)^2$$

$$= \sum_{i=1}^{k} n_i [(\mu + \alpha_i + \bar{E}_{i.}) - (\mu + \bar{E}..)]^2$$

$$= \sum_{i=1}^{k} n_i (\alpha_i + \bar{E}_{i.} - \bar{E}..)^2$$

$$= \sum_{i=1}^{k} n_i \alpha_i^2 + 2 \sum_{i=1}^{k} n_i \alpha_i \bar{E}_{i.} + \sum_{i=1}^{k} n_i \bar{E}_{i.}^2 - N\bar{E}..^2$$

Taking the expected value of each term, we get

$$E[SS_{\text{Tr}}] = \sum_{i=1}^{k} n_i \alpha_i^2 + 2 \sum_{i=1}^{k} n_i \alpha_i E[\bar{E}_{i.}] + \sum_{i=1}^{k} n_i E[\bar{E}_{i.}^2] - NE[\bar{E}..^2]$$

In Exercise 1 we outline the proof that $E[\bar{E}_{i.}^2] = \sigma^2/n_i$ for each i. A similar argument shows that $E[\bar{E}..^2] = \sigma^2/N$. It is easy to see that $E[\bar{E}_{i.}] = 0$. Substituting, we see that

$$E[SS_{\text{Tr}}] = \sum_{i=1}^{k} n_i \alpha_i^2 + \sum_{i=1}^{k} n_i \sigma^2/n_i - N\sigma^2/N$$

$$= (k - 1)\sigma^2 + \sum_{i=1}^{k} n_i \alpha_i^2$$

By dividing SS_{Tr} by $k - 1$, we obtain a statistic called the *treatment mean square,* which we denote by MS_{Tr}. That is,

> **Treatment mean square**
> $$MS_{\text{Tr}} = SS_{\text{Tr}}/(k - 1)$$

It is easy to see that $E[MS_{\text{Tr}}] = \sigma^2 + \sum_{i=1}^{k} n_i \alpha_i^2/(k - 1)$. Recall that, in the regression context, the residual sum of squares helped to estimate σ^2. The same is true here. To obtain an unbiased estimator for σ^2 we divide the residual sum of squares SS_E by $N - k$. This estimator is called the *error mean square* and is denoted by MS_E. That is,

> **Error mean square**
> $$MS_E = SS_E/(N - k)$$

How can we use MS_{Tr} and MS_E to test H_0? To answer this question, we need only note that if H_0 is true, then $\alpha_1 = \alpha_2 = \cdots = \alpha_k = 0$, and hence $\Sigma_{i=1}^k n_i \alpha_i^2 / (k-1) = 0$. If H_0 is not true, then this term will be positive. Thus if H_0 is true, we would *expect* MS_{Tr} and MS_E to be close in value, since both of them estimate σ^2; if H_0 is not true, we would *expect* MS_{Tr} to be somewhat larger than MS_E. This suggests the ratio

> **Test statistic H_0: $\mu_1 = \mu_2 = \cdots \mu_k$**
>
> $$F_{k-1, N-k} = MS_{Tr}/MS_E$$

as a logical test statistic. If H_0 is true, its value is expected to be close to 1; otherwise it is expected to be larger than 1. This ratio can be used as a test statistic, since if H_0 is true, it is known to have an F distribution with $k-1$ and $N-k$ degrees of freedom. The test is always a right-tailed test, with rejection of H_0 occurring for values of the $F_{k-1, N-k}$ random variable that appear to be too large to have occurred by chance. Values of $F = MS_{Tr}/MS_E$ *can* be less than 1, since F is a random variable. Such an outcome can occur by chance alone or because the assumed linear model is incorrect.

Although in practice analysis of variance is usually performed via the computer, some computational shortcuts are available. We leave it as an exercise to show that

> **Computational shortcuts**
>
> $$SS_{Tot} = \sum_{i=1}^k \sum_{j=1}^{n_i} Y_{ij}^2 - \frac{T_{..}^2}{N}$$
>
> $$SS_{Tr} = \sum_{i=1}^k \frac{T_{i\cdot}^2}{n_i} - \frac{T_{..}^2}{N}$$
>
> $$SS_E = SS_{Tot} - SS_{Tr}$$

The theoretical ideas behind the analysis of variance procedure for the one-way classification fixed-effects model are summarized in Table 11.2. This type of table is called an *analysis of variance (ANOVA) table*.

Table 11.2

ANOVA table for the one-way classification design with fixed effects

Source of variation	Degrees of freedom (DF)	Sum of squares (SS)	Mean square (MS)	Expected mean square	F
Treatment or level	$k-1$	$\sum_{i=1}^k \dfrac{T_{i\cdot}^2}{n_i} - \dfrac{T_{..}^2}{N}$ (SS_{Tr})	$\dfrac{SS_{Tr}}{k-1}$	$\sigma^2 + \sum_{i=1}^k \dfrac{n_i \alpha_i^2}{k-1}$	$\dfrac{MS_{Tr}}{MS_E}$
Error or residual	$N-k$	Subtraction (SS_E)	$\dfrac{SS_E}{N-k}$	σ^2	
Total	$N-1$	$\sum_{i=1}^k \sum_{j=1}^{n_i} Y_{ij}^2 - \dfrac{T_{..}^2}{N}$			

To illustrate the use of the F ratio, we continue the analysis of the coal data begun in Example 11.1.2.

Example 11.1.3. We are testing

$$H_0: \mu_1 = \mu_2 = \mu_3 = \mu_4 = \mu_5$$

or

$$H_0: \alpha_1 = \alpha_2 = \alpha_3 = \alpha_4 = \alpha_5 = 0$$

based on our previous data. Recall that μ_i, $i = 1, 2, 3, 4, 5$, denote the mean sulfur content of the five major coal seams in a particular geographical region. We have these summary statistics available:

$$T_1. = 11.62 \qquad T_4. = 7.04 \qquad n_1 = 7 \qquad n_4 = 8$$
$$T_2. = 9.36 \qquad T_5. = 8.8 \qquad n_2 = 8 \qquad n_5 = 10$$
$$T_3. = 13.14 \qquad T.. = 49.96 \qquad n_3 = 9 \qquad N = 42$$

The only new statistic needed is $\sum_{i=1}^{5}\sum_{j=1}^{n_i}Y_{ij}^2$. For the data given in Example 11.1.2 this statistic assumes the value 67.861. Substituting into the computational formulas, we obtain

$$SS_{\text{Tot}} = \sum_{i=1}^{5}\sum_{j=1}^{n_i}Y_{ij}^2 - T..^2/N = 67.861 - \frac{(49.96)^2}{42} = 8.432$$

$$SS_{\text{Tr}} = \sum_{i=1}^{5}\frac{T_i.^2}{n_i} - \frac{T..^2}{N}$$

$$= \frac{(11.62)^2}{7} + \frac{(9.36)^2}{8} + \frac{(13.14)^2}{9} + \frac{(7.04)^2}{8} + \frac{(8.8)^2}{10} - \frac{(49.96)^2}{42}$$

$$= 3.935$$

$$SS_E = SS_{\text{Tot}} - SS_{\text{Tr}}$$

$$= 8.432 - 3.935 = 4.497$$

$$MS_{\text{Tr}} = \frac{SS_{\text{Tr}}}{k-1} = \frac{3.935}{4} = .984$$

$$MS_E = \frac{SS_E}{N-k} = \frac{4.497}{37} = .122$$

The observed value of the $F_{k-1, N-k} = F_{4,37}$ test statistic is

$$F_{4,37} = \frac{MS_{\text{Tr}}}{MS_E} = \frac{.984}{.122} = 8.066$$

Since $f_{.05}(4, 37) \doteq 2.626$, we can reject H_0 with $P < .05$. We do have statistical evidence that at least two of the coal seams differ in mean sulfur content. The ANOVA table for these data is shown in Table 11.3.

Table 11.3
ANOVA for coal seam data

Source of variation	Degrees of freedom (*DF*)	Sum of squares (*SS*)	Mean square (*MS*)	*F*
Treatments	4	3.935	.984	8.066
Error	37	4.497	.122	
Total	41	8.432		

The corresponding SAS printout is given below.

ONE WAY ANOVA

GENERAL LINEAR MODELS PROCEDURE

DEPENDENT VARIABLE: SULFUR

SOURCE	DF	SUM OF SQUARES	MEAN SQUARE	F VALUE
MODEL	4	3.93539048	0.98384762	8.09
ERROR	37	4.49700000	0.12154054	PR > F
CORRECTED TOTAL	41	8.43239048		0.0001

Before concluding this section, recall that we assume that each of the k independent samples is drawn from normally distributed populations with equal variances σ^2. If the sample sizes are reasonably large, the test is quite robust to departures from normality in the sense that P values reported, although approximate, are fairly accurate. However, the test can be quite sensitive to departures from the assumption of equal variances. This is particularly true if the respective sample sizes differ considerably. When possible, it is advantageous to design experiments so that sample sizes are equal. A method for testing for equality of variances is given in the next section.

11.2   COMPARING VARIANCES

As indicated earlier, the F test for testing equality of means is sensitive to the violation of the assumption of equal variances. This is especially true when sample sizes differ greatly. Before performing an analysis of variance, we need to test the hypothesis

$$H_0: \sigma_1^2 = \sigma_2^2 = \cdots = \sigma_k^2$$
$$H_1: \sigma_i^2 \neq \sigma_j^2 \quad \text{for some } i \text{ and } j$$

(at least two of the variances are not equal)

If H_0 is rejected, then either a nonparametric analysis should be used or else the data should be transformed in hopes of stabilizing the variances. Variance stabilization transformations are found in analysis of variance texts [14].

The most frequently used test for testing the null hypothesis of equal variances is called *Bartlett's test*. The statistic used in this test can be shown to follow an approximate chi-squared distribution with $k - 1$ degree of freedom when sampling from normal populations.

To conduct Bartlett's test, we compute the sample variances $S_1^2, S_2^2, \ldots, S_k^2$ for each of the k samples. We also compute the error mean square, the pooled estimate of σ^2 under the assumption that H_0 is true. In this context it is convenient to compute MS_E directly from the individual sample variances by means of the formula

$$MS_E = S_p^2 = \sum_{i=1}^{k} \frac{(n_i - 1)S_i^2}{N - k}$$

It is left as an exercise to verify that this equation holds. (See Exercise 8.) We next form the statistic Q defined by

$$Q = (N - k)\log_{10} S_p^2 - \sum_{i=1}^{k} (n_i - 1)\log S_i^2$$

The observed value of this statistic is large when the sample variances S_i^2, $i = 1$, $2, \ldots, k$ are quite different; it is near 0 when these sample variances are close in value. The Bartlett statistic is defined by

> **Test statistic H_0: $\sigma_1^2 = \sigma_2^2 = \cdots \sigma_k^2$**
> $$B = 2.3026Q/h$$

where

$$h = 1 + \frac{1}{3(k-1)} \left(\sum_{i=1}^{k} \frac{1}{n_i - 1} - \frac{1}{N-k} \right)$$

An example should demonstrate the use of Bartlett's test.

Example 11.2.1.   Let us return to the coal seam data given in Example 11.1.2. The sample variances and their logarithms must be found for each of the five factor levels. The results of the calculations are summarized below:

Coal seam	Sample variance (s_i^2)	$\text{Log}_{10}\, s_i^2$	Sample size (n_i)
1	.175	−.757	7
2	.144	−.842	8
3	.115	−.939	9
4	.123	−.910	8
5	.074	−1.131	10

The pooled estimate for σ^2 is

$$MS_E = s_p^2 = \sum_{i=1}^{k} \frac{(n_i - 1)s_i^2}{N-k}$$

$$= \frac{\begin{aligned}(7-1)(.175) + (8-1)(.144) + (9-1)(.115) \\ + (8-1)(.123) + (10-1)(.074)\end{aligned}}{42-5}$$

$$= .122$$

Note that this value agrees with the value obtained for MS_E in Example 11.1.3. By substitution

$$q = (N-k)\log_{10} s_p^2 - \sum_{i=1}^{k} (n_i - 1)\log_{10} s_i^2$$

$$= 37\log_{10}.122 - [6(-.757) + 7(-.842) \\ + 8(-.939) + 7(-.910) + 9(-1.131)]$$

$$= .692$$

and

$$h = 1 + \frac{1}{3(k-1)} \left[\sum_{i=1}^{k} \frac{1}{n_i - 1} - \frac{1}{N-k} \right]$$

$$= 1 + \frac{1}{3(4)} \left[\frac{1}{6} + \frac{1}{7} + \frac{1}{8} + \frac{1}{7} + \frac{1}{9} - \frac{1}{37} \right]$$

$$= 1.055$$

The observed value of the Bartlett statistic is

$$b = 2.3026q/h$$

$$= 2.3026(.692)/1.055$$

$$= 1.510$$

Based on the $X^2_{k-1} = X^2_4$ distribution, the P value lies between .75 and .9. Since this value is large, we are unable to reject

$$H_0: \sigma_1^2 = \sigma_2^2 = \sigma_3^2 = \sigma_4^2 = \sigma_5^2$$

We have no reason to doubt that the assumption of equal population variances is not valid.

11.3   PAIRWISE COMPARISONS

In testing for equality of means in the one-way classification model, we either reject H_0 or fail to do so. If H_0 is rejected, we conclude that at least two of the population means differ in value. Unfortunately, the analysis of variance procedure does not tell us which of the k population means may be regarded as being different from the others.

There are many different techniques that have been suggested for comparing pairs of means. Several of these procedures do not require that the F test in the overall analysis of variance table be conducted prior to pairwise comparisons. However, this procedure is usually followed in practice. As pointed out at the end of this section, requiring a significant F test in the analysis of variance prior to conducting pairwise comparisons gives added protection against falsely declaring pairs of means significantly different.

Lentner and Bishop [31] present a good overview of most of the pairwise comparison procedures. Here we discuss three possibilities. These are the Bonferroni T tests, the Duncan's multiple range test, and Tukey's test.

Bonferroni T Tests

Consider a set of k population means. There are $\binom{k}{2} = k(k-1)/2$ possible pairs of means. Thus there are $k(k-1)/2$ possible tests of the form

$$H_0: \mu_i = \mu_j$$
$$H_1: \mu_i \neq \mu_j$$

that can be conducted. Since one of the model assumptions is that the population variances are equal, each of these hypotheses can be tested by using a pooled T test as described in Chap. 9. In such a test the test statistic is

$$T_{n_i + n_j - 2} = \frac{\bar{Y}_{i\cdot} - \bar{Y}_{j\cdot}}{\sqrt{S_p^2 \left(\dfrac{1}{n_i} + \dfrac{1}{n_j} \right)}}$$

where S_p^2 is the pooled estimator for the common population variance based on samples drawn from populations i and j. In this case, another estimator of σ^2 is available, namely, $\sigma^2 = MS_E$. Since this estimator is based on all available data, the T test can be improved by using

> **Bonferroni test statistic H_0: $\mu_i = \mu_j$**
>
> $$T_{N-k} = \frac{\overline{Y}_{i\cdot} - \overline{Y}_{j\cdot}}{\sqrt{MS_E\left(\dfrac{1}{n_i} + \dfrac{1}{n_j}\right)}}$$

as the test statistic. The critical point for the two-tailed test at the α level of significance is

> **Bonferroni critical point**
>
> $$cp = t_{N-k,\,1-\alpha/2}\sqrt{MS_E\left(\frac{1}{n_i} + \frac{1}{n_j}\right)}$$

We reject H_0 whenever $|\overline{Y}_{i\cdot} - \overline{Y}_{j\cdot}|$ exceeds the computed critical point. Note that when sample sizes are unequal, it might be necessary to compute a different critical point for each test; if sample sizes are the same, then a single critical point will suffice. Performing $k(k-1)/2$ individual T tests is laborious, but it can be done. However, this procedure has a more serious drawback that must be handled with care.

To understand the problem, suppose that we ask, "What is the probability of making at least one incorrect rejection and therefore drawing at least one incorrect conclusion?" If the $= k(k-1)/2$ tests are independent, then this probability, called the *experimentwise error rate* and denoted by α', can be calculated as follows:

$$P[\text{at least one incorrect rejection}] = 1 - P[\text{no incorrect rejections}]$$

If each test is conducted at the α level of significance, then the probability of making an incorrect rejection in each case is α; the probability that a rejection is not incorrect is $1 - \alpha$. If the tests are independent, then the definition of independence guarantees that if c tests are run then

$$P[\text{no incorrect rejections}] = (1 - \alpha)^c$$

Hence in the case of independent tests α' is given by $1 - (1 - \alpha)^c$. The problem in the case of multiple comparisons is that many tests are being conducted on the same set of data. For this reason, these tests are not independent. However, it can be shown that the above expression provides an upper bound for α' in this setting. In any hypothesis-testing setting in which more than one test is conducted we can say that

$$\text{Experimentwise error rate} = \alpha' \leq 1 - (1 - \alpha)^c$$

Here α is the common level of significance per test, called the *comparisonwise error rate,* and c is the number of tests run.

As an example for $k = 5$, there are $\binom{5}{2} = 10$ possible paired comparisons. If each test is conducted at the $\alpha = .05$ level, then an upper bound on the probability of making at least one incorrect rejection is $1 - (1 - .05)^{10} = .40$. It is easy to see that as k increases, the overall probability of error may become unacceptably high.

To compensate for this problem, we suggest that only those tests of real interest to the researcher be conducted and that some reasonably small target upper bound, b, be chosen. We then conduct each T test at the b/c level of significance, where c denotes the actual number of tests run. It can be shown that if $\alpha = b/c$, then $\alpha' \leq b$. (See Exercise 14.) For example, if we want α' to be at most .10 and we run all possible tests for $k = 5$ groups, we would conduct each of our paired comparisons at the $\alpha = .10/10 = .01$ level of significance. The numerical value of the bound b is chosen at the discretion of the researcher. Its value is somewhat dependent on the number of tests being run. Notice that if b is small, then $\alpha = b/c$ is even smaller. If we try to force the experimentwise error rate to be too small, then α becomes so small that it becomes extremely difficult to reject H_0: $\mu_i = \mu_j$, even when this hypothesis should be rejected. That is, the price paid for setting too small a bound on α' is lack of power. For this reason, values of α' as high as .15 or .20 are not unreasonable.

The technique presented here is equivalent to what is called the *Fisher's least significant difference* (LSD) procedure run with $\alpha = b/c$. For an excellent discussion of LSD you are referred to [31]. Since the procedure is based on an inequality known as the Bonferroni inequality, it bears the same name. Bonferroni T tests are available on SAS.

Duncan's Multiple Range Test

Duncan's multiple range test was developed by D. B. Duncan in the early 1950s. It was one of the first methods suggested for doing pairwise comparisons of means, and you will see references to this test in the literature.

To understand Duncan's idea, let us compare what is done to the Bonferroni technique just discussed. First, the Duncan procedure initially assumes that sample sizes are equal. That is, it is assumed that we have available a collection of k sample means $\bar{Y}_1, \bar{Y}_2, \bar{Y}_3, \ldots, \bar{Y}_k$, each based on a sample of size n. Notice that in this setting there is a common critical point cp, that applies to all Bonferroni type T tests conducted. We reject H_0: $\mu_i = \mu_j$ whenever $|\bar{Y}_i - \bar{Y}_j| > cp$. No attempt is made to account for the relative positions of $\bar{Y}_i - \bar{Y}_j$ in the ordered list of sample means. The same critical point is used whether these means are side by side, or are separated by several other means, or are the extremes in the ordered list. In the Duncan procedure an attempt is made to adjust the critical point to account for positioning. Thus in the Duncan procedure, multiple critical points are used which lead to the name "multiple range" test. If two means are adjacent on the ordered list of sample means, then the critical point used is the same as that of an ordinary T test at the specified α level; however, if the sample means being compared are not adjacent, then this critical value is increased according to the span of the means in the ordered set of means. In other words, critical points are chosen so that sample means that lie close together are not required to exhibit as much difference as those that are more widely separated in order to declare the corresponding population means to be "significantly different." Duncan's contribution was to develop tables from which these adjusted critical points can be calculated.

The test was first developed under the assumption that sample sizes are equal. However, C. Y. Kramer adapted the procedure to the case of unequal sample sizes. The Duncan test is conducted as follows:

1. Linearly order the k sample means.
2. Find the value of the least significant "studentized range," r_p, for each $p = 2$, $3, \ldots, k$. This value is given in Table XI of App. A for α levels of .1, .01, or

.05. In this table γ denotes the number of degrees of freedom associated with MS_E, the error mean square in the original analysis of variance.

3. For each $p = 2, 3, \ldots, k$ find the shortest or least significant range, SSR_p. This value is given by

$$SSR_p = r_p \sqrt{\frac{MS_E}{n}} \qquad \text{if the sample sizes are all equal with value } n$$

$$SSR_p = r_p \sqrt{MS_E} \qquad \text{if the sample sizes are unequal}$$

4. Consider any subset of p adjacent sample means. Let $|\bar{Y}_i. - \bar{Y}_j.|$ denote the range of the means in this subgroup. The population means, of span p, μ_i, and μ_j are considered to be different if

$$|\bar{Y}_i. - \bar{Y}_j.| > SSR_p \qquad \text{for equal sample sizes}$$

or

$$|\bar{Y}_i. - \bar{Y}_j.| \sqrt{\frac{2 n_i n_j}{n_i + n_j}} > SSR_p \qquad \text{for unequal sample sizes}$$

5. Summarize your results by underlining any subset of adjacent sample means that are not considered to be significantly different at your chosen α level.

Although this sounds complicated, it is not! An example should make the idea clear.

Example 11.3.1. In Example 11.1.3 we rejected

$$H_0: \mu_1 = \mu_2 = \mu_3 = \mu_4 = \mu_5$$

and concluded that at least two of the coal seams sampled differ in mean sulfur content. To pinpoint the differences, we run the Duncan's multiple range test. The sample means in linear order are

$\bar{Y}_4.$	$\bar{Y}_5.$	$\bar{Y}_2.$	$\bar{Y}_3.$	$\bar{Y}_1.$
.88	.88	1.17	1.46	1.66

The values of r_p for span $p = 2, 3, 4, 5$ for an $\alpha = .01$ level test based on 37 degrees of freedom are found in Table XI of App. A. These values are

p	2	3	4	5
r_p	3.825	3.988	4.098	4.180

The error mean square found earlier is $MS_E = .122$. Since the sample sizes $n_1 = 7$, $n_2 = 8$, $n_3 = 9$, $n_4 = 8$, and $n_5 = 10$ are unequal, the shortest significant range for each p is given by

$$SSR_p = r_p \sqrt{MS_E}$$

These values are

p	2	3	4	5
r_p	3.825	3.988	4.098	4.180
SSR_p	1.336	1.393	1.431	1.460

A pair of means μ_i and μ_j of span p are considered to differ if

$$|\bar{Y}_i. - \bar{Y}_j.| \sqrt{\frac{2 n_i n_j}{n_i + n_j}} > SSR_p$$

For five populations there are $\binom{5}{2} = 10$ possible comparisons. We first compare the largest to the smallest sample mean. In this case we compare $\bar{Y}_1.$ to $\bar{Y}_4..$ Since these means span the entire set of sample means, $p = 5$ and $SSR_p = 1.460$. The observed value of the test statistic is

$$|\bar{Y}_1. - \bar{Y}_4.|\sqrt{\frac{2n_1 n_4}{n_1 + n_4}} = (1.66 - .88)\sqrt{\frac{2(7)(8)}{7 + 8}}$$

$$= 2.131$$

This value exceeds SSR_p, and hence we conclude that μ_1 and μ_4 are significantly different. The results of other comparisons are as follows:

Treatment pair	p	Value of test statistic	SSR_p	Reject $\mu_i = \mu_j$?
4-1	5	2.131	1.460	Yes
4-3	4	1.688	1.431	Yes
4-2	3	.820	1.393	No

$$\overline{\bar{Y}_4. \ \bar{Y}_5. \ \bar{Y}_2. \ \bar{Y}_3. \ \bar{Y}_1.}$$

4-5	2	0	1.336	No
5-1	4	2.238	1.431	Yes
5-3	3	1.785	1.393	Yes
5-2	2	.8646	1.336	No
2-1	3	1.339	1.393	No

$$\bar{Y}_4. \ \bar{Y}_5. \ \overline{\bar{Y}_2. \ \bar{Y}_3. \ \bar{Y}_1.}$$

2-3	2	.8440	1.336	No
3-1	2	.5612	1.336	No

In summary, we can conclude that

$$\mu_1 \neq \mu_4$$
$$\mu_1 \neq \mu_5$$
$$\mu_3 \neq \mu_4$$
$$\mu_3 \neq \mu_5$$

Each statement has probability .01 of being in error.

If sample sizes are all the same, then we might not have to make all $\binom{k}{2}$ comparisons. This is due to the fact that when sample sizes are equal, whenever the most extreme pair of means is found not to be significantly different then all means within the subgroup are assumed to be equal with no further testing required.

Duncan's test is widely used and available on SAS and other computer packages. However, it should be emphasized that the significance level given does *not* refer to α', the overall probability of making at least one incorrect rejection. Therefore a Bonferroni test run at the $\alpha' = .01$ or $.05$ level can give results that differ from that of a Duncan's test at the same level. In using SAS, you must be aware of this fact. In an actual study only one procedure should be run. The choice of which one to run is yours.

The Bonferroni technique is analytically attractive, since it is easy to calculate an upper bound for the experimentwise error rate. However, this technique tends to be very conservative, and therefore it may not detect differences in means when they exist. Duncan's procedure attempts to control the experimentwise error rate but yet not be too conservative. The trade-off is that Duncan's procedure may yield large experimentwise error rates when the number of comparisons is relatively large.

An approach that is sometimes recommended is to use what can be called a *protected* Duncan's procedure. That is, do not use Duncan's technique unless the ANOVA F test is first significant at level α. This then assures the experimenter that if Duncan's technique is also used at level α, then the experimentwise error rate $\alpha' \le \alpha$. Duncan's standard usage does not require a preliminary significant F test from the ANOVA.

Tukey's Test

Tukey's test is based on the studentized range distribution and has the advantage of allowing simultaneous confidence intervals for the paired mean difference. Let $q(\alpha, k, v)$ denote the upper-tail α-level critical value of the studentized range, where α represents the significance level, k denotes the number of treatment groups, and v denotes the number of degrees of freedom for the mean squared error (MS_E) in the ANOVA table used to conduct the overall F test. Values of $q(\alpha, k, v)$ are given in Table XIII for $\alpha = .05$.

The simultaneous confidence intervals are given by (11.1) and (11.2) for equal sample sizes and unequal sample sizes, respectively. For equal sample sizes, n, in each treatment group, we have

$$(\bar{Y}_{i.} - \bar{Y}_{j.}) \pm q(\alpha, k, v)\sqrt{MS_E/n} \qquad (11.1)$$

for all pairs of means $\bar{Y}_{i.}$ and $\bar{Y}_{j.}$ and for $i \ne j$.

When the sample sizes are not equal, the Tukey procedure for simultaneous confidence intervals can be approximated as follows:

Let $q_{ij} = q(\alpha, k, v)\sqrt{MS_E\left(\dfrac{1}{n_i} + \dfrac{1}{n_j}\right)}$, where n_i and n_j denote the sample sizes

for $\bar{Y}_{i.}$ and $\bar{Y}_{j.}$, respectively. Then the confidence interval is given by

$$(\bar{Y}_{i.} - \bar{Y}_{j.}) \pm q_{ij} \qquad (11.2)$$

for all pairs of means $\bar{Y}_{i.}$ and $\bar{Y}_{j.}$, and for $i \ne j$.

The interpretation for both (11.1) and (11.2) is to conclude that the true means μ_i and μ_j differ significantly at level α if the confidence interval does not include zero.

11.4 ALTERNATIVE NONPARAMETRIC METHODS

As seen in Secs. 7.7 and 9.6, there are nonparametric analogs to the normal theory T tests for two independent samples and for paired samples when testing location differences. The multiple sample extensions of the normal theory two-independent-sample and paired-sample problems were seen in this chapter to be the one-way analysis of variance. This procedure assumed underlying normally distributed populations with rather restrictive assumptions on the population variances. Fortunately, nonparametric analogs are available when the normal theory test assumptions are not met. The nonparametric test for the one-way classification analysis of variance is the *Kruskal-Wallis test*, which is discussed in this section.

Kruskal-Wallis Test

Assume that k independent random samples of sizes $n_1, n_2, \ldots, n_k$ are drawn from continuously distributed populations. The Kruskal-Wallis procedure tests for hypothesis that each of the k samples has been drawn from identical populations. However, the test is particularly sensitive to location differences, and therefore the null hypothesis is usually stated in terms of equality of population medians. Thus the null and alternative hypotheses can be stated as

$$H_0: M_1 = M_2 = \cdots = M_k$$

$$H_1: \text{at least two population medians are not equal}$$

To perform the test, the $N = n_1 + n_2 + \cdots + n_k$ sample observations are pooled and ranked from the smallest to the largest, retaining group identity. As was done for the Wilcoxon tests, ties are assigned the average rank for their group. Let T_i, $i = 1, 2, \ldots, k$, denote the sum of the ranks associated with the observations from the ith population. The Kruskal-Wallis test statistic is given by

$$H = \frac{12}{N(N+1)} \sum_{i=1}^{k} n_i \left(\bar{T}_i - \frac{N+1}{2} \right)^2$$

where $\bar{T}_i = T_i/n_i$ denotes the average of the ranks assigned to the ith group. If the null hypothesis is true, it can be shown that $E(\bar{T}_i) = (N+1)/2$. Therefore the Kruskal-Wallis test statistic is a measure of the deviations of the observed average ranks for the k groups from the value expected if the null hypothesis is true. Large deviations lead to relatively large values of H, and hence to rejection of H_0. Although exact tables are available for small k and n_i, it has been shown that H approximately follows a chi-squared distribution with $k - 1$ degrees of freedom if all $n_i \geq 5$. Hence approximate critical values for H can be obtained from the chi-squared distribution given in Table IV of App. A. The reader can easily verify that an equivalent but computationally easier form for the test statistic H is

---

Test statistic $H_0: M_1 = M_2 = \cdots M_k$

$$H = \left[\frac{12}{N(N+1)} \sum_{i=1}^{k} \frac{T_i^2}{n_i} \right] - 3(N+1)$$

Example 11.4.1. An experiment was conducted to compare the amount of pressure needed to compress three types of materials. Random samples of sizes 7, 7, and 10 were obtained for materials labeled A, B, and C, respectively. The pressure measurements and corresponding ranks (in parentheses) of the combined sample of $7 + 7 + 10 = 24$ observations are given below:

Material A	Material B	Material C
207 (14)	194 (11)	288 (12.5)
150 (5)	146 (3)	269 (20)
197 (12)	175 (8)	288 (21.5)
173 (7)	186 (9)	358 (24)
147 (4)	223 (17)	229 (18)
144 (2)	143 (1)	249 (19)
192 (10)	170 (6)	346 (23)
		217 (16)
		203 (13)
		214 (15)

The rank sums are

$$T_1 = 14 + 5 + 12 + \cdots + 2 + 10 = 54$$
$$T_2 = 11 + 3 + 8 + \cdots + 1 + 6 = 55$$
$$T_3 = 21.5 + 20 + 21.5 + \cdots + 13 + 15 = 191$$

Calculating the Kruskal-Wallis test statistic, we obtain

$$H = \left[\frac{12}{N(N+1)} \sum_{i=1}^{3} \frac{T_i^2}{n_i} \right] - 3(N+1)$$
$$= \frac{12}{24(25)} \left(\frac{54^2}{7} + \frac{55^2}{7} + \frac{191^2}{10} \right) - 3(25)$$
$$= 14.94$$

From Table IV for $k - 1 = 3 - 1 = 2$ degrees of freedom we see that the critical value for a significance level of .005 is 10.6. Hence we conclude that the amount of pressure required to compress the materials is significantly different for at least two of the materials tested.

Like the Wilcoxon rank-sum test, the Kruskal-Wallis test is very robust relative to the usual normal theory F test. In addition, various multiple comparison procedures are available in standard texts on nonparametric statistics. The reader is referred to [23], [6], and [30].

CHAPTER SUMMARY

The fundamentals of analysis of variance were presented for the one-way classification model. Multiple comparison methods were presented for fixed effects. General analysis of variance (ANOVA) tables, useful for computational purposes and for selecting appropriate mean square error ratios for hypothesis testing, were given in the chapter. We attempted to be careful about assumptions necessary for the model. A test for equality of variances was given, and alternative nonparametric methods were discussed for possible use when the normal theory assumptions are questioned.

We also introduced and defined important terms that you should know. These are:

One-way classification	Bonferroni inequality
Homogeneity of variances	Multiple comparisons
Randomized complete block design	Fixed effects
Random effects	Nonparametric analysis of variance
Variance components	Duncan's multiple range test
Comparisonwise error rate	Tukey's test
Experimentwise error rate	

REAL WORLD APPLICATION OF ANALYSIS OF VARIANCE

According to the data from the Brazilian Institute of Geography and Statistics, there were about 8.5 million people 15 and 24 years old who engaged in full-time jobs and also enrolled in evening courses in Brazil. A study on the factors associated with time spent in class among working college students was carried out from August 2007 to June 2008. A sample of 78 students from the University of São Paulo was selected. The students filled in a comprehensive questionnaire on living and working conditions. Time of activity and rest were continuously monitored by wrist actigraph, which provide information on the beginning and end of the sleep phase, sleep length, sleep efficiency, sleep latency, nocturnal awakenings, and daytime napping. The students were also requested to record their daily activities, including work time, commuting time, extracurricular pursuits, time in classes at college, leisure time and sleep.

One of results from the study was the effect of weekly working hours on workday sleep length. The working hours were classified into three categories: less than 30 hours/week, from 30 to 40 hours/week, and greater than 40 hours/week. Sleep length was defined as time from sleep onset to wake.

The sample size and sample means of three categories of working hours were shown in the following table.

	Less than 30 hours/week	From 30 to 40 hours/week	Greater than 40 hours/week
Size	29	32	17
Mean of sleep length (hour)	6.25	6.49	5.56

One-way ANOVA was used to test the means of sleep lengths of the students in three working hours categories. The value of the test statistic was $F = 4.35$. The P value of the test was 0.016. We have sufficient evidence to reject the null hypothesis of the same population means of three categories. There was a significant association between weekly working hours and sleep length.

Tukey's test was then used as post hoc test. It showed that those who worked longer than 40 hours/week had a shorter sleep length compared to the others.

EXERCISES

Section 11.1

1. Show that $E[\overline{E}_{i\cdot}^2] = \sigma^2/n_i$ for each $i = 1, 2, \ldots, k$. *Hint:*

$$E[\overline{E}_{i\cdot}^2] = E\left[\left(\sum_{j=1}^{n_i} E_{ij}\Big/n_i\right)^2\right]$$

$$= E\left[\left(\sum_{j=1}^{n_i} E_{ij}\right)^2\right]\Big/n_i^2$$

Argue that due to the independence of the terms E_{ij}, all cross-product terms $E_{ij} E_{if}$, $j \neq f$, have expectation 0. Argue also that $E[E_{ij}^2] = \sigma^2$.

2. Show that

$$SS_{\text{Tot}} = \sum_{i=1}^{k} \sum_{j=1}^{n_i} Y_{ij}^2 - T_{..}^2/N$$

3. Show that

$$SS_{\text{Tr}} = \sum_{i=1}^{k} T_{i\cdot}^2/n_i - T_{..}^2/N$$

4. Experiments were conducted to study whether commercial processing of various foods changes the concentration of essential elements for human consumption. One such experiment was to study the concentration of zinc in green beans. A batch of green beans was divided into four groups. The four groups were then

randomly assigned to be measured for zinc as follows: group 1 measured raw; group 2 measured before blanching; group 3 measured after blanching; and group 4 measured after the final processing step. Independent measurements were taken from the four groups (treatments), yielding the following observations:

Zinc concentration

Group 1	Group 2	Group 3	Group 4
2.23	3.71	2.53	5.46
2.20	4.67	2.87	5.19
2.44	3.45	2.83	5.51
2.11	2.73	2.33	4.82
2.30	2.58	2.19	6.63
1.72	1.85	1.80	2.39
1.78	1.81	1.75	2.09
2.36	2.32	1.83	2.27
2.91	2.50	1.97	2.39

(*a*) State the appropriate null and alternative hypotheses.
(*b*) Test your hypothesis for significance at the 5% level.
(*c*) Verbally state your conclusion.

5. It was known that a toxic material was dumped in a river leading into a large salt water commercial fishing area. Civil engineers studied the way the water carried the toxic material by measuring the amount of the material (in parts per million) found in oysters harvested at three different locations, ranging from the estuary out into the bay where the majority of commercial fishing was carried out. The resulting data are given below:

Site 1	Site 2	Site 3
15	19	22
26	15	26
20	10	24
20	26	26
29	11	15
28	20	17
21	13	24
26	15	
	18	

(*a*) Test whether there is a significant difference in the average parts per million of toxic material found in oysters harvested at the three sites. Use $\alpha = .05$.
(*b*) Would the means be significantly different at the .01 level of significance? (Use a computer package to find the P value.)
(*c*) Do your answers in parts (*a*) and (*b*) contradict each other? Justify your answer.

6. As landfills begin to fill up and land for such operations becomes more scarce, recycling will become more and more necessary. Some fast-food chains have become aware of this problem and are beginning to use biodegradable packaging or to recycle their packaging material. A study is conducted to investigate the amount of money received per year for recycled goods by various sized businesses. These data, in thousands of dollars, are obtained (based on figures in "New Ways to Take Out the Trash," Tom Andel, *Transportation and Distribution,* May 1993):

Company Size					
101–250 employees		251–500 employees		Over 500 employees	
3.5	3.8	13.4	14.1	9.0	10.8
3.4	3.0	14.0	14.7	10.3	10.7
3.7	2.8	14.5	15.0	10.1	9.5
3.3	3.2	14.7	13.6	10.6	11.2
2.9	3.9	15.2	14.3	10.4	9.7
2.6	4.1	15.5	14.6	10.5	10.0
3.5	4.0	14.2	14.8	10.3	11.3
4.2	4.9	15.3	15.6	10.7	11.8

(a) Construct stem-and-leaf diagrams for each of these samples, and comment on the reasonableness of the normality assumption.

(b) Estimate the variance of each of the populations from which these data are drawn, and comment on the reasonableness of the assumption of equal variances. (Do not try to test anything.) What has been done on this study to guard against a possible violation of the assumption of equal variances?

(c) Test H_0: $\mu_1 = \mu_2 = \mu_3$, and report the P value of your test. What practical conclusion can be drawn?

7. Scroll speed is an important consideration in the development of color graphics cards. A study is conducted to compare the time, in seconds, required to scroll one screen of WORD documents using five different color graphics cards with 24-inch monitors. The test conducted was a standard Hydra Quick Draw performance test. These data are obtained (based on information found in "Gauging Video Speed," *MAC WORLD,* June 1993, p. 28):

Graphics cards				
A	B	C	D	E
30.5	48.3	79.2	51.6	79.0
32.4	42.1	84.7	59.4	85.3
27.2	43.5	85.0	57.3	86.2
26.3	40.6	88.2	59.0	82.0
25.1	38.6	76.3	58.7	87.2
38.2	32.1	83.1	68.1	81.7
30.6	41.6	92.6	64.8	93.5
33.7	38.8	88.5	55.5	89.1

Use these data to test for equality of means. State your conclusion and report the P value of the test.

Section 11.2

8. Recall that MS_E was defined as the ratio $(SS_E/N - k)$. Show that

$$MS_E = S_p^2 = \frac{\sum_{i=1}^{k}(n_i - 1)S_i^2}{N - k}$$

for the one-way classification fixed-effects model.

9. In Exercise 4, use Bartlett's test to determine whether it is reasonable to assume homogeneity of variances for the four treatment groups. If not, what type of alternative procedure is available to analyze these data?

10. Test for equality of variances by using the data in Exercise 6. Be sure to state your conclusion and to report the P value of your test.

11. Test for equality of variances by using the data of Exercise 7.

Section 11.3

12. In each case, determine the number of paired comparisons possible.
    (*a*) $k = 3$
    (*b*) $k = 6$
    (*c*) $k = 10$

13. Let $\alpha = .05$, and assume that all possible paired comparisons are to be made. For each value of k given in Exercise 12, use the Bonferroni procedure to find an upper bound for α', the overall probability of making at least one incorrect rejection. In each case, what level should be used to guarantee that $\alpha' \leq .10$?

14. Consider a situation in which c tests are to be run. Let b denote the target upper bound for α'. In each case, show that $1 - [1 - (b/c)]^c \leq b$, thus guaranteeing that the choice of $\alpha = b/c$ implies that $\alpha' \leq b$.
    (*a*) $c = 2$
    (*b*) $c = 3$
    In each case, find the value of $1 - [1 - (b/c)]^c$ for $b = .10$.

15. (*Bonferroni inequality.*) The following inequality provides the basis for determining the comparisonwise error rate needed to control the experimentwise error rate at some specified level.

Let $A_1, A_2, \ldots, A_c$ be events. Then

$$P[A_1 \cap A_2 \cap \cdots \cap A_c] \geq 1 - [P[A_1'] + P[A_2'] + \cdots + P[A_c']]$$

    (*a*) Prove this result for $c = 2$. *Hint:* Notice that by the general addition rule, $P[A_1' \cup A_2'] \leq P[A_1'] + P[A_2']$. Use the fact that $A_1 \cap A_2 = (A_1' \cup A_2')'$ to complete the proof.
    (*b*) Let A_i denote the event that no incorrect rejection is made on the ith comparison, $i = 1, 2, \ldots, c$. In set notation, express the probability that no incorrect rejections are made. In terms of this probability, what is the probability that at least one incorrect rejection is made? If $\alpha = P[A_i']$ for $i = 1, 2, \ldots, c$, then use the Bonferroni inequality to show that $\alpha' \leq \alpha c$. Argue that if a target upper bound b is chosen and then c tests are run at $\alpha = b/c$, then $\alpha' \leq b$.

16. Consider the study described in Exercise 6. If Duncan's multiple comparisons are run at $\alpha = .05$, what is an upper bound for the experimentwise error rate? Conduct these tests, and discuss your conclusions.

17. Reconsider the data of Exercise 5.
    (*a*) How many paired comparisons are possible?
    (*b*) Suppose that these comparisons are to be made using Bonferroni T tests. If we want $\alpha' \leq .15$, at what α level should each test be conducted?
    (*c*) Conduct each of the tests indicated in part (*b*), and discuss your results.

18. Use Bonferroni T tests with $\alpha' \leq .10$ to pinpoint differences among groups in Exercise 4.

19. Use Bonferroni T tests with $\alpha' \leq .20$ to pinpoint differences among graphics cards based on the data of Exercise 7.

20. Scientists concerned with treatment of tar sand wastewater studied three treatment methods for the removal of organic carbon. (Based on "Statistical Planning and Analysis for Treatments of Tar Sand Wastewater," W. R. Pirie, Technical Report, Virginia Tech University). The three treatment methods used were air flotation (A.F.), foam separation (F.S.), and ferric-chloride coagulation (F.C.C.). The organic carbon material measurements for the three treatments yielded the following data:

A.F.	F.S.	F.C.C.
34.6	38.8	26.7
35.1	39.0	26.7
35.3	40.1	27.0
35.8	40.9	27.1
36.1	41.0	27.5
36.5	43.2	28.1
36.8	44.9	28.1
37.2	46.9	28.7
37.4	51.6	30.7
37.7	53.6	31.2

Test for differences among means. Be sure to report the P value of your test and state your conclusion.

21. If you found significant differences in Exercise 20, determine which treatment methods differ from each other.

22. A study on the tensile strength of aluminum rods is conducted. Forty identical rods are randomly divided into four groups, each of size 10. Each group is subjected to a different heat treatment, and the tensile strength, in thousands of pounds per square inch, of each rod is determined. The following data result:

	Treatment		
1	2	3	4
18.9	18.3	21.3	15.9
20.0	19.2	21.5	16.0
20.5	17.8	19.9	17.2
20.6	18.4	20.2	17.5
19.3	18.8	21.9	17.9
19.5	18.6	21.8	16.8
21.0	19.9	23.0	17.7
22.1	17.5	22.5	18.1
20.8	16.9	21.7	17.4
20.7	18.0	21.9	19.0

(a) Construct a stem-and-leaf diagram for each data set. Does the assumption that each sample is drawn from a normal distribution appear reasonable?

(b) Test the null hypothesis of equal treatment means by using a one-way ANOVA.

(c) Compare all possible pairs of means by using Bonferroni T tests with an overall level of significance that is at most .05.

(d) Conduct a Duncan's multiple test at the .05 significance level.

(e) Conduct a Tukey's test at the .05 significance level.

(f) Do the tests in (c), (d), and (e) lead you to the same conclusion?

(g) Explain differences if they exist.

Section 11.4

23. For a certain manufacturing plant, filters used to remove solid pollutants must be replaced as soon as they fail due to cracking or holes in the filter. An experiment was conducted to test five types of filters made from different fabrics. Six filters of each type were used under the same conditions, with the number of hours until failure recorded for each. The experiment yielded the following information:

Filter type (hours until failure)

1	2	3	4	5
261.1	221.9	201.4	600.9	160.6
186.2	188.7	146.1	301.2	135.0
239.1	167.6	96.8	608.9	455.1
243.3	224.9	173.9	283.3	402.3
296.8	178.8	280.8	193.3	457.9
270.5	147.9	100.3	159.4	559.6

(a) Use the Kruskal-Wallis test to determine whether there is significant evidence that the median time to failure among the filter types is different at the .05 level of significance.

(b) Repeat this test by using the appropriate normal theory test procedure.

24. Following a major accidental spill from a chemical manufacturing plant near a river, a study was conducted to determine whether certain species of fish caught from the river differ in terms of the amounts of the chemical absorbed. If differences are found, regulations on human consumption may be recommended. Samples from catches of three major species were measured in parts per million. The resulting data are given below:

Species

A	B	C
18.1	29.1	26.6
16.5	15.8	16.1
21.0	20.4	18.8
18.7	23.5	25.0
7.4	18.5	21.8
12.4	21.3	15.4
16.1	23.1	19.9
17.9	23.8	15.5
	20.1	21.1
	11.9	25.5

Test whether the median amounts of chemical absorbed by the three species of fish differ at the .05 level of significance.

REVIEW EXERCISES

Carbon dioxide is known to have a critical effect on microbiological growth. Small amounts of CO_2 stimulate growth of some organisms, whereas high concentrations inhibit the growth of most. The latter effect is used commercially when perishable food products are stored. A study is conducted to investigate the effect of CO_2 on the growth rate of *Pseudomonas fragi,* a food-spoiling organism. Carbon dioxide is administered at five predetermined different atmospheric pressures. The response measured was the percentage change in cell mass after a 1-hour growing time. Ten cultures were used at each atmospheric pressure level, resulting in the following data:

Factor level (CO_2 pressure)

0.0	.083	.29	.50	.86
62.6	50.9	45.5	29.5	24.9
59.6	44.3	41.1	22.8	17.2
64.5	47.5	29.8	19.2	7.8
59.3	49.5	38.3	20.6	10.5
58.6	48.5	40.2	29.2	17.8
64.6	50.4	38.5	24.1	22.1
50.9	35.2	30.2	22.6	22.6
56.2	49.9	27.0	32.7	16.8
52.3	42.6	40.0	24.4	15.9
62.8	41.6	33.9	29.6	8.8

Exercises 25 through 27 refer to these data.

25. State the assumptions required to test the null hypothesis

$$H_0: \mu_1 = \mu_2 = \cdots = \mu_5$$

26. (a) Numerically complete the appropriate analysis of variance (ANOVA) table.
 (b) Is there sufficient evidence to reject H_0 at the $\alpha = .05$ level of significance?

27. Use Bartlett's test to test for homogeneity of variances. Does this test lend support to the assumptions given in Exercise 25?

28. Three treatments were randomly selected from a large population of possible treatments. Ten randomly selected observations were then obtained from each treatment selected.
 (a) State an appropriate null hypothesis to be tested, and list all assumptions necessary to make this test for the described experiment.
 (b) The data yielded the following partial analysis of variance table:

ANOVA

Source	DF	SS	MS	F	EMS
Treatment	2	110.6			
Error	27				
Total	29	608.3			

Complete the ANOVA table, and test your null hypothesis in part (a) at the .05 level of significance.
 (c) Estimate the proportions of total variability due to error and treatments, respectively.

29. An experiment is conducted to compare the energy requirements of three physical activities: running, walking, and bicycle riding. The variable of interest is the number of kilocalories expended per kilometer traveled. To control for possible metabolic differences, eight subjects were selected and then randomly assigned (in terms of order) each of the three tasks, with ample rest between tasks to eliminate fatigue. Each activity is monitored exactly once for each individual. The resulting data are given below:

	Task		
Individual	Running	Walking	Bicycling
1	1.4	1.1	0.7
2	1.5	1.2	0.8
3	1.8	1.3	0.7
4	1.7	1.3	0.8
5	1.6	0.7	0.1
6	1.5	1.2	0.7
7	1.7	1.1	0.4
8	2.0	1.3	0.6

(*a*) Test for possible differences in average kilocalories expended among the three tasks. State an approximate *P* value.

(*b*) Use Duncan's multiple range test to determine which energy task, if any, is different from the other tasks in terms of energy expended at the .05 significance level.

(*c*) Repeat part (*c*) using Tukey's test, and compare differences in results found, if any.

30. The OPEC oil embargo made it evident that fuel economy in automobiles needed to be improved. Newer lightweight materials were sought for use in automobile engines. Comparisons were made among test samples of steel, aluminum, and phenolic thermoset composites containing glass fibers. Two variables, density (g/cm^3) and tensile strength (ksi), were considered. These data were obtained (based on information found in "Phenolics Creep Up on Engine Applications," John Arimind and William Ayles, *Advanced Materials and Processes,* vol. 143, no. 6, June 1993; pp. 34–36):

Material (density)					
Steel		Aluminum		Phenolics	
7.60	7.82	2.90	2.71	1.79	1.73
7.80	7.90	2.65	2.77	1.74	1.72
7.81	7.75	2.67	2.78	1.69	1.66
7.65	7.73	2.75	2.72	1.68	1.67
7.72	7.75	2.80	2.86	1.50	1.70
7.71	7.80	2.81	2.73	1.67	1.72
7.68	7.87	2.85	2.75	1.80	1.71
7.79	7.89	2.72	2.79	1.78	1.63
7.76	7.78	2.60	2.81	1.62	1.62

Material (tensile strength)					
Steel		Aluminum		Phenolics	
60	100	17	40	8	13
73	122	25	19	17	12
87	86	35	29	15	10
98	112	38	22	9	11
175	77	27	26	10	16

Analyze each of these data sets by using whatever tests you believe are appropriate. Write a report to summarize your results. Explain your choice of test statistics.

31. Studies are conducted to investigate the use of slag in road pavement base, subbase, and surfacing. The composition of slag obtained from various sources is of interest. These data are obtained on the percentage of S_1O_2 in samples from five different sources (based on information from "Steel Plant Slag in Road Pavements," B. S. Heaton, *Australian Civil Engineering Transactions,* March 1993, pp. 49–53):

<table>
<tr><th colspan="5">Source</th></tr>
<tr><th>Blast furnace slag</th><th>Steel slag</th><th>Portland cement</th><th>Fly ash</th><th>Natural basalt</th></tr>
<tr><td>35.1</td><td>16.0</td><td>20.1</td><td>58.3</td><td>45.9</td></tr>
<tr><td>34.7</td><td>15.6</td><td>22.0</td><td>57.6</td><td>46.3</td></tr>
<tr><td>34.8</td><td>17.2</td><td>23.1</td><td>55.0</td><td>44.5</td></tr>
<tr><td>33.2</td><td>16.2</td><td>19.7</td><td>60.1</td><td>45.2</td></tr>
<tr><td>33.6</td><td>16.3</td><td>19.5</td><td>61.2</td><td>44.5</td></tr>
<tr><td>36.8</td><td>14.7</td><td>16.2</td><td>58.2</td><td>49.6</td></tr>
</table>

Analyze these data, and write a report summarizing your results. Be sure to defend your choice of test statistics.

It is well known that power surges or line "spikes" can damage sensitive electronic equipment. A study of the surges was conducted to ascertain whether or not there are differences in the average frequency of the surges among the seven days of the week. Over a randomly selected ten-week period, the number of spikes was observed for ten 24-hour periods for each of the seven days in the week. The resulting data are as follows:

Mon.	Tues.	Wed.	Thurs.	Fri.	Sat.	Sun.
25	24	16	22	33	28	35
21	20	20	17	28	28	20
20	19	19	19	44	31	29
20	16	21	21	33	21	19
21	21	17	19	22	33	26
20	19	13	13	36	22	23
25	24	23	27	28	22	26
21	15	23	15	27	20	30
18	20	23	20	22	22	29
22	14	23	20	16	26	29

Exercises 32 through 35 refer to these data.

32. Using analysis of variance, state the correct hypothesis for testing equality of means and conduct the appropriate ANOVA test. State your conclusion.

33. If some of the means in Exercise 32 were found to be different, use Duncan's procedure to test for any significant differences in pairs of means. Test at $\alpha = .05$.

34. Repeat Exercise 33 using Tukey's test. Do you get different results in Exercises 33 and 34? If so, explain the reasons for these differences.

35. Repeat Exercise 32, but use an appropriate nonparametric test.

CHAPTER 12

Categorical Data

In this chapter we are concerned with the analysis of data characterized by the fact that each observation in the data set can be classed as falling into exactly one of several mutually exclusive "cells" or categories. Interest centers on the number of observations falling into each category. The statistical problem is to determine whether the observed category frequencies tend to refute a stated hypothesis. We are concerned with three problems in particular. These are:

1. Testing to see whether a set of observations was drawn from a specified probability distribution
2. Testing for independence between two variables used for classification purposes
3. Comparing proportions

The statistical procedures used in much of the work to come are based on the *multinomial distribution.* We begin by describing this distribution.

12.1 MULTINOMIAL DISTRIBUTION

To develop the definition of a multinomial random variable, we need to consider first the idea of a multinomial trial.

> **Definition 12.1.1 (Multinomial trial).** A multinomial trial with parameters $p_1, p_2, \ldots, p_k$ is a trial that can result in exactly one of k possible outcomes. The probability that outcome i will occur on a given trial is p_i for $i = 1, 2, 3, \ldots, k$.

Note that since $p_1, p_2, \ldots, p_k$ are probabilities, each of them lies between 0 and 1 inclusive. Furthermore, since each trial results in exactly one of the k possible outcomes, these probabilities sum to 1.

Example 12.1.1. It is noted that 1% of the items coming off a production line are defective and nonsalvageable, 5% are defective but salvageable, and the rest are nondefective. One item is selected at random and classified. Since exactly one of three possible outcomes can result, this experiment can be viewed as constituting a single multinomial trial with parameters $p_1 = .01, p_2 = .05$, and $p_3 = .94$.

The multinomial random variable arises quite naturally whenever we observe a series of independent and identical multinomial trials. This multivariate random variable is defined as follows:

> **Definition 12.1.2 (Multinomial random variable).** Let an experiment consist of n independent and identical multinomial trials with parameters $p_1, p_2, \ldots, p_k$. Let X_i denote the number of trials that result in outcome i for $i = 1, 2, \ldots, k$. The k-tuple $(X_1, X_2, \ldots, X_k)$ is called a *multinomial random variable* with parameters $n, p_1, p_2, \ldots, p_k$.

Example 12.1.2. Assume that a random sample of 100 items is selected from the production line described in Example 12.1.1. This experiment can be viewed as consisting of $n = 100$ independent multinomial trials, each with parameters $p_1 = .01$, $p_2 = .05$, and $p_3 = .94$. Let X_1 denote the number of defective and nonsalvageable items selected, X_2 the number of defective but salvageable items selected, and X_3 the number of nondefective items selected. The triple or 3-tuple (X_1, X_2, X_3) is a multinomial random variable with parameters 100, .01, .05, and .94. For example, if we observe 2 defective items that cannot be salvaged, 6 that are defective and can be salvaged, and 92 that are nondefective, then the multinomial random variable (X_1, X_2, X_3) assumes the observed value of $(2, 6, 92)$.

Although it is not hard to derive the density for a multinomial random variable, we shall not need to do so here. However, we do need to determine the expected value of each of the random variables $X_1, X_2, \ldots, X_k$. This is easy to do. Consider a single multinomial trial and any fixed outcome i. This trial either does or does not result in outcome i. If outcome i does occur, we consider the trial a success; otherwise it is a failure. The probability of success is p_i, the probability that outcome i will result on a given trial; the probability of failure is $1 - p_i$. Consider now a series of n independent and identical multinomial trials. Let X_i denote the number of trials that result in outcome i. Note that X_i also denotes the number of successes in n independent trials, each with probability of success p_i. Therefore X_i is a binomial random variable with parameters n and p_i. From our discussion in Chap. 3 we know that for each i, $E[X_i] = np_i$. This result plays an important role in analyzing categorical data.

Example 12.1.3. In a random sample of 100 items selected from the production line described in Example 12.1.1 the expected number of items falling into each category is

Expected number of defective and nonsalvageable items	$= E[X_1] = np_1 = 100(.01) = 1$
Expected number of defective but salvageable items	$= E[X_2] = np_2 = 100(.05) = 5$
Expected number of nondefective items	$= E[X_3] = np_3 = 100(.94) = 94$

When the sampling is complete, we observe that $x_1 = 2$, $x_2 = 6$, and $x_3 = 92$. These values do not coincide exactly with the expected values, but they do not seem to differ drastically from them. For this reason, the values do not lead us to suspect the accuracy of the stated category probabilities of .01, .05, and .94, respectively.

The previous example Illustrates the basic idea of categorical data analysis. In analyzing count data, we compare the observed category frequencies to those expected under a stated null hypothesis. If these agree fairly well we do not reject H_0; if there are substantial disagreements, we do reject H_0. In the sections that follow we develop the statistics needed to determine when the differences are extreme enough to warrant rejection of the stated hypothesis.

12.2   CHI-SQUARED GOODNESS OF FIT TESTS

The purpose of the chi-squared goodness of fit test is to test the null hypothesis that a given set of observations is drawn from, or "fits," a specified probability distribution. We consider the situation in which the hypothesized distribution is completely specified before the sampling is done. The procedure for handling this case is based on the next theorem, which is offered without proof.

> **Theorem 12.2.1.** Let $(X_1, X_2, \ldots, X_k)$ be a multinomial random variable with parameters $n, p_1, p_2, \ldots, p_k$. For large n the random variable
>
> $$\sum_{i=1}^{k} \frac{(X_i - np_i)^2}{np_i}$$
>
> follows an approximate chi-squared distribution with $k - 1$ degrees of freedom.

We make two notational changes to make this random variable easier to remember. Since in the multinomial context X_i is the actual or "observed" number of trials resulting in outcome i or falling into category i, we denote X_i by O_i. Recall that np_i is the theoretical expected number of trials resulting in outcome i, and so we let $np_i = E[X_i] = E_i$. Thus Theorem 12.2.1 states that the statistic,

$$\sum_{i=1}^{k} \frac{(O_i - E_i)^2}{E_i} = \sum_{i=1}^{k} \frac{[(\text{observed frequency}) - (\text{expected frequency})]^2}{\text{expected frequency}}$$

is, for large n, approximately chi-squared with $k - 1$ degrees of freedom, where k is the number of mutually exclusive categories involved. This naturally brings up the question, "How large is large?" There are various opinions as to the answer to this question. However, it is usually felt that n should be large enough that *no expected frequency is less than 1 and no more than 20% of the expected frequencies are less than 5.* If this condition is not met, either categories should be combined or redefined or the sample size should be increased so that the expected frequencies will be of adequate size.

This random variable serves quite logically as a test statistic for testing a null hypothesis that a given set of observations is drawn from a specified probability distribution. If H_0 is true, the value of p_i will be known for each i, and hence E_i can be computed easily. In effect, the above statistic compares the observed number of observations per category with the number expected under H_0. If these figures agree fairly well (there is a good fit), then the term $(O_i - E_i)^2$ will be small for each i, $\sum_{i=1}^{k}[(O_i - E_i)^2]/E_i$ will be small, and H_0 should not be rejected. If the observed and expected frequencies differ greatly, then $(O_i - E_i)^2$ will be large for some i, $\sum_{i=1}^{k}[(O_i - E_i)^2]/E_i$ will be large, and H_0 should be rejected. The use of this test statistic is illustrated in the next example.

Example 12.2.1. Computer systems crash for many reasons, among them software failure, hardware failure, operator error, and system overloading. It is thought that 10% of the crashes are due to software failure, 5% to hardware failure, 25% to operator error, 40% to system overloading, and the rest to other causes. Over an extended study period 150 crashes are observed, and each is classified according to its probable cause. It is found that 13 are due to software failure, 10 to hardware failure, 42 to operator error, 65 to system overloading, and the rest to other causes. Do these data lead us to suspect the accuracy of the stated percentages? To answer this question, we test

$$H_0: p_1 = .10, p_2 = .05, p_3 = .25, p_4 = .40, p_5 = .20$$
$$H_1: p_i \text{ is not as stated for some } i = 1, 2, 3, 4, 5$$

If H_0 is true, then

$$E[X_1] = E_1 = np_1 = 150(.10) = 15$$
$$E[X_2] = E_2 = np_2 = 150(.05) = 7.5$$
$$E[X_3] = E_3 = np_3 = 150(.25) = 37.5$$
$$E[X_4] = E_4 = np_4 = 150(.40) = 60$$
$$E[X_5] = E_5 = np_5 = 150(.20) = 30$$

The situation is summarized in Table 12.1. Note that the expected and observed frequencies do not agree exactly. The question to be answered is, "Do they differ enough to cause us to reject H_0?" Since $E_i > 5$ in each case, the test statistic

$$\sum_{i=1}^{5} \frac{(O_i - E_i)^2}{E_i}$$

follows an approximate $X_{k-1}^2 = X_4^2$ distribution. The observed value of this statistic is

$$\frac{(13 - 15)^2}{15} + \frac{(10 - 7.5)^2}{7.5} + \frac{(42 - 37.5)^2}{37.5} + \frac{(65 - 60)^2}{60} + \frac{(20 - 30)^2}{30} = 5.39$$

Is this value large enough to cause us to reject H_0? From Table IV of App. A we see that the probability of observing a value of 5.39 or larger is .25. That is, if we reject H_0, the P value of the test is .25. Since this probability is large, we do not reject H_0. The data gathered are not sufficient to allow us to conclude that the stated percentages are incorrect.

12.3 TESTING FOR INDEPENDENCE

In this section we discuss a problem involving categorical data that is somewhat different from that considered in the last section. However, the idea behind the test procedure used is identical to that studied earlier. Namely, the test statistic compares

Table 12.1

Category	Software failure 1	Hardware failure 2	Operator error 3	System overloading 4	Other 5
Observed frequency, O_i	13	10	42	65	20
Expected frequency, E_i	15	7.5	37.5	60	30

observed category frequencies with those expected under the assumption that the stated null hypothesis is true, with rejection coming if these differ too much to have occurred by chance.

Here we consider experiments in which two random variables are being studied. The purpose of the study is to test these random variables for independence. For example, a highway engineer may be interested in seeing whether the extent of an injury is independent of the type of restraint being used by an accident victim; a manufacturer may want to see if the quality of an item produced is independent of the day of the week on which it was made; a cancer researcher may want to see whether the development of lung cancer is independent of exposure to airborne asbestos.

We illustrate the test for independence in the context of what are called 2×2 *contingency* tables. These tables arise in experiments in which each of the two random variables being considered is studied at two levels. This naturally defines 2×2 or 4 mutually exclusive "cells" or categories. The data analysis is based on an examination of the number of observations falling into each cell.

Example 12.3.1. A cancer researcher performs what is called a prospective study by selecting a large group of individuals at random and following their progress for a long period of time. At the end of the study period each individual is classified according to whether or not lung cancer is present and according to whether or not the individual has been exposed to an identifiable source of airborne asbestos. Let C denote the presence of lung cancer, and let A denote the fact that the individual has been exposed to airborne asbestos.

These four mutually exclusive categories result:

$$C \cap A: \text{has cancer and exposed to asbestos}$$
$$C \cap A': \text{has cancer but not exposed to asbestos}$$
$$C' \cap A: \text{no cancer but exposed to asbestos}$$
$$C' \cap A': \text{no cancer and not exposed to asbestos}$$

Each individual in the study falls into exactly one of these cells.

Since we are concerned with the number of observations falling into each cell, we need a notational convention for the cell frequencies. We also need a notational convention to indicate the number of observations falling into each level of each of the two classification variables. We use the following:

$$n_{11} = \text{number of observations falling into cell in row 1 and column 1}$$
$$n_{12} = \text{number of observations falling into cell in row 1 and column 2}$$
$$n_{21} = \text{number of observations falling into cell in row 2 and column 1}$$
$$n_{22} = \text{number of observations falling into cell in row 2 and column 2}$$
$$n_{1.} = n_{11} + n_{12} = \text{number of observations in row 1}$$
$$n_{2.} = n_{21} + n_{22} = \text{number of observations in row 2}$$
$$n_{.1} = n_{11} + n_{21} = \text{number of observations in column 1}$$
$$n_{.2} = n_{12} + n_{22} = \text{number of observations in column 2}$$
$$n = \text{total number of observations}$$

This notational convention is illustrated in Example 12.3.2.

Example 12.3.2. When the study of Example 12.3.1 is completed it is found that 50 of the 5000 persons involved had developed lung cancer. Of these, 10 had been exposed to an identifiable source of airborne asbestos. A total of 500 persons in the study had been exposed to an identifiable airborne asbestos source. These data are summarized in Table 12.2. Note that $n._1$ and $n._2$ are column totals that appear along the margins of the 2×2 table. They are called *marginal column totals.* Similarly, $n_1.$ and $n_2.$ are called *marginal row totals.*

The general null hypothesis to be tested via a contingency table is that there is "no association" between the two classification variables. The alternative is that there is an association. The tables studied in this section are characterized by the fact that *only* the overall sample size n is fixed by the researcher. Prior to data collection all other entries, including the row and column marginal totals, are *free to vary.* In this sort of study the null hypothesis of "no association" is equivalent to a null hypothesis of independence.

To develop the general test, let A and B denote the classification variables. We want to test

H_0: A and B are independent

H_1: A and B are not independent

If H_0 is true, then knowledge of the classification level of an object relative to characteristic A has no bearing on its level relative to characteristic B. To express this idea mathematically, we use the table of probabilities given in Table 12.3. Note that p_{11} denotes the probability that a randomly selected object has characteristics A *and* B, $p_1.$ denotes the probability that it has characteristic A, and $p._1$ denotes the probability that it has characteristic B. Recall that A and B are independent if and only if

$$P[A \cap B] = P[A] \cdot P[B]$$

Thus the null hypothesis that A and B are independent can be expressed as

$$H_0: p_{11} = p_1. p._1$$

Table 12.2

	A	A'	
C	$10 = n_{11}$	$40 = n_{12}$	$50 = n_1.$
C'	$490 = n_{21}$	$4460 = n_{22}$	$4950 = n_2.$
	$500 = n._1$	$4500 = n._2$	$5000 = n$

Table 12.3

	B	B'	
A	p_{11}	p_{12}	$p_1.$
A'	p_{21}	p_{22}	$p_2.$
	$p._1$	$p._2$	1

This implies that $p_{ij} = p_{i\cdot} \, p_{\cdot j}$ for $i = 1, 2$ and $j = 1, 2$. That is, A and B are independent if and only if the cell probability for any cell can be found by multiplying the corresponding row and column probabilities.

Since in a 2×2 table each observation falls into exactly one of four mutually exclusive categories, a random sample of size n can be viewed as constituting a series of n independent multinomial trials, each with parameters $p_{11}, p_{12}, p_{21},$ and p_{22}. Hence the set $(n_{11}, n_{12}, n_{21}, n_{22})$ of observed cell frequencies is a multinomial random variable with parameters $n, p_{11}, p_{12}, p_{21},$ and p_{22}. Thus the expected cell frequencies are given by

$$E_{ij} = np_{ij}$$

where p_{ij} is the probability of an observation falling into the (ij)th cell and n is the sample size. These probabilities are not known and must be estimated from the data under the assumption that the null hypothesis is true. How can this be done? Quite simply! Note, for instance, that if H_0 is true and characteristics A and B are independent, then

$$p_{11} = p_{1\cdot} \, p_{\cdot 1}$$

Since $p_{1\cdot}$ is the probability of an observation falling into row 1, it is logical to estimate $p_{1\cdot}$ by

$$\hat{p}_{1\cdot} = \frac{\text{number of elements in row 1}}{\text{sample size}} = \frac{n_{1\cdot}}{n}$$

Similarly, since $p_{\cdot 1}$ is the probability of an observation falling into column 1, we estimate $p_{\cdot 1}$ by

$$\hat{p}_{\cdot 1} = \frac{\text{number of elements in column 1}}{n} = \frac{n_{\cdot 1}}{n}$$

Thus

$$\hat{p}_{11} = \hat{p}_{1\cdot} \, \hat{p}_{\cdot 1} = \frac{n_{1\cdot}}{n} \frac{n_{\cdot 1}}{n}$$

This in turn implies that

$$\hat{E}_{11} = \hat{p}_{11} n = \frac{n_{1\cdot}}{n} \frac{n_{\cdot 1}}{n} n$$

$$= \frac{n_{1\cdot} \, n_{\cdot 1}}{n} = \frac{(\text{marginal row total}) \, (\text{marginal column total})}{\text{sample size}}$$

A similar argument holds for other cell expectations. Thus we conclude that for each i and j

$$\hat{E}_{ij} = \frac{n_{i\cdot} \, n_{\cdot j}}{n} = \frac{(\text{marginal row total}) \, (\text{marginal column total})}{\text{sample size}}$$

Recall that for large samples

$$\sum_{i=1}^{2} \sum_{j=1}^{2} \frac{(O_{ij} - \hat{E}_{ij})^2}{\hat{E}_{ij}} = \sum_{i=1}^{2} \sum_{j=1}^{2} \frac{(n_{ij} - \hat{E}_{ij})^2}{\hat{E}_{ij}}$$

follows an approximate chi-squared distribution. The number of degrees of freedom is $k - 1 - m$, where m is the number of parameters estimated from the data used in computing the expected cell frequencies. Note that we actually need estimate only

$p_1.$ and $p._1$ from the data, since $p._2 = 1 - p._1$ and $p_2. = 1 - p_1..$ Hence the number of degrees of freedom associated with the test statistic is

$$k - 1 - m = 4 - 1 - 2 = 1$$

In this case, to satisfy the rule that no expected frequency should be less than 1 and no more than 20% should be less than 5, we must, in fact, have *no expected frequency less than 5*. If this rule cannot be satisfied, then the data should be analyzed by a procedure called Fisher's exact test [6]. Let us now complete the analysis of the data of Example 12.3.2.

Example 12.3.3. We want to see if there is evidence that the development of lung cancer (C) is not independent of the exposure of the individual to airborne asbestos (A). We shall test

$$H_0: C \text{ and } A \text{ are independent}$$
$$H_1: C \text{ and } A \text{ are not independent}$$

Using the data given in Table 12.2, we see that the expected cell frequencies under H_0 are given by

$$\hat{E}_{11} = \frac{n_1.n._1}{n} = \frac{50(500)}{5000} = 5$$

$$\hat{E}_{12} = \frac{n_1.n._2}{n} = \frac{50(4500)}{5000} = 45$$

$$\hat{E}_{21} = \frac{n_2.n._1}{n} = \frac{4950(500)}{5000} = 495$$

$$\hat{E}_{22} = \frac{n_2.n._2}{n} = \frac{4950(4500)}{5000} = 4455$$

The situation is summarized in Table 12.4. Note that there are some differences between what is expected if H_0 is true, listed in parentheses, and what is actually observed. Are those differences too large to have occurred strictly by chance? Note that no expected cell frequency is less than 5, as required. The observed value of the test statistic is given by

$$\sum_{i=1}^{2} \sum_{j=1}^{2} \frac{(n_{ij} - \hat{E}_{ij})^2}{\hat{E}_{ij}} = \frac{(10-5)^2}{5} + \frac{(40-45)^2}{45} + \frac{(490-495)^2}{495}$$
$$+ \frac{(4460-4455)^2}{4455}$$
$$= 5.61$$

The number of degrees of freedom associated with this chi-squared statistic is 1. Since $\chi^2_{.01} = 6.63$ and $\chi^2_{.025} = 5.02$, the P value of the test lies between .01 and .025.

Table 12.4

	A		A'		
C	10		40		50
		(5)		(45)	
C'	490		4460		4950
		(495)		(4455)	
	500		4500		5000

Since these probabilities are small, we reject H_0 and conclude that the development of lung cancer is associated with exposure to airborne asbestos. An inspection of Table 12.4 reveals that the number of cancers observed among those exposed to asbestos is higher than that expected if no association exists.

$r \times c$ Test for Independence

We have illustrated the test for independence in the case in which each variable is studied at two levels. This results in a 2×2 contingency table. In general, we may study one variable at r levels and the other at c levels, leading to what is called an $r \times c$ contingency table. The data layout and associated probabilities are shown in Tables 12.5(a) and 12.5(b), respectively.

The null hypothesis of independence is stated mathematically as

$$H_0: p_{ij} = p_{i.}p_{.j} \qquad i = 1, 2, 3, \ldots, r$$
$$j = 1, 2, 3, \ldots, c$$

The alternative is that $p_{ij} \neq p_{i.}p_{.j}$ for at least one i and j. The test statistic is

Test statistic H_0: no association between variables A and B

$$\sum_{i=1}^{r} \sum_{j=1}^{c} \frac{(n_{ij} - \hat{E}_{ij})^2}{\hat{E}_{ij}}$$

$$\text{where } \hat{E}_{ij} = \frac{(\text{marginal row total}) (\text{column row total})}{\text{sample size}}$$

Table 12.5(a)

	Variable B					
Variable A	1	2	3	$\cdots$	c	
1	n_{11}	n_{12}	n_{13}	$\cdots$	n_{1c}	$n_{1.}$
2	n_{21}	n_{22}	n_{23}	$\cdots$	n_{2c}	$n_{2.}$
3	n_{31}	n_{32}	n_{33}	$\cdots$	n_{3c}	$n_{3.}$
$\vdots$	$\vdots$	$\vdots$	$\vdots$			$\vdots$
r	n_{r1}	n_{r2}	n_{r3}	$\cdots$	n_{rc}	$n_{r.}$
	$n_{.1}$	$n_{.2}$	$n_{.3}$	$\cdots$	$n_{.c}$	n

Table 12.5(b)

	Variable B					
Variable A	1	2	3	$\cdots$	c	
1	p_{11}	p_{12}	p_{13}	$\cdots$	p_{1c}	$p_{1.}$
2	p_{21}	p_{22}	p_{23}	$\cdots$	p_{2c}	$p_{2.}$
3	p_{31}	p_{32}	p_{33}	$\cdots$	p_{3c}	$p_{3.}$
$\vdots$	$\vdots$	$\vdots$	$\vdots$			$\vdots$
r	p_{r1}	p_{r2}	p_{r3}	$\cdots$	p_{rc}	$p_{r.}$
	$p_{.1}$	$p_{.2}$	$p_{.3}$	$\cdots$	$p_{.c}$	1

Table 12.6

Extent of injury	Seat belt only		Seat belt and harness		None		
			Type of restraint				
None	75	(70.0)	60	(50.0)	65	(80.0)	200
Minor	160	(157.5)	115	(112.5)	175	(180.0)	450
Major	100	(105.0)	65	(75.0)	135	(120.0)	300
Death	15	(17.5)	10	(12.5)	25	(20.0)	50
	350		250		400		1000

The only question to answer is, "How many degrees of freedom are associated with this test statistic?" We must estimate the $r - 1$ probabilities $p_{1\cdot}, p_{2\cdot}, \ldots, p_{(r-1)\cdot}$ and the $c - 1$ probabilities $p_{\cdot 1}, p_{\cdot 2}, \ldots, p_{\cdot(c-1)}$ from the data in order to compute the expected cell frequencies. Recall that the number of degrees of freedom is given by $k - 1 - m$, where k is the number of cells in the table and m is the number of parameters estimated from the data used to compute expected frequencies. In this case the number of degrees of freedom is

$$k - 1 - m = rc - 1 - (r - 1 + c - 1)$$
$$= rc - r - c + 1$$
$$= (r - 1)(c - 1)$$

These ideas are illustrated in the next example.

> **Example 12.3.4.** To try to convince the public to use safety equipment in automobiles, a random sample of 1000 accidents is chosen from the records. Each accident is classed according to the type of safety restraint used by the occupants and the severity of the injuries received. The data given in Table 12.6 results.
>
> Expected cell frequencies are given in parentheses. Do these indicate an association between the type of restraint used and the extent of injury? The observed value of the $X^2_{(r-1)(c-1)} = X^2_{3\cdot2} = X^2_6$ statistic is
>
> $$\frac{(75 - 70)^2}{70} + \frac{(60 - 50)^2}{50} + \frac{(65 - 80)^2}{80} + \cdots + \frac{(25 - 20)^2}{20} = 10.96$$
>
> The probability of observing a value of 10.96 or greater is between .05 ($\chi^2_{.95} = 12.6$) and .1 ($\chi^2_{.90} = 10.6$). That is, the P value of the test is between .05 and .1.
>
> Since this probability is fairly small we shall reject the null hypothesis and conclude that the extent of an individual's injury is not independent of the type of safety restraint being used at the time of the accident.

Keep in mind the fact that when *only* the sample size is fixed by the experimenter and the remainder of the entries in the contingency table are free to vary, we are testing for independence. Other types of tests are considered in the next section.

12.4 COMPARING PROPORTIONS

In this section we consider the use of the chi-squared statistic in comparing proportions. Once again, we begin by describing an experiment that results in a 2×2 table. This example will demonstrate an important difference between the problems presented in this section and those considered earlier.

Table 12.7

	D (has disorder)	D' (does not have disorder)	
E (exposed)	n_{11}	n_{12}	$n_1. = 300$ (fixed)
E' (not exposed)	n_{21}	n_{22}	$n_2. = 320$ (fixed)
	$n_{.1}$	$n_{.2}$	$n = 620$ (fixed)

Example 12.4.1. A large number of people living in a particular community have been exposed over the last 10 years to radioactivity from an atomic waste storage dump. A study is to be run to find out whether there is any association between this exposure and the development of a particular blood disorder. To conduct the experiment, random samples will be chosen of 300 persons from the community who have been exposed to the hazard and 320 persons who have not been so exposed. Each subject will be screened to determine whether or not the blood disorder is present. This experiment generates a table of the form given in Table 12.7. Note that although this 2 × 2 table looks exactly like those studied earlier, there is a difference. In particular, the marginal row totals are *fixed* at 300 and 320 *prior* to conducting the field study. That is, these marginal totals as well as the overall sample size are predetermined by the experimenter. All other entries in the table are free to vary.

In experiments such as this where either the row or column totals, but not both, are fixed by the researcher, the null hypothesis of "no association" is stated in terms of proportions. To see how this is done, let A and B denote the classification variables and assume that the marginal totals for the levels of A are fixed. Thus we essentially have two independent random samples: one from the population of objects with trait A and the other from the population of objects without trait A. We want to test

H_0: proportion of objects with trait B among those with

trait A = proportion of objects with trait B among

those without trait A

This implies that characteristic B is no more prevalent among those with characteristic A than among those without characteristic A. Hence there is no apparent association between A and B. In terms of our example we want to test

H_0: proportion of individuals with the blood disorder

among those exposed to the hazard = proportion of

individuals with the blood disorder among those

not exposed to the hazard

In other words, there is no apparent association of the blood disorder with exposure to the hazardous waste material.

The null hypothesis can be expressed mathematically with the aid of Table 12.8. Since p_{11} denotes the proportion of objects with trait B among those with trait A and p_{21} denotes the proportion with trait B among those that do not have trait A, we are testing

$$H_0: p_{11} = p_{21}$$

Note that since $p_{12} = 1 - p_{11}$ and $p_{22} = 1 - p_{21}$, we are also testing $p_{12} = p_{22}$. It is convenient to state the null hypothesis as

$$H_0: p_{1j} = p_{2j} \qquad j = 1, 2$$

Table 12.8

	B	B'	
A	p_{11}	$p_{12} = 1 - p_{11}$	1 (fixed)
A'	p_{21}	$p_{22} = 1 - p_{21}$	1 (fixed)

When the null hypothesis is expressed in this form, the test is referred to as a test of *homogeneity*.

To understand the logic behind the test, we need to look closer at the structure of a 2×2 table with the marginal row totals fixed. In particular, note that we can view the data in row 1 as constituting a random sample of size $n_1.$ drawn from a binomial distribution with probability of success p_{11}. Here success is finding an object with trait B. Similarly, the data in row 2 constitutes a random sample of size $n_2.$ from a binomial distribution with parameter p_{21}. Thus when we test

$$H_0: p_{11} = p_{21}$$

we are actually comparing two population proportions as we did in Chap. 8. In fact, the test statistic that we develop here is simply the square of the Z statistic used earlier! The important point to note here is that since the number of objects in each group with trait B is binomially distributed,

$$E_{11} = n_1.p_{11}$$

and
$$E_{21} = n_2.p_{21}$$

Thus to estimate E_{11} and E_{21} from the contingency table, we need to find only a logical way to estimate p_{11} and p_{21}. This is not hard to do. Note that if H_0 is true, $p_{11} = p_{21}$. We denote this common population proportion by p. Furthermore, if the proportion of objects with trait B is the same for both populations, then the overall proportion of objects in the two populations combined will also be p. A logical estimator for the overall proportion of objects with trait B is

$$\hat{p} = \frac{\text{number of objects in column 1}}{\text{overall sample size}} = \frac{n_{.1}}{n}$$

Since we are assuming that $p_{11} = p_{21} = p$, we can also use $\hat{p}$ as an estimator for p_{11} and p_{21}. Substituting, we see that the estimated expected cell frequencies under H_0 are

$$\hat{E}_{11} = n_1.\hat{p}_{11} = n_1.\frac{n_{.1}}{n}$$

$$= \frac{(\text{marginal row total}) (\text{marginal column total})}{\text{sample size}}$$

$$\hat{E}_{21} = n_2.\hat{p}_{21} = n_2.\frac{n_{.1}}{n}$$

$$= \frac{(\text{marginal row total}) (\text{marginal column total})}{\text{sample size}}$$

We leave it to you to verify that

$$\hat{E}_{12} = \frac{n_1.n_{.2}}{n} \qquad \text{and} \qquad \hat{E}_{22} = \frac{n_2.n_{.2}}{n}$$

These expectations are exactly the same as those used in testing for independence. From this point on the test for homogeneity is identical to that for independence. We illustrate the idea by reconsidering the experiment described in Example 12.4.1.

Example 12.4.2. When the experiment of Example 12.4.1 is conducted, the data of Table 12.9 results. The expected cell frequencies are given in parentheses. The observed value of the X_1^2 statistic is

$$\frac{(52 - 48.39)^2}{48.39} + \frac{(248 - 251.61)^2}{251.61} + \frac{(48 - 51.61)^2}{51.61} + \frac{(272 - 268.39)^2}{268.39} = .62$$

This value is not significant even at the $\alpha = .25$ level ($\chi_{.25}^2 = 1.32$). These data do not allow us to conclude that there is an association between this particular blood disorder and exposure to this source of radioactivity.

$r \times c$ Test for Homogeneity

As in the test for independence, we can test for homogeneity via an $r \times c$ table. In this case we are dealing with two variables, one of which is studied at r levels and the other at c levels. The marginal totals for exactly one of these variables is fixed by the researcher prior to data gathering. To illustrate the idea, consider Example 12.4.3.

Example 12.4.3. A study is to be conducted to consider the association between the sulfur dioxide (SO_2) level in the air and the mean number of chloroplasts per leaf cell of trees in the area. Three regions are selected for study. One is known to have a high SO_2 concentration, one to have a normal level of SO_2, and the third to have a low SO_2 level. Twenty trees are to be randomly selected from within each area, and the mean number of chloroplasts per leaf cell is to be determined for each tree. On this basis each tree will be classified as having a low, normal, or high chloroplast count. This experiment generates a table of the form given in Table 12.10. Note that by fixing the row totals prior to experimentation, we are essentially selecting

Table 12.9

	D (has disorder)		D' (does not have disorder)		
E (exposed)	52	(48.39)	248	(251.61)	300
E' (not exposed)	48	(51.61)	272	(268.39)	320
	100		520		620

Table 12.10

	Chloroplast level			
SO_2 level	High	Normal	Low	
High	n_{11}	n_{12}	n_{13}	$n_{1.} = 20$ (fixed)
Normal	n_{21}	n_{22}	n_{23}	$n_{2.} = 20$ (fixed)
Low	n_{31}	n_{32}	n_{33}	$n_{3.} = 20$ (fixed)
	$n_{.1}$	$n_{.2}$	$n_{.3}$	n

three independent random samples. One sample is selected from the population of trees exposed to a high SO_2 concentration, one is selected from the population of trees exposed to a normal SO_2 level, and the third is selected from the population of trees with a low SO_2 exposure.

To express the null hypothesis of "no association" when one set of marginal totals is fixed in an $r \times c$ table, let us assume that the row totals are fixed. Consider the probabilities shown in Table 12.11. Note that p_{ij} denotes the proportion of objects in the ith level relative to variable A that are in the jth level relative to variable B. The null hypothesis of no association essentially states that within each column, no row classification is more prevalent than any other. The alternative is that for some columns this is not the case. Statistically, this null hypothesis takes the form

$$H_0\text{: } p_{1j} = p_{2j} = p_{3j} = \cdots = p_{rj} \qquad j = 1, 2, 3, \ldots, c$$

We can think of this null hypothesis as testing to see whether or not r multinomial populations are identical. For instance, in our last example we are testing the null hypothesis of no association between chloroplast level and level of exposure to SO_2. We want to see whether or not the proportions of trees falling into each chloroplast level are identical regardless of the level of SO_2 to which the trees are exposed. The null hypothesis of homogeneity is tested in exactly the same way as the null hypothesis of independence.

Example 12.4.4. When the study described in Example 12.4.3 is conducted, the data shown in Table 12.12 are obtained. Again, expected cell frequencies are given in parentheses. The observed value of the $X^2_{(r-1)(c-1)} = X^2_4$ statistic is

$$\frac{(3-5)^2}{5} + \frac{(4-8.33)^2}{8.33} + \cdots + \frac{(2-6.67)^2}{6.67} = 14.74$$

Table 12.11

Variable A	Variable B					
	1	2	3	$\cdots$	c	
1	p_{11}	p_{12}	p_{13}	$\cdots$	p_{1c}	1
2	p_{21}	p_{22}	p_{23}	$\cdots$	p_{2c}	1
3	p_{31}	p_{32}	p_{33}	$\cdots$	p_{3c}	1
$\vdots$	$\vdots$	$\vdots$	$\vdots$		$\vdots$	$\vdots$
r	p_{r1}	p_{r2}	p_{r3}	$\cdots$	p_{rc}	1

Table 12.12

SO_2 level	Chloroplast level			
	High	Normal	Low	
High	3 (5)	4 (8.33)	13 (6.67)	20
Normal	5 (5)	10 (8.33)	5 (6.67)	20
Low	7 (5)	11 (8.33)	2 (6.67)	20
	15	25	20	60

Since $\chi^2_{.01} = 13.3$ and $\chi^2_{.005} = 14.9$, we can reject H_0 with $.005 < P < .01$. We have strong evidence that there is an association between the SO_2 concentration in the area and the chloroplast level in the leaf cells of the trees. To see the association more clearly, note that if there is no association, the proportion of trees with a low chloroplast count should be the same in each of the three regions. However, it is easy to see that this is not the case. Based on the data of Table 12.12, the estimated proportions of trees with a low chloroplast count are $13/20 = .65$, $5/20 = .25$, and $2/20 = .10$, respectively. These proportions suggest that a high SO_2 level tends to suppress the chloroplast count.

Comparing Proportions with Paired Data: McNemar's Test

Before ending our discussion of categorical data, let us consider one additional type of problem. Note that thus far we have been concerned with the problem of comparing proportions based on *independent* samples drawn from two or more populations. Occasionally there is a need to compare two proportions when the samples drawn are *not* independent. In this case neither the methods presented thus far in this section nor those discussed in Chap. 8 are applicable. However, a method of comparison based on a chi-squared statistic can be used. This technique, called *McNemar's test* is illustrated now.

Example 12.4.5. One problem that concerns the industrial engineer is that of the economical storage of small items that are distributed in less than case lot quantities. Two schemes for storing items are being studied. The first, called *alphameric placement,* stores items in strict alphameric order. The second, the selection density factor (SDF) method, uses a numerical factor computed for each stock item to determine its position relative to the distributor's workstation. We want to see whether the two schemes result in the same proportion of items being placed within 10 feet (ft) of the workstation. To decide, 100 items are classified first by alphameric placement and then via SDF. In this way each item generates a pair of observations. Although we have a sample of 100 observations from each population, the samples are *not* independent. Rather, they are matched. We record the data obtained in the format shown in Table 12.13. Note that if there is a difference in the proportions of objects placed within 10 ft of the workstation by two schemes, then this difference will be reflected in the cells in which the methods disagree on the placement of an item. Thus we are interested only in the starred cells of Table 12.13. Altogether there are 35 observations in these two cells, if the schemes place the same proportion of objects within 10 ft of the workstation, then we expect half or 17.5 of these observations to fall into each of the two cells. Just as before, we now compare the observed cell frequencies to those expected under the assumption that the

Table 12.13

SDF	Alphameric		
	Within 10 ft	10 ft or farther	
Within 10 ft	4	33*	37
10 ft or farther	2*	61	63
	6	94	100

proportions are the same via a chi-squared statistic with 1 degree of freedom. For these data we obtain

$$\frac{(2 - 17.5)^2}{17.5} + \frac{(33 - 17.5)^2}{17.5} = 27.46$$

Since this value is significant even at the $\alpha = .005$ level ($\chi^2_{.005} = 7.88$), we can reject the null hypothesis. We do have evidence that the two proportions are not the same. An inspection of Table 12.13 shows that the SDF method tends to place a higher proportion of objects closer to the workstation than does the alphameric procedure.

In this chapter we have covered some of the most frequently used tests for data of a categorical nature. We should say that there is a large body of literature on categorical data analysis. For further study we refer the reader to [6] and [15].

CHAPTER SUMMARY

In this chapter we discussed problems involving categorical data. These are data characterized by the fact that each observation can be classed as falling into exactly one of several mutually exclusive categories or cells. All the procedures presented involve comparing the actual number of observations in a cell to that expected if a specified null hypothesis were true. We reject H_0 if the differences observed are too large to have occurred by chance. The theory behind the test statistic used is based on the multinomial distribution. For this reason, we began our study of categorical data by considering this important multivariate distribution.

We first learned how to test to see if a data set was drawn from a specified distribution. We next considered experiments in which two random variables are being studied. The purpose of the study is to test these random variables for independence. We examined 2×2 tables in which each variable is studied at two classification levels in some detail. We later extended the ideas presented to include $r \times c$ tables. In testing for independence, we see that the only fixed entry in the contingency table is n, the overall sample size. All other entries are random variables. The null and alternative hypotheses assume these forms:

$$H_0: p_{ij} = p_{i \cdot} p_{\cdot j} \qquad \text{for all } i \text{ and } j$$
$$H_1: p_{ij} \neq p_{i \cdot} p_{\cdot j} \qquad \text{for some } i \text{ and } j$$

The test procedure used does require a fairly large sample. Our guideline for using the procedure requires that no expected cell frequency should be less than 1 and that no more than 20% should be less than 5.

We next considered tests of homogeneity. These are tests that compare two binomial populations via a 2×2 contingency table or r multinomial populations via an $r \times c$ contingency table. The null hypothesis takes this form:

$$H_0: p_{1j} = p_{2j} = \cdots = p_{rj} \qquad j = 1, 2, 3, \ldots, c$$

We saw that in such tables the marginal row totals as well as the overall sample size are fixed by the experimenter. However, despite these differences, the mathematical analysis is mechanically the same as that used in testing for independence.

The last test that we considered, called McNemar's test, is used to test for equality of two population proportions based on paired data.

We also introduced and defined important terms that you should know. These are:

Categorical data   Goodness of fit test
Cell   Marginal row total
Multinomial trial   Marginal column total
Multinomial random variable   McNemar's test

REAL WORLD APPLICATION OF COMPARING PROPORTIONS

Smoking is hazardous to health. It may also be hazardous to fetal health. A study on the effect of smoking on pregnancy was carried out between 1952 and 1958. A sample of 5659 obstetrical patients at the Palo Alto Medical Clinic was selected. The final outcomes of pregnancies, which were full term, premature, and abortion, were noted for the smokers and nonsmokers. The following table shows the contingency table for premature delivery.

	Smokers	Nonsmokers	
Premature	88	66	154
Abortion or full term	2542	2963	5505
	2630	3029	5659

The observed value of the $X^2_{(2-1)(2-1)} = X^2_1$ statistic is 7.68. The probability of observing a value of 7.68 is between 0.005 and 0.01. Since the P value is very small, the null hypothesis of the same proportion was rejected. It was concluded that maternal smoking is associated with premature delivery.

EXERCISES

Section 12.1

1. A study is run to determine whether the general public favors the construction of a dam for the generation of electricity and flood control. It is thought that 40% favor dam construction, 30% are neutral, 20% oppose the dam, and the rest have given the issue no thought. A random sample of 150 individuals in the affected area is selected and interviewed. If the above figures accurately reflect public opinion, how many individuals are expected in each category? If in the sample 42 are in favor, 61 are neutral, 33 are opposed, and the rest have given the issue no thought, do you think, on an intuitive basis, that the proposed percentages are correct or incorrect? Explain.

2. It is assumed that the labor pool for a particular industry consists of 40% white males, 30% white females, 5% black females, 15% black males, and 10% others. Ideally the work force should reflect these percentages. To see if this is the case, a random sample of 200 workers is selected and each worker is placed into exactly one of the above categories. If the work force reflects the labor pool, how many workers are expected in each category? When the sampling is complete, it is observed that there are 95 white males, 50 white females, 2 black females, 20 black males, and 33 others employed. Do these data lead you to suspect that the work force does not reflect the percentages in the labor pool very well? Explain.

3. A random digit generator should produce the digits 0 to 9 inclusive with equal probability. If such a generator is activated 100 times, how many of each digit is expected? If we observe 10 zeros, 8 ones, 9 twos, 11 threes, 12 fours, 7 fives, 10 sixes, 13 sevens, 9 eights, and the rest nines, do you think that there is reason to suspect that the generator does not produce the digits with equal frequency in the long run?

Section 12.2

4. Use the data of Exercise 1 to test

$$H_0: p_1 = .4, \qquad p_2 = .3, \qquad p_3 = .2, \qquad p_4 = .1$$

Is there evidence to support the contention that the stated probabilities are incorrect? Explain based on the P value of the test.

5. Use the data of Exercise 2 to test the null hypothesis that the work force reflects the percentages in the labor pool.

6. Use the data of Exercise 3 to test the null hypothesis that the random digit generator produces the digits 0 to 9 inclusive with equal frequency.

7. Select a random sample of 50 one-digit random numbers from Table III of App. A. Test the null hypothesis that the digits 0 to 9 inclusive occur with equal frequency in this table.

Section 12.3

8. A study is conducted to see if there is an association between age and willingness to use computerized banking systems. The data shown in Table 12.14 are obtained in a survey of 500 randomly selected customers of a bank that has been offering computerized banking for over a year. Is there evidence of an association between these two variables? Explain, based on the P value of your test and inspection of the table.

9. It is suspected that the tendency of an automobile to catch fire in a rear-end collision is not independent of the make of the car. To support this contention, a random sample of 200 cars involved in rear-end collisions is selected from past records. Each car is classified as to make and whether or not it is one of the cars suspected of being especially susceptible to fire under these circumstances. The data gathered is shown in Table 12.15. Is there evidence of an association between this make of car and the presence of fire when involved in a rear-end collision? Explain.

10. A study is conducted to test for independence between air quality and air temperature. These data are obtained from records on 200 randomly selected days over the last few years. (See Table 12.16.) Do these data indicate an association between these variables? Explain, based on the P value of the test.

Table 12.14

| Age | Use computerized banking | |
	Yes	No
Under 40	150	75
40 or over	150	125
		500

Table 12.15

	Suspect make	
Fire	**Yes**	**No**
Yes	9	31
No	16	144
		200

Table 12.16

	Air quality		
Temperature	**Poor**	**Fair**	**Good**
Below average	1	3	24
Average	12	28	76
Above average	12	14	30
			200

11. In a study of the association between color and the effectiveness of a graphical display 100 graphs are randomly selected from among current scientific journals. Each is classified as to whether or not color is used. Each is also rated as to its effectiveness in making its point. Resulting data are given in Table 12.17. Is there evidence that the effectiveness of a graphical display is not independent of color? Explain, based on the P value of the test.

12. It is suspected that there is an association between the day of the week on which an item is produced and the quality of the item. To support this contention, a random sample of 500 items is selected from stock and each item is classified as to the day on which it was produced via its lot number. The item is also rated for quality. The data gathered are shown in Table 12.18.

(a) Our guideline on expected cell frequencies states that no more than 20% can be less than 5 and none can be less than 1. Is this criterion satisfied in this case?

(b) To satisfy the criterion, combine the quality of categories "Fair" and "Poor" to form a new table with three rows and five columns. Use this table to test for independence.

(c) Has an association between quality and day of production been established? Explain.

Table 12.17

	Color present	
Effective	**Yes**	**No**
Excellent	7	4
Good	10	19
Fair	9	26
Poor	4	21
		100

Table 12.18

Quality	Day produced				
	M	T	W	Th	F
Excellent	44	74	79	72	31
Good	14	25	27	24	10
Fair	15	20	20	23	9
Poor	3	5	5	0	0
					500

Section 12.4

13. A study is conducted to assess the effectiveness of a new computerized system of filling orders in a particular industry. Random samples of 100 customers served via the old system and 100 served via the new system are selected. Each customer is contacted to determine whether or not the order was filled satisfactorily within 2 weeks. Table 12.19 gives the results of the study. Test the null hypothesis that the proportion of satisfied customers among those served by the new system is the same as that among those served by the old system at the $\alpha = .05$ level.

14. Although many jobs in the airline industry entail stress, it is thought that air traffic controllers are particularly susceptible to stress-related disorders such as heart problems, high blood pressure, and ulcers. To support this contention, a random sample of 500 air traffic controllers is selected and surveyed. For comparative purposes a sample of 700 workers from other areas of the airline industry is also selected and surveyed. The data obtained is presented in Table 12.20. Test the null hypothesis that the proportion of air traffic controllers with stress-related disorders is the same as that of other workers in the airline industry. Explain your results in a practical sense based on the P value of the test and inspection of Table 12.20.

Table 12.19

	Satisfied		
	Yes	No	
New	82	18	100
Old	70	30	100
			200

Table 12.20

	Stress-related disorder present		
	Yes	No	
Controllers	115	385	500
Others	125	575	700
			1200

Table 12.21

Method	Quality				
	Excellent	**Good**	**Fair**	**Poor**	
High pressure (old)	113	34	21	32	200
Reactive ion (old)	117	31	25	27	200
Magnetron (new)	130	40	20	10	200
					600

15. A new method for etching semiconductors is being studied. The quality of the etch is to be compared to that obtained using two older techniques. The results of the study are given in Table 12.21. State the null hypothesis of homogeneity mathematically. Test this hypothesis at the $\alpha = .05$ level. Interpret your result in a practical sense.

16. A study of the salary gains by workers in research, development, and quality control is conducted. The data in Table 12.22 gives a breakdown of the percentage salary increases over last year of men and women working in these areas. The study is based on a sample of 300 men and 150 women randomly selected from among these workers. Raises were classified according to their integer value. For example, a raise of 5.75% is classified in the category 2–5%. Do these data tend to support the claim that there is an association between the percentage increase in the salary of the worker and the worker's sex? Explain, based on the P value of your test. Interpret your result in a practical sense by inspecting the data of Table 12.22.

17. A recent study claims that an increasing proportion of engineering firms are purchasing liability insurance. This claim is based on a survey of 753 engineering firms. The status of each firm is recorded for the current and for the previous year. The data on which the claim is based are shown in Table 12.23. Do the data support the claim? Explain, based on the P value of McNemar's test.

Table 12.22

	% increase					
	< 2%	**2–5%**	**6–9%**	**10–13%**	**> 14%**	
Male	50	47	103	76	24	300
Female	21	27	50	35	17	150
						450

Table 12.23

Last year	This year		
	Insured	**Uninsured**	
Insured	650	5	655
Uninsured	28	70	98
	678	75	753

Table 12.24

	Rate under 60	
Rate 60 or over	**Command accepted**	**Command rejected**
Command accepted	14	1
Command rejected	28	7
		50

18. A study is conducted of the association between the rate at which words are spoken and the ability of a "talking computer" to recognize commands that it is programmed to accept. A random sample of 50 commands is spoken at a rate under 60 words per minute and the response of the computer is noted. The same commands are repeated at a rate of 60 words per minute or faster and the response is again noted. The data gathered are shown in Table 12.24. Is there a difference in the proportion of commands accepted at the two speaking speeds? Explain, based on the P value of the McNemar test.

19. A study of packaging of "over-the-counter" drugs is conducted. The purpose of the study is to determine whether or not the proportion of drugs in tamper-resistant packages is the same this year as last. A sample of 100 products is selected, and the manner of packaging for each year is determined. Table 12.25 gives the results of the study. Is there evidence that the proportions differ? Explain.

Table 12.25

	Current year, tamper-resistant	
Previous year, tamper-resistant	**Yes**	**No**
Yes	30	3
No	52	15
		100

20. Show that when testing for homogeneity in a 2×2 table, $\hat{E}_{12} = (n_1 . n_{.2})/n$ and $\hat{E}_{22} = (n_2 . n_{.2})/n$. *Hint:* $\hat{E}_{12} = n_1 . - \hat{E}_{11}$.

21. Rework Exercise 13 using the pooled Z test statistic given in Chap. 8. Show that the square of this Z statistic is identical to the observed chi-squared value obtained in Exercise 13.

22. Rework Exercise 14 using the pooled Z statistic given in Chap. 8. Show that the square of this Z statistic is identical to the observed chi-squared value obtained in Exercise 14.

REVIEW EXERCISES

23. The industrial robot is a programmable mechanism designed to do work in a limited space. Spray-painting robots are used in the automobile industry. Their main advantage is that they can work in areas with ventilation levels that would

be unhealthy for human workers. Robots are highly efficient but not infallible, and they, like humans, occasionally produce a paint job with heavy edges or thin spots. In a study of these robots 50 car hoods painted by a robot are randomly selected and classified as to whether or not the paint job is flawed. A second sample of 50 hoods painted by a skilled painter is also studied. The resulting data are given in Table 12.26. Do these data support the contention that robots produce a smaller proportion of flawed hoods than do humans? Explain, based on the P value of the appropriate test.

Table 12.26

	Flawed		
	Yes	No	
Robot	2	48	50
Human	4	46	50
	6	94	100

24. Anaerobic bacteria are microbes that cannot grow in the presence of oxygen. These organisms are now recognized as a major causative agent in infectious disease. In the past their presence was often overlooked in the clinical laboratory. New methods for detection have been devised recently. A study of 826 specimens yields the data of Table 12.27. Do these data indicate that the presence of anaerobic bacteria is not independent of the presence of aerobic bacteria? Explain, based on the P value of the appropriate test.

25. Scientists have suggested that animals use the earth's magnetic field as a clue to their orientation. An experiment to investigate this theory is conducted by using homing pigeons. A pair of coils is placed around each pigeon and a magnetic field that reverses the earth's field is applied. This could disorient the bird. Each day for 118 consecutive days a single bird is released. The bird's orientation and the type of day is noted. Do the data of Table 12.28 indicate

Table 12.27

		Anaerobic bacteria present		
		Yes	No	
Aerobic bacteria present	Yes	322	286	(Random)
	No	81	137	(Random)
		(Random)	(Random)	826

Table 12.28

	Sunny		
Orient home	Yes	No	
Yes	79	5	
No	16	18	
			118

that the bird's orientation is not independent of the cloud cover? Explain, based on the P value of the appropriate test.

26. Two types of coatings are being compared for use as a rust preventive. Fifty pieces of pipe, each of the same type and size, are used in the equipment. Half of each pipe is coated with a .5-mil layer of compound A; the other half receives a .5-mil layer of compound B. Each pipe is then subjected to 1000 hours of salt fog. At the end of the experiment an impartial judge compares the two compounds for effectiveness in preventing rust. The data gathered are shown in Table 12.29. Is there a difference in the proportion of pipes that are deemed effective for the two compounds? Explain, based on the P value of McNemar's test.

Table 12.29

	A effective	
B effective	Yes	No
Yes	35	5
No	8	2
		50

27. A study of a computer-based haul truck dispatching system for open-pit mines is conducted. A simulation of truck availability is included in the study. Simulation is done in such a way that each truck should be deemed fully operable 50% of the time, partially operable 25% of the time, and inoperable the rest of the time.

   (a) The condition of a particular truck is simulated 350 times. How many of these simulations are expected to result in the truck being classified as fully operable? Partially operable? Inoperable?

   (b) When the simulation is complete, it is found that 168 trucks are classified as fully operable, 94 are classified as partially operable, and the rest are classified as inoperable. Do these data lead you to suspect that the simulator is not functioning properly? Explain.

28. In Bangladesh some flooding occurs yearly, and abnormally high floods occur every 2 or 3 years. Poldering is a flood control technique by which the flood level is regulated inside an enclosed area by means of drainage regulators such as sluice gates and pumping stations. A study of the perceived effectiveness of one such polder is conducted. Samples were taken of 259 longtime residents of the area within the polder and 121 residents of the region outside the polder after a recent abnormal flood. Each person was asked to estimate the depth of the water in and around his or her residence. Data obtained are shown in Table 12.30. (Based on a discussion found in "Poldering vs. Compartmentalization:

Table 12.30

	Water depth (d) (in meters)			
	Very shallow ($d < .5$)	Shallow ($.5 \leq d < 1$)	Moderate ($1 \leq d < 1.5$)	Deep $d \geq 1.5$
Inside polder	25	145	81	8
Outside polder	0	13	43	65

The Choice of Flood Control in Bangladesh," Harun Rasid and Azim Mallik, *Environmental Management,* vol. 17, no. 1, January 1993, pp. 59–71.)

(*a*) From the description of the sampling scheme, is the test for association a test of independence or of homogeneity?

(*b*) Has an association been detected between location and perceived severity of the flood in question? Explain, based on your statistical test.

References

1. Amin, R. W., M. R. Reynolds, Jr., and J. C. Arnold: "Cusum Charts with Variable Sampling Intervals," *Technometrics* Vol. 32, No. 4, pp. 371–384, 1990.
2. Beyer, William, ed.: *Handbook of Tables for Probability and Statistics,* 2d ed., CRC Press, Boca Raton, Fla., 1968.
3. Bishop, Y., S. Feinberg, and P. Holland: *Discrete Multivariate Analysis: Theory and Practice,* MIT Press, Cambridge, Mass., 1975.
4. Bowker, A., and G. Lieberman: *Engineering Statistics,* 2d ed., Prentice-Hall, Englewood Cliffs, N.J., 1972.
5. Box, G. E. P.: "Signal-to-Noise Ratios, Performance Criteria, and Transformation," *Technometrics,* Vol. 30, pp. 1–17, 1988.
6. Bradley, J. V.: *Distribution Free Statistical Tests,* Prentice-Hall, Englewood Cliffs, N.J., 1968.
7. Burr, I. W.: *Engineering Statistics and Quality Control,* McGraw-Hill, New York, 1953.
8. Chengalur, I. N., J. C. Arnold, and M. R. Reynolds, Jr.: "Variable Sampling Intervals for Multiparameter Shewhart Charts," *Commun. Statist.—Theory Meth.,* Vol. 18, No. 5, pp. 1769–1792, 1989.
9. Conover, W. J.: *Practical Nonparametric Statistics,* Wiley, New York, 1971.
10. Daniel, W.: *Applied Nonparametric Statistics,* PWS-Kent, Boston, Mass., 1990.
11. Dodge, H., and H. Romig: *Sampling Inspection Tables: Single and Double Sampling,* 2d ed., Wiley, New York, 1959.
12. Downing, George C., and W. E. Chapman: "Smoking and Pregnancy—A Statistical Study of 5,659 Patients," *Calif Med.,* Vol. 104(3), pp. 187, 1966 March.
13. Draper, N., and H. Smith: *Applied Regression Analysis,* 2d ed., Wiley, New York, 1981.
14. Duncan, A. J.: *Quality Control and Industrial Statistics,* 4th ed., Irwin, Homewood, Ill., 1974.
15. Dunn, O., and V. Clark: *Applied Statistics: Analysis of Variance and Regression,* Wiley, New York, 1974.
16. Fisher, R., and F. Yates: *Statistical Tables for Biological Agriculture and Medical Research,* 6th ed., Oliver and Boyd Ltd., London, 1963.
17. Fleiss, J. L.: *Statistical Methods for Rates and Proportions,* 2d ed., Wiley, New York, 1981.
18. Gibbons, J., and J. Pratt: "*P* Values: Interpretation and Methodology," *Amer. Statist.,* Vol. 29, No. 29, 1, 1975.
19. Grant, E., and R. Leavenworth: *Statistical Quality Control,* 4th ed., McGraw-Hill, New York, 1972.
20. Graybill, F.: *Theory and Application of the Linear Model,* Duxbury Press, North Scituate, Mass., 1976.
21. Gupta, S.: "An Asymptotically Nonparametric Test for Symmetry," *Ann. Math. Statist.,* Vol. 38, pp. 849–866,1967.
22. Hicks, C.: *Fundamental Concepts in the Design of Experiments,* Holt, Rinehart, and Winston, New York, 1965.
23. Hoel, P.: *Introduction to Mathematical Statistics,* 3rd ed., Wiley, New York, 1984.
24. Hollander, M., and D. Wolf: *Nonparametric Statistical Methods,* Wiley, New York, 1973.
25. Hong, K., Z. Li, H-J. Wang, R. Elashoff, and D. Heber: "Analysis of Weight Loss Outcomes Using VLCD in Black and White Overweight and Obese Women with and without Metabolic Syndrome," *International Journal of Obesity*, Vol. 29, pp. 436–442, 2005.

26. Hunter, J. S.: "Statistical Design Applied to Production Design," *J. Qual. Tech.,* Vol. 17, pp. 210–221, 1985.

27. Iman, R.: "Graphs for Use with the Lilliefors Test for Normal and Exponential Distributions," *Amer. Statist.,* Vol. 36, No. 2, 1982.

28. Kackar, R. N.: "Off-line Quality Control, Parameter Design, and the Taguchi Method," *J. Qual. Tech.,* Vol.17, pp. 176–221, 1985.

29. Kempthorne, O.: *Design and Analysis of Experiments,* Wiley, New York, 1952.

30. Kramer, C.: "Extension of Multiple Range Tests to Group Means with Unequal Numbers of Replications," *Biometrics,* Vol. 12, p. 307, 1956.

31. Lehmann, E.: *Nonparametric: Statistical Methods Based on Ranks,* Holden-Day, San Francisco, 1975.

32. Lentner, M., and T. Bishop: *Experimental Design and Analysis,* Valley Book Company, Blacksburg, Va., 1986.

33. Lentner, M., J. Arnold, and K. Hinkelmann: "How to Use the Ratio of MS(Blocks) and MS(Error) Correctly," *Amer. Statist.,* Vol. 43, No. 2, 1989.

34. Leon, R. V., A. C. Shoemaker, and R. N. Kackar: "Performance Measures Independent of Adjustment," *Technometrics,* Vol. 29, pp. 253–286, 1987.

35. *Military Standard 105D, Sampling Procedures and Tables for Inspection by Attributes,* Superintendent of Documents, Government Printing Office, Washington, D.C., 1963.

36. Miller, R.: *Simultaneous Statistical Inference,* McGraw-Hill, New York, 1966.

37. Milton, J., and J. Tsokos: *Statistical Methods in the Biological and Health Sciences,* McGraw-Hill Companies, Inc., 1999.

38. Milton, J., and R. Myers: *A First Course in the Theory of Linear Statistical Models,* McGraw-Hill Companies, Inc., 1999.

39. Milton, J., J. Corbet, and P. McTeer: *Introduction to Statistics,* McGraw-Hill Companies, Inc., 1996.

40. Montgomery, D., and E. Peck: *Linear Regression Analysis,* Wiley, New York, 1982.

41. Montgomery, D. C.: *Statistical Quality Control,* Wiley, New York, 1985.

42. Mood, A., F. Graybill, and D. Boes: *Introduction to the Theory of Statistics,* McGraw-Hill, New York, 1974.

43. Nagai-Manelli, R., A. Lowden, C. R. de Castro Moreno, L. R. Teixeira, A. A. da Luz, M. H. Mussi, A. B. Conceição, and F. M. Fischer: "Sleep Length, Working Hours and Socio-Demographic Variables are Associated with Time Attending Evening Classes Among Working College Students," *Sleep and Biological Rhythms*, Vol. 10, pp. 53–60, 2012.

44. Olmsted, J.: *Real Variables,* Appleton-Century-Crofts, New York, 1956.

45. Reynolds, M. R., Jr., and J. C. Arnold: "Optimal One-Sided Shewhart Control Charts with Variable Sampling Intervals," *Seq. Anal.,* Vol. 8, pp. 51–57, 1989.

46. Reynolds, M. R., Jr., R. W. Amin, J. C. Arnold, and J. A. Nachlas: "$\overline{X}$-Charts with Variable Sampling Intervals," *Technometrics,* Vol. 30, pp. 181–192, 1988.

47. *SAS User's Guide,* SAS Institute Inc., Raleigh, N.C., 1982.

48. Snedecor, G., and W. Cochran: *Statistical Methods,* 7th ed., Iowa State University Press, Ames, 1980.

49. Stephens, M.: *Three Mile Island,* Random House, New York, 1981.

50. Taguchi, G.: *Introduction to Quality Engineering,* Asian Productivity Organization, Tokyo, 1986.

51. Taguchi, G., and Y. Wu: *Introduction to Off-line Quality Control: ehpl,* Central Japan Quality Control Association, Tokyo, 1980.

52. Tsokos, C.: *Probability Distributions: An Introduction to Probability Theory with Applications,* Duxbury Press, Belmont, Calif., 1972.

53. Tukey, J.: *Exploratory Data Analysis,* Addison-Wesley, Reading, Mass., 1977.

54. Wald, A., and J. Wolfowitz: "Sampling Inspection Plans for Continuous Production which Ensure a Prescribed Limit on the Outgoing Quality," *Ann. Math. Stat.,* Vol. 16, p. 30, 1945.

APPENDIX A

Statistical Tables

Table I

Cumulative binomial distribution

$$F_X(t) = P[X \le t] = \sum_{x \le t} \binom{n}{x} p^x (1-p)^{n-x}$$

n	t	0.1	0.2	0.25	0.3	0.4	0.5	0.6	0.7	0.75	0.8	0.9
5	0	0.5905	0.3277	0.2373	0.1681	0.0778	0.0312	0.0102	0.0024	0.0010	0.0003	0.0000
	1	0.9185	0.7373	0.6328	0.5282	0.3370	0.1875	0.0870	0.0308	0.0156	0.0067	0.0005
	2	0.9914	0.9421	0.8965	0.8369	0.6826	0.5000	0.3174	0.1631	0.1035	0.0579	0.0086
	3	0.9995	0.9933	0.9844	0.9692	0.9130	0.8125	0.6630	0.4718	0.3672	0.2627	0.0815
	4	1.0000	0.9997	0.9990	0.9976	0.9898	0.9688	0.9222	0.8319	0.7627	0.6723	0.4095
	5	1.0000	1.0000	1.0000	1.0000	1.0000	1.0000	1.0000	1.0000	1.0000	1.0000	1.0000
6	0	0.5314	0.2621	0.1780	0.1176	0.0467	0.0156	0.0041	0.0007	0.0002	0.0001	0.0000
	1	0.8857	0.6554	0.5339	0.4202	0.2333	0.1094	0.0410	0.0109	0.0046	0.0016	0.0001
	2	0.9841	0.9011	0.8306	0.7443	0.5443	0.3437	0.1792	0.0705	0.0376	0.0170	0.0013
	3	0.9987	0.9830	0.9624	0.9295	0.8208	0.6562	0.4557	0.2557	0.1694	0.0989	0.0159
	4	0.9999	0.9984	0.9954	0.9891	0.9590	0.8906	0.7667	0.5798	0.4661	0.3446	0.1143
	5	1.0000	0.9999	0.9998	0.9993	0.9959	0.9844	0.9533	0.8824	0.8220	0.7379	0.4686
	6	1.0000	1.0000	1.0000	1.0000	1.0000	1.0000	1.0000	1.0000	1.0000	1.0000	1.0000
7	0	0.4783	0.2097	0.1335	0.0824	0.0280	0.0078	0.0016	0.0002	0.0001	0.0000	0.0000
	1	0.8503	0.5767	0.4449	0.3294	0.1586	0.0625	0.0188	0.0038	0.0013	0.0004	0.0000
	2	0.9743	0.8520	0.7564	0.6471	0.4199	0.2266	0.0963	0.0288	0.0129	0.0047	0.0002
	3	0.9973	0.9667	0.9294	0.8740	0.7102	0.5000	0.2898	0.1260	0.0706	0.0333	0.0027
	4	0.9998	0.9953	0.9871	0.9712	0.9037	0.7734	0.5801	0.3529	0.2436	0.1480	0.0257
	5	1.0000	0.9996	0.9987	0.9962	0.9812	0.9375	0.8414	0.6706	0.5551	0.4233	0.1497
	6	1.0000	1.0000	0.9999	0.9998	0.9984	0.9922	0.9720	0.9176	0.8665	0.7903	0.5217
	7	1.0000	1.0000	1.0000	1.0000	1.0000	1.0000	1.0000	1.0000	1.0000	1.0000	1.0000
8	0	0.4305	0.1678	0.1001	0.0576	0.0168	0.0039	0.0007	0.0001	0.0000	0.0000	0.0000
	1	0.8131	0.5033	0.3671	0.2553	0.1064	0.0352	0.0085	0.0013	0.0004	0.0001	0.0000
	2	0.9619	0.7969	0.6785	0.5518	0.3154	0.1445	0.0498	0.0113	0.0042	0.0012	0.0000
	3	0.9950	0.9437	0.8862	0.8059	0.5941	0.3633	0.1737	0.0580	0.0273	0.0104	0.0004
	4	0.9996	0.9896	0.9727	0.9420	0.8263	0.6367	0.4059	0.1941	0.1138	0.0563	0.0050
	5	1.0000	0.9988	0.9958	0.9887	0.9502	0.8555	0.6846	0.4482	0.3215	0.2031	0.0381
	6	1.0000	0.9999	0.9996	0.9987	0.9915	0.9648	0.8936	0.7447	0.6329	0.4967	0.1869
	7	1.0000	1.0000	1.0000	0.9999	0.9993	0.9961	0.9832	0.9424	0.8999	0.8322	0.5695
	8	1.0000	1.0000	1.0000	1.0000	1.0000	1.0000	1.0000	1.0000	1.0000	1.0000	1.0000
9	0	0.3874	0.1342	0.0751	0.0404	0.0101	0.0020	0.0003	0.0000	0.0000	0.0000	0.0000
	1	0.7748	0.4362	0.3003	0.1960	0.0705	0.0195	0.0038	0.0004	0.0001	0.0000	0.0000
	2	0.9470	0.7382	0.6007	0.4628	0.2318	0.0898	0.0250	0.0043	0.0013	0.0003	0.0000
	3	0.9917	0.9144	0.8343	0.7297	0.4826	0.2539	0.0994	0.0253	0.0100	0.0031	0.0001
	4	0.9991	0.9804	0.9511	0.9012	0.7334	0.5000	0.2666	0.0988	0.0489	0.0196	0.0009
	5	0.9999	0.9969	0.9900	0.9747	0.9006	0.7461	0.5174	0.2703	0.1657	0.0856	0.0083
	6	1.0000	0.9997	0.9987	0.9957	0.9750	0.9102	0.7682	0.5372	0.3993	0.2618	0.0530
	7	1.0000	1.0000	0.9999	0.9996	0.9962	0.9805	0.9295	0.8040	0.6997	0.5638	0.2252
	8	1.0000	1.0000	1.0000	1.0000	0.9997	0.9980	0.9899	0.9596	0.9249	0.8658	0.6126
	9	1.0000	1.0000	1.0000	1.0000	1.0000	1.0000	1.0000	1.0000	1.0000	1.0000	1.0000

Table I
Cumulative binomial distribution *(continued)*

							p					
n	t	0.1	0.2	0.25	0.3	0.4	0.5	0.6	0.7	0.75	0.8	0.9
10	0	0.3487	0.1074	0.0563	0.0282	0.0060	0.0010	0.0001	0.0000	0.0000	0.0000	0.0000
	1	0.7361	0.3758	0.2440	0.1493	0.0464	0.0107	0.0017	0.0001	0.0000	0.0000	0.0000
	2	0.9298	0.6778	0.5256	0.3828	0.1673	0.0547	0.0123	0.0016	0.0004	0.0001	0.0000
	3	0.9872	0.8791	0.7759	0.6496	0.3823	0.1719	0.0548	0.0106	0.0035	0.0009	0.0000
	4	0.9984	0.9672	0.9219	0.8497	0.6331	0.3770	0.1662	0.0473	0.0197	0.0064	0.0001
	5	0.9999	0.9936	0.9803	0.9527	0.8338	0.6230	0.3669	0.1503	0.0781	0.0328	0.0016
	6	1.0000	0.9991	0.9965	0.9894	0.9452	0.8281	0.6177	0.3504	0.2241	0.1209	0.0128
	7	1.0000	0.9999	0.9996	0.9984	0.9877	0.9453	0.8327	0.6172	0.4744	0.3222	0.0702
	8	1.0000	1.0000	1.0000	0.9999	0.9983	0.9893	0.9536	0.8507	0.7560	0.6242	0.2639
	9	1.0000	1.0000	1.0000	1.0000	0.9999	0.9990	0.9940	0.9718	0.9437	0.8926	0.6513
	10	1.0000	1.0000	1.0000	1.0000	1.0000	1.0000	1.0000	1.0000	1.0000	1.0000	1.0000
11	0	0.3138	0.0859	0.0422	0.0198	0.0036	0.0005	0.0000	0.0000	0.0000	0.0000	0.0000
	1	0.6974	0.3221	0.1971	0.1130	0.0302	0.0059	0.0007	0.0000	0.0000	0.0000	0.0000
	2	0.9104	0.6174	0.4552	0.3127	0.1189	0.0327	0.0059	0.0006	0.0001	0.0000	0.0000
	3	0.9815	0.8389	0.7133	0.5696	0.2963	0.1133	0.0293	0.0043	0.0012	0.0002	0.0000
	4	0.9972	0.9496	0.8854	0.7897	0.5328	0.2744	0.0994	0.0216	0.0076	0.0020	0.0000
	5	0.9997	0.9883	0.9657	0.9218	0.7535	0.5000	0.2465	0.0782	0.0343	0.0117	0.0003
	6	1.0000	0.9980	0.9924	0.9784	0.9006	0.7256	0.4672	0.2103	0.1146	0.0504	0.0028
	7	1.0000	0.9998	0.9988	0.9957	0.9707	0.8867	0.7037	0.4304	0.2867	0.1611	0.0185
	8	1.0000	1.0000	0.9999	0.9994	0.9941	0.9673	0.8811	0.6873	0.5448	0.3826	0.0896
	9	1.0000	1.0000	1.0000	1.0000	0.9993	0.9941	0.9698	0.8870	0.8029	0.6779	0.3026
	10	1.0000	1.0000	1.0000	1.0000	1.0000	0.9995	0.9964	0.9802	0.9578	0.9141	0.6862
	11	1.0000	1.0000	1.0000	1.0000	1.0000	1.0000	1.0000	1.0000	1.0000	1.0000	1.0000
12	0	0.2824	0.0687	0.0317	0.0138	0.0022	0.0002	0.0000	0.0000	0.0000	0.0000	0.0000
	1	0.6590	0.2749	0.1584	0.0850	0.0196	0.0032	0.0003	0.0000	0.0000	0.0000	0.0000
	2	0.8891	0.5583	0.3907	0.2528	0.0834	0.0193	0.0028	0.0002	0.0000	0.0000	0.0000
	3	0.9744	0.7946	0.6488	0.4925	0.2253	0.0730	0.0153	0.0017	0.0004	0.0001	0.0000
	4	0.9957	0.9274	0.8424	0.7237	0.4382	0.1938	0.0573	0.0095	0.0028	0.0006	0.0000
	5	0.9995	0.9806	0.9456	0.8822	0.6652	0.3872	0.1582	0.0386	0.0143	0.0039	0.0001
	6	0.9999	0.9961	0.9857	0.9614	0.8418	0.6128	0.3348	0.1178	0.0544	0.0194	0.0005
	7	1.0000	0.9994	0.9972	0.9905	0.9427	0.8062	0.5618	0.2763	0.1576	0.0726	0.0043
	8	1.0000	0.9999	0.9996	0.9983	0.9847	0.9270	0.7747	0.5075	0.3512	0.2054	0.0256
	9	1.0000	1.0000	1.0000	0.9998	0.9972	0.9807	0.9166	0.7472	0.6093	0.4417	0.1109
	10	1.0000	1.0000	1.0000	1.0000	0.9997	0.9968	0.9804	0.9150	0.8416	0.7251	0.3410
	11	1.0000	1.0000	1.0000	1.0000	1.0000	0.9998	0.9978	0.9862	0.9683	0.9313	0.7176
	12	1.0000	1.0000	1.0000	1.0000	1.0000	1.0000	1.0000	1.0000	1.0000	1.0000	1.0000
13	0	0.2542	0.0550	0.0238	0.0097	0.0013	0.0001	0.0000	0.0000	0.0000	0.0000	0.0000
	1	0.6213	0.2336	0.1267	0.0637	0.0126	0.0017	0.0001	0.0000	0.0000	0.0000	0.0000
	2	0.8661	0.5017	0.3326	0.2025	0.0579	0.0112	0.0013	0.0001	0.0000	0.0000	0.0000
	3	0.9658	0.7473	0.5843	0.4206	0.1686	0.0461	0.0078	0.0007	0.0001	0.0000	0.0000
	4	0.9935	0.9009	0.7940	0.6543	0.3530	0.1334	0.0321	0.0040	0.0010	0.0002	0.0000
	5	0.9991	0.9700	0.9198	0.8346	0.5744	0.2905	0.0977	0.0182	0.0056	0.0012	0.0000
	6	0.9999	0.9930	0.9757	0.9376	0.7712	0.5000	0.2288	0.0624	0.0243	0.0070	0.0001
	7	1.0000	0.9988	0.9944	0.9818	0.9023	0.7095	0.4256	0.1654	0.0802	0.0300	0.0009
	8	1.0000	0.9998	0.9990	0.9960	0.9679	0.8666	0.6470	0.3457	0.2060	0.0991	0.0065

Table I
Cumulative binomial distribution *(continued)*

							p					
n	*t*	0.1	0.2	0.25	0.3	0.4	0.5	0.6	0.7	0.75	0.8	0.9
	9	1.0000	1.0000	0.9999	0.9993	0.9922	0.9539	0.8314	0.5794	0.4157	0.2527	0.0342
	10	1.0000	1.0000	1.0000	0.9999	0.9987	0.9888	0.9421	0.7975	0.6674	0.4983	0.1339
	11	1.0000	1.0000	1.0000	1.0000	0.9999	0.9983	0.9874	0.9363	0.8733	0.7664	0.3787
	12	1.0000	1.0000	1.0000	1.0000	1.0000	0.9999	0.9987	0.9903	0.9762	0.9540	0.7458
	13	1.0000	1.0000	1.0000	1.0000	1.0000	1.0000	1.0000	1.0000	1.0000	1.0000	1.0000
14	0	0.2288	0.0440	0.0178	0.0068	0.0008	0.0001	0.0000	0.0000	0.0000	0.0000	0.0000
	1	0.5846	0.1979	0.1010	0.0475	0.0081	0.0009	0.0001	0.0000	0.0000	0.0000	0.0000
	2	0.8416	0.4481	0.2811	0.1608	0.0398	0.0065	0.0006	0.0000	0.0000	0.0000	0.0000
	3	0.9559	0.6982	0.5213	0.3552	0.1243	0.0287	0.0039	0.0002	0.0000	0.0000	0.0000
	4	0.9908	0.8702	0.7415	0.5842	0.2793	0.0898	0.0175	0.0017	0.0003	0.0000	0.0000
	5	0.9985	0.9561	0.8883	0.7805	0.4859	0.2120	0.0583	0.0083	0.0022	0.0004	0.0000
	6	0.9998	0.9884	0.9617	0.9067	0.6925	0.3953	0.1501	0.0315	0.0103	0.0024	0.0000
	7	1.0000	0.9976	0.9897	0.9685	0.8499	0.6047	0.3075	0.0933	0.0383	0.0116	0.0002
	8	1.0000	0.9996	0.9978	0.9917	0.9417	0.7880	0.5141	0.2195	0.1117	0.0439	0.0015
	9	1.0000	1.0000	0.9997	0.9983	0.9825	0.9102	0.7207	0.4158	0.2585	0.1298	0.0092
	10	1.0000	1.0000	1.0000	0.9998	0.9961	0.9713	0.8757	0.6448	0.4787	0.3018	0.0441
	11	1.0000	1.0000	1.0000	1.0000	0.9994	0.9935	0.9602	0.8392	0.7189	0.5519	0.1584
	12	1.0000	1.0000	1.0000	1.0000	0.9999	0.9991	0.9919	0.9525	0.8990	0.8021	0.4154
	13	1.0000	1.0000	1.0000	1.0000	1.0000	0.9999	0.9992	0.9932	0.9822	0.9560	0.7712
	14	1.0000	1.0000	1.0000	1.0000	1.0000	1.0000	1.0000	1.0000	1.0000	1.0000	1.0000
15	0	0.2059	0.0352	0.0134	0.0047	0.0005	0.0000	0.0000	0.0000	0.0000	0.0000	0.0000
	1	0.5490	0.1671	0.0802	0.0353	0.0052	0.0005	0.0000	0.0000	0.0000	0.0000	0.0000
	2	0.8159	0.3980	0.2361	0.1268	0.0271	0.0037	0.0003	0.0000	0.0000	0.0000	0.0000
	3	0.9444	0.6482	0.4613	0.2969	0.0905	0.0176	0.0019	0.0001	0.0000	0.0000	0.0000
	4	0.9873	0.8358	0.6865	0.5155	0.2173	0.0592	0.0093	0.0007	0.0001	0.0000	0.0000
	5	0.9978	0.9389	0.8516	0.7216	0.4032	0.1509	0.0338	0.0037	0.0008	0.0001	0.0000
	6	0.9997	0.9819	0.9434	0.8689	0.6098	0.3036	0.0950	0.0152	0.0042	0.0008	0.0000
	7	1.0000	0.9958	0.9827	0.9500	0.7869	0.5000	0.2131	0.0500	0.0173	0.0042	0.0000
	8	1.0000	0.9992	0.9958	0.9848	0.9050	0.6964	0.3902	0.1311	0.0566	0.0181	0.0003
	9	1.0000	0.9999	0.9992	0.9963	0.9662	0.8491	0.5968	0.2784	0.1484	0.0611	0.0022
	10	1.0000	1.0000	0.9999	0.9993	0.9907	0.9408	0.7827	0.4845	0.3135	0.1642	0.0127
	11	1.0000	1.0000	1.0000	0.9999	0.9981	0.9824	0.9095	0.7031	0.5387	0.3518	0.0556
	12	1.0000	1.0000	1.0000	1.0000	0.9997	0.9963	0.9729	0.8732	0.7639	0.6020	0.1841
	13	1.0000	1.0000	1.0000	1.0000	1.0000	0.9995	0.9948	0.9647	0.9198	0.8329	0.4510
	14	1.0000	1.0000	1.0000	1.0000	1.0000	1.0000	0.9995	0.9953	0.9866	0.9648	0.7941
	15	1.0000	1.0000	1.0000	1.0000	1.0000	1.0000	1.0000	1.0000	1.0000	1.0000	1.0000
16	0	0.1853	0.0281	0.0100	0.0033	0.0003	0.0000	0.0000	0.0000	0.0000	0.0000	0.0000
	1	0.5147	0.1407	0.0635	0.0261	0.0033	0.0003	0.0000	0.0000	0.0000	0.0000	0.0000
	2	0.7892	0.3518	0.1971	0.0994	0.0183	0.0021	0.0001	0.0000	0.0000	0.0000	0.0000
	3	0.9316	0.5981	0.4050	0.2459	0.0651	0.0106	0.0009	0.0000	0.0000	0.0000	0.0000
	4	0.9830	0.7982	0.6302	0.4499	0.1666	0.0384	0.0049	0.0003	0.0000	0.0000	0.0000
	5	0.9967	0.9183	0.8103	0.6598	0.3288	0.1051	0.0191	0.0016	0.0003	0.0000	0.0000

Table I
Cumulative binomial distribution *(continued)*

							p					
n	t	0.1	0.2	0.25	0.3	0.4	0.5	0.6	0.7	0.75	0.8	0.9
	6	0.9995	0.9733	0.9204	0.8247	0.5272	0.2272	0.0583	0.0071	0.0016	0.0002	0.0000
	7	0.9999	0.9930	0.9729	0.9256	0.7161	0.4018	0.1423	0.0257	0.0075	0.0015	0.0000
	8	1.0000	0.9985	0.9925	0.9743	0.8577	0.5982	0.2839	0.0744	0.0271	0.0070	0.0001
	9	1.0000	0.9998	0.9984	0.9929	0.9417	0.7728	0.4728	0.1753	0.0796	0.0267	0.0005
	10	1.0000	1.0000	0.9997	0.9984	0.9809	0.8949	0.6712	0.3402	0.1897	0.0817	0.0033
	11	1.0000	1.0000	1.0000	0.9997	0.9951	0.9616	0.8334	0.5501	0.3698	0.2018	0.0170
	12	1.0000	1.0000	1.0000	1.0000	0.9991	0.9894	0.9349	0.7541	0.5950	0.4019	0.0684
	13	1.0000	1.0000	1.0000	1.0000	0.9999	0.9979	0.9817	0.9006	0.8029	0.6482	0.2108
	14	1.0000	1.0000	1.0000	1.0000	1.0000	0.9997	0.9967	0.9739	0.9365	0.8593	0.4853
	15	1.0000	1.0000	1.0000	1.0000	1.0000	1.0000	0.9997	0.9967	0.9900	0.9719	0.8147
	16	1.0000	1.0000	1.0000	1.0000	1.0000	1.0000	1.0000	1.0000	1.0000	1.0000	1.0000
17	0	0.1668	0.0225	0.0075	0.0023	0.0002	0.0000	0.0000	0.0000	0.0000	0.0000	0.0000
	1	0.4818	0.1182	0.0501	0.0193	0.0021	0.0001	0.0000	0.0000	0.0000	0.0000	0.0000
	2	0.7618	0.3096	0.1637	0.0774	0.0123	0.0012	0.0001	0.0000	0.0000	0.0000	0.0000
	3	0.9174	0.5489	0.3530	0.2019	0.0464	0.0064	0.0005	0.0000	0.0000	0.0000	0.0000
	4	0.9779	0.7582	0.5739	0.3887	0.1260	0.0245	0.0025	0.0001	0.0000	0.0000	0.0000
	5	0.9953	0.8943	0.7653	0.5968	0.2639	0.0717	0.0106	0.0007	0.0001	0.0000	0.0000
	6	0.9992	0.9623	0.8929	0.7752	0.4478	0.1662	0.0348	0.0032	0.0006	0.0001	0.0000
	7	0.9999	0.9891	0.9598	0.8954	0.6405	0.3145	0.0919	0.0127	0.0031	0.0005	0.0000
	8	1.0000	0.9974	0.9876	0.9597	0.8011	0.5000	0.1989	0.0403	0.0124	0.0026	0.0000
	9	1.0000	0.9995	0.9969	0.9873	0.9081	0.6855	0.3595	0.1046	0.0402	0.0109	0.0001
	10	1.0000	0.9999	0.9994	0.9968	0.9652	0.8338	0.5522	0.2248	0.1071	0.0377	0.0008
	11	1.0000	1.0000	0.9999	0.9993	0.9894	0.9283	0.7361	0.4032	0.2347	0.1057	0.0047
	12	1.0000	1.0000	1.0000	0.9999	0.9975	0.9755	0.8740	0.6113	0.4261	0.2418	0.0221
	13	1.0000	1.0000	1.0000	1.0000	0.9995	0.9936	0.9536	0.7981	0.6470	0.4511	0.0826
	14	1.0000	1.0000	1.0000	1.0000	0.9999	0.9988	0.9877	0.9226	0.8363	0.6904	0.2382
	15	1.0000	1.0000	1.0000	1.0000	1.0000	0.9999	0.9979	0.9807	0.9499	0.8818	0.5182
	16	1.0000	1.0000	1.0000	1.0000	1.0000	1.0000	0.9998	0.9977	0.9925	0.9775	0.8332
	17	1.0000	1.0000	1.0000	1.0000	1.0000	1.0000	1.0000	1.0000	1.0000	1.0000	1.0000
18	0	0.1501	0.0180	0.0056	0.0016	0.0001	0.0000	0.0000	0.0000	0.0000	0.0000	0.0000
	1	0.4503	0.0991	0.0395	0.0142	0.0013	0.0001	0.0000	0.0000	0.0000	0.0000	0.0000
	2	0.7338	0.2713	0.1353	0.0600	0.0082	0.0007	0.0000	0.0000	0.0000	0.0000	0.0000
	3	0.9018	0.5010	0.3057	0.1646	0.0328	0.0038	0.0002	0.0000	0.0000	0.0000	0.0000
	4	0.9718	0.7164	0.5187	0.3327	0.0942	0.0154	0.0013	0.0000	0.0000	0.0000	0.0000
	5	0.9936	0.8671	0.7175	0.5344	0.2088	0.0481	0.0058	0.0003	0.0000	0.0000	0.0000
	6	0.9988	0.9487	0.8610	0.7217	0.3743	0.1189	0.0203	0.0014	0.0002	0.0000	0.0000
	7	0.9998	0.9837	0.9431	0.8593	0.5634	0.2403	0.0576	0.0061	0.0012	0.0002	0.0000
	8	1.0000	0.9957	0.9807	0.9404	0.7368	0.4073	0.1347	0.0210	0.0054	0.0009	0.0000
	9	1.0000	0.9991	0.9946	0.9790	0.8653	0.5927	0.2632	0.0596	0.0193	0.0043	0.0000
	10	1.0000	0.9998	0.9988	0.9939	0.9424	0.7597	0.4366	0.1407	0.0569	0.0163	0.0002
	11	1.0000	1.0000	0.9998	0.9986	0.9797	0.8811	0.6257	0.2783	0.1390	0.0513	0.0012
	12	1.0000	1.0000	1.0000	0.9997	0.9942	0.9519	0.7912	0.4656	0.2825	0.1329	0.0064
	13	1.0000	1.0000	1.0000	1.0000	0.9987	0.9846	0.9058	0.6673	0.4813	0.2836	0.0282
	14	1.0000	1.0000	1.0000	1.0000	0.9998	0.9962	0.9672	0.8354	0.6943	0.4990	0.0982
	15	1.0000	1.0000	1.0000	1.0000	1.0000	0.9993	0.9918	0.9400	0.8647	0.7287	0.2662
	16	1.0000	1.0000	1.0000	1.0000	1.0000	0.9999	0.9987	0.9858	0.9605	0.9009	0.5497
	17	1.0000	1.0000	1.0000	1.0000	1.0000	1.0000	0.9999	0.9984	0.9944	0.9820	0.8499
	18	1.0000	1.0000	1.0000	1.0000	1.0000	1.0000	1.0000	1.0000	1.0000	1.0000	1.0000

Table I
Cumulative binomial distribution *(concluded)*

n	t	0.1	0.2	0.25	0.3	0.4	0.5	0.6	0.7	0.75	0.8	0.9
19	0	0.1351	0.0144	0.0042	0.0011	0.0001	0.0000	0.0000	0.0000	0.0000	0.0000	0.0000
	1	0.4203	0.0829	0.0310	0.0104	0.0008	0.0000	0.0000	0.0000	0.0000	0.0000	0.0000
	2	0.7054	0.2369	0.1113	0.0462	0.0055	0.0004	0.0000	0.0000	0.0000	0.0000	0.0000
	3	0.8850	0.4551	0.2631	0.1332	0.0230	0.0022	0.0001	0.0000	0.0000	0.0000	0.0000
	4	0.9648	0.6733	0.4654	0.2822	0.0696	0.0096	0.0006	0.0000	0.0000	0.0000	0.0000
	5	0.9914	0.8369	0.6678	0.4739	0.1629	0.0318	0.0031	0.0001	0.0000	0.0000	0.0000
	6	0.9983	0.9324	0.8251	0.6655	0.3081	0.0835	0.0116	0.0006	0.0001	0.0000	0.0000
	7	0.9997	0.9767	0.9225	0.8180	0.4878	0.1796	0.0352	0.0028	0.0005	0.0000	0.0000
	8	1.0000	0.9933	0.9713	0.9161	0.6675	0.3238	0.0885	0.0105	0.0023	0.0003	0.0000
	9	1.0000	0.9984	0.9911	0.9674	0.8139	0.5000	0.0861	0.0326	0.0089	0.0016	0.0000
	10	1.0000	0.9997	0.9977	0.9895	0.9115	0.6762	0.3325	0.0839	0.0287	0.0067	0.0000
	11	1.0000	1.0000	0.9995	0.9972	0.9648	0.8204	0.5122	0.1820	0.0775	0.0233	0.0003
	12	1.0000	1.0000	0.9999	0.9994	0.9884	0.9165	0.6919	0.3345	0.1749	0.0676	0.0017
	13	1.0000	1.0000	1.0000	0.9999	0.9969	0.9682	0.8371	0.5261	0.3322	0.1631	0.0086
	14	1.0000	1.0000	1.0000	1.0000	0.9994	0.9904	0.9304	0.7178	0.5346	0.3267	0.0352
	15	1.0000	1.0000	1.0000	1.0000	0.9999	0.9978	0.9770	0.8668	0.7369	0.5449	0.1150
	16	1.0000	1.0000	1.0000	1.0000	1.0000	0.9996	0.9945	0.9538	0.8887	0.7631	0.2946
	17	1.0000	1.0000	1.0000	1.0000	1.0000	1.0000	0.9992	0.9896	0.9690	0.9171	0.5797
	18	1.0000	1.0000	1.0000	1.0000	1.0000	1.0000	0.9999	0.9989	0.9958	0.9856	0.8649
	19	1.0000	1.0000	1.0000	1.0000	1.0000	1.0000	1.0000	1.0000	1.0000	1.0000	1.0000
20	0	0.1216	0.0115	0.0032	0.0008	0.0000	0.0000	0.0000	0.0000	0.0000	0.0000	0.0000
	1	0.3917	0.0692	0.0243	0.0076	0.0005	0.0000	0.0000	0.0000	0.0000	0.0000	0.0000
	2	0.6769	0.2061	0.0913	0.0355	0.0036	0.0002	0.0000	0.0000	0.0000	0.0000	0.0000
	3	0.8670	0.4114	0.2252	0.1071	0.0160	0.0013	0.0000	0.0000	0.0000	0.0000	0.0000
	4	0.9568	0.6296	0.4148	0.2375	0.0510	0.0059	0.0003	0.0000	0.0000	0.0000	0.0000
	5	0.9887	0.8042	0.6172	0.4164	0.1256	0.0207	0.0016	0.0000	0.0000	0.0000	0.0000
	6	0.9976	0.9133	0.7858	0.6080	0.2500	0.0577	0.0065	0.0003	0.0000	0.0000	0.0000
	7	0.9996	0.9679	0.8982	0.7723	0.4159	0.1316	0.0210	0.0013	0.0002	0.0000	0.0000
	8	0.9999	0.9900	0.9591	0.8867	0.5956	0.2517	0.0565	0.0051	0.0009	0.0001	0.0000
	9	1.0000	0.9974	0.9861	0.9520	0.7553	0.4119	0.1275	0.0171	0.0039	0.0006	0.0000
	10	1.0000	0.9994	0.9961	0.9829	0.8725	0.5881	0.2447	0.0480	0.0139	0.0026	0.0000
	11	1.0000	0.9999	0.9991	0.9949	0.9435	0.7483	0.4044	0.1133	0.0409	0.0100	0.0001
	12	1.0000	1.0000	0.9998	0.9987	0.9790	0.8684	0.5841	0.2277	0.1018	0.0321	0.0004
	13	1.0000	1.0000	1.0000	0.9997	0.9935	0.9423	0.7500	0.3920	0.2142	0.0867	0.0024
	14	1.0000	1.0000	1.0000	1.0000	0.9984	0.9793	0.8744	0.5836	0.3828	0.1958	0.0113
	15	1.0000	1.0000	1.0000	1.0000	0.9997	0.9941	0.9490	0.7625	0.5852	0.3704	0.0432
	16	1.0000	1.0000	1.0000	1.0000	1.0000	0.9987	0.9840	0.8929	0.7748	0.5886	0.1330
	17	1.0000	1.0000	1.0000	1.0000	1.0000	0.9998	0.9964	0.9645	0.9087	0.7939	0.3231
	18	1.0000	1.0000	1.0000	1.0000	1.0000	1.0000	0.9995	0.9924	0.9757	0.9308	0.6083
	19	1.0000	1.0000	1.0000	1.0000	1.0000	1.0000	1.0000	0.9992	0.9968	0.9885	0.8784
	20	1.0000	1.0000	1.0000	1.0000	1.0000	1.0000	1.0000	1.0000	1.0000	1.0000	1.0000

Table II
Poisson distribution function

$$F_X(t) = P[X \le t] = \sum_{x \le t} e^{-\lambda s}(\lambda s)^x / x!$$

t	0.5	1.0	2.0	3.0	4.0	5.0	6.0	7.0	8.0	9.0	10.0	11.0	12.0	13.0	14.0	15.0
0	0.607	0.368	0.135	0.050	0.018	0.007	0.002	0.001	0.000	0.000	0.000	0.000	0.000	0.000	0.000	0.000
1	0.910	0.736	0.406	0.199	0.092	0.040	0.017	0.007	0.003	0.001	0.000	0.000	0.000	0.000	0.000	0.000
2	0.986	0.920	0.677	0.423	0.238	0.125	0.062	0.030	0.014	0.006	0.003	0.001	0.001	0.000	0.000	0.000
3	0.998	0.981	0.857	0.647	0.433	0.265	0.151	0.082	0.042	0.021	0.010	0.005	0.002	0.001	0.000	0.000
4	1.000	0.996	0.947	0.815	0.629	0.440	0.285	0.173	0.100	0.055	0.029	0.015	0.008	0.004	0.002	0.001
5	1.000	0.999	0.983	0.916	0.785	0.616	0.446	0.301	0.191	0.116	0.067	0.038	0.020	0.011	0.006	0.003
6	1.000	1.000	0.995	0.966	0.889	0.762	0.606	0.450	0.313	0.207	0.130	0.079	0.046	0.026	0.014	0.008
7	1.000	1.000	0.999	0.988	0.949	0.867	0.744	0.599	0.453	0.324	0.220	0.143	0.090	0.054	0.032	0.018
8	1.000	1.000	1.000	0.996	0.979	0.932	0.847	0.729	0.593	0.456	0.333	0.232	0.155	0.100	0.062	0.037
9	1.000	1.000	1.000	0.999	0.992	0.968	0.916	0.830	0.717	0.587	0.458	0.341	0.242	0.166	0.109	0.070
10	1.000	1.000	1.000	1.000	0.997	0.986	0.957	0.901	0.816	0.706	0.583	0.460	0.347	0.252	0.176	0.118
11	1.000	1.000	1.000	1.000	0.999	0.995	0.980	0.947	0.888	0.803	0.697	0.579	0.462	0.353	0.260	0.185
12	1.000	1.000	1.000	1.000	1.000	0.998	0.991	0.973	0.936	0.876	0.792	0.689	0.576	0.463	0.358	0.268
13	1.000	1.000	1.000	1.000	1.000	0.999	0.996	0.987	0.966	0.926	0.864	0.781	0.682	0.573	0.464	0.363
14	1.000	1.000	1.000	1.000	1.000	1.000	0.999	0.994	0.983	0.959	0.917	0.854	0.772	0.675	0.570	0.466
15	1.000	1.000	1.000	1.000	1.000	1.000	0.999	0.998	0.992	0.978	0.951	0.907	0.844	0.764	0.669	0.568
16	1.000	1.000	1.000	1.000	1.000	1.000	1.000	0.999	0.996	0.989	0.973	0.944	0.899	0.835	0.756	0.664
17	1.000	1.000	1.000	1.000	1.000	1.000	1.000	1.000	0.998	0.995	0.986	0.968	0.937	0.890	0.827	0.749
18	1.000	1.000	1.000	1.000	1.000	1.000	1.000	1.000	0.999	0.998	0.993	0.982	0.963	0.930	0.883	0.819
19	1.000	1.000	1.000	1.000	1.000	1.000	1.000	1.000	1.000	0.999	0.997	0.991	0.979	0.957	0.923	0.875
20	1.000	1.000	1.000	1.000	1.000	1.000	1.000	1.000	1.000	1.000	0.998	0.995	0.988	0.975	0.952	0.917
21											0.999	0.998	0.994	0.986	0.971	0.947
22											1.000	0.999	0.997	0.992	0.983	0.967
23											1.000	1.000	0.999	0.996	0.991	0.981
24											1.000	1.000	0.999	0.998	0.995	0.989
25											1.000	1.000	1.000	0.999	0.997	0.994
26											1.000	1.000	1.000	1.000	0.999	0.997
27											1.000	1.000	1.000	1.000	0.999	0.998
28											1.000	1.000	1.000	1.000	1.000	0.999
29											1.000	1.000	1.000	1.000	1.000	1.000

Table III
A table of random digits

Line/Col.	1	2	3	4	5	6	7	8	9	10	11	12	13	14
1	10480	15011	01536	02011	81647	91646	69179	14194	62590	36207	20969	99570	91291	90700
2	22368	46573	25595	85393	30995	89198	27982	53402	93965	34095	52666	19174	39615	99505
3	24130	48360	22527	97265	76393	64809	15179	24830	49340	32081	30680	19655	63348	58629
4	42167	93093	06243	61680	07856	16376	39440	53537	71341	57004	00849	74917	97758	16379
5	37570	39975	81837	16656	06121	91782	60468	81305	49684	60672	14110	06927	01263	54613
6	77921	06907	11008	42751	27756	53498	18602	70659	90655	15053	21916	81825	44394	42880
7	99562	72905	56420	69994	98872	31016	71194	18738	44013	48840	63213	21069	10634	12952
8	96301	91977	05463	07972	18876	20922	94595	56869	69014	60045	18425	84903	42508	32307
9	89579	14342	63661	10281	17453	18103	57740	84378	25331	12566	58678	44947	05585	56941
10	85485	36857	43342	53988	53060	59533	38867	62300	08158	17983	16439	11458	18593	64952
11	28918	69578	88231	33276	70997	79936	56865	05859	90106	31595	01547	85590	91610	78188
12	63553	40961	48235	03427	49626	69445	18663	72695	52180	20847	12234	90511	33703	90322
13	09429	93969	52636	92737	88974	33488	36320	17617	30015	08272	84115	27156	30613	74952
14	10365	61129	87529	85689	48237	52267	67689	92294	01511	26358	85104	20285	29975	89868
15	07119	97336	71048	08178	77233	13916	47564	81056	97735	85977	29372	74461	28551	90707
16	51085	12765	51821	51259	77452	16308	60756	92144	49442	53900	70960	63990	75601	40719
17	02368	21382	52404	60268	89368	19885	55322	44819	01188	65255	64835	44919	05944	55157
18	01011	54092	33362	94904	31273	04146	18594	29852	71585	85030	51132	01915	92747	64951
19	52162	53916	46369	58586	23216	14513	83149	98736	23495	64350	94738	17752	35156	35749
20	07056	97628	33787	09998	42698	06691	76988	13602	51851	46104	88916	19509	25625	58104
21	48663	91245	85828	14346	09172	30168	90229	04734	59193	22178	30421	61666	99904	32812
22	54164	58492	22421	74103	47070	25306	76468	26384	58151	06646	21524	15227	96909	44592
23	32639	32363	05597	24200	13363	38005	94342	28728	35806	06912	17012	64161	18296	22851
24	29334	27001	87637	87308	58731	00256	45834	15398	46557	41135	10367	07684	36188	18510
25	02488	33062	28834	07351	19731	92420	60952	61280	50001	67658	32586	86679	50720	94953

Table III

A table of random digits (*concluded*)

Line/Col.	1	2	3	4	5	6	7	8	9	10	11	12	13	14
26	81525	72295	04839	96423	24878	82651	66566	14778	76797	14780	13300	87074	79666	95725
27	29676	20591	68086	26432	46901	20849	89768	81536	86645	12659	92259	57102	80428	25280
28	00742	57392	39064	66432	84673	40027	32832	61362	98947	96067	64760	64584	96096	98253
29	05366	04213	25669	26422	44407	44048	37937	63904	45766	66134	75470	66520	34693	90449
30	91921	26418	64117	94305	26766	25940	39972	22209	71500	64568	91402	42416	07844	69618
31	00582	04711	87917	77341	42206	35126	74087	99547	81817	42607	43808	76655	62028	76630
32	00725	69884	62797	56170	86324	88072	76222	36086	84637	93161	76038	65855	77919	88006
33	69011	65797	95876	55293	18988	27354	26575	08625	40801	59920	29841	80150	12777	48501
34	25976	57948	29888	88604	67917	48708	18912	82271	65424	69774	33611	54262	85963	03547
35	09763	83473	73577	12908	30883	18317	28290	35797	05998	41688	34952	37888	38917	88050
36	91567	42595	27958	30134	04024	86385	29880	99730	55536	84855	29080	09250	79656	73211
37	17955	56349	90999	49127	20044	59931	06115	20542	18059	02008	73708	83517	36103	42791
38	46503	18584	18845	49618	02304	51038	20655	58727	28168	15475	56942	53389	20562	87338
39	92157	89634	94824	78171	84610	82834	09922	25417	44137	48413	25555	21246	35509	20468
40	14577	62765	35605	81263	39667	47358	56873	56307	61607	49518	89656	20103	77490	18062
41	98427	07523	33362	64270	01638	92477	66969	98420	04880	45585	46565	04102	46880	45709
42	34914	63976	88720	82765	34476	17032	87589	40836	32427	70002	70663	88863	77775	69348
43	70060	28277	39475	46473	23219	53416	94970	25832	69975	94884	19661	72828	00102	66794
44	53976	54914	06990	67245	68350	82948	11398	42878	80287	88267	47363	46634	06541	97809
45	76072	29515	40980	07391	58745	25774	22987	80059	39911	96189	41151	14222	60697	59583
46	90725	52210	83974	29992	65831	38857	50490	83765	55657	14361	31720	57375	56228	41546
47	64364	67412	33339	31926	14883	24413	59744	92351	97473	89286	35931	04110	23726	51900
48	08962	00358	31662	25388	61642	34072	81249	35648	56891	69352	48373	45578	78547	81788
49	95012	68379	93526	70765	10593	04542	76463	54328	02349	17247	28865	14777	62730	92277
50	15664	10493	20492	38391	91132	21999	59516	81652	27195	48223	46751	22923	32261	85653

From W. H. Beyer (ed.), in *CRC Handbook of Tables for Probability and Statistics*, 2d ed., 1968. Copyright CRC Press, Inc., Boca Raton, Fla.

Table IV

Cumulative chi-squared distribution

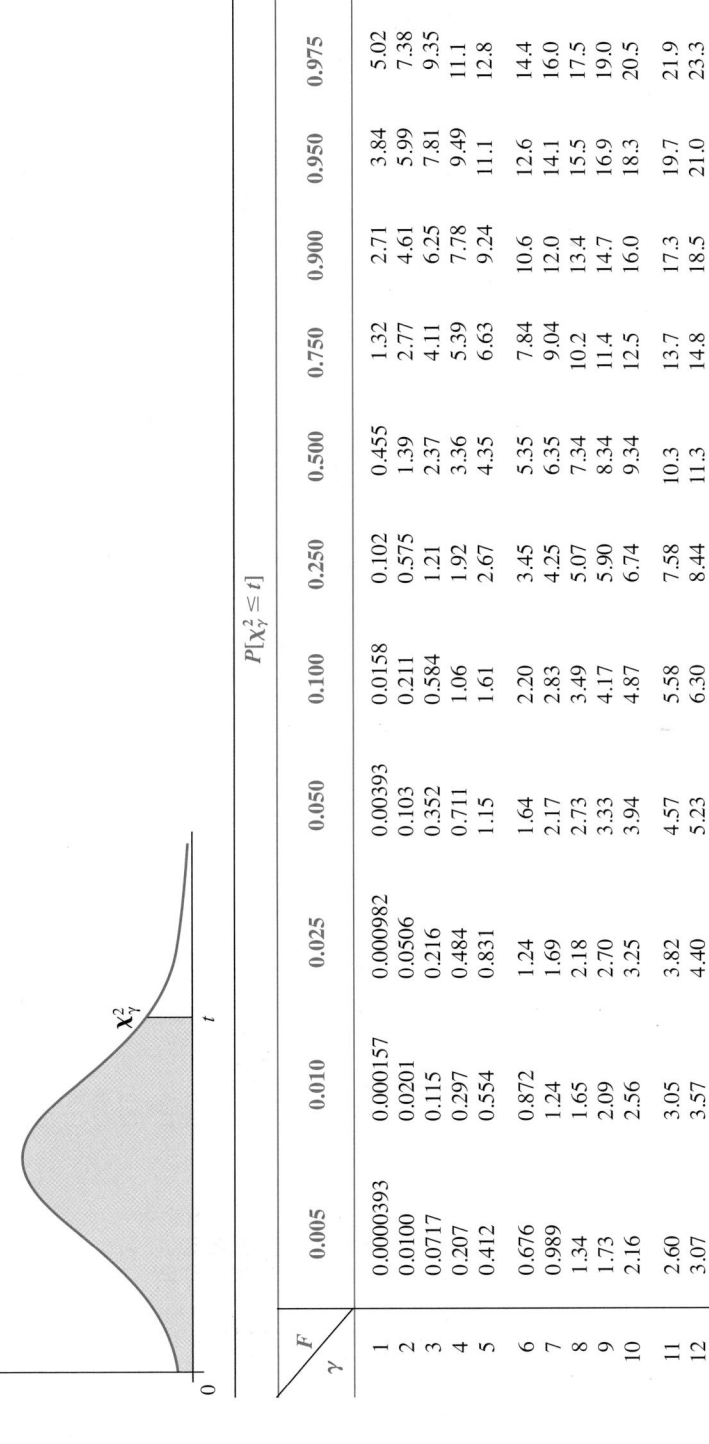

$$P[\chi_\gamma^2 \le t]$$

$\frac{F}{\gamma}$	0.005	0.010	0.025	0.050	0.100	0.250	0.500	0.750	0.900	0.950	0.975	0.990	0.995
1	0.0000393	0.000157	0.000982	0.00393	0.0158	0.102	0.455	1.32	2.71	3.84	5.02	6.63	7.88
2	0.0100	0.0201	0.0506	0.103	0.211	0.575	1.39	2.77	4.61	5.99	7.38	9.21	10.6
3	0.0717	0.115	0.216	0.352	0.584	1.21	2.37	4.11	6.25	7.81	9.35	11.3	12.8
4	0.207	0.297	0.484	0.711	1.06	1.92	3.36	5.39	7.78	9.49	11.1	13.3	14.9
5	0.412	0.554	0.831	1.15	1.61	2.67	4.35	6.63	9.24	11.1	12.8	15.1	16.7
6	0.676	0.872	1.24	1.64	2.20	3.45	5.35	7.84	10.6	12.6	14.4	16.8	18.5
7	0.989	1.24	1.69	2.17	2.83	4.25	6.35	9.04	12.0	14.1	16.0	18.5	20.3
8	1.34	1.65	2.18	2.73	3.49	5.07	7.34	10.2	13.4	15.5	17.5	20.1	22.0
9	1.73	2.09	2.70	3.33	4.17	5.90	8.34	11.4	14.7	16.9	19.0	21.7	23.6
10	2.16	2.56	3.25	3.94	4.87	6.74	9.34	12.5	16.0	18.3	20.5	23.2	25.2
11	2.60	3.05	3.82	4.57	5.58	7.58	10.3	13.7	17.3	19.7	21.9	24.7	26.8
12	3.07	3.57	4.40	5.23	6.30	8.44	11.3	14.8	18.5	21.0	23.3	26.2	28.3

Table IV

Cumulative chi-squared distribution (concluded)

$$P[\chi_\gamma^2 \le t]$$

$F \backslash \gamma$	0.005	0.010	0.025	0.050	0.100	0.250	0.500	0.750	0.900	0.950	0.975	0.990	0.995
13	3.57	4.11	5.01	5.89	7.04	9.30	12.3	16.0	19.8	22.4	24.7	27.7	29.8
14	4.07	4.66	5.63	6.57	7.79	10.2	13.3	17.1	21.1	23.7	26.1	29.1	31.3
15	4.60	5.23	6.26	7.26	8.55	11.0	14.3	18.2	22.3	25.0	27.5	30.6	32.8
16	5.14	5.81	6.91	7.96	9.31	11.9	15.3	19.4	23.5	26.3	28.8	32.0	34.3
17	5.70	6.41	7.56	8.67	10.1	12.8	16.3	20.5	24.8	27.6	30.2	33.4	35.7
18	6.26	7.01	8.23	9.39	10.9	13.7	17.3	21.6	26.0	28.9	31.5	34.8	37.2
19	6.84	7.63	8.91	10.1	11.7	14.6	18.3	22.7	27.2	30.1	32.9	36.2	38.6
20	7.43	8.26	9.59	10.9	12.4	15.5	19.3	23.8	28.4	31.4	34.2	37.6	40.0
21	8.03	8.90	10.3	11.6	13.2	16.3	20.3	24.9	29.6	32.7	35.5	38.9	41.4
22	8.64	9.54	11.0	12.3	14.0	17.2	21.3	26.0	30.8	33.9	36.8	40.3	42.8
23	9.26	10.2	11.7	13.1	14.8	18.1	22.3	27.1	32.0	35.2	38.1	41.6	44.2
24	9.89	10.9	12.4	13.8	15.7	19.0	23.3	28.2	33.2	36.4	39.4	43.0	45.6
25	10.5	11.5	13.1	14.6	16.5	19.9	24.3	29.3	34.4	37.7	40.6	44.3	46.9
26	11.2	12.2	13.8	15.4	17.3	20.8	25.3	30.4	35.6	38.9	41.9	45.6	48.3
27	11.8	12.9	14.6	16.2	18.1	21.7	26.3	31.5	36.7	40.1	43.2	47.0	49.6
28	12.5	13.6	15.3	16.9	18.9	22.7	27.3	32.6	37.9	41.3	44.5	48.3	51.0
29	13.1	14.3	16.0	17.7	19.8	23.6	28.3	33.7	39.1	42.6	45.7	49.6	52.3
30	13.8	15.0	16.8	18.5	20.6	24.5	29.3	34.8	40.3	43.8	47.0	50.9	53.7

From Beyer, W. H. (ed.), in *CRC Handbook of Tables for Probability and Statistics*, 2d ed., 1968. Copyright CRC Press, Inc., Boca Raton, Fla.

Table V
Cumulative distribution: Standard normal

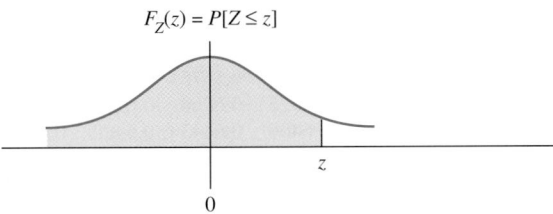

$F_Z(z) = P[Z \leq z]$

$F_Z(z) = P[Z \leq z]$

z	0.00	0.01	0.02	0.03	0.04	0.05	0.06	0.07	0.08	0.09
−3.4	0.0003	0.0003	0.0003	0.0003	0.0003	0.0003	0.0003	0.0003	0.0003	0.0002
−3.3	0.0005	0.0005	0.0005	0.0004	0.0004	0.0004	0.0004	0.0004	0.0004	0.0003
−3.2	0.0007	0.0007	0.0006	0.0006	0.0006	0.0006	0.0006	0.0005	0.0005	0.0005
−3.1	0.0010	0.0009	0.0009	0.0009	0.0008	0.0008	0.0008	0.0008	0.0007	0.0007
−3.0	0.0013	0.0013	0.0013	0.0012	0.0012	0.0011	0.0011	0.0011	0.0010	0.0010
−2.9	0.0019	0.0018	0.0018	0.0017	0.0016	0.0016	0.0015	0.0015	0.0014	0.0014
−2.8	0.0026	0.0025	0.0024	0.0023	0.0023	0.0022	0.0021	0.0021	0.0020	0.0019
−2.7	0.0035	0.0034	0.0033	0.0032	0.0031	0.0030	0.0029	0.0028	0.0027	0.0026
−2.6	0.0047	0.0045	0.0044	0.0043	0.0041	0.0040	0.0039	0.0038	0.0037	0.0036
−2.5	0.0062	0.0060	0.0059	0.0057	0.0055	0.0054	0.0052	0.0051	0.0049	0.0048
−2.4	0.0082	0.0080	0.0078	0.0075	0.0073	0.0071	0.0069	0.0068	0.0066	0.0064
−2.3	0.0107	0.0104	0.0102	0.0099	0.0096	0.0094	0.0091	0.0089	0.0087	0.0084
−2.2	0.0139	0.0136	0.0132	0.0129	0.0125	0.0122	0.0119	0.0116	0.0113	0.0110
−2.1	0.0179	0.0174	0.0170	0.0166	0.0162	0.0158	0.0154	0.0150	0.0146	0.0143
−2.0	0.0228	0.0222	0.0217	0.0212	0.0207	0.0202	0.0197	0.0192	0.0188	0.0183
−1.9	0.0287	0.0281	0.0274	0.0268	0.0262	0.0256	0.0250	0.0244	0.0239	0.0233
−1.8	0.0359	0.0351	0.0344	0.0336	0.0329	0.0322	0.0314	0.0307	0.0301	0.0294
−1.7	0.0446	0.0436	0.0427	0.0418	0.0409	0.0401	0.0392	0.0384	0.0375	0.0367
−1.6	0.0548	0.0537	0.0526	0.0516	0.0505	0.0495	0.0485	0.0475	0.0465	0.0455
−1.5	0.0668	0.0655	0.0643	0.0630	0.0618	0.0606	0.0594	0.0582	0.0571	0.0559
−1.4	0.0808	0.0793	0.0778	0.0764	0.0749	0.0735	0.0721	0.0708	0.0694	0.0681
−1.3	0.0968	0.0951	0.0934	0.0918	0.0901	0.0885	0.0869	0.0853	0.0838	0.0823
−1.2	0.1151	0.1131	0.1112	0.1093	0.1075	0.1056	0.1038	0.1020	0.1003	0.0985
−1.1	0.1357	0.1335	0.1314	0.1292	0.1271	0.1251	0.1230	0.1210	0.1190	0.1170
−1.0	0.1587	0.1562	0.1539	0.1515	0.1492	0.1469	0.1446	0.1423	0.1401	0.1379
−0.9	0.1841	0.1814	0.1788	0.1762	0.1736	0.1711	0.1685	0.1660	0.1635	0.1611
−0.8	0.2119	0.2090	0.2061	0.2033	0.2005	0.1977	0.1949	0.1921	0.1894	0.1867
−0.7	0.2420	0.2389	0.2358	0.2327	0.2296	0.2266	0.2236	0.2206	0.2177	0.2148
−0.6	0.2743	0.2709	0.2676	0.2643	0.2611	0.2578	0.2546	0.2514	0.2483	0.2451
−0.5	0.3085	0.3050	0.3015	0.2981	0.2946	0.2912	0.2877	0.2843	0.2810	0.2776
−0.4	0.3446	0.3409	0.3372	0.3336	0.3300	0.3264	0.3228	0.3192	0.3156	0.3121
−0.3	0.3821	0.3783	0.3745	0.3707	0.3669	0.3632	0.3594	0.3557	0.3520	0.3483
−0.2	0.4207	0.4168	0.4129	0.4090	0.4052	0.4013	0.3974	0.3936	0.3897	0.3859
−0.1	0.4602	0.4562	0.4522	0.4483	0.4443	0.4404	0.4364	0.4325	0.4286	0.4247
−0.0	0.5000	0.4960	0.4920	0.4880	0.4840	0.4801	0.4761	0.4721	0.4681	0.4641

Table V
Cumulative distribution: Standard normal *(concluded)*

z	0.00	0.01	0.02	0.03	0.04	0.05	0.06	0.07	0.08	0.09
0.0	0.5000	0.5040	0.5080	0.5120	0.5160	0.5199	0.5239	0.5279	0.5319	0.5359
0.1	0.5398	0.5438	0.5478	0.5517	0.5557	0.5596	0.5636	0.5675	0.5714	0.5753
0.2	0.5793	0.5832	0.5871	0.5910	0.5948	0.5987	0.6026	0.6064	0.6103	0.6141
0.3	0.6179	0.6217	0.6255	0.6293	0.6331	0.6368	0.6406	0.6443	0.6480	0.6517
0.4	0.6554	0.6591	0.6628	0.6664	0.6700	0.6736	0.6772	0.6808	0.6844	0.6879
0.5	0.6915	0.6950	0.6985	0.7019	0.7054	0.7088	0.7123	0.7157	0.7190	0.7224
0.6	0.7257	0.7291	0.7324	0.7357	0.7389	0.7422	0.7454	0.7486	0.7517	0.7549
0.7	0.7580	0.7611	0.7642	0.7673	0.7704	0.7734	0.7764	0.7794	0.7823	0.7852
0.8	0.7881	0.7910	0.7939	0.7967	0.7995	0.8023	0.8051	0.8078	0.8106	0.8133
0.9	0.8159	0.8186	0.8212	0.8238	0.8264	0.8289	0.8315	0.8340	0.8365	0.8389
1.0	0.8413	0.8438	0.8461	0.8485	0.8508	0.8531	0.8554	0.8577	0.8599	0.8621
1.1	0.8643	0.8665	0.8686	0.8708	0.8729	0.8749	0.8770	0.8790	0.8810	0.8830
1.2	0.8849	0.8869	0.8888	0.8907	0.8925	0.8944	0.8962	0.8980	0.8997	0.9015
1.3	0.9032	0.9049	0.9066	0.9082	0.9099	0.9115	0.9131	0.9147	0.9162	0.9177
1.4	0.9192	0.9207	0.9222	0.9236	0.9251	0.9265	0.9279	0.9292	0.9306	0.9319
1.5	0.9332	0.9345	0.9357	0.9370	0.9382	0.9394	0.9406	0.9418	0.9429	0.9441
1.6	0.9452	0.9463	0.9474	0.9484	0.9495	0.9505	0.9515	0.9525	0.9535	0.9545
1.7	0.9554	0.9564	0.9573	0.9582	0.9591	0.9599	0.9608	0.9616	0.9625	0.9633
1.8	0.9641	0.9649	0.9656	0.9664	0.9671	0.9678	0.9686	0.9693	0.9699	0.9706
1.9	0.9713	0.9719	0.9726	0.9732	0.9738	0.9744	0.9750	0.9756	0.9761	0.9767
2.0	0.9772	0.9778	0.9783	0.9788	0.9793	0.9798	0.9803	0.9808	0.9812	0.9817
2.1	0.9821	0.9826	0.9830	0.9834	0.9838	0.9842	0.9846	0.9850	0.9854	0.9857
2.2	0.9861	0.9864	0.9868	0.9871	0.9875	0.9878	0.9881	0.9884	0.9887	0.9890
2.3	0.9893	0.9896	0.9898	0.9901	0.9904	0.9906	0.9909	0.9911	0.9913	0.9916
2.4	0.9918	0.9920	0.9922	0.9925	0.9927	0.9929	0.9931	0.9932	0.9934	0.9936
2.5	0.9938	0.9940	0.9941	0.9943	0.9945	0.9946	0.9948	0.9949	0.9951	0.9952
2.6	0.9953	0.9955	0.9956	0.9957	0.9959	0.9960	0.9961	0.9962	0.9963	0.9964
2.7	0.9965	0.9966	0.9967	0.9968	0.9969	0.9970	0.9971	0.9972	0.9973	0.9974
2.8	0.9974	0.9975	0.9976	0.9977	0.9977	0.9978	0.9979	0.9979	0.9980	0.9981
2.9	0.9981	0.9982	0.9982	0.9983	0.9984	0.9984	0.9985	0.9985	0.9986	0.9986
3.0	0.9987	0.9987	0.9987	0.9988	0.9988	0.9989	0.9989	0.9989	0.9990	0.9990
3.1	0.9990	0.9991	0.9991	0.9991	0.9992	0.9992	0.9992	0.9992	0.9993	0.9993
3.2	0.9993	0.9993	0.9994	0.9994	0.9994	0.9994	0.9994	0.9995	0.9995	0.9995
3.3	0.9995	0.9995	0.9995	0.9996	0.9996	0.9996	0.9996	0.9996	0.9996	0.9997
3.4	0.9997	0.9997	0.9997	0.9997	0.9997	0.9997	0.9997	0.9997	0.9997	0.9998

Table VI

T distribution

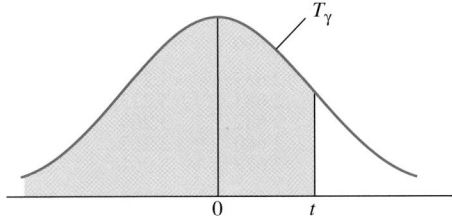

Column heading = cumulative probability

Row heading = degrees of freedom

Row ∞ = standard normal values

$$P[T_\gamma \leq t]$$

γ	.6	.75	.9	.95	.975	.99	.995	.999	.9995
1	0.325	1.000	3.078	6.314	12.706	31.821	63.657	318.317	636.607
2	0.289	0.816	1.886	2.920	4.303	6.965	9.925	22.327	31.598
3	0.277	0.765	1.638	2.353	3.182	4.541	5.841	10.215	12.924
4	0.271	0.741	1.533	2.132	2.776	3.747	4.604	7.173	8.610
5	0.267	0.727	1.476	2.015	2.571	3.365	4.032	5.893	6.869
6	0.265	0.718	1.440	1.943	2.447	3.143	3.707	5.208	5.959
7	0.263	0.711	1.415	1.895	2.365	2.998	3.499	4.785	5.408
8	0.262	0.706	1.397	1.860	2.306	2.896	3.355	4.501	5.041
9	0.261	0.703	1.383	1.833	2.262	2.821	3.250	4.297	4.781
10	0.260	0.700	1.372	1.812	2.228	2.764	3.169	4.144	4.587
11	0.260	0.697	1.363	1.796	2.201	2.718	3.106	4.025	4.437
12	0.259	0.695	1.356	1.782	2.179	2.681	3.055	3.930	4.318
13	0.259	0.694	1.350	1.771	2.160	2.650	3.012	3.852	4.221
14	0.258	0.692	1.345	1.761	2.145	2.624	2.977	3.787	4.140
15	0.258	0.691	1.341	1.753	2.131	2.602	2.947	3.733	4.073
16	0.258	0.690	1.337	1.746	2.120	2.583	2.921	3.686	4.015
17	0.257	0.689	1.333	1.740	2.110	2.567	2.898	3.646	3.965
18	0.257	0.688	1.330	1.734	2.101	2.552	2.878	3.611	3.922
19	0.257	0.688	1.328	1.729	2.093	2.539	2.861	3.579	3.883
20	0.257	0.687	1.325	1.725	2.086	2.528	2.845	3.552	3.850
21	0.257	0.686	1.323	1.721	2.080	2.518	2.831	3.527	3.819
22	0.256	0.686	1.321	1.717	2.074	2.508	2.819	3.505	3.792
23	0.256	0.685	1.319	1.714	2.069	2.500	2.807	3.485	3.768
24	0.256	0.685	1.318	1.711	2.064	2.492	2.797	3.467	3.745
25	0.256	0.684	1.316	1.708	2.060	2.485	2.787	3.450	3.725
26	0.256	0.684	1.315	1.706	2.056	2.479	2.779	3.435	3.707
27	0.256	0.684	1.314	1.703	2.052	2.473	2.771	3.421	3.690
28	0.256	0.683	1.313	1.701	2.048	2.467	2.763	3.408	3.674
29	0.256	0.683	1.311	1.699	2.045	2.462	2.756	3.396	3.659
30	0.256	0.683	1.310	1.697	2.042	2.457	2.750	3.385	3.646
31	0.256	0.682	1.309	1.696	2.040	2.453	2.744	3.375	3.633
32	0.255	0.682	1.309	1.694	2.037	2.449	2.738	3.365	3.622
33	0.255	0.682	1.308	1.692	2.035	2.445	2.733	3.356	3.611
34	0.255	0.682	1.307	1.691	2.032	2.441	2.728	3.348	3.601
35	0.255	0.682	1.306	1.690	2.030	2.438	2.724	3.340	3.591
36	0.255	0.681	1.306	1.688	2.028	2.434	2.719	3.333	3.582
37	0.255	0.681	1.305	1.687	2.026	2.431	2.715	3.326	3.574
38	0.255	0.681	1.304	1.686	2.024	2.429	2.712	3.319	3.566
39	0.255	0.681	1.304	1.685	2.023	2.426	2.708	3.313	3.558
40	0.255	0.681	1.303	1.684	2.021	2.423	2.704	3.307	3.551
41	0.255	0.681	1.303	1.683	2.020	2.421	2.701	3.301	3.544
42	0.255	0.680	1.302	1.682	2.018	2.418	2.698	3.296	3.538
43	0.255	0.680	1.302	1.681	2.017	2.416	2.695	3.291	3.532
44	0.255	0.680	1.301	1.680	2.015	2.414	2.692	3.286	3.526

Table VI
T distribution *(concluded)*

γ	.6	.75	.9	.95	.975	.99	.995	.999	.9995
45	0.255	0.680	1.301	1.679	2.014	2.412	2.690	3.281	3.520
46	0.255	0.680	1.300	1.679	2.013	2.410	2.687	3.277	3.515
47	0.255	0.680	1.300	1.678	2.012	2.408	2.685	3.273	3.510
48	0.255	0.680	1.299	1.677	2.011	2.407	2.682	3.269	3.505
49	0.255	0.680	1.299	1.677	2.010	2.405	2.680	3.265	3.500
50	0.255	0.679	1.299	1.676	2.009	2.403	2.678	3.261	3.496
51	0.255	0.679	1.298	1.675	2.008	2.402	2.676	3.258	3.492
52	0.255	0.679	1.298	1.675	2.007	2.400	2.674	3.255	3.488
53	0.255	0.679	1.298	1.674	2.006	2.399	2.672	3.251	3.484
54	0.255	0.679	1.297	1.674	2.005	2.397	2.670	3.248	3.480
55	0.255	0.679	1.297	1.673	2.004	2.396	2.668	3.245	3.476
56	0.255	0.679	1.297	1.673	2.003	2.395	2.667	3.242	3.473
57	0.255	0.679	1.297	1.672	2.002	2.394	2.665	3.239	3.470
58	0.255	0.679	1.296	1.672	2.002	2.392	2.663	3.237	3.466
59	0.254	0.679	1.296	1.671	2.001	2.391	2.662	3.234	3.463
60	0.254	0.679	1.296	1.671	2.000	2.390	2.660	3.232	3.460
61	0.254	0.679	1.296	1.670	2.000	2.389	2.659	3.229	3.457
62	0.254	0.678	1.295	1.670	1.999	2.388	2.658	3.227	3.455
63	0.254	0.678	1.295	1.669	1.998	2.387	2.656	3.225	3.452
64	0.254	0.678	1.295	1.669	1.998	2.386	2.655	3.223	3.449
65	0.254	0.678	1.295	1.669	1.997	2.385	2.654	3.221	3.447
66	0.254	0.678	1.295	1.668	1.997	2.384	2.652	3.218	3.444
67	0.254	0.678	1.294	1.668	1.996	2.383	2.651	3.217	3.442
68	0.254	0.678	1.294	1.668	1.995	2.382	2.650	3.215	3.440
69	0.254	0.678	1.294	1.667	1.995	2.382	2.649	3.213	3.437
70	0.254	0.678	1.294	1.667	1.994	2.381	2.648	3.211	3.435
71	0.254	0.678	1.294	1.667	1.994	2.380	2.647	3.209	3.433
72	0.254	0.678	1.293	1.666	1.993	2.379	2.646	3.207	3.431
73	0.254	0.678	1.293	1.666	1.993	2.379	2.645	3.206	3.429
74	0.254	0.678	1.293	1.666	1.993	2.378	2.644	3.204	3.427
75	0.254	0.678	1.293	1.665	1.992	2.377	2.643	3.203	3.425
76	0.254	0.678	1.293	1.665	1.992	2.376	2.642	3.201	3.423
77	0.254	0.678	1.293	1.665	1.991	2.376	2.641	3.200	3.422
78	0.254	0.678	1.292	1.665	1.991	2.375	2.640	3.198	3.420
79	0.254	0.678	1.292	1.664	1.990	2.375	2.640	3.197	3.418
80	0.254	0.678	1.292	1.664	1.990	2.374	2.639	3.195	3.416
81	0.254	0.678	1.292	1.664	1.990	2.373	2.638	3.194	3.415
82	0.254	0.677	1.292	1.664	1.989	2.373	2.637	3.193	3.413
83	0.254	0.677	1.292	1.663	1.989	2.372	2.636	3.191	3.412
84	0.254	0.677	1.292	1.663	1.989	2.372	2.636	3.190	3.410
85	0.254	0.677	1.292	1.663	1.988	2.371	2.635	3.189	3.409
86	0.254	0.677	1.291	1.663	1.988	2.371	2.634	3.188	3.407
87	0.254	0.677	1.291	1.663	1.988	2.370	2.634	3.187	3.406
88	0.254	0.677	1.291	1.662	1.987	2.369	2.633	3.186	3.405
89	0.254	0.677	1.291	1.662	1.987	2.369	2.632	3.184	3.403
90	0.254	0.677	1.291	1.662	1.987	2.369	2.632	3.183	3.402
91	0.254	0.677	1.291	1.662	1.986	2.368	2.631	3.182	3.401
92	0.254	0.677	1.291	1.662	1.986	2.368	2.630	3.181	3.400
93	0.254	0.677	1.291	1.661	1.986	2.367	2.630	3.180	3.398
94	0.254	0.677	1.291	1.661	1.986	2.367	2.629	3.179	3.397
95	0.254	0.677	1.291	1.661	1.985	2.366	2.629	3.178	3.396
96	0.254	0.677	1.290	1.661	1.985	2.366	2.628	3.177	3.395
97	0.254	0.677	1.290	1.661	1.985	2.365	2.627	3.176	3.394
98	0.254	0.677	1.290	1.661	1.984	2.365	2.627	3.176	3.393
99	0.254	0.677	1.290	1.660	1.984	2.365	2.626	3.175	3.392
100	0.254	0.677	1.290	1.660	1.984	2.364	2.626	3.174	3.391
∞	0.253	0.674	1.282	1.645	1.960	2.326	2.576	3.090	3.291

Table VII

Sample size for estimating the mean

Level of *t*-test

Single-sided test → Double-sided test	α = 0.005 / α = 0.01					α = 0.01 / α = 0.02					α = 0.025 / α = 0.05					α = 0.05 / α = 0.1					
$\beta =$	0.01	0.05	0.1	0.2	0.5	0.01	0.05	0.1	0.2	0.5	0.01	0.05	0.1	0.2	0.5	0.01	0.05	0.1	0.2	0.5	Value of $\Delta = (\mu - \mu_0)/\sigma$
0.05																					0.05
0.10																					0.10
0.15																				122	0.15
0.20										139					99					70	0.20
0.25					110					90				128	64			139	101	45	0.25
0.30				134	78				115	63			119	90	45		122	97	71	32	0.30
0.35			125	99	58			109	85	47		109	88	67	34		90	72	52	24	0.35
0.40		115	97	77	45		101	85	66	37	117	84	68	51	26	101	70	55	40	19	0.40
0.45		92	77	62	37	110	81	68	53	30	93	67	54	41	21	80	55	44	33	15	0.45
0.50	100	75	63	51	30	90	66	55	43	25	76	54	44	34	18	65	45	36	27	13	0.50
0.55	83	63	53	42	26	75	55	46	36	21	63	45	37	28	15	54	38	30	22	11	0.55
0.60	71	53	45	36	22	63	47	39	31	18	53	38	32	24	13	46	32	26	19	9	0.60
0.65	61	46	39	31	20	55	41	34	27	16	46	33	27	21	12	39	28	22	17	8	0.65
0.70	53	40	34	28	17	47	35	30	24	14	40	29	24	19	10	34	24	19	15	8	0.70
0.75	47	36	30	25	16	42	31	27	21	13	35	26	21	16	9	30	21	17	13	7	0.75
0.80	41	32	27	22	14	37	28	24	19	12	31	22	19	15	9	27	19	15	12	6	0.80
0.85	37	29	24	20	13	33	25	21	17	11	28	21	17	13	8	24	17	14	11	6	0.85
0.90	34	26	22	18	12	29	23	19	16	10	25	19	16	12	7	21	15	13	10	5	0.90
0.95	31	24	20	17	11	27	21	18	14	9	23	17	14	11	7	19	14	11	9	5	0.95
1.00	28	22	19	16	10	25	19	16	13	9	21	16	13	10	6	18	13	11	8	5	1.00

Table VII
Sample size for estimating the mean (concluded)

	Level of t-test																			
Single-sided test	$\alpha = 0.005$					$\alpha = 0.01$					$\alpha = 0.025$					$\alpha = 0.05$				
Double-sided test	$\alpha = 0.01$					$\alpha = 0.02$					$\alpha = 0.05$					$\alpha = 0.1$				
$\beta =$	0.01	0.05	0.1	0.2	0.5	0.01	0.05	0.1	0.2	0.5	0.01	0.05	0.1	0.2	0.5	0.01	0.05	0.1	0.2	0.5
1.1	24	19	16	14	9	21	16	14	12	8	18	13	11	9	6	15	11	9	7	
1.2	21	16	14	12	8	18	14	12	10	7	15	12	10	8	5	13	10	8	6	
1.3	18	15	13	11	8	16	13	11	9	6	14	10	9	7		11	8	7	6	
1.4	16	13	12	10	7	14	11	10	9	6	12	9	8	7		10	8	7	5	
1.5	15	12	11	9	7	13	10	9	8	6	11	8	7	6		9	7	6		
1.6	13	11	10	8	6	12	10	9	7	5	10	8	7	6		8	6	6		
1.7	12	10	9	8	6	11	9	8	7		9	7	6	5		8	6	5		
1.8	12	10	9	8	6	10	8	7	7		8	7	6			7	6			
1.9	11	9	8	7	6	10	8	7	6		8	6	6			7	5			
2.0	10	8	8	7	5	9	7	7	6		7	6	5			6				
2.1	10	8	7	7		8	7	6	6		7	6				6				
2.2	9	8	7	6		8	7	6	5		7	6				6				
2.3	9	7	7	6		8	6	6			6	5				5				
2.4	8	7	7	6		7	6	6			6									
2.5	8	7	6	6		7	6	6			6									
3.0	7	6	6	5		6	5	5			5									
3.5	6	5	5			5														
4.0	6																			

Value of Δ $= (\mu - \mu_0)/\sigma$

From Beyer, W. H. (ed.), in *CRC Handbook of Tables for Probability and Statistics*, 2d ed., 1968. Copyright CRC Press, Inc., Boca Raton, Fla.

Table VIII
Wilcoxon signed-rank test

					n = 5(1)50		
One-sided	Two-sided	*n* = 5	*n* = 6	*n* = 7	*n* = 8	*n* = 9	*n* = 10
P = 0.05	*P* = 0.10	1	2	4	6	8	11
P = 0.025	*P* = 0.05		1	2	4	6	8
P = 0.01	*P* = 0.02			0	2	3	5
P = 0.005	*P* = 0.01				0	2	3
One-sided	Two-sided	*n* = 11	*n* = 12	*n* = 13	*n* = 14	*n* = 15	*n* = 16
P = 0.05	*P* = 0.10	14	17	21	26	30	36
P = 0.025	*P* = 0.05	11	14	17	21	25	30
P = 0.01	*P* = 0.02	7	10	13	16	20	24
P = 0.005	*P* = 0.01	5	7	10	13	16	19
One-sided	Two-sided	*n* = 17	*n* = 18	*n* = 19	*n* = 20	*n* = 21	*n* = 22
P = 0.05	*P* = 0.10	41	47	54	60	68	75
P = 0.025	*P* = 0.05	35	40	46	52	59	66
P = 0.01	*P* = 0.02	28	33	38	43	49	56
P = 0.005	*P* = 0.01	23	28	32	37	43	49
One-sided	Two-sided	*n* = 23	*n* = 24	*n* = 25	*n* = 26	*n* = 27	*n* = 28
P = 0.05	*P* = 0.10	83	92	101	110	120	130
P = 0.025	*P* = 0.05	73	81	90	98	107	117
P = 0.01	*P* = 0.02	62	69	77	85	93	102
P = 0.005	*P* = 0.01	55	61	68	76	84	92
One-sided	Two-sided	*n* = 29	*n* = 30	*n* = 31	*n* = 32	*n* = 33	*n* = 34
P = 0.05	*P* = 0.10	141	152	163	175	188	201
P = 0.025	*P* = 0.05	127	137	148	159	171	183
P = 0.01	*P* = 0.02	111	120	130	141	151	162
P = 0.005	*P* = 0.01	100	109	118	128	138	149
One-sided	Two-sided	*n* = 35	*n* = 36	*n* = 37	*n* = 38	*n* = 39	
P = 0.05	*P* = 0.10	214	228	242	256	271	
P = 0.025	*P* = 0.05	195	208	222	235	250	
P = 0.01	*P* = 0.02	174	186	198	211	224	
P = 0.005	*P* = 0.01	160	171	183	195	208	
One-sided	Two-sided	*n* = 40	*n* = 41	*n* = 42	*n* = 43	*n* = 44	*n* = 45
P = 0.05	*P* = 0.10	287	303	319	336	353	371
P = 0.025	*P* = 0.05	264	279	295	311	327	344
P = 0.01	*P* = 0.02	238	252	267	281	297	313
P = 0.005	*P* = 0.01	221	234	248	262	277	292
One-sided	Two-sided	*n* = 46	*n* = 47	*n* = 48	*n* = 49	*n* = 50	
P = 0.05	*P* = 0.10	389	408	427	446	466	
P = 0.025	*P* = 0.05	361	379	397	415	434	
P = 0.01	*P* = 0.02	329	345	362	380	398	
P = 0.005	*P* = 0.01	307	323	339	356	373	

Table IX
F distribution

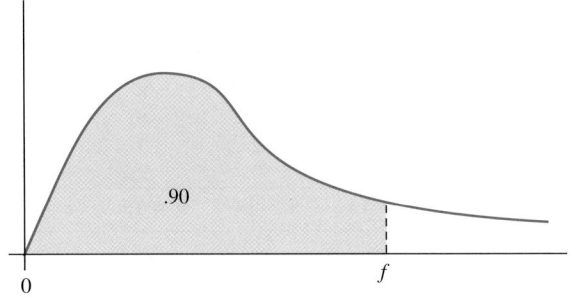

Column heading = numerator
degrees of freedom

Row heading = denominator
degrees of freedom

Points given are $f_{.10}$ points

For degrees of freedom > 120,
use row or column 120

$P[F_{\gamma_1, \gamma_2} \leq f] = .90$

γ_2 \ γ_1	1	2	3	4	5	6	7	8	9	10
1	39.862	49.500	53.593	55.833	57.240	58.204	58.906	59.439	59.857	60.195
2	8.526	9.000	9.162	9.243	9.293	9.326	9.349	9.367	9.381	9.392
3	5.538	5.462	5.391	5.343	5.309	5.285	5.266	5.252	5.240	5.231
4	4.545	4.325	4.191	4.107	4.051	4.010	3.979	3.955	3.936	3.920
5	4.060	3.780	3.619	3.520	3.453	3.405	3.368	3.339	3.316	3.297
6	3.776	3.463	3.289	3.181	3.108	3.055	3.014	2.983	2.958	2.937
7	3.589	3.257	3.074	2.961	2.883	2.827	2.785	2.752	2.725	2.703
8	3.458	3.113	2.924	2.806	2.726	2.668	2.624	2.589	2.561	2.538
9	3.360	3.006	2.813	2.693	2.611	2.551	2.505	2.469	2.440	2.416
10	3.285	2.924	2.728	2.605	2.522	2.461	2.414	2.377	2.347	2.323
11	3.225	2.860	2.660	2.536	2.451	2.389	2.342	2.304	2.274	2.248
12	3.177	2.807	2.606	2.480	2.394	2.331	2.283	2.245	2.214	2.188
13	3.136	2.763	2.560	2.434	2.347	2.283	2.234	2.195	2.164	2.138
14	3.102	2.726	2.522	2.395	2.307	2.243	2.193	2.154	2.122	2.095
15	3.073	2.695	2.490	2.361	2.273	2.208	2.158	2.119	2.086	2.059
16	3.048	2.668	2.462	2.333	2.244	2.178	2.128	2.088	2.055	2.028
17	3.026	2.645	2.437	2.308	2.218	2.152	2.102	2.061	2.028	2.001
18	3.007	2.624	2.416	2.286	2.196	2.130	2.079	2.038	2.005	1.977
19	2.990	2.606	2.397	2.266	2.176	2.109	2.058	2.017	1.984	1.956
20	2.975	2.589	2.380	2.249	2.158	2.091	2.040	1.999	1.965	1.937
21	2.961	2.575	2.365	2.233	2.142	2.075	2.023	1.982	1.948	1.920
22	2.949	2.561	2.351	2.219	2.128	2.061	2.008	1.967	1.933	1.904
23	2.937	2.549	2.339	2.207	2.115	2.047	1.995	1.953	1.919	1.890
24	2.927	2.538	2.327	2.195	2.103	2.035	1.983	1.941	1.906	1.877
25	2.918	2.528	2.317	2.184	2.092	2.024	1.971	1.929	1.895	1.866
26	2.909	2.519	2.307	2.174	2.082	2.014	1.961	1.919	1.884	1.855
27	2.901	2.511	2.299	2.165	2.073	2.005	1.952	1.909	1.874	1.845
28	2.894	2.503	2.291	2.157	2.064	1.996	1.943	1.900	1.865	1.836
29	2.887	2.495	2.283	2.149	2.057	1.988	1.935	1.892	1.857	1.827
30	2.881	2.489	2.276	2.142	2.049	1.980	1.927	1.884	1.849	1.819
31	2.875	2.482	2.270	2.136	2.042	1.973	1.920	1.877	1.842	1.812
32	2.869	2.477	2.263	2.129	2.036	1.967	1.913	1.870	1.835	1.805
33	2.864	2.471	2.258	2.123	2.030	1.961	1.907	1.864	1.828	1.799
34	2.859	2.466	2.252	2.118	2.024	1.955	1.901	1.858	1.822	1.793
35	2.855	2.461	2.247	2.113	2.019	1.950	1.896	1.852	1.817	1.787
36	2.850	2.456	2.243	2.108	2.014	1.945	1.891	1.847	1.811	1.781
37	2.846	2.452	2.238	2.103	2.009	1.940	1.886	1.842	1.806	1.776
38	2.842	2.448	2.234	2.099	2.005	1.935	1.881	1.838	1.802	1.772
39	2.839	2.444	2.230	2.095	2.001	1.931	1.877	1.833	1.797	1.767
40	2.835	2.440	2.226	2.091	1.997	1.927	1.873	1.829	1.793	1.763
50	2.809	2.412	2.197	2.061	1.966	1.895	1.840	1.796	1.760	1.729
60	2.791	2.393	2.177	2.041	1.946	1.875	1.819	1.775	1.738	1.707
120	2.748	2.347	2.130	1.992	1.896	1.824	1.767	1.722	1.684	1.652

Table IX
F **distribution** (*continued*)

γ_2 \ γ_1	11	12	13	14	15	16	17	18	19	20
1	60.473	60.705	60.903	61.072	61.220	61.350	61.464	61.566	61.658	61.740
2	9.401	9.408	9.414	9.420	9.425	9.429	9.432	9.435	9.438	9.441
3	5.223	5.216	5.210	5.205	5.200	5.196	5.193	5.190	5.187	5.185
4	3.907	3.896	3.886	3.878	3.870	3.864	3.858	3.853	3.849	3.844
5	3.282	3.268	3.257	3.247	3.238	3.230	3.223	3.217	3.212	3.207
6	2.920	2.905	2.892	2.881	2.871	2.863	2.855	2.848	2.842	2.836
7	2.684	2.668	2.654	2.643	2.632	2.623	2.615	2.607	2.601	2.595
8	2.519	2.502	2.488	2.475	2.464	2.455	2.446	2.438	2.431	2.425
9	2.396	2.379	2.364	2.351	2.340	2.329	2.320	2.312	2.305	2.298
10	2.302	2.284	2.269	2.255	2.244	2.233	2.224	2.215	2.208	2.201
11	2.227	2.209	2.193	2.179	2.167	2.156	2.147	2.138	2.130	2.123
12	2.166	2.147	2.131	2.117	2.105	2.094	2.084	2.075	2.067	2.060
13	2.116	2.097	2.080	2.066	2.053	2.042	2.032	2.023	2.014	2.007
14	2.073	2.054	2.037	2.022	2.010	1.998	1.988	1.979	1.970	1.962
15	2.037	2.017	2.000	1.985	1.972	1.961	1.950	1.941	1.932	1.924
16	2.005	1.985	1.968	1.953	1.940	1.928	1.917	1.908	1.899	1.891
17	1.978	1.958	1.940	1.925	1.912	1.900	1.889	1.879	1.870	1.862
18	1.954	1.933	1.916	1.900	1.887	1.875	1.864	1.854	1.845	1.837
19	1.932	1.912	1.894	1.878	1.865	1.852	1.841	1.831	1.822	1.814
20	1.913	1.892	1.875	1.859	1.845	1.833	1.821	1.811	1.802	1.794
21	1.896	1.875	1.857	1.841	1.827	1.815	1.803	1.793	1.784	1.776
22	1.880	1.859	1.841	1.825	1.811	1.798	1.787	1.777	1.768	1.759
23	1.866	1.845	1.827	1.811	1.796	1.784	1.772	1.762	1.753	1.744
24	1.853	1.832	1.814	1.797	1.783	1.770	1.759	1.748	1.739	1.730
25	1.841	1.820	1.802	1.785	1.771	1.758	1.746	1.736	1.726	1.718
26	1.830	1.809	1.790	1.774	1.760	1.747	1.735	1.724	1.715	1.706
27	1.820	1.799	1.780	1.764	1.749	1.736	1.724	1.714	1.704	1.695
28	1.811	1.790	1.771	1.754	1.740	1.726	1.715	1.704	1.694	1.685
29	1.802	1.781	1.762	1.745	1.731	1.717	1.705	1.695	1.685	1.676
30	1.794	1.773	1.754	1.737	1.722	1.709	1.697	1.686	1.676	1.667
31	1.787	1.765	1.746	7.729	1.714	1.701	1.689	1.678	1.668	1.659
32	1.780	1.758	1.739	1.722	1.707	1.694	1.682	1.671	1.661	1.652
33	1.773	1.751	1.732	1.715	1.700	1.687	1.675	1.664	1.654	1.645
34	1.767	1.745	1.726	1.709	1.694	1.680	1.668	1.657	1.647	1.638
35	1.761	1.739	1.720	1.703	1.688	1.674	1.662	1.651	1.641	1.632
36	1.756	1.734	1.715	1.697	1.682	1.669	1.656	1.645	1.635	1.626
37	1.751	1.729	1.709	1.692	1.677	1.663	1.651	1.640	1.630	1.620
38	1.746	1.724	1.704	1.687	1.672	1.658	1.646	1.635	1.624	1.615
39	1.741	1.719	1.700	1.682	1.667	1.653	1.641	1.630	1.619	1.610
40	1.737	1.715	1.695	1.678	1.662	1.649	1.636	1.625	1.615	1.605
50	1.703	1.680	1.660	1.643	1.627	1.613	1.600	1.588	1.578	1.568
60	1.680	1.657	1.637	1.619	1.603	1.589	1.576	1.564	1.553	1.543
120	1.625	1.601	1.580	1.562	1.545	1.530	1.516	1.504	1.493	1.482

Table IX
F distribution *(continued)*

γ_2 \ γ_1	21	22	23	24	25	26	27	28	29	30
1	61.815	61.882	61.945	62.002	62.054	62.103	62.148	62.189	62.228	62.265
2	9.444	9.446	9.448	9.450	9.451	9.453	9.454	9.456	9.457	9.458
3	5.182	5.180	5.178	5.176	5.175	5.173	5.172	5.170	5.169	1.168
4	3.841	3.837	3.834	3.831	3.828	3.826	3.824	3.821	3.819	3.817
5	3.202	3.198	3.194	3.191	3.187	3.184	3.181	3.179	3.176	3.174
6	2.831	2.827	2.822	2.818	2.815	2.811	2.808	2.805	2.803	2.800
7	2.589	2.584	2.580	2.575	2.571	2.568	2.564	2.561	2.558	2.555
8	2.419	2.414	2.409	2.404	2.400	2.396	2.392	2.389	2.386	2.383
9	2.292	2.287	2.282	2.277	2.272	2.268	2.265	2.261	2.258	2.255
10	2.194	2.189	2.183	2.178	2.174	2.170	2.166	2.162	2.159	2.155
11	2.117	2.111	2.105	2.100	2.095	2.091	2.087	2.083	2.080	2.076
12	2.053	2.047	2.041	2.036	2.031	2.027	2.022	2.019	2.015	2.011
13	2.000	1.994	1.988	1.983	1.978	1.973	1.969	1.965	1.961	1.958
14	1.955	1.949	1.943	1.938	1.933	1.928	1.923	1.919	1.916	1.912
15	1.917	1.911	1.905	1.899	1.894	1.889	1.885	1.880	1.876	1.873
16	1.884	1.877	1.871	1.866	1.860	1.855	1.851	1.847	1.843	1.839
17	1.855	1.848	1.842	1.836	1.831	1.826	1.821	1.817	1.813	1.809
18	1.829	1.823	1.816	1.810	1.805	1.800	1.795	1.791	1.787	1.783
19	1.807	1.800	1.793	1.787	1.782	1.777	1.772	1.767	1.763	1.759
20	1.786	1.779	1.773	1.767	1.761	1.756	1.751	1.746	1.742	1.738
21	1.768	1.761	1.754	1.748	1.742	1.737	1.732	1.728	1.723	1.719
22	1.751	1.744	1.737	1.731	1.726	1.720	1.715	1.711	1.706	1.702
23	1.736	1.729	1.722	1.716	1.710	1.705	1.700	1.695	1.691	1.686
24	1.722	1.715	1.708	1.702	1.696	1.691	1.686	1.681	1.676	1.672
25	1.710	1.702	1.695	1.689	1.683	1.678	1.672	1.668	1.663	1.659
26	1.698	1.690	1.684	1.677	1.671	1.666	1.660	1.656	1.651	1.647
27	1.687	1.680	1.673	1.666	1.660	1.655	1.649	1.645	1.640	1.636
28	1.677	1.669	1.662	1.656	1.650	1.644	1.639	1.634	1.630	1.625
29	1.668	1.660	1.653	1.647	1.640	1.635	1.630	1.625	1.620	1.616
30	1.659	1.651	1.644	1.638	1.632	1.626	1.621	1.616	1.611	1.606
31	1.651	1.643	1.636	1.630	1.623	1.618	1.612	1.607	1.602	1.598
32	1.643	1.636	1.628	1.622	1.616	1.610	1.604	1.599	1.595	1.590
33	1.636	1.628	1.621	1.615	1.608	1.603	1.597	1.592	1.587	1.583
34	1.630	1.622	1.614	1.608	1.601	1.596	1.590	1.585	1.580	1.576
35	1.623	1.615	1.608	1.601	1.595	1.589	1.584	1.579	1.574	1.569
36	1.617	1.609	1.602	1.595	1.589	1.583	1.578	1.572	1.567	1.563
37	1.612	1.604	1.596	1.590	1.583	1.577	1.572	1.567	1.562	1.557
38	1.606	1.598	1.591	1.584	1.578	1.572	1.566	1.561	1.556	1.551
39	1.601	1.593	1.586	1.579	1.573	1.567	1.561	1.556	1.551	1.546
40	1.596	1.588	1.581	1.574	1.568	1.562	1.556	1.551	1.546	1.541
50	1.559	1.551	1.543	1.536	1.529	1.523	1.517	1.512	1.507	1.502
60	1.534	1.526	1.518	1.511	1.504	1.498	1.492	1.486	1.481	1.476
120	1.472	1.463	1.455	1.447	1.440	1.433	1.427	1.421	1.415	1.409

Table IX
F distribution *(continued)*

γ_1 / γ_2	31	32	33	34	35	36	37	38	39	40
1	62.298	62.330	62.361	62.389	62.415	62.441	62.464	62.487	62.508	62.528
2	9.459	9.460	9.461	9.462	9.463	9.463	9.464	9.465	9.466	9.466
3	5.167	5.166	5.165	5.164	5.163	5.163	5.162	5.161	5.160	5.160
4	3.816	3.814	3.812	3.811	3.810	3.808	3.807	3.806	3.805	3.804
5	3.172	3.170	3.168	3.166	3.165	3.163	3.161	3.160	3.159	3.157
6	2.798	2.795	2.793	2.791	2.789	2.787	2.786	2.784	2.783	2.781
7	2.553	2.550	2.548	2.546	2.544	2.542	2.540	2.538	2.537	2.535
8	2.380	2.378	2.375	2.373	2.371	2.369	2.367	2.365	2.363	2.361
9	2.252	2.249	2.247	2.244	2.242	2.240	2.238	2.236	2.234	2.232
10	2.152	2.150	2.147	2.144	2.142	2.140	2.138	2.135	2.134	2.132
11	2.073	2.070	2.067	2.065	2.062	2.060	2.058	2.056	2.054	2.052
12	2.008	2.005	2.002	2.000	1.997	1.995	1.992	1.990	1.988	1.986
13	1.954	1.951	1.948	1.945	1.943	1.940	1.938	1.936	1.934	1.931
14	1.909	1.905	1.902	1.899	1.897	1.894	1.892	1.889	1.887	1.885
15	1.869	1.866	1.863	1.860	1.857	1.855	1.852	1.850	1.848	1.845
16	1.835	1.832	1.829	1.826	1.823	1.820	1.818	1.815	1.813	1.811
17	1.805	1.802	1.799	1.796	1.793	1.790	1.788	1.785	1.783	1.781
18	1.779	1.776	1.772	1.769	1.766	1.764	1.761	1.758	1.756	1.754
19	1.756	1.752	1.749	1.746	1.743	1.740	1.737	1.735	1.732	1.730
20	1.734	1.731	1.728	1.724	1.721	1.718	1.716	1.713	1.711	1.708
21	1.715	1.712	1.708	1.705	1.702	1.699	1.696	1.694	1.691	1.689
22	1.698	1.695	1.691	1.688	1.685	1.682	1.679	1.676	1.674	1.671
23	1.683	1.679	1.675	1.672	1.669	1.666	1.663	1.660	1.658	1.655
24	1.668	1.664	1.661	1.658	1.654	1.651	1.648	1.646	1.643	1.641
25	1.655	1.651	1.648	1.644	1.641	1.638	1.635	1.632	1.630	1.627
26	1.643	1.639	1.635	1.632	1.629	1.626	1.623	1.620	1.617	1.615
27	1.632	1.628	1.624	1.621	1.617	1.614	1.611	1.608	1.606	1.603
28	1.621	1.617	1.614	1.610	1.607	1.604	1.601	1.598	1.595	1.592
29	1.611	1.607	1.604	1.600	1.597	1.594	1.591	1.588	1.585	1.583
30	1.602	1.598	1.595	1.591	1.588	1.585	1.582	1.579	1.576	1.573
31	1.594	1.590	1.586	1.583	1.579	1.576	1.573	1.570	1.567	1.565
32	1.586	1.582	1.578	1.575	1.571	1.568	1.565	1.562	1.559	1.556
33	1.578	1.574	1.571	1.567	1.564	1.560	1.557	1.554	1.551	1.549
34	1.571	1.567	1.564	1.560	1.556	1.553	1.550	1.547	1.544	1.541
35	1.565	1.561	1.557	1.553	1.550	1.546	1.543	1.540	1.537	1.535
36	1.559	1.554	1.551	1.547	1.543	1.540	1.537	1.534	1.531	1.528
37	1.553	1.548	1.545	1.541	1.537	1.534	1.531	1.528	1.525	1.522
38	1.547	1.543	1.539	1.535	1.532	1.528	1.525	1.522	1.519	1.516
39	1.542	1.538	1.534	1.530	1.526	1.523	1.520	1.517	1.514	1.511
40	1.537	1.532	1.528	1.525	1.521	1.518	1.515	1.511	1.508	1.506
50	1.497	1.493	1.489	1.485	1.481	1.477	1.474	1.471	1.468	1.465
60	1.471	1.466	1.462	1.458	1.454	1.450	1.447	1.444	1.440	1.437
120	1.404	1.399	1.395	1.390	1.386	1.382	1.378	1.375	1.371	1.368

Table IX
F distribution *(continued)*

γ_2 \ γ_1	50	60	120
1	62.688	62.793	63.060
2	9.471	9.475	9.483
3	5.155	5.151	5.143
4	3.795	3.790	3.775
5	3.147	3.140	3.123
6	2.770	2.762	2.742
7	2.523	2.514	2.493
8	2.348	2.339	2.316
9	2.218	2.208	2.184
10	2.117	2.107	2.082
11	2.036	2.026	2.000
12	1.970	1.960	1.932
13	1.915	1.904	1.876
14	1.869	1.857	1.828
15	1.828	1.817	1.787
16	1.793	1.782	1.751
17	1.763	1.751	1.791
18	1.736	1.723	1.691
19	1.711	1.699	1.666
20	1.690	1.677	1.643
21	1.670	1.657	1.623
22	1.652	1.639	1.604
23	1.636	1.622	1.587
24	1.621	1.607	1.571
25	1.607	1.593	1.557
26	1.594	1.581	1.544
27	1.583	1.569	1.531
28	1.572	1.558	1.520
29	1.562	1.547	1.509
30	1.552	1.538	1.499
31	1.543	1.529	1.489
32	1.535	1.520	1.481
33	1.527	1.512	1.472
34	1.520	1.505	1.464
35	1.513	1.497	1.457
36	1.506	1.491	1.450
37	1.500	1.484	1.443
38	1.494	1.478	1.437
39	1.488	1.473	1.431
40	1.483	1.467	1.425
50	1.441	1.424	1.379
60	1.413	1.395	1.348
120	1.340	1.320	1.265

Table IX
F distribution (*continued*)

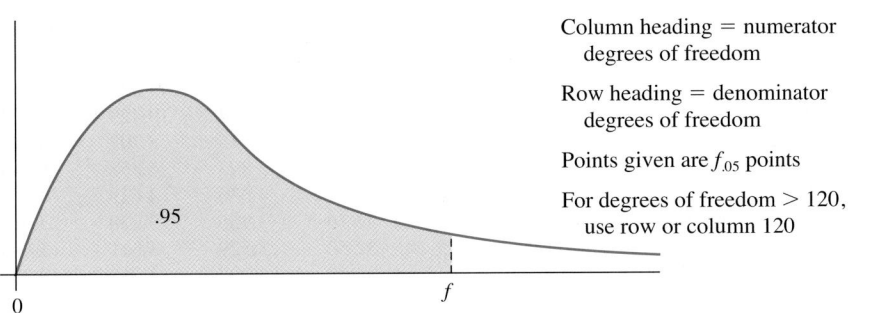

Column heading = numerator
degrees of freedom

Row heading = denominator
degrees of freedom

Points given are $f_{.05}$ points

For degrees of freedom > 120,
use row or column 120

$$P[F_{\gamma_1, \gamma_2} \leq f] = .95$$

γ_2 \ γ_1	1	2	3	4	5	6	7	8
1	161.448	199.500	215.707	224.583	230.161	233.985	236.768	238.882
2	18.513	19.000	19.164	19.247	19.296	19.329	19.353	19.371
3	10.128	9.552	9.277	9.117	9.013	8.941	8.887	8.845
4	7.709	6.944	6.591	6.388	6.256	6.163	6.094	6.041
5	6.608	5.786	5.409	5.192	5.050	4.950	4.876	4.818
6	5.987	5.143	4.757	4.534	4.387	4.284	4.207	4.147
7	5.591	4.737	4.347	4.120	3.972	3.866	3.787	3.726
8	5.318	4.459	4.066	3.838	3.687	3.581	3.500	3.438
9	5.117	4.256	3.863	3.633	3.482	3.374	3.293	3.230
10	4.965	4.103	3.708	3.478	3.326	3.217	3.135	3.072
11	4.844	3.982	3.587	3.357	3.204	3.095	3.012	2.948
12	4.747	3.885	3.490	3.259	3.106	2.996	2.913	2.849
13	4.667	3.806	3.411	3.179	3.025	2.915	2.832	2.767
14	4.600	3.739	3.344	3.112	2.958	2.848	2.764	2.699
15	4.543	3.682	3.287	3.056	2.901	2.790	2.707	2.641
16	4.494	3.634	3.239	3.007	2.852	2.741	2.657	2.591
17	4.451	3.592	3.197	2.965	2.810	2.699	2.614	2.548
18	4.414	3.555	3.160	2.928	2.773	2.661	2.577	2.510
19	4.381	3.522	3.127	2.895	2.740	2.628	2.544	2.477
20	4.351	3.493	3.098	2.866	2.711	2.599	2.514	2.447
21	4.325	3.467	3.072	2.840	2.685	2.573	2.488	2.420
22	4.301	3.443	3.049	2.817	2.661	2.549	2.464	2.397
23	4.279	3.422	3.028	2.796	2.640	2.528	2.442	2.375
24	4.260	3.403	3.009	2.776	2.621	2.508	2.423	2.355
25	4.242	3.385	2.991	2.759	2.603	2.490	2.405	2.337
26	4.225	3.369	2.975	2.743	2.587	2.474	2.388	2.321
27	4.210	3.354	2.960	2.728	2.572	2.459	2.373	2.305
28	4.196	3.340	2.947	2.714	2.558	2.445	2.359	2.291
29	4.183	3.328	2.934	2.701	2.545	2.432	2.346	2.278
30	4.171	3.316	2.922	2.690	2.534	2.421	2.334	2.266
31	4.160	3.305	2.911	2.679	2.523	2.409	2.323	2.255
32	4.149	3.295	2.901	2.668	2.512	2.399	2.313	2.244
33	4.139	3.285	2.892	2.659	2.503	2.389	2.303	2.235
34	4.130	3.276	2.883	2.650	2.494	2.380	2.294	2.225
35	4.121	3.267	2.874	2.641	2.485	2.372	2.285	2.217
36	4.113	3.259	2.866	2.634	2.477	2.364	2.277	2.209
37	4.105	3.252	2.859	2.626	2.470	2.356	2.270	2.201
38	4.098	3.245	2.852	2.619	2.463	2.349	2.262	2.194
39	4.091	3.238	2.845	2.612	2.456	2.342	2.255	2.187
40	4.085	3.232	2.839	2.606	2.449	2.336	2.249	2.180
50	4.034	3.183	2.790	2.557	2.400	2.286	2.199	2.130
60	4.001	3.150	2.758	2.525	2.368	2.254	2.167	2.097
120	3.9301	3.072	2.6681	2.447	2.290	2.175	2.087	2.016

Table IX
F **distribution** *(continued)*

γ_2 \ γ_1	9	10	11	12	13	14	15	16
1	240.543	241.881	242.983	243.905	244.689	245.363	245.949	246.462
2	19.385	19.396	19.405	19.412	19.419	19.424	19.429	19.433
3	8.812	8.786	8.763	8.745	8.729	8.715	8.703	8.692
4	5.999	5.964	5.936	5.912	5.891	5.873	5.858	5.844
5	4.772	4.735	4.704	4.678	4.655	4.636	4.619	4.604
6	4.099	4.060	4.027	4.000	3.976	3.956	3.938	3.922
7	3.677	3.637	3.603	3.575	3.550	3.529	3.511	3.494
8	3.388	3.347	3.313	3.284	3.259	3.237	3.218	3.202
9	3.179	3.137	3.102	3.073	3.048	3.025	3.006	2.989
10	3.020	2.978	2.943	2.913	2.887	2.865	2.845	2.828
11	2.896	2.854	2.818	2.788	2.761	2.739	2.719	2.701
12	2.796	2.753	2.717	2.687	2.660	2.637	2.617	2.599
13	2.714	2.671	2.635	2.604	2.577	2.554	2.533	2.515
14	2.646	2.602	2.566	2.534	2.507	2.484	2.463	2.445
15	2.588	2.544	2.507	2.475	2.448	2.424	2.403	2.385
16	2.538	2.494	2.456	2.425	2.397	2.373	2.352	2.333
17	2.494	2.450	2.413	2.381	2.353	2.329	2.308	2.289
18	2.456	2.412	2.374	2.342	2.314	2.290	2.269	2.250
19	2.423	2.378	2.340	2.308	2.280	2.256	2.234	2.215
20	2.393	2.348	2.310	2.278	2.250	2.225	2.203	2.184
21	2.366	2.321	2.283	2.250	2.222	2.197	2.176	2.156
22	2.342	2.297	2.259	2.226	2.198	2.173	2.151	2.131
23	2.320	2.275	2.236	2.204	2.175	2.150	2.128	2.109
24	2.300	2.255	2.216	2.183	2.155	2.130	2.108	2.088
25	2.282	2.236	2.198	2.165	2.136	2.111	2.089	2.069
26	2.265	2.220	2.181	2.148	2.119	2.094	2.072	2.052
27	2.250	2.204	2.166	2.132	2.103	2.078	2.056	2.036
28	2.236	2.190	2.151	2.118	2.089	2.064	2.041	2.021
29	2.223	2.177	2.138	2.105	2.075	2.050	2.027	2.007
30	2.211	2.165	2.126	2.092	2.063	2.037	2.015	1.995
31	2.199	2.153	2.114	2.080	2.051	2.026	2.003	1.983
32	2.189	2.142	2.103	2.070	2.040	2.015	1.992	1.972
33	2.179	2.133	2.093	2.060	2.030	2.004	1.982	1.961
34	2.170	2.123	2.084	2.050	2.021	1.995	1.972	1.952
35	2.161	2.114	2.075	2.041	2.012	1.986	1.963	1.942
36	2.153	2.106	2.067	2.033	2.003	1.977	1.954	1.934
37	2.145	2.098	2.059	2.025	1.995	1.969	1.946	1.926
38	2.138	2.091	2.051	2.017	1.988	1.962	1.939	1.918
39	2.131	2.084	2.044	2.010	1.981	1.954	1.931	1.911
40	2.124	2.077	2.038	2.003	1.974	1.948	1.924	1.904
50	2.073	2.026	1.986	1.952	1.921	1.895	1.871	1.850
60	2.040	1.993	1.952	1.917	1.887	1.860	1.836	1.815
120	1.959	1.910	1.869	1.834	1.803	1.775	1.750	1.728

Table IX
F distribution (continued)

γ_2 \ γ_1	17	18	19	20	21	22	23	24
1	246.917	247.322	247.685	248.012	248.308	248.577	248.824	249.051
2	19.437	19.440	19.443	19.446	19.448	19.450	19.452	19.454
3	8.683	8.675	8.667	8.660	8.654	8.648	8.643	8.639
4	5.832	5.821	5.811	5.803	5.795	5.787	5.781	5.774
5	4.590	4.579	4.568	4.558	4.549	4.541	4.534	4.527
6	3.908	3.896	3.884	3.874	3.865	3.856	3.849	3.841
7	3.480	3.467	3.455	3.445	3.435	3.426	3.418	3.411
8	3.187	3.173	3.161	3.150	3.140	3.131	3.123	3.115
9	2.974	2.960	2.948	2.936	2.926	2.917	2.908	2.900
10	2.812	2.798	2.785	2.774	2.764	2.754	2.745	2.737
11	2.685	2.671	2.658	2.646	2.636	2.626	2.617	2.609
12	2.583	2.568	2.555	2.544	2.533	2.523	2.514	2.505
13	2.499	2.484	2.471	2.459	2.448	2.438	2.429	2.420
14	2.428	2.413	2.400	2.388	2.377	2.367	2.357	2.349
15	2.368	2.353	2.340	2.328	2.316	2.306	2.297	2.288
16	2.317	2.302	2.288	2.276	2.264	2.254	2.244	2.235
17	2.272	2.257	2.243	2.230	2.219	2.208	2.199	2.190
18	2.233	2.217	2.203	2.191	2.179	2.168	2.159	2.150
19	2.198	2.182	2.168	2.156	2.144	2.133	2.123	2.114
20	2.167	2.151	2.137	2.124	2.112	2.102	2.092	2.082
21	2.139	2.123	2.109	2.096	2.084	2.073	2.063	2.054
22	2.114	2.098	2.084	2.071	2.059	2.048	2.038	2.028
23	2.091	2.075	2.061	2.048	2.036	2.025	2.014	2.005
24	2.070	2.054	2.040	2.027	2.015	2.003	1.993	1.984
25	2.051	2.035	2.021	2.007	1.995	1.984	1.974	1.964
26	2.034	2.018	2.003	1.990	1.978	1.966	1.956	1.946
27	2.018	2.002	1.987	1.974	1.961	1.950	1.940	1.930
28	2.003	1.987	1.972	1.959	1.946	1.935	1.924	1.915
29	1.989	1.973	1.958	1.945	1.932	1.921	1.910	1.901
30	1.976	1.960	1.945	1.932	1.919	1.908	1.897	1.887
31	1.965	1.948	1.933	1.920	1.907	1.896	1.885	1.875
32	1.953	1.937	1.922	1.908	1.896	1.884	1.873	1.864
33	1.943	1.926	1.911	1.898	1.885	1.873	1.863	1.853
34	1.933	1.917	1.902	1.888	1.875	1.863	1.853	1.843
35	1.924	1.907	1.892	1.878	1.866	1.854	1.843	1.833
36	1.915	1.899	1.883	1.870	1.857	1.845	1.834	1.824
37	1.907	1.890	1.875	1.861	1.848	1.837	1.826	1.816
38	1.899	1.883	1.867	1.853	1.841	1.829	1.818	1.808
39	1.892	1.875	1.860	1.846	1.833	1.821	1.810	1.800
40	1.885	1.868	1.853	1.839	1.826	1.814	1.803	1.793
50	1.831	1.814	1.798	1.784	1.771	1.759	1.748	1.737
60	1.796	1.778	1.763	1.748	1.735	1.722	1.711	1.700
120	1.709	1.690	1.674	1.659	1.645	1.632	1.620	1.608

Table IX
F distribution (continued)

γ_2 \ γ_1	25	26	27	28	29	30	31	32
1	249.258	249.451	249.629	249.795	249.949	250.093	250.228	250.355
2	19.456	19.457	19.459	19.460	19.461	19.462	19.463	19.464
3	8.634	8.630	8.626	8.623	8.620	8.617	8.614	8.611
4	5.769	5.764	5.759	5.754	5.750	5.746	5.742	5.739
5	4.521	4.515	4.510	4.505	4.500	4.496	4.492	4.488
6	3.835	3.829	3.823	3.818	3.813	3.808	3.804	3.800
7	3.404	3.397	3.391	3.386	3.381	3.376	3.371	3.367
8	3.108	3.102	3.095	3.090	3.084	3.079	3.075	3.070
9	2.893	2.886	2.880	2.874	2.869	2.864	2.859	2.854
10	2.730	2.723	2.716	2.710	2.705	2.700	2.695	2.690
11	2.601	2.594	2.588	2.582	2.576	2.570	2.565	2.561
12	2.498	2.491	2.484	2.478	2.472	2.466	2.461	2.456
13	2.412	2.405	2.398	2.392	2.386	2.380	2.375	2.370
14	2.341	2.333	2.326	2.320	2.314	2.308	2.303	2.298
15	2.280	2.272	2.265	2.259	2.253	2.247	2.241	2.236
16	2.227	2.220	2.212	2.206	2.200	2.194	2.188	2.183
17	2.181	2.174	2.167	2.160	2.154	2.148	2.142	2.137
18	2.141	2.134	2.126	2.119	2.113	2.107	2.102	2.096
19	2.106	2.098	2.090	2.084	2.077	2.071	2.066	2.060
20	2.074	2.066	2.059	2.052	2.045	2.039	2.033	2.028
21	2.045	2.037	2.030	2.023	2.016	2.010	2.004	1.999
22	2.020	2.012	2.004	1.997	1.990	1.984	1.978	1.973
23	1.996	1.988	1.981	1.973	1.967	1.961	1.955	1.949
24	1.975	1.967	1.959	1.952	1.945	1.939	1.933	1.927
25	1.955	1.947	1.939	1.932	1.926	1.919	1.913	1.908
26	1.938	1.929	1.921	1.914	1.907	1.901	1.895	1.889
27	1.921	1.913	1.905	1.898	1.891	1.884	1.878	1.872
28	1.906	1.897	1.889	1.882	1.875	1.869	1.863	1.857
29	1.891	1.883	1.875	1.868	1.861	1.854	1.848	1.842
30	1.878	1.870	1.862	1.854	1.847	1.841	1.835	1.829
31	1.866	1.857	1.849	1.842	1.835	1.828	1.822	1.816
32	1.854	1.846	1.838	1.830	1.823	1.817	1.810	1.804
33	1.844	1.835	1.827	1.819	1.812	1.806	1.799	1.793
34	1.833	1.825	1.817	1.809	1.802	1.795	1.789	1.783
35	1.824	1.815	1.807	1.799	1.792	1.786	1.779	1.773
36	1.815	1.806	1.798	1.790	1.783	1.776	1.770	1.764
37	1.806	1.798	1.789	1.782	1.775	1.768	1.761	1.755
38	1.798	1.790	1.781	1.774	1.766	1.760	1.753	1.747
39	1.791	1.782	1.774	1.766	1.759	1.752	1.745	1.739
40	1.783	1.775	1.766	1.759	1.751	1.744	1.738	1.732
50	1.727	1.718	1.710	1.702	1.694	1.687	1.680	1.674
60	1.690	1.681	1.672	1.664	1.656	1.649	1.642	1.636
120	1.598	1.588	1.579	1.570	1.562	1.554	1.547	1.540

Table IX
F distribution (continued)

γ_1 / γ_2	33	34	35	36	37	38	39	40
1	250.474	250.586	250.691	250.793	250.886	250.975	251.061	251.141
2	19.465	19.466	19.467	19.468	19.468	19.469	19.470	19.470
3	8.609	8.606	8.604	8.602	8.600	8.598	8.596	8.594
4	5.735	5.732	5.729	5.727	5.724	5.722	5.719	5.717
5	4.484	4.481	4.478	4.474	4.472	4.469	4.466	4.464
6	3.796	3.792	3.789	3.786	3.783	3.780	3.777	3.774
7	3.363	3.359	3.356	3.352	3.349	3.346	3.343	3.340
8	3.066	3.062	3.059	3.055	3.052	3.049	3.046	3.043
9	2.850	2.846	2.842	2.839	2.835	2.832	2.829	2.826
10	2.686	2.681	2.678	2.674	2.670	2.667	2.664	2.661
11	2.556	2.552	2.548	2.544	2.541	2.537	2.534	2.531
12	2.452	2.447	2.443	2.439	2.436	2.432	2.429	2.426
13	2.366	2.361	2.357	2.353	2.349	2.346	2.342	2.339
14	2.293	2.289	2.284	2.280	2.277	2.273	2.270	2.266
15	2.232	2.227	2.223	2.219	2.215	2.211	2.208	2.204
16	2.178	2.174	2.169	2.165	2.161	2.158	2.154	2.151
17	2.132	2.127	2.123	2.119	2.115	2.111	2.107	2.104
18	2.091	2.087	2.082	2.078	2.074	2.070	2.066	2.063
19	2.055	2.050	2.046	2.042	2.037	2.034	2.030	2.026
20	2.023	2.018	2.013	2.009	2.005	2.001	1.997	1.994
21	1.994	1.989	1.984	1.980	1.976	1.972	1.968	1.965
22	1.968	1.963	1.958	1.954	1.949	1.945	1.942	1.938
23	1.944	1.939	1.934	1.930	1.925	1.921	1.918	1.914
24	1.922	1.917	1.912	1.908	1.904	1.900	1.896	1.892
25	1.902	1.897	1.892	1.888	1.884	1.879	1.876	1.872
26	1.884	1.879	1.874	1.869	1.865	1.861	1.857	1.853
27	1.867	1.862	1.857	1.852	1.848	1.844	1.840	1.836
28	1.851	1.846	1.841	1.837	1.832	1.828	1.824	1.820
29	1.837	1.832	1.827	1.822	1.818	1.813	1.809	1.806
30	1.823	1.818	1.813	1.808	1.804	1.800	1.796	1.792
31	1.811	1.805	1.800	1.796	1.791	1.787	1.783	1.779
32	1.799	1.794	1.789	1.784	1.779	1.775	1.771	1.767
33	1.788	1.783	1.777	1.773	1.768	1.764	1.760	1.756
34	1.777	1.772	1.767	1.762	1.758	1.753	1.749	1.745
35	1.768	1.762	1.757	1.752	1.748	1.743	1.739	1.735
36	1.758	1.753	1.748	1.743	1.738	1.734	1.730	1.726
37	1.750	1.744	1.739	1.734	1.730	1.725	1.721	1.717
38	1.741	1.736	1.731	1.726	1.721	1.717	1.712	1.708
39	1.733	1.728	1.723	1.718	1.713	1.709	1.704	1.700
40	1.726	1.721	1.715	1.710	1.706	1.701	1.697	1.693
50	1.668	1.662	1.657	1.652	1.647	1.642	1.638	1.634
60	1.630	1.624	1.618	1.613	1.608	1.603	1.599	1.594
120	1.534	1.527	1.521	1.516	1.510	1.505	1.500	1.495

Table IX
F distribution (*concluded*)

γ_2 \ γ_1	50	60	120
1	251.773	252.191	253.252
2	19.475	19.478	19.487
3	8.581	8.572	8.549
4	5.699	5.688	5.658
5	4.444	4.431	4.398
6	3.754	3.740	3.705
7	3.319	3.304	3.267
8	3.020	3.005	2.967
9	2.803	2.787	2.748
10	2.637	2.621	2.580
11	2.507	2.490	2.448
12	2.401	2.384	2.341
13	2.314	2.297	2.252
14	2.241	2.223	2.178
15	2.178	2.160	2.114
16	2.124	2.106	2.059
17	2.077	2.058	2.011
18	2.035	2.017	1.968
19	1.999	1.980	1.930
20	1.966	1.946	1.896
21	1.936	1.916	1.866
22	1.909	1.889	1.838
23	1.885	1.865	1.813
24	1.863	1.842	1.790
25	1.842	1.822	1.768
26	1.823	1.803	1.749
27	1.806	1.785	1.731
28	1.790	1.769	1.714
29	1.775	1.754	1.698
30	1.761	1.740	1.683
31	1.748	1.726	1.670
32	1.736	1.714	1.657
33	1.724	1.702	1.645
34	1.713	1.691	1.633
35	1.703	1.681	1.623
36	1.694	1.671	1.612
37	1.685	1.662	1.603
38	1.676	1.653	1.594
39	1.668	1.645	1.585
40	1.660	1.637	1.577
50	1.599	1.576	1.511
60	1.559	1.534	1.467
120	1.457	1.429	1.352

Table X
Wilcoxon rank-sum test

$m = 3(1)25$ and $n = m(1)m + 25$
$P = .025$ one-sided; $P = .05$ two-sided

n	$m = 3$	$m = 4$	$m = 5$	$m = 6$	$m = 7$	$m = 8$	$m = 9$	$m = 10$	$m = 11$	$m = 12$	$m = 13$	$m = 14$
$n = m$	5,16	11,25	18,37	26,52	37,68	49,87	63,108	79,131	96,157	116,184	137,214	160,246
$n = m + 1$	6,18	12,28	19,41	28,56	39,73	51,93	66,114	82,138	100,164	120,192	141,223	165,255
$n = m + 2$	6,21	12,32	20,45	29,61	41,78	54,98	68,121	85,145	103,172	124,200	146,231	170,264
$n = m + 3$	7,23	13,35	21,49	31,65	43,83	56,104	71,127	88,152	107,179	128,208	150,240	174,274
$n = m + 4$	7,26	14,38	22,53	32,70	45,88	58,110	74,133	91,159	110,187	131,217	154,249	179,283
$n = m + 5$	8,28	15,41	24,56	34,74	46,94	61,115	77,139	94,116	114,194	135,225	159,257	184,292
$n = m + 6$	8,31	16,44	25,60	36,78	48,99	63,121	79,146	97,173	118,201	139,233	163,266	189,301
$n = m + 7$	9,33	17,47	26,64	37,83	50,104	65,127	82,152	101,179	121,209	143,241	168,274	194,310
$n = m + 8$	10,35	17,51	27,68	39,87	52,109	68,132	85,158	104,186	125,216	147,249	172,283	198,320
$n = m + 9$	10,38	18,54	29,71	41,91	54,114	70,138	88,164	107,193	128,224	151,257	176,292	203,329
$n = m + 10$	11,40	19,57	30,75	42,96	56,119	72,144	90,171	110,200	132,231	155,265	181,300	208,338
$n = m + 11$	11,43	20,60	31,79	44,100	58,124	75,149	93,177	113,207	135,239	159,273	185,309	213,347
$n = m + 12$	12,45	21,63	32,83	45,105	60,129	77,155	96,183	117,213	139,246	163,281	190,317	218,356
$n = m + 13$	12,48	22,66	33,87	47,109	62,134	80,160	99,189	120,220	143,253	167,289	194,326	222,366
$n = m + 14$	13,50	23,69	35,90	49,113	64,139	82,166	101,196	123,227	146,261	171,297	198,335	227,375
$n = m + 15$	13,53	24,72	36,94	50,118	66,144	84,172	104,202	126,234	150,268	175,305	203,343	232,384
$n = m + 16$	14,55	24,76	37,98	52,122	68,149	87,177	107,208	129,241	153,276	179,313	207,352	237,393
$n = m + 17$	14,58	25,79	37,102	53,127	70,154	89,183	110,214	132,248	157,283	183,321	212,360	242,402
$n = m + 18$	15,60	26,82	38,105	55,131	72,159	91,188	113,220	136,254	161,290	187,329	216,369	247,411
$n = m + 19$	15,63	27,85	40,105	57,135	74,164	94,194	115,227	139,261	164,298	191,337	221,377	252,420
$n = m + 20$	16,65	28,88	42,113	58,140	76,169	96,200	118,233	142,268	168,305	195,345	225,286	256,430
$n = m + 21$	16,68	29,91	43,117	60,144	78,174	99,205	121,239	145,275	171,313	199,353	229,395	261,439
$n = m + 22$	17,70	30,94	45,120	61,149	80,179	101,211	124,245	148,282	175,320	203,361	234,403	266,448
$n = m + 23$	17,73	31,97	46,124	63,153	82,184	103,217	127,251	152,288	179,327	207,369	238,412	271,457
$n = m + 24$	18,75	31,101	47,128	65,157	84,189	106,222	129,258	155,295	182,335	211,377	243,420	276,466
$n = m + 25$	18,78	32,104	48,132	66,162	86,194	108,228	132,264	158,302	186,342	216,384	247,429	281,475

Table X
Wilcoxon rank-sum test (*continued*)

$$m = 3(1)25 \text{ and } n = m(1)m + 25$$
$$P = .025 \text{ one-sided}; P = .05 \text{ two-sided}$$

n	m = 15	m = 16	m = 17	m = 18	m = 19	m = 20	m = 21	m = 22	m = 23	m = 24	m = 25
n = m	185,280	212,316	240,355	271,395	303,438	337,483	373,530	411,579	451,630	493,683	536,739
n = m + 1	190,290	217,327	246,366	277,407	310,450	345,495	381,543	419,593	460,644	502,698	546,754
n = m + 2	195,300	223,337	252,377	284,418	317,462	352,508	389,556	428,606	468,659	511,713	555,770
n = m + 3	201,309	229,347	258,388	290,430	324,474	359,521	397,569	436,620	477,673	520,728	565,785
n = m + 4	206,319	234,358	264,399	297,441	331,486	367,533	404,583	444,634	486,687	529,743	574,801
n = m + 5	211,329	240,368	271,409	303,453	338,498	374,546	412,596	452,648	494,702	538,758	584,816
n = m + 6	216,339	245,379	277,420	310,464	345,510	381,559	420,609	460,662	503,716	547,773	593,832
n = m + 7	221,349	251,389	283,431	316,476	351,523	389,571	428,622	469,675	512,730	556,788	603,847
n = m + 8	227,358	257,399	289,442	323,487	358,535	396,584	436,635	477,689	520,745	565,803	612,863
n = m + 9	232,368	262,410	295,453	329,499	365,547	403,597	443,649	485,703	529,759	575,817	622,878
n = m + 10	237,378	268,420	301,464	336,510	372,559	411,609	451,662	493,717	538,773	584,832	632,893
n = m + 11	242,388	274,430	307,475	342,522	379,571	418,622	459,675	502,730	546,788	593,847	641,909
n = m + 12	248,397	279,441	313,486	349,533	386,583	426,634	467,688	510,744	555,802	602,862	651,924
n = m + 13	253,407	285,451	319,497	355,545	393,595	433,647	475,701	518,758	564,816	611,877	660,940
n = m + 14	258,417	291,461	325,508	362,556	400,607	440,660	482,715	526,772	572,831	620,892	670,955
n = m + 15	263,427	296,472	331,519	368,568	407,619	448,672	490,728	535,785	581,845	629,907	679,971
n = m + 16	269,436	302,482	338,529	375,579	414,631	455,685	498,741	543,799	590,859	638,922	689,986
n = m + 17	274,446	308,492	344,540	381,591	421,643	463,697	506,754	551,813	599,873	648,936	699,1001
n = m + 18	279,456	314,502	350,551	388,602	428,655	470,710	514,767	560,826	607,888	657,951	708,1017
n = m + 19	284,466	319,513	356,562	395,613	435,667	477,723	522,780	568,840	616,902	666,966	718,1032
n = m + 20	290,475	325,523	362,573	401,625	442,679	485,735	530,793	576,854	625,916	675,981	727,1048
n = m + 21	295,485	331,533	368,584	408,636	449,691	492,748	537,807	584,868	633,931	684,996	737,1063
n = m + 22	300,495	336,544	374,595	414,648	456,703	500,760	545,820	593,881	642,945	693,1011	747,1078
n = m + 23	306,504	342,554	380,606	421,659	463,715	507,773	553,833	601,895	651,959	703,1025	756,1094
n = m + 24	311,514	348,564	387,616	427,671	470,727	515,785	561,846	609,909	660,973	712,1040	766,1109
n = m + 25	316,524	353,575	393,627	434,682	477,739	522,798	569,859	618,922	668,988	721,1055	775,1125

Table X

Wilcoxon rank-sum test (continued)

$$m = 3(1)25 \text{ and } n = m(1)m + 25$$
$$P = .05 \text{ one-sided}; P = .10 \text{ two-sided}$$

n	$m = 3$	$m = 4$	$m = 5$	$m = 6$	$m = 7$	$m = 8$	$m = 9$	$m = 10$	$m = 11$	$m = 12$	$m = 13$	$m = 14$
$n = m$	6,15	12,24	19,36	28,50	39,66	52,84	66,105	83,127	101,152	121,179	143,208	167,239
$n = m + 1$	7,17	13,27	20,40	30,54	41,71	54,90	69,111	86,134	105,159	125,187	148,216	172,248
$n = m + 2$	7,20	14,30	22,43	32,58	43,76	57,95	72,117	89,141	109,166	129,195	152,225	177,257
$n = m + 3$	8,22	15,33	24,46	33,63	46,80	60,100	75,123	93,147	112,174	134,202	157,233	182,266
$n = m + 4$	9,24	16,36	25,50	35,67	48,85	62,106	78,129	96,154	116,181	138,210	162,241	187,275
$n = m + 5$	9,27	17,39	26,54	37,71	50,90	65,111	81,135	100,160	120,188	142,218	166,250	192,284
$n = m + 6$	10,29	18,42	27,58	39,75	52,95	67,117	84,141	103,167	124,195	147,225	171,258	197,293
$n = m + 7$	11,31	19,45	29,61	41,79	54,100	70,122	87,147	107,173	128,202	151,233	176,266	203,301
$n = m + 8$	11,34	20,48	30,65	42,84	57,104	73,127	90,153	110,180	132,209	155,241	181,274	208,310
$n = m + 9$	12,36	21,51	32,68	44,88	59,109	75,133	93,159	114,186	136,216	159,249	185,283	213,319
$n = m + 10$	13,38	22,54	33,72	46,92	61,114	78,138	96,165	117,193	139,224	164,256	190,291	218,328
$n = m + 11$	13,41	23,57	34,76	48,96	63,119	80,144	100,170	120,200	143,231	168,264	195,299	223,337
$n = m + 12$	14,43	24,60	36,79	50,100	65,124	83,149	103,176	124,206	147,238	172,272	199,308	228,346
$n = m + 13$	15,45	25,63	37,83	52,104	68,128	86,154	106,182	127,213	151,245	177,279	204,316	234,355
$n = m + 14$	15,48	26,66	39,86	53,109	70,133	88,160	109,188	131,219	155,252	181,287	209,324	239,363
$n = m + 15$	16,50	27,69	40,90	55,113	72,138	91,165	112,194	134,226	159,259	185,295	214,332	244,372
$n = m + 16$	17,52	28,72	42,93	57,117	74,143	94,170	115,200	138,232	163,266	190,302	218,341	249,381
$n = m + 17$	17,55	29,75	43,97	59,121	77,147	96,176	118,206	141,239	167,273	194,310	223,349	254,390
$n = m + 18$	18,57	30,78	44,101	61,125	79,152	99,181	121,212	145,245	171,280	198,318	228,357	260,398
$n = m + 19$	19,59	31,81	46,104	62,130	81,157	102,186	124,218	148,252	175,287	203,325	233,365	265,407
$n = m + 20$	19,62	32,84	47,108	64,134	83,162	104,192	127,224	152,258	178,295	207,333	237,374	270,416
$n = m + 21$	20,64	33,87	49,111	66,138	86,166	107,197	130,230	155,265	182,302	211,341	242,382	275,425
$n = m + 22$	21,66	34,90	50,115	68,142	88,171	109,203	133,236	159,271	186,309	216,348	247,390	280,434
$n = m + 23$	21,69	35,93	52,118	70,146	90,176	112,208	136,242	162,278	190,316	220,356	252,398	285,443
$n = m + 24$	22,71	37,95	53,122	72,150	92,181	115,213	139,248	166,284	194,323	224,364	257,406	291,451
$n = m + 25$	23,73	38,98	54,126	73,155	94,186	117,219	142,254	169,291	198,330	229,371	261,415	296,460

Table X
Wilcoxon rank-sum test *(concluded)*

$m = 3(1)25$ and $n = m(1)m + 25$
$P = .05$ one-sided; $P = .10$ two-sided

n	$m = 15$	$m = 16$	$m = 17$	$m = 18$	$m = 19$	$m = 20$	$m = 21$	$m = 22$	$m = 23$	$m = 24$	$m = 25$
$n = m$	192,273	220,308	249,346	280,386	314,427	349,471	386,517	424,566	465,616	508,668	552,723
$n = m + 1$	198,282	226,318	256,356	287,397	321,439	356,484	394,530	433,579	474,630	517,683	562,738
$n = m + 2$	203,292	232,328	262,367	294,408	328,451	364,496	402,543	442,592	483,644	527,697	572,753
$n = m + 3$	209,301	238,338	268,378	301,419	336,462	372,508	410,556	450,606	492,658	536,712	582,768
$n = m + 4$	215,310	244,348	275,388	308,430	343,474	380,520	418,569	459,619	501,672	546,726	592,783
$n = m + 5$	220,320	250,358	281,399	315,441	350,486	387,533	427,581	468,632	511,685	555,741	602,798
$n = m + 6$	226,329	256,368	288,409	322,452	358,497	395,545	435,594	476,646	520,699	565,755	612,813
$n = m + 7$	231,339	262,378	294,420	329,463	365,509	403,557	443,607	485,659	529,713	574,770	622,828
$n = m + 8$	237,348	268,388	301,430	336,474	372,521	411,569	451,620	494,672	538,727	584,784	632,843
$n = m + 9$	242,358	274,398	307,441	342,486	380,532	419,581	459,633	502,686	547,741	594,798	642,858
$n = m + 10$	248,367	280,408	314,451	349,497	387,544	426,594	468,645	511,699	556,755	603,813	652,873
$n = m + 11$	254,376	286,418	320,462	356,508	394,556	434,606	476,658	520,712	565,769	613,827	662,888
$n = m + 12$	259,386	292,428	327,472	363,519	402,567	442,618	484,671	528,726	574,783	622,842	672,903
$n = m + 13$	265,395	298,438	333,483	370,530	409,579	450,630	492,684	537,739	584,796	632,856	682,918
$n = m + 14$	270,405	304,448	340,493	377,541	416,591	458,642	501,696	546,752	593,810	642,870	692,933
$n = m + 15$	276,414	310,458	346,504	384,552	424,602	465,655	509,709	554,766	602,824	651,885	702,948
$n = m + 16$	282,423	316,468	353,514	391,563	431,614	473,667	517,722	563,779	611,838	661,899	712,963
$n = m + 17$	287,433	322,478	359,525	398,574	438,626	481,679	526,734	572,792	620,852	670,914	723,977
$n = m + 18$	293,442	328,488	366,535	405,585	446,637	489,691	534,747	581,805	629,866	680,928	733,992
$n = m + 19$	299,451	334,498	372,546	412,596	453,649	497,703	542,760	589,819	639,879	690,942	743,1007
$n = m + 20$	304,461	340,508	379,556	419,607	461,660	505,715	550,773	598,832	648,893	699,957	753,1022
$n = m + 21$	310,470	347,517	385,568	426,618	468,672	512,728	559,785	607,845	657,907	709,971	763,1037
$n = m + 22$	315,480	353,527	392,577	433,629	475,684	520,740	567,798	615,859	666,921	718,986	773,1052
$n = m + 23$	321,489	359,537	398,588	439,641	483,695	528,752	575,811	624,872	675,935	728,100	783,1067
$n + m + 24$	327,498	365,547	405,598	446,652	490,707	536,764	583,824	633,885	684,949	738,1014	793,1082
$n = m + 25$	332,508	371,557	411,609	453,663	498,718	544,776	592,836	642,898	694,962	747,1029	803,1097

From Beyer, W. H. (ed.), in *CRC Handbook of Tables for Probability and Statistics*, 2nd ed., 1968. Copyright CRC Press, Inc., Boca Raton, Fla.

Table XI
Least significant studentized ranges r_p

Least significant studentized ranges r_p — $\alpha = 0.05$, P

r	2	3	4	5	6
1	17.97	17.97	17.97	17.97	17.97
2	6.085	6.085	6.085	6.085	6.085
3	4.501	4.516	4.516	4.516	4.516
4	3.927	4.013	4.033	4.033	4.033
5	3.635	3.749	3.797	3.814	3.814
6	3.461	3.587	3.649	3.680	3.694
7	3.344	3.477	3.548	3.588	3.611
8	3.261	3.399	3.475	3.521	3.549
9	3.199	3.399	3.420	3.470	3.502
10	3.151	3.293	3.376	3.430	3.465
11	3.113	3.256	3.342	3.397	3.435
12	3.082	3.225	3.313	3.370	3.410
13	3.055	3.200	3.289	3.348	3.389
14	3.033	3.178	3.268	3.329	3.372
15	3.014	3.160	3.250	3.312	3.356
16	2.998	3.144	3.235	3.298	3.343
17	2.984	3.130	3.222	3.285	3.331
18	2.971	3.118	3.210	3.274	3.321
19	2.960	3.107	3.199	3.264	3.311
20	2.950	3.097	3.190	3.255	3.303
24	2.919	3.066	3.160	3.226	3.276
30	2.888	3.035	3.131	3.199	3.250
40	2.858	3.006	3.102	3.171	3.224
60	2.829	2.976	3.073	4.143	3.198
120	2.800	2.947	3.045	3.116	3.172
∞	2.772	2.918	3.017	3.089	3.146

Least significant studentized ranges r_p — $\alpha = 0.01$, P

r	2	3	4	5	6
1	90.03	90.03	90.03	90.03	90.03
2	14.04	14.04	14.04	14.04	14.04
3	8.261	8.321	8.321	8.321	8.321
4	6.512	6.667	6.740	6.756	6.756
5	5.702	5.893	5.898	6.040	6.065
6	5.243	5.439	5.549	5.614	5.655
7	4.949	5.145	5.260	5.334	5.383
8	4.746	4.939	5.057	5.135	5.189
9	4.596	4.787	4.906	4.986	5.043
10	4.482	4.671	4.790	4.871	4.931
11	4.392	4.579	4.697	4.780	4.841
12	4.320	4.504	4.622	4.706	4.767
13	4.260	4.442	4.560	4.644	4.706
14	4.210	4.391	4.508	4.591	4.654
15	4.168	4.347	4.463	4.547	4.610
16	4.131	4.309	4.425	4.509	4.572
17	4.099	4.275	4.391	4.475	4.539
18	4.071	4.246	4.362	4.445	4.509
19	4.046	4.220	4.335	4.419	4.483
20	4.024	4.197	4.312	4.395	4.459
24	3.956	4.126	4.239	4.322	4.386
30	2.889	4.506	4.168	4.250	4.314
40	3.825	3.988	4.098	4.180	4.244
60	3.762	3.922	4.031	4.111	4.174
120	3.702	3.858	3.965	4.044	4.107
∞	3.643	3.796	3.900	3.978	4.040

Least significant studentized ranges r_p — $\alpha = 0.1$, P

r	2	3	4	5	6
1	8.929	8.929	8.929	8.929	8.929
2	4.130	4.130	4.130	4.130	4.130
3	3.328	3.330	3.330	3.330	3.330
4	3.015	3.074	3.081	3.081	3.081
5	2.850	2.934	2.964	2.970	2.970
6	2.748	2.846	2.890	2.908	2.911
7	2.680	2.785	2.838	2.864	2.876
8	2.630	2.742	2.800	2.832	2.849
9	2.592	2.708	2.771	2.808	2.829
10	2.563	2.682	2.748	2.788	2.813
11	2.540	2.660	2.730	2.772	2.799
12	2.521	2.643	2.714	2.759	2.879
13	2.505	2.628	2.701	2.748	2.779
14	2.491	2.616	2.690	2.739	2.771
15	2.479	2.605	2.681	2.731	2.765
16	2.469	2.596	2.673	2.723	2.759
17	2.460	2.588	2.665	2.717	2.753
18	2.452	2.580	2.659	2.712	2.749
19	2.445	2.574	2.653	2.707	2.745
20	2.439	2.568	2.648	2.702	2.741
24	2.420	2.550	2.632	2.688	2.729
30	2.400	2.532	2.615	2.674	2.717
40	2.381	2.514	2.600	2.660	2.705
60	2.363	2.497	2.584	2.646	2.694
120	2.344	2.479	2.568	2.632	2.682
∞	2.326	2.462	2.552	2.619	2.670

Abridgment of H. L. Harter's "Critical Values for Duncan's New Multiple Range Test," Biometrics, Vol. 16, no. 4, 1960. With permission from the Biometric Society.

Table XII
**Upper percentage points of the studentized range distribution:
values of $q(0.05; k, v)$**

Degrees of Freedom, v	Number of Treatments k								
	2	3	4	5	6	7	8	9	10
3	4.50	5.91	6.83	7.51	8.04	8.47	8.85	9.18	9.46
4	3.93	5.04	5.76	6.29	6.71	7.06	7.35	7.60	7.83
5	3.64	4.60	5.22	5.67	6.03	6.33	6.58	6.80	6.99
6	3.46	4.34	4.90	5.31	5.63	5.89	6.12	6.32	6.49
7	3.34	4.16	4.68	5.06	5.35	5.59	5.80	5.99	6.15
8	3.26	4.04	4.53	4.89	5.17	5.40	5.60	5.77	5.92
9	3.20	3.95	4.42	4.76	5.02	5.24	5.43	5.60	5.74
10	3.15	3.88	4.33	4.66	4.91	5.12	5.30	5.46	5.60
11	3.11	3.82	4.26	4.58	4.82	5.03	5.20	5.35	5.49
12	3.08	3.77	4.20	4.51	4.75	4.95	5.12	5.27	5.40
13	3.06	3.73	4.15	4.46	4.69	4.88	5.05	5.19	5.32
14	3.03	3.70	4.11	4.41	4.64	4.83	4.99	5.13	5.25
15	3.01	3.67	4.08	4.37	4.59	4.78	4.94	5.08	5.20
16	3.00	3.65	4.05	4.34	4.56	4.74	4.90	5.03	5.15
17	2.98	3.62	4.02	4.31	4.52	4.70	4.86	4.99	5.11
18	2.97	3.61	4.00	4.28	4.49	4.67	4.83	4.96	5.07
19	2.96	3.59	3.98	4.26	4.47	4.64	4.79	4.92	5.04
20	2.95	3.58	3.96	4.24	4.45	4.62	4.77	4.90	5.01
24	2.92	3.53	3.90	4.17	4.37	4.54	4.68	4.81	4.92
30	2.89	3.48	3.84	4.11	4.30	4.46	4.60	4.72	4.83
40	2.86	3.44	3.79	4.04	4.23	4.39	4.52	4.63	4.74
60	2.83	3.40	3.74	3.98	4.16	4.31	4.44	4.55	4.65
120	2.80	3.36	3.69	3.92	4.10	4.24	4.36	4.47	4.56
∞	2.77	3.32	3.63	3.86	4.03	4.17	4.29	4.39	4.47

Answers to Selected Problems

Section 1.1

1. .3; relative frequency
3. .25; classical

Section 1.2

5. (*a*)

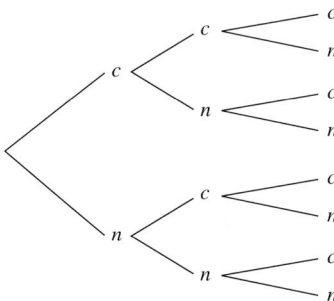

(*b*) {*ccc, ccn, cnc, cnn, ncc, ncn, nnc, nnn*}
(*c*) A_1 = {*ccc, ccn, cnc, cnn, ncc, ncn, nnc*}
A_2 = {*ccc*}
A_3 = {*nnn*}
(*d*) no; yes; yes; no
(*e*) no; the eight sample points are not equally likely

7. (*a*)

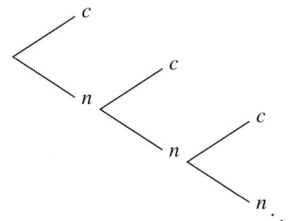

(b) no

(c) {c, nc, nnc, nnnc, . . . }; the list cannot be completed

(d) {c, nc, nnc, nnnc} = A

(e) A_1 = contact is made on the first try = {c}

A_2 = contact is made on the second try = {nc}

CHAPTER 1 review exercises

9. (a)

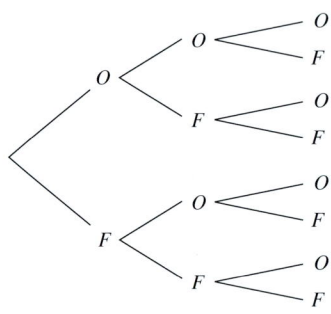

(b) {OOO, OOF, OFO, OFF, FOO, FOF, FFO, FFF}

(c) A = {OOO, OOF, OFO, OFF, FOO, FOF, FFO}

B = {OOO, OOF, OFO, OFF}

C = {FFF}

D = ∅

(d) no; yes; yes

(e) impossible event

(f) 1/8

10. (a)

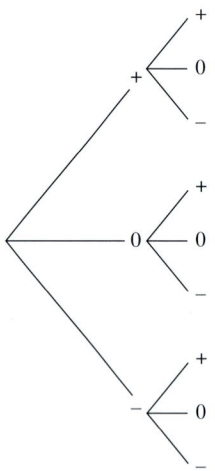

(b) {+ +, +0, + −, 0 + , 00, 0 −, − +, −0, − −}

(c) A = {− +, −0, − −}

B = {+ +, 00, − −}

C = {+0, + −, 0 −}

(*d*) no; yes

(*e*) The first item selected is not of inferior quality, and both items are of the same quality ($A' \cap B = \{+ +, 00\}$); the first item selected is of inferior quality, but the items are not of the same quality ($A \cap B' = \{- +, -0\}$); the first item selected is not of inferior quality, and the two items are not of the same quality ($A' \cap B' = \{0 +, 0 -, +0, + -\}$); the first item selected is of inferior quality, the quality of the first does not exceed that of the second, and the items are of the same quality ($A \cap C' \cap B = \{- -\}$)

(*f*) The argument is invalid because the nine outcomes are not equally likely.

11. (*a*)

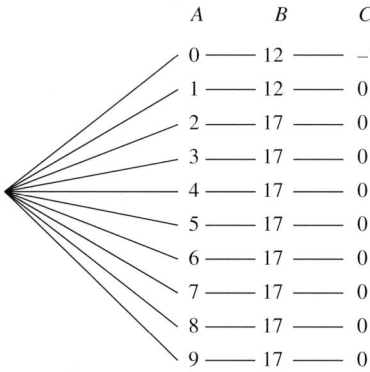

(*b*) $\{(0, 12, -1), (1, 12, 0), (2, 17, 0), (3, 17, 0), (4, 17, 0), (5, 17, 0), (6, 17, 0),$
$(7, 17, 0), (8, 17, 0), (9, 17, 0)\}$

(*c*) yes

(*d*) 5/10

(*e*) 1/10

(*f*) 9/10

(*g*) 1

Section 2.1

1. 12/13

3. .45; .13; .46

5. .58; .28; .1

7. .4

9. .42; .46

11. .004

Section 2.2

13. (*a*) 30/58

(*b*) 28/58

(*c*) Theorem 2.1.2

(*d*) 10/42

(*e*) no; exposure to the lethal dose should increase the probability of death

15. 5/35; 35/40

17. (*a*) 1/5
  (*b*) 1/80
  (*c*) .04
  (*d*) 4/20
  (*e*) .84

Section 2.3

19. no; $P[A_1 \cap A_2] = .2 \neq P[A_1]P[A_2]$
21. $(.39)^2 \doteq .15$
23. (*a*) .21
  (*b*) 21/23
25. .085
27. $.0144(.67) + .0012(.33) = .010044$
29. no; $P[B|T] \neq P[B]$
31. .931

Section 2.4

35. $.85(.10)/[.85(.10) + .04(.90)] = .7025$
37. .9999

CHAPTER 2 review exercises

38. (*a*) .85
  (*b*) .15
  (*c*) 5/20
  (*d*) 5/10
39. (*a*) 1/2
  (*b*) 1/8
40. .3529; .2353; .2647; .1471
41. .24; .6; .16
42. (*a*) .5
  (*b*) .35
  (*c*) .50
  (*d*) 1/3
  (*e*) 35/85
43. (*a*) .0008
  (*b*) .0002
  (*c*) .2
44. $(.99)^3(.01) = .00970299$; $.01 + .99(.01) + (.99)^2(.01) + (.99)^3(.01)$

Section 3.1

1. not discrete
3. discrete
5. not discrete

Section 3.2

7. (*a*) .01

(*b*)

x	0	1	2	3	4	5
$F(x)$	.7	.9	.95	.98	.99	1.00

(*c*) .98; .1

(*d*) .03

9. (*a*)

x	0	1	2	3
$f(x)$	$(.1)^3$	$3(.9)(.1)^2$	$3(.9)^2(.1)$	$(.9)^3$

(*b*) $k(x) = \dfrac{3!}{x!(3-x)!}$

(*c*)

x	0	1	2	3
$F(x)$	.001	.028	.271	1.00

(*d*) .999

(*e*) .028

11. (*a*)

x	0	1	2	3	4
$f(x)$	16/31	8/31	4/31	2/31	1/31

(*b*) $F(x) = 0$ for any $x < 0$

(*c*) $F(x) = 1$ for any $x > 4$

13. The sum of probabilities is the sum of numbers greater than or equal to 0.

Section 3.3

15. (*a*) 4.96; 26.34

(*b*) 1.7384; 1.3185

(*c*) holes per bit

17. $1/.7 = 10/7$; $1/p$

21. (*a*) 11

(*b*) -17

(*c*) 16

(*d*) 4

(*e*) 64

(*f*) 8

(*g*) 208

(*h*) 640

(*i*) 0; 1

(*j*) 0; 1

(*k*) $E\left[\dfrac{X-\mu}{\sigma}\right] = 0$ and $\text{Var}\left[\dfrac{X-\mu}{\sigma}\right] = 1$

23. (*e*) $163/60 = 2.7167$

(*f*) 2.7007

(*g*) $E[X_{100}] = 94.7953$

Section 3.4

25. 2; .6778; yes, $P[X \geq 5] = .0328$

27. (a) $p = 1/4$

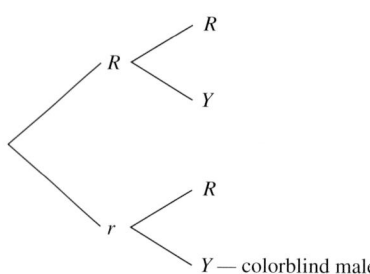

(b) 1.25; .1035

29. (a) .3758

(b) $(.3758)^5$

31. (a) $m_X(t) = (pe^t + q)^n$

32. (a) X = number of successes in one trial

$$f(x) = \binom{1}{x} p^x q^{1-x} \qquad x = 0,1$$

(b) $f(x) = p^x q^{1-x} \qquad x = 0, 1$

(c) $m_X(t) = pe^t + q$

(d) p; pq

(e) .14

Section 3.5

34. (a) 10

(b) 10

(c) $\sqrt{10}$

(d) $f(x) = \dfrac{e^{-10}10^x}{x!} \qquad x = 0, 1, 2, \ldots$

(e) .029

(f) .010

(g) .019

(h) .99

(i) .448

36. .010

38. yes; $P[X > 10] = .043$

40. yes; $P[X < 2] = .04$

43. $(.0247)^5 = .00000000919$

CHAPTER 3 review exercises

47. .118; .882

48. (a) 5/4; 15/16

(b) $(3/4)^5$

(c) $(3/4)^{10}$

49. (*a*) X is binomial with $n = 100$ and $p = .1$; $E[X] = 10$
  (*b*) Poisson
  (*c*) $P[X \geq 17] \doteq .027$
50. 5; 61/125; 64/125
51. 35/210; 105/210
52. $(999/1000)^{12,000} \doteq e^{-12}12^0/0! \doteq .000006; .999994$
53. yes; $P[X \geq 5] = .0127$
54. (*b*) $36/14$; $98/14$
  (*c*) $76/196$; $\sqrt{76/196}$
55. (*a*) $f(x) = (.99)^{x-1}(.01)$ $x = 1, 2, 3, \ldots$
  (*b*) 100
  (*c*) $F(x) = 1 - .99^x$ x a positive integer
  (*d*) $P[X \leq 90] = 1 - (.99)^{90} \doteq .595$

Section 4.1

1. (*a*) 1/6
  (*b*) 11/48
  (*c*) 0
  (*d*) 11/48
3. (*b*) $1 - e^{-.7} = .5034$; .4966; 0
  (*c*) yes, $P[1 \leq X \leq 2] = e^{-.1} - e^{-.2} \doteq .086$
5. (*d*) .5
  (*e*) equal; probabilities are constant or "uniform" over intervals of equal lengths
9. (*a*)

$$F(x) = \begin{cases} 0 & x < 2 \\ x^2/12 - 1/3 & 2 \leq x \leq 4 \\ 1 & x > 4 \end{cases}$$

  (*c*) yes; yes; 0; 1; yes
  (*d*) $\dfrac{dF(x)}{dx} = f(x)$

11. (*a*)

$$F(x) = \begin{cases} 0 & x < 0 \\ x/2\pi & 0 \leq x < 2\pi \\ 1 & x \geq 2\pi \end{cases}$$

  (*b*) yes; yes; 0; 1; yes
  (*c*) $\dfrac{dF(x)}{dx} = f(x)$

13.

$$F(x) = \begin{cases} 0 & x < 25 \\ (\ln x - \ln 25)/\ln 2 & 25 \leq x \leq 50 \\ 1 & x > 50 \end{cases}$$

Section 4.2

15. (*a*) 56/18
  (*b*) 10
  (*c*) 104/324; $\sqrt{104/324}$

17. (a) $m_X(t) = (1 - 10t)^{-1}$ $t < 1/10$
   (b) 10 minutes
   (c) 100; 10 minutes

19. π; $\pi^2/3$; $\pi/\sqrt{3}$

21. 10; 10; X

23. (b) $F(x) = \dfrac{25}{24} - \dfrac{25}{6}x^{-2}, 2 < x < 10$; .78125
   (c) 1.033
   (d) 5.364; 4.297

Section 4.3

24. (a) .9418
   (b) .9418
   (c) 0
   (d) .0582
   (e) .8543
   (f) 1.28
   (g) -1.28
   (h) 1.96
   (i) 1.645

25. (a) .9544
   (b) 1.24%
   (c) 128 parsecs
   (d) $m_X(t) = e^{5000t^2}$

28. (a) .9525
   (b) 5.065

Section 4.4

34. $P[-2\sigma < X - \mu < 2\,\sigma] = .95$

36. (a) no
   (b) $P[|X - \mu| < .5\sigma] \geq -3, P[|X - \mu| < 1\sigma] \geq 0,$
      $P[|X - \mu| < .2\sigma] \geq .75, P[|X - \mu| < 3\sigma] \geq .89$; $k = 2$

Section 4.5

38. (a) yes; .9956
   (b) .0668

40. (a) yes
   (b) 37.5
   (c) .0322
   (d) .0392

42. .9484; .8413

CHAPTER 4 review exercises

46. (a) 1/18
   (b) 0; 5.4
   (c) 5.4; $\sqrt{5.4}$
   (d) 35/54; 9/54; 26/54

(*e*)

$$F(x) = \begin{cases} 0 & x < -3 \\ (x^3 + 27)/54 & -3 \le x \le 3 \\ 1 & x > 3 \end{cases}$$

47. $\Gamma(11) = 10! = 3,628,800$

48. (*a*) $\alpha = 9$ and $\beta = 2$
   (*b*) .01; .10; .725 (chi-squared with 18 degrees of freedom)

49. (*a*) $f(x) = \dfrac{1}{\sqrt{2\pi}\,(5)} \exp\left[-\dfrac{1}{2}\left(\dfrac{x-15}{5}\right)^2\right]$

   (*b*) .0082
   (*c*) no; $P[10 \le X \le 20] \doteq .68$ by the normal probability rule
   (*d*) yes; $P[X \ge 30] = .0013$

50. .7904; $e^{-6.25(6)}$; .042

53. 8; .2912; no; $P[X \le 5] \doteq .1788$

54. $f(x) = (3/2)x^2 + x, \qquad 0 \le x \le 1$

55. .4892

Section 5.1

1. yes; set of all days from past, present, and future
3. no
5. yes; the 50,000 workers affected

Section 5.2

7. (*a*) 58.8
   (*b*) 8.5
   (*c*) 16.25
   (*d*) 16.25 to 24.75
       24.75 to 33.25
       33.25 to 41.75
       41.75 to 50.25
       50.25 to 58.75
       58.75 to 67.25
       67.25 to 75.75

9. (*a*)

0	2
0	5 7
1	1 2 2 3 4 4 4 4
1	5 5 6 6 7 9
2	0 1 1 2 3
2	5 6 8 9
3	0 0 1
3	7 7
4	0 1
4	5
5	1
5	8

   (*b*) yes; right

11. (*d*)

Category	Boundaries	Frequency	Relative frequency	Relative cumulative frequency
1	.45 to 1.25	7	7/50	7/50
2	1.25 to 2.05	15	15/50	22/50
3	2.05 to 2.85	15	15/50	37/50
4	2.85 to 3.65	8	8/50	45/50
5	3.65 to 4.45	3	3/50	48/50
6	4.45 to 5.25	2	2/50	50/50

the gamma distribution might be appropriate

13. (*a*) $P[X < p_{25/100}] \leq 25/100$ and $P[X \leq p_{25/100}] \geq 25/100$
    (*b*) 8
    (*c*) $-\ln .75 \doteq .288$
15. (*a*) approximately .044
    (*b*) approximately 7.7
    (*c*) approximately 2.2

Section 5.3

17. (*a*) group I   3; 3
        group II   3; 2.5
    (*b*) 4, 4
    (*c*) group I   1.5; 1.2
        group II   2.91; 1.7
    (*d*) yes
19. .8; 1.07
21. 7.94; 1.221; 1.11
23. (*a*) $\bar{x} = 2.31, \tilde{x} = 2.05$
    (*b*) $s = 1.29, s^2 = 1.6745$
    (*c*) $\bar{x}, \tilde{x}, s$ are measured in minutes; s^2 is unitless
25. (*a*) yes
    (*b*) 1.86
    (*c*) 1.975

Section 5.4

27. (*b*) 1.34σ

29. (*a*)

```
1 | 1
1 | 5 8
2 | 4 3
2 | 6 5 7 9
3 | 2 0
3 | 6 9 5 7 8 7 5
4 | 0 0 2
4 | 7 9
5 | 0 1
5 | 6
6 | 0
```

(*b*) $\tilde{x} = 3.65$ $iqr = 1.45$ no outliers
    $q_1 = 2.65$ $f_1 = .475$
    $q_3 = 4.10$ $f_2 = 6.275$

(*c*)
```
2 | 0
  |
4 |
4 | 2 3
4 | 5 4
4 | 6 7 6
4 | 9 8
5 | 0 1 0
5 | 2 2 3 2 3
5 | 4 4
5 | 6
5 | 8 9
  |
7 | 8
```

(*d*) $\tilde{x} = 5.05$ $iqr = .75$ $F_1 = 2.35$
    $q_1 = 4.6$ $f_1 = 3.475$ $F_3 = 7.6$
    $q_3 = 5.35$ $f_3 = 6.475$
    2 is an extreme outlier; 7.8 is an extreme outlier

(*e*)
```
3 |
3 | 9
4 | 2 4 3 1
4 | 5 7 9 8 7 6 8
5 | 0 1 1 2 0 4
5 | 9 6
6 | 1
6 |
```
    58 has a misplaced decimal and should read 5.8; $\bar{x} = 7.29$ (using bad point); $\bar{x} = 4.91$ (using corrected point)

CHAPTER 5 review exercises

26. (*a*) Poisson
    (*b*) 1.9; 3.36; 1.8; 1
    (*c*) no
    (*d*) 6.24; 1.852; 1.36; 6.3
31. (*a*) range = 5.73

Category	Boundaries	Frequency	Relative frequency	Relative cumulative frequency
1	.235 to 1.195	2	2/50	2/50
2	1.195 to 2.155	18	18/50	20/50
3	2.155 to 3.115	13	13/50	33/50
4	3.115 to 4.075	7	7/50	40/50
5	4.075 to 5.035	7	7/50	47/50
6	5.035 to 5.995	3	3/50	50/50

    (*b*) 2.788; 1.8594; 1.364
    (*c*) approximately 2.644; about 16%

(d) p; 3/50; 3/50

(e) $p(1 - p)$; (3/50)(47/50) = .0564; .0576; no; .0576

(f) $\sigma_X \doteq .955$

32. (a)

```
1 | 1
1 | 3
1 | 5
1 |
1 | 9 8
2 | 1 0
2 | 2 2 2 3 3
2 | 5 5 5 5 4 5 4 4 5 5 4
2 | 6 7 7 6 7 7 6
2 | 9 8 8
3 | 0
3 |
3 |
3 |
3 |
3 |
```

(b) $\tilde{x} = 25$ $iqr = 4$ $F_1 = 10$

   $q_1 = 22$ $f_1 = 16$ $F_3 = 38$

   $q_3 = 26$ $f_3 = 32$

   the values 11, 13, 14, and 15 are flagged as outliers

(c) there is a left skew; gamma; no; outliers are flagged, assuming normality

(d)

```
3 |
3 | 3
3 | 5 5
3 | 6 7 7 6 6 6 7 7
3 | 8 8 9 9 8 9 8 9 9
4 | 0 0 0 0 0 1 1
4 | 2 3 3
4 | 5 5
4 | 6 6
4 |
```

   yes

(e) $\tilde{x} = 3.9$ $f_1 = 3.1$ no outliers

    $q_1 = 3.7$ $f_3 = 4.7$

    $q_3 = 4.1$ $F_1 = 2.5$

    $iqr = .4$ $F_3 = 5.3$

33. (a)

```
15 | 0
16 | 0 0
17 | 0 0 0 0 0 0
18 | 0 0 0 0 0
19 | 0 0 0
20 | 0 0
21 | 0
22 | 0
```

   yes

(b) $\bar{x} = 18.0$; $\tilde{x} = 18$

(c) $s = 1.7$; $s^2 = 2.95$ (from TI83)

(*d*) $q_1 = 17, q_3 = 19, iqr = q_3 - q_1 = 2$
The TI83 and MINITAB give these same results.

34. (*a*) 275.87, 30.57
(*b*)

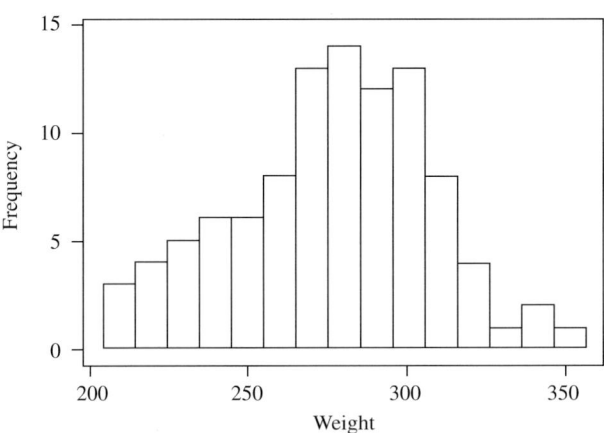

(*c*) yes
(*d*) 250; 295; 275
(*e*) TI83 results: $q_1 = 255.85, q_3 = 298.25, \tilde{x} = 277.85$
MINITAB results: $q_1 = 255.63, q_3 = 298.28, \tilde{x} = 277.85$

35. (*a*) The distributions of life span for both lamp types are skewed to the right
(*b*)

Variable	Lamp type	Mean	Median	StDev	Variance
Life span	1	38.38	28.78	40.48	20.07
	2	28.97	17.14	30.22	913.25

mean > median

(*c*)

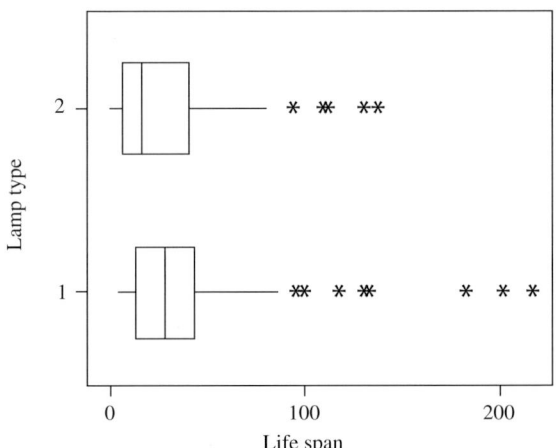

(*d*)

Variable	Lamp type	Mean	Median	StDev	Variance
Life span	1	28.92	23.02	22.04	485.76
	2	25.04	15.28	23.61	557.43

36. (*a*)

Day	Mean	Median	StDev	Min	Max	Q_1	Q_3
sn	25.9	25.5	5.39	1.21	19	21.25	30.75
m	20	20	4.255	12	31	17.25	22
t	18.55	19	4.032	10	24	16.25	21.75
w	21.15	22	3.76	13	28	19.25	23
th	22.4	21	5.35	13	34	19	27
f	27.45	27.5	6.44	16	44	22	32.25

(*b*)

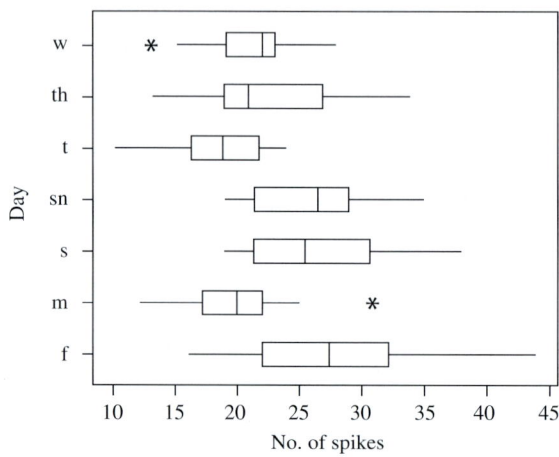

Section 6.1

1. 8; 5/20

3. (*a*) 5.25

   (*b*) 5.25

   (*c*) 5.25/9

5. (*b*) .44

7. (*a*) 9.7

   (*b*) 3.57

   (*c*) 19.4

   (*d*) 1.9; no

9. (*b*) .817; no, the sample sizes are very different

   (*d*) .708

13. (*a*)

16	0 3 2 0 2 3 1 2 4 3
16	5 7 8 7 9
17	0 2 3
17	6 9 8
18	0
18	5
19	
19	6

   skewed right

   (*b*) $\tilde{x} = 167$ $f_1 = 148$

         $q_1 = 163$ $f_3 = 188$

         $q_3 = 173$ $F_1 = 133$

        $iqr = 10$ $F_3 = 203$

196 is flagged; the distribution is not bell-shaped, so the "outlier" is probably not unusual

(c) $\bar{x} = 169.0\ (169.04)$

$s^2 = 81.54$

Section 6.3

16. (a) .643

(b) $.643 \pm .0039$

(c) shorter; $.643 \pm .0033$

(d) longer; $.643 \pm .0052$

18. (a) no, X is discrete

(b) 2.8

(c) $2.8 \pm .2036$; Central Limit Theorem

(d) no; 3.0 lies in this interval—it has not been ruled out as a possible value for μ

CHAPTER 6 review exercises

20. (a) normal with mean μ and variance 4/16

(b) 42.883

(c) $42.883 \pm .98$; no, 42.2 lies in the confidence interval

21. (a) \$3.698

(b) .0487

(c) .221; no

(d) .0438; no

22. (a) point binomial with probability of success p

(b) normal with mean p and variance $p(1 - p)/n$

(c) .05

23. (a) 7.1

(b) normal with mean μ and variance 25/36

(c) 7.1 ± 1.9

(d) yes; 10 lies outside the 99% confidence interval

24. (a) standard normal

(b) X_1^2

(c) X_{10}^2; $\bar{y} \sim$ normal; $\mu = \dfrac{\lambda s}{n}$; $\sigma^2 = \dfrac{\lambda s}{n^2}$

25. (a)

x	1	2	3	4	5	6
$f(x)$	1/6	1/6	1/6	1/6	1/6	1/6

$$E[X] = 3.5$$
$$E[X^2] = 91/6$$
$$\sigma^2 = 35/12 = 2.92$$

(b) $E[\bar{X}] = 3.5$; Var $\bar{X} = 2.92/20 = .146$

(c) bell

(d) 3.5

(e) .146

(f) 24

Section 7.1

1. (*b*) .0129
   (*c*) [.0082, .0234]
   (*d*) [.091, .153]
   (*e*) yes; .2 is not in the confidence interval
3. (*b*) 2.455
   (*c*) [1.322, 5.699]
   (*d*) [1.15, 2.39], reduce the confidence
5. (*b*) .00000375
   (*c*) [0, .00000573]; [0, .00239]
   (*d*) yes; σ appears to be at most .00239
7. [5.49, 11.83]

Section 7.2

9. (*a*) 1.86
   (*b*) -1.86
   (*c*) -2.179
   (*d*) 2.179
   (*e*) 1.645
   (*f*) 1.645
   (*g*) 1.708
   (*h*) 2.060
   (*i*) 1.753
   (*j*) 1.325
   (*k*) 1.746
   (*l*) 1.310
11. (*a*) 1.2896; .0000123; .0035
   (*b*) $1.2896 \pm .0016$
   (*c*) no, 1.29 is contained in the confidence interval
13. (*a*) 2.35; .89
   (*b*) $2.35 \pm .45$
   (*c*) yes, we are 99% confident that the new mean time is at most 2.80 seconds
15. (*a*)

2	9						
3	8	5	9				
4	2	1	7	8			
5	1	3	5	6	7	5	
6	1	8	1	0	2	5	7	3
7	9	7	3	2			
8	1	0					
9	2						
10	0						

   (*b*) $\tilde{x} = 605$ $f_1 = 120$ no outliers
        $q_1 = 480$ $f_3 = 1080$
        $q_3 = 720$
        $iqr = 240$
   (*c*) $\bar{x} = 602.3$; $s = 169.1$; 602.3 ± 85.1
   (*d*) lower the confidence
17. $[37.11, \infty)$
19. (*b*) 385
   (*c*) 153

Section 7.3

21. (*a*) H_0: $\mu \geq .08$
H_1: $\mu < .08$
(*b*) We shall conclude that the average percentage of metal in household wastes has been reduced when, in fact, it has not been reduced.
(*c*) We shall be unable to detect the fact that the mean percentage of metal in household waste has been reduced.
(*d*) We have a 5% chance of having committed a Type I error.

23. (*a*) We shall conclude that the model is not credible when, in fact, it is a valid model.
(*b*) We shall be unable to detect the fact that the proposed model is not credible.

25. (*a*) $C = \{10, 11, 12, 13, 14, 15\}$, $\alpha = .0338$
(*b*) yes; Type I

27. (*a*) H_0: $p \leq .5$
H_1: $p > .5$
(*b*) 7.5
(*c*) .0592
(*d*) .7827; .4845; .1642; .0127
(*e*) .2173; 5155; 8358; 9873
(*f*) yes; Type I
(*g*) no; Type II

29. .0065; .0003; 0; 0; yes

Section 7.4

31. (*a*) H_0: $\mu \leq .05$
H_1: $\mu > .05$
(*b*) We shall assume that the percentage titanium exceeds 5% when, in fact, it does not; we shall be unable to detect a situation in which the percentage titanium exceeds 5%.
(*c*) .1056; debatable, a P value of .1056 might be considered small by some and large by others

33. (*a*) H_0: $p \leq .15$
H_1: $p > .15$
(*b*) .1335; no, this probability is not unusually small; Type II

Section 7.5

35. (*a*) -1.711
(*b*) -1.282
(*c*) 2.093
(*d*) 2.602
(*e*) ± 1.729
(*f*) ± 2.045

37. (*a*) H_0: $\mu = .12$
H_1: $\mu > .12$
(*b*) 2.462
(*c*) yes, $t = 2.738$
(*d*) that X is at least approximately normal

39. (*a*) H_0: $\mu = 2.5$
H_1: $\mu < 2.5$
(*b*) $t = -3.5$; $.001 < P < .005$; yes, P seems to be small; at least approximate normality
(*c*) conclude that the mean noise level is below 2.5 db; we shall assume that the new product reduces noise when, in fact, it does not
41. (*a*) H_0: $\mu = 4.8$
H_1: $\mu < 4.8$
(*b*) $t = -2.828$; $.001 < P < .005$; yes
43. (*a*) H_1: $\mu < 5$
(*b*) $t = -3.47$; $.001 < P < .005$; reject H_0
(*c*) yes, because $P < .05$
(*d*) probably not since $\bar{x} = 4.28$; there is still a large accumulation there
45. H_1: $\mu < 7$, $P = .27$, no

Section 7.6

47. (*a*) unable to reject H_0
(*b*) $t = 1.154$, critical points $= \pm 2.145$; unable to reject H_0
(*c*) $\chi^2 = 17.81$, critical point $= 23.7$, unable to reject H_0
49. (*a*) unable to reject H_0
(*b*) $t = 1.22$, critical point $= 1.729$, unable to reject H_0
(*c*) $\chi^2 = 9.297$, critical point $= 11.7$, reject H_0; yes

Section 7.7

51. (*a*) yes, .0037
(*b*) yes, .0207
(*c*) yes, .0207
(*d*) yes, .0107
(*e*) no, .0547
(*f*) yes, .0074
(*g*) yes, .0352
53. no, $P = .3770$
55. (*a*) 110/4
(*b*) yes, $|W_-| = 8.5$, critical point $= 11$
57. (*a*) 100(101)/4
(*b*) 100(101)(201)/24
(*c*) H_0: $M = 2$
H_1: $M > 2$
(*d*) $P \doteq .0007$, yes

CHAPTER 7 review exercises

58. (*a*) 44
(*b*) no
(*c*) 3.10; .1213; .348, no
(*d*) $\chi^2_{.95} \doteq 33.65$; [.08998, .1766]; [.298, .418]
$\chi^2_{.05} \doteq 66.05$
(*e*) $3.10 \pm .082$

59. (*a*) $H_0: \mu = 3$
   $H_1: \mu < 3$
   (*c*) 13
   (*d*) $t_{.95} = -1.753$; $t \doteq -3.71$; reject H_0; yes, the product should be marketed
60. (*a*) $H_0: p \leq .5$
   $H_1: p > .5$
   (*b*) $C = \{15, 16, 17, 18, 19, 20\}$
   (*c*) no; Type II
   (*d*) .8744; .5836; .1958; .0113
   (*e*) .1256; .4164; .8042; .9887
61. yes; $t = -2.69$; $.005 < P < .01$
62. $.10 < P < .25$
63. (*a*)

$-.3$	0
$-.2$	0 0 0 0
$-.1$	0 0 0 0 0 0 0 0
0	0 0 0 0 0 0 0 0 0 0 0 0
.1	0 0 0 0 0 0
.2	0 0 0 0
.3	0
.4	
.5	
.6	
.7	
.8	0

   yes; .8 looks like an outlier
   (*b*) $\tilde{x} = 0$ $f_1 = -.4$
     $q_1 = -.1$ $f_3 = .4$
     $q_3 = .1$ $F_1 = -.7$
     $iqr = .2$ $F_3 = .7$
     .8 is an extreme outlier
   (*c*) $\bar{x} = -.008$; $s = .198$; $-.008 \pm .066$
64. (*a*) The number of trials is at most 3000; in a geometric setting there is no a priori number of trials.
   (*b*) 20
   (*c*) exceed
   (*d*) 59; $P[X \geq 59] = 0.51$; see Sec. 3.4
   (*e*) no; Type II; we think that the system will not crash when, in fact, it will
   (*f*) reject H_0 and conclude that the system will crash; Type I; we shall stop the system unnecessarily
65. (*b*) $8.1 \pm .8$
   (*c*) [4.36, 10.56]
   (*d*) 3.3 ± 10.7; [.80, 3.17]; [.90, 1.8]
   (*e*) no
   (*f*) no
66. (*a*) H_1: hollow arrows are faster than those made of solid aluminum
     $H_1: \mu < 0$
   (*b*) H_0: hollow arrows are no faster than those made of solid aluminum
     $H_0: \mu = 0$
   (*c*) $\bar{x} = -20.41$; $s = .89$; negative sign means that the time with the hollow arrow was better than that with the solid arrow
   (*d*) $t = -102.5$; $P < .0005$

67. (*a*) **e:**
```
0 | 6
1 | 4 5 9
2 | 3 4 4 6 6 6 7 9 9
3 | 0 1 2 4 5 6 6 6 9 9 9
4 | 0 2 3 4 4 4 6 6 6 6 6 7 7 7 8 8
5 | 1 1 2 2 2 3 3 3 4 9 9
6 | 0 0 1 2 2 3 4 5 6 6 8 9 9
7 | 1 1 4 4 5 8
8 | 0 0 1
9 | 6
```

wb:
```
 2 | 4 8
 3 | 1 9 9
 4 | 0 5 5 9
 5 | 0 0 1 2 3 6 8 8 9
 6 | 3 4 6 7 7 7 8 8 9
 7 | 0 0 1 1 3 4 5 5 6
 8 | 0 0 1 1 1 2 3 3 3 4 5 6 8 9
 9 | 2 4 4 5 5 5 5 6 6 7 8 8 9
10 | 2 3 3 5 6
11 | 2 3 8 9
12 | 5
13 | 7
14 | 1
```

(*b*) **e:** [4.615, 5.359]
wb: [74.112, 83.608]
absolutely; the confidence intervals do not come close to overlapping

(*c*)

Company/Accident	N	Mean	Median	StDev
ge	25	4.782	4.790	1.527
gwb	25	77.750	73.250	24.450
me	25	4.829	4.750	2.195
mwb	25	69.370	70.200	25.820
pe	25	5.352	5.180	2.046
pwb	25	89.460	84.600	20.120

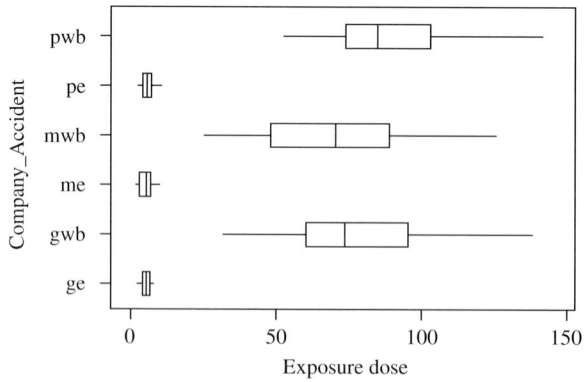

The differences are obviously due to the differences among accident types.

Section 8.1

1. (*a*) .9
   (*b*) .9 ± .069
   (*c*) 609
3. (*a*) .6 ± .025
   (*b*) .9 ± .019
5. 1068
7. 6766
9. (*a*) $1 - 2\hat{p}$
   (*b*) 1/2
   (*c*) -2
   (*d*) 1/4

Section 8.2

11. (*a*) $H_0\!: p = .99$
    $H_1\!: p > .99$
    (*b*) $z = .57$; no, $P = .2843$
    (*c*) We are unable to show that the new network is compatible with more than 99% of the equipment already in use.
13. (*a*) $H_0\!: p = .6$
    $H_1\!: p > .6$
    (*b*) 1.645
    (*c*) $z = .842$, no; Type II
    (*d*) We shall not be able to show that more than 60% of the business offices in the United States have a mainframe when, in fact, this is true.
15. (*a*) ± 1.96
    (*b*) $z = 1.38$, no; Type II

Section 8.3

17. (*a*) .356, .489, $-.133$
    (*b*) $-.133 ± .062$
    (*c*) yes, 0 is not contained in the confidence interval
19. (*a*) .31, .40, $-.09$
    (*b*) $-.09 ± .079$
    (*c*) yes, 0 is not contained in the confidence interval
23. 1527
25. 542

Section 8.4

27. (*a*) $H_0\!: p_1 = p_2$
    $H_1\!: p_1 < p_2$
    (*b*) -1.28
    (*c*) .0015, .0025, $-.001$; $\hat{p} = .002$, $z = -.71$, no
29. (*a*) $H_0\!: p_1 = p_2$
    $H_1\!: p_1 > p_2$
    (*b*) $\hat{p} = .334$, $z = 12.35$, yes, $P \doteq 0$

CHAPTER 8 review exercises

33. (*a*) 533

   (*b*) .21; .21 ± .034

34. (*a*) $H_0: p = .5$

   $H_1: p > .5$

   (*b*) $z = .2828$; no, $P \doteq .39$

35. (*a*) $H_0: p = .5$

   $H_1: p > .5$

   (*b*) 1.645

   (*c*) no; the observed value of the test statistic is 1.00

36. (*a*) $- .02$

   (*b*) $- .02 \pm .047$

   (*c*) no; 0 lies in the interval

   (*d*) no; $n \doteq 2229$, using $\hat{p}_1 = .02$ and $\hat{p}_2 = .04$ as prior estimates

37. $H_0: p_1 = p_2$

   $H_1: p_1 > p_2$

   where p_1 denotes the proportion of customers reordering during the current year; $\hat{p} = .6, z = 1.87$, yes, $P = .0307$

38. (*a*) $H_0: p_1 - p_2 = .1$

   $H_1: p_1 = p_2 > .1$

   (*b*) 1.645

   (*c*) $z = .25$; no, Type II

39. (*a*) .2 ± .078

   (*b*) 75,000 ± 29,400

   (*c*) 1537

40. (*a*) .25

   (*b*) .25 ± .003

   (*c*) [2468, 2532]

Section 9.1

1. 4.2; 2; 2.2

3. 8.008; 7.609; .399

Section 9.2

5. (*a*) .9

   (*b*) 2.416

   (*c*) .05

   (*d*) 1.841

   (*e*) .05

   (*f*) 5.803

7. (*a*) $f = 5$, upper critical point = 2.725, reject H_0

   (*b*) $f = 2$, upper critical point = 2.308, unable to reject H_0

   (*c*) $f = 1.36$, upper critical point = 1.352, reject H_0

9. (*a*) $H_0: \sigma_1^2 = \sigma_2^2$

   $H_1: \sigma_1^2 < \sigma_2^2$

   (*b*) $f = .9151$; $P = .4485$; fail to reject H_0

11. $P = .7213$; no; no

Section 9.3

13. (a) $s_1^2 = .034$, $s_2^2 = .0525$, $f = 1.54$, upper critical point $= 2.147$, unable to reject H_0
 (b) .0433
 (c) $- .67 \pm .14$
 (d) yes, the confidence interval does not contain 0
15. (a) $f = 1.23$, upper critical point $= 2.086$, unable to reject H_0
 (b) 7.93
 (c) 7.88 ± 3.18
 (d) yes, 0 is not in the interval
17. no; $f = 1.79$, $P = .421$
19. (a) yes, $f = 1.599$, $P = .2598$
 (b) [.42 and 20.36]; night hours
 (c) 99%; decrease confidence level

Section 9.4

21. (a) H_0: $\mu_1 = \mu_2$ $\mu_1 =$ mean level for children
     H_1: $\mu_1 > \mu_2$
 (b) $f = 119.01$, upper critical point $(\alpha = .2) = 1.348$, do not pool
 (c) $t = 20.00$, $\gamma = 123$, reject H_0: $P < .0005$; conclude that the average strontium level in children is higher than that in adults
23. (a) H_0: $\mu_1 = \mu_2$
     H_1: $\mu_1 > \mu_2$ (premium $>$ regular)
 (b) $t = 1.583$, reject H_0, $.05 < P < .1$
 (c) conclude that the average mileage using premium gasoline is higher than that using regular gasoline
25. (a) H_0: $\mu_1 = \mu_2$
     H_1: $\mu_1 < \mu_2$ (acrylic $<$ butyl)
 (b) $f = 2.75$, critical point $(\alpha = .1) = 1.972$, do not pool
 (c) $t = -5.88$, $\gamma = 24$, reject H_0, $P < .0005$; conclude that the average tensile strength for butyl coating is higher than that for acrylic coatings
27. $[-.3116, -.1664]$; yes, all values in the interval are negative
29. (a) $f = 3.32$, critical point $(\alpha = .10) = 1.984$, do not pool
 (b) 2.8 ± 1.94, $\gamma = 37$
 (c) yes, 0 is not in the interval

Section 9.5

31. (b) $- .412, 1.14$
 (c) $- .412 \pm .723$, no, the confidence interval contains 0, so the results are inconclusive
33. $.04 \pm .585$; no, 0 lies in the interval
35. H_0: $\mu_X = \mu_Y$ $t = 3.04$, reject H_0: $.0005 < P < .005$
 H_1: $\mu_X > \mu_Y$ $X =$ travel lane

Section 9.6

37. H_0: $M_I = M_P$, $W_m = 101.5$, upper critical point $= 98$; reject H_0
H_1: $M_I < M_P$

39. $W_m = 20$, critical point $= 22$, reject H_0

41. (*a*) 6600
(*b*) 110000
(*c*) $z = -1.999, P = .0228$, yes

43. $|W_-| = 7$, can reject at $\alpha = .05$ level (critical point $= 11$); no, $P[Q_- \leq 3] = .1719$, the sign test ignores the magnitude of the differences involved

CHAPTER 9 review exercises

45. (*a*) H_0: $\mu_1 = \mu_2$ where $\mu_1 =$ mean temperature setting required by
H_1: $\mu_1 < \mu_2$ using the computerized system
H_0: $\sigma_1^2 = \sigma_2^2$ where $\sigma_1^2 =$ variance in temperature settings required
H_1: $\sigma_1^2 < \sigma_2^2$ by using the computerized system
(*b*) $s_2^2/s_1^2 = 25$; reject H_0: $\sigma_1^2 = \sigma_2^2$; $P < .05$; do not pool; $\gamma \doteq 25$; $t = -8.73$; reject H_0: $\mu_1 = \mu_2$; $P < .0005$; yes, both claims are supported

46. (*a*) H_0: $\mu_1 - \mu_2 = 15$ where $\mu_1 =$ mean amount of dross obtained by using
H_1: $\mu_1 - \mu_2 > 15$ the old method
(*b*) $s_1^2/s_2^2 = 25$; reject H_0: $\sigma_1^2 = \sigma_2^2$; $P < .10$; do not pool; $\gamma \doteq 9$; $t = 4.96$; reject H_0; $P < .0005$; yes, it appears that the new process will be profitable

47. (*a*) yes
(*b*) 2.6 ± 1.0; yes, the confidence interval consists entirely of positive values

48. $s_1^2/s_2^2 = 1.14$; pool; 900 ± 58.2; yes, the confidence interval consists of positive values

49. yes, $t \doteq 7.56$; $P < .0005$

50. (*a*) $\bar{d} = -14.43$
(*b*) $-14.43 \pm .40$; $s_d = .72$
(*c*) yes; 0 is not contained in the confidence interval

51. (*a*)

	A		R	
$\tilde{x} = 130$	$f_1 = 90.25$	$\tilde{x} = 253$	$f_1 = 188.5$
$q_1 = 119.5$	$f_2 = 168.25$	$q_1 = 238$	$f_3 = 320.5$
$q_3 = 139$	$F_1 = 61$	$q_3 = 271$	$F_1 = 139$
$iqr = 19.5$	$F_3 = 197.5$	$iqr = 33$	$F_3 = 370$

12.4 and 200 are extreme   516 is an extreme outlier
extreme outliers; 170 is a
mild outlier
(*b*) $f = 8.83$; upper critical point 1.908; reject at .20 level; conclude that $\sigma_A^2 \neq \sigma_R^2$
(*c*) $(267.4 - 133.9) \pm 2.093 \sqrt{4448.89/17 + 503.83/19}$; $df \doteq 19$
(*d*) yes, the confidence interval is negative

52. (*a*)

nonsmokers								smokers						
15	1							15	1 7 3 0 2 9					
16	2 9							16	8 1 0 3					
17	2 5 6 4							17	7 2 9					
18	9 8 1 3							18	1 0 3					
19	8 5 7 9 8 2 3							19	4 9 9					
20	0 6 1 5 4 1 2 7							20	5					
21	2 1 8 1 4 3							21	2 3 6					
22	6 1 1 4							22	8 4 1					
23	0 6 3							23	2 1 8 0					
24	1 9							24	3 1 1 9 8					
25	0							25	8 7 1 2 2 0 1					

           nonsmokers   smokers

    (*b*) yes; there is a difference in shape, and also variability may be different
    (*c*) use the Wilcoxon rank-sum test to test location

53.

bank										other										
1	2									3	4									
1	8 9 7 9 5 7									3	7 9 7								
2	0 1 2 0 4 3 1 3 3									4	3 2 0 0 1								
2	9 7 8 5 9 7 6 5 5									4	7 9 8								
3	1 3 0 2 1 4									5	1 4 2 0 1 3 2 1 2 0								
3	8									5	9 7 9 8								
										6	4 3 2								
										6	9								

    bank   other

    a two-sample *t* test should apply; $f = 2.19$; variances are unequal ($\alpha = .10$);
    $t = -13.17$; $df \doteq 50$

54. (*a*) Site 1: 17; Site 2: 13
    (*b*) 28
    (*c*) 26
    (*d*) yes; 17 is not too different from 13 AND $s_1 \approx s_2$
    (*e*) two-tailed
    (*f*) $P = .058$
    (*g*) no; $P > .05$; yes, since $P < .01$

55. (*a*) 1: observed value of test statistic
         2: *P* value for two-tailed test
         3: standard deviation of the difference $X - Y$, s_d
         4: Mean difference, $\bar{d}$
         5: Standard error of the mean difference, $s_{\bar{d}} = s_d / \sqrt{n}$
    (*b*) yes; yes

Section 10.2

7. (*b*) $\hat{\mu}_{Y|x} = 0.2177 + 0.0957x$
    (*c*) 5.004 for both estimates
    (*d*) increase .1914 $\times 10^8$ Btu
    (*e*) decrease .1914 $\times 10^8$ Btu

9. (*b*) $\hat{\mu}_{Y|x} = -.1064 + 33.5178x$
    (*d*) .8995
    (*e*) .8995
    (*f*) yes, yes, no

15. .1814

Section 10.3

17. significant ($P < .001$)
19. $t = 4.025$, significant at $\alpha = .01$
21. significant ($P < .001$)
23. $\hat{\mu}_{Y|x} = 14.491 + 1.498x$
25. $t = 1.197$, not significant
27. $\hat{\mu}_{Y|x=15} = 89.588$ (Regression not significant.)
29. (a) $\hat{\mu}_{Y|x} = 911.667 - 49.667x$
    (b) $t = -19.431$, significant at $P < .0001$
    (c) no
31. $\hat{\sigma}^2 = .0109$
33. (b) $\hat{\mu}_{Y|x} = 6.375 + 2.943x$
    (c) 41.987
    (d) $t = 6.378$, significant at .0002

Section 10.5

39. (b) $\hat{\rho} = .887$
41. $.657 \le \rho \le .966$
43. (b) $\hat{\rho} = .586$
    (c) $-.068 \le \rho \le .888$
49. 96.9%
51. $R^2 \doteq 0$

CHAPTER 10 review exercises

60. (b) $\hat{\mu}_{Y|x} = 145.667 + 6.20x$
61. (a) no, $f = 27.91$, significant at $\alpha = .05$
    (b) yes
62. $\hat{\mu}_{Y|x} = 99.383 - .0052x$
63. significant at $\alpha < .001$
64. $R^2 = .967$
65. $99.1603 \le \beta_0 \le 99.6057$
66. $-.00602 \le \beta_1 \le -.00431$
67. $\hat{\mu}_{Y|x} = .8233 - .0589x$
68. $t = -9.264$, significant at $P < .0001$
69. (a) $\hat{\mu}_{Y|x=3.25} = .6318$; $(\hat{Y}|x = 3.25) = .6318$
    (b) $.6121 \le \mu_{Y|x=3.25} \le .6516$
    (c) $.5703 \le (Y|x = 3.25) \le .6933$
70. (b) $\hat{\rho} = -.959$
71. significant at $P < .0001$
72. $-.866 \le \rho \le -.988$
73. (a) $t = .889, P > .20$
    (b) $t = 2.179, P < .05$
75. (a) $\hat{\rho} = .989$
    (b) $.953 \le \rho \le .997$
    (c) significant; yes
    (d) 97.86%
        A coefficient of determination of .9786 implies that 97.86% of the variance in the dependent variable Y is explained by the linear regression model.

78. (*a*) As pH increases to 4.0, weight loss decreases.
   (*b*) $r = .7259$
   (*c*) $t = 4.22$; $P < .001$; there is a significant correlation between pH and weight loss
   (*d*) [.392, .891]
79. (*a*) no, the relationship is not linear
   (*b*) yes
   (*c*) fit a quadratic or other nonlinear model
80. (*a*) $\hat{\mu}_{Y|x} = 6.97 - 0.120x$
   (*b*) $t = -14.96$; $P \doteq 0 < .05$; reject H_0
   (*c*) 8.77
   (*d*) [8.36, 9.18]
81. (*a*) $\hat{\mu}_{Y|x} = -5.94 + 0.0973x$
   (*b*) $t = 9.85$; $P \doteq 0 < .05$; reject H_0
   (*c*) 2.817
   (*d*) [0, 9.2]

Section 11.1

5. (*a*) $f = 4.95$, $P = .0174$, significant at $\alpha = .05$
   (*b*) no
   (*c*) no

Section 11.2

7. $f_{4, 35} = 216.34$, $P < .001$
9. $b = 23.187$ with 3 df, significant at $\alpha < .005$
   not reasonable to assume homogeneity of variances; use a nonparametric test
11. $b = .354$, not significant

Section 11.3

13. (*a*) $\alpha' \leq .1426, .0333$
   (*b*) $\alpha' \leq .5367, .0067$
   (*c*) $\alpha' \leq .9006, .0022$
17. (*a*) 3
   (*b*) .05
   (*c*) $\mu_1 \neq \mu_2$
21. $\mu_1 \neq \mu_2, \mu_1 \neq \mu_3, \mu_2 \neq \mu_3$ at $\alpha = .01$

Section 11.4

23. (*a*) $H = 9.05$, not significant at $\alpha = .05$ ($P \doteq .06$)
   (*b*) $f = 3.30$, significant at $\alpha = .05$ ($P \doteq .026$)

CHAPTER 11 review exercises

26. (*a*)

Source	Df	SS	MS
CO_2 level	4	11,274.32	2818.50
Error	45	1,248.04	27.73
Corrected total	49	12,522.36	

   (*b*) yes, $f = 101.63$

27. $b = 1.07$, not significant; yes

28. (b)

Source	df	SS	MS	F	EMS
Treatment	2	110.6	55.3	3.0	$\sigma^2 + 10\sigma_{Tr}^2$
Error	27	497.7	18.433		σ^2
Total	29	608.3			

    $f = 3.0$ with 2 and 27 df; not significant at $\alpha = .05$

   (c) due to error, 83.3%; due to treatments, 16.7%

29. (a) $f = 79.9$, significant at $\alpha < .0001$

   (b) all three are significantly different

   (c) all three are significantly different

31. $f = 579.61, P \doteq 0$

Section 12.1

1. $60, 45, 30, 15$; yes, these seem to differ quite a bit from the expected numbers

3. 10; questionable, the observed values are not drastically different from those expected

Section 12.2

5. $\chi^2 = 22.66$; reject H_0, $P < .005$ based on the X_4^2 distribution

Section 12.3

9. $\chi^2 = 4.57$; reject H_0, $.025 < P < .05$ based on the X_1^2 distribution

11. $\chi^2 = 8.84$; reject H_0, $.025 < P < .05$ based on the X_3^2 distribution

Section 12.4

13. $\chi^2 = 3.95$; reject H_0, critical point $= 3.84$

15. $H_0: p_{11} = p_{21} = p_{31}$ $\chi^2 = 14.72$; reject H_0, $.01 < P < .025$ based

        $p_{12} = p_{22} = p_{32}$ on the X_6^2 distribution

        $p_{13} = p_{23} = p_{33}$

        $p_{14} = p_{24} = p_{34}$

17. $\chi^2 = 16.03$; reject H_0; $P < .005$

19. $\chi^2 = 43.65$; reject H_0; $P < .005$

CHAPTER 12 review exercises

23. $H_0: p_{11} = p_{21}$; $\chi^2 = .709$; no, $.25 < P < .50$

24. yes; $\chi^2 \doteq 16.04$; $P < .005$

25. yes; $\chi^2 \doteq 34.05$; $P < .005$

26. no; $\chi^2 \doteq .69$; $.25 < P < .5$

27. (a) $175; 87.5; 87.5$

   (b) no; $\chi^2 \doteq .77$; $.50 < P < .75$

Selected Derivations

I **Theorem 4.5.2 (Chebyshev's inequality).**
Let X be a random variable with mean μ, and standard deviation σ. Then for any positive number k,

$$P[|X - \mu| < k\sigma] \geq 1 - \frac{1}{k^2}$$

Proof.
Assume that X is continuous with mean μ, and standard deviation σ, and density f. By definition,

$$\sigma^2 = \text{Var}(X) = \int_{-\infty}^{\infty} (x - \mu)^2 f(x)\, dx$$

Let $k > 0$, and $c = k^2\sigma^2$, and note that

$$\sigma^2 = \int_{-\infty}^{\mu - \sqrt{c}} (x - \mu)^2 f(x)\, dx + \int_{\mu - \sqrt{c}}^{\mu + \sqrt{c}} (x - \mu)^2 f(x)\, dx$$

$$+ \int_{\mu + \sqrt{c}}^{\infty} (x - \mu)^2 f(x)\, dx$$

Since $(x - \mu)^2 f(x) \geq 0$,

$$\int_{\mu - \sqrt{c}}^{\mu + \sqrt{c}} (x - \mu)^2 f(x)\, dx \geq 0$$

and thus

$$\sigma^2 \geq \int_{-\infty}^{\mu - \sqrt{c}} (x - \mu)^2 f(x)\, dx + \int_{\mu + \sqrt{c}}^{\infty} (x - \mu)^2 f(x)\, dx$$

Note that over both regions of integration $(x - \mu)^2 \geq c$, and so it can be concluded that

$$\sigma^2 \geq \int_{-\infty}^{\mu - \sqrt{c}} cf(x)\, dx + \int_{\mu + \sqrt{c}}^{\infty} cf(x)\, dx$$

In terms of probabilities,

$$\sigma^2 \geq cP[X \leq \mu - \sqrt{c}] + cP[X \geq \mu + \sqrt{c}]$$

or

$$\sigma^2 \geq c\{P[X - \mu \leq -\sqrt{c}] + P[X - \mu \geq \sqrt{c}]\}$$

This inequality can be rewritten to conclude that

$$P[X - \mu \leq -\sqrt{c}] + P[X - \mu \geq \sqrt{c}] \leq \frac{\sigma^2}{c}$$

or that

$$P[-\sqrt{c} \leq X - \mu \leq \sqrt{c}] \geq 1 - \frac{\sigma^2}{c}$$

Since $c = k^2\sigma^2$ where k and c are each nonnegative, $\sqrt{c} = k\sigma$. Substitution yields

$$P[-k\sigma \leq X - \mu \leq k\sigma] \geq 1 - \frac{1}{k^2}$$

or

$$P[|X - \mu| \leq k\sigma] \geq 1 - \frac{1}{k^2}$$

Since X is continuous, we can conclude that

$$P[|X - \mu| < k\sigma] \geq 1 - \frac{1}{k^2}$$

as claimed. The proof in the discrete case is similar with summation replacing integration.

II   **Theorem 7.1.3.**

Let S^2 be the sample variance based on a random sample of size n from a distribution with mean μ and variance σ^2. S^2 is an unbiased estimator for σ^2.

Proof.
By definition,

$$
\begin{aligned}
E[S^2] &= E\left[\sum_{i=1}^{n} \frac{(X_i - \overline{X})^2}{n-1}\right] \\[2mm]
&= \frac{1}{n-1} E\left[\sum_{i=1}^{n} (X_i - \mu + \mu - \overline{X})^2\right] \\[2mm]
&= \frac{1}{n-1} E\left[\sum_{i=1}^{n} (X_i - \mu)^2 - 2(\overline{X} - \mu)\sum_{i=1}^{n}(X_i - \mu) + \sum_{i=1}^{n}(\overline{X} - \mu)^2\right] \\[2mm]
&= \frac{1}{n-1} E\left[\sum_{i=1}^{n} (X_i - \mu)^2 - 2(\overline{X} - \mu)\frac{n\left(\sum X_i - n\mu\right)}{n} + n(\overline{X} - \mu)^2\right] \\[2mm]
&= \frac{1}{n-1} E\left[\sum_{i=1}^{n} (X_i - \mu)^2 - 2n(\overline{X} - \mu)^2 + n(\overline{X} - \mu)^2\right] \\[2mm]
&= \frac{1}{n-1} E\left[\sum_{i=1}^{n} (X_i - \mu)^2 - n(\overline{X} - \mu)^2\right] \\[2mm]
&= \frac{1}{n-1}\left[\sum_{i=1}^{n} E[(X_i - \mu)^2] - nE[(\overline{X} - \mu)^2]\right]
\end{aligned}
$$

Note that since $X_1, X_2, X_3, \ldots, X_n$ is a random sample from a distribution with variance σ^2, $E[(X_i - \mu)^2] = \sigma^2$ for each $i = 1, 2, 3, \ldots, n$. Note that by Theorems 7.1.2 and 7.1.1, $\text{Var }\overline{X} = E[(\overline{X} - \mu)^2] = \sigma^2/n$. By substitution, we obtain

$$
\begin{aligned}
E[S^2] &= \frac{1}{n-1}\left[\sum_{i=1}^{n} \sigma^2 - n\sigma^2/n\right] \\[2mm]
&= \frac{1}{n-1}(n\sigma^2 - \sigma^2) = \sigma^2
\end{aligned}
$$

and the proof is complete.

III   **Theorem 8.1.1.**

[Distribution of $(n-1)S^2/\sigma^2$]. Let $X_1, X_2, X_3, \ldots, X_n$ be a random sample from a normal distribution with mean μ and variance σ^2. The random variable

$$(n-1)S^2/\sigma^2 = \sum_{i=1}^{n}(X_i - \overline{X})^2/\sigma^2$$

has a chi-squared distribution with $n-1$ degrees of freedom.

Proof.
This argument does not constitute a rigorous proof of our theorem. However, it does suggest that the distribution of the random variable $(n-1)S^2/\sigma^2$ is as stated. We begin by rewriting the random variable as a difference between two chi-squared random variables:

$$(n-1)S^2/\sigma^2 = \sum_{i=1}^{n} \frac{(X_i - \overline{X})^2}{\sigma^2} = \sum_{i=1}^{n} \frac{[(X_i - \mu) - (\overline{X} - \mu)]^2}{\sigma^2}$$

$$= \sum_{i=1}^{n} \frac{(X_i - \mu)^2}{\sigma^2} - 2(\overline{X} - \mu) \sum_{i=1}^{n} \frac{(X_i - \mu)}{\sigma^2} + \frac{n(\overline{X} - \mu)^2}{\sigma^2}$$

$$= \sum_{i=1}^{n} \frac{(X_i - \mu)^2}{\sigma^2} - 2(\overline{X} - \mu) \frac{\left(\sum\limits_{i=1}^{n} X_i - n\mu\right)}{\sigma^2} + \frac{n(\overline{X} - \mu)^2}{\sigma^2}$$

$$= \sum_{i=1}^{n} \frac{(X_i - \mu)^2}{\sigma^2} - \frac{2n(\overline{X} - \mu)^2}{\sigma^2} + \frac{n(\overline{X} - \mu)^2}{\sigma^2}$$

$$= \sum_{i=1}^{n} \frac{(X_i - \mu)^2}{\sigma^2} - \left(\frac{\overline{X} - \mu}{\sigma/\sqrt{n}}\right)^2$$

We now see that

$$(n-1)S^2/\sigma^2 + \left(\frac{\overline{X} - \mu}{\sigma/\sqrt{n}}\right)^2 = \sum_{i=1}^{n} \frac{(X_i - \mu)^2}{\sigma^2}$$

Note that the random variable $(\overline{X} - \mu)/(\sigma/\sqrt{n})$ is standard normal. By Exercise 45 of Chap. 7 $[(\overline{X} - \mu)/(\sigma/\sqrt{n})]^2$ has a chi-squared distribution with 1 degree of freedom, and $\sum_{i=1}^{n}[(X_i - \mu)^2/\sigma^2]$ has a chi-squared distribution with n degrees of freedom. Since the sum of independent chi-squared random variables is also a chi-squared random variable (see Exercise 44 of Chap. 7), it is logical to assume that the random variable $(n-1)S^2/\sigma^2$ has a chi-squared distribution with $n-1$ degrees of freedom as claimed.

Index